T0142160

Information Fusion and Data Science

This book series provides a forum to systematically summarize recent developments, discoveries and progress on multi-sensor, multi-source/multi-level data and information fusion along with its connection to data-enabled science. Emphasis is also placed on fundamental theories, algorithms and real-world applications of massive data as well as information processing, analysis, fusion and knowledge generation. The aim of this book series is to provide the most up-to-date research results and tutorial materials on current topics in this growing field as well as to stimulate further research interest by transmitting the knowledge to the next generation of scientists and engineers in the corresponding fields. The target audiences are graduate students, academic scientists as well as researchers in industry and government, related to computational sciences and engineering, complex systems and artificial intelligence. Formats suitable for the series are contributed volumes, monographs and lecture notes.

More information about this series at http://www.springer.com/series/15462

Jun Zhao • Wei Wang • Chunyang Sheng

Data-Driven Prediction for Industrial Processes and Their Applications

Jun Zhao
Dalian University of Technology
Dalian, China

Wei Wang
Dalian University of Technology
Dalian, China

Chunyang Sheng
Shandong University of Science
and Technology
Qingdao, China

ISSN 2510-1528 ISSN 2510-1536 (electronic)
Information Fusion and Data Science
ISBN 978-3-030-06785-4 ISBN 978-3-319-94051-9 (eBook)
https://doi.org/10.1007/978-3-319-94051-9

Printed on acid-free paper

This Springer imprint is published by the registered company Springer Nature Switzerland AG
The registered company address is: Gewerbestrasse 11, 6330 Cham, Switzerland

Preface

This book will introduce a wealth of data-driven prediction approaches for indicators (variables) in industrial processes. The motivation stems from decision-making demands in real production for control, optimization, and scheduling, among others. In an optimal control task, the estimated future output is fed back to the input in real time to correct and optimize the system. Therefore, accurate prediction is undoubtedly beneficial to solve such problems. Typically, the indicators can be divided into two categories: 1) indices that directly describe the state of the production process (e.g., temperature, pressure) and 2) manually defined indicators in a production process (e.g., the grade of concentrated ore in an ore production process), which have a complex relationship with variables of other processes. Considering the nonlinear and hysteretic nature of generic industrial production, it can be quite time consuming or even impossible to obtain the value of these indicators. Thus, effective real-time models for their estimation are helpful.

Traditional prediction methods are mostly derived from analyses of manufacturing mechanisms involving utilization of the material balance principle and thermodynamics, among others. However, due to the complex characteristics of the dynamics, instability, and nonlinearity of a real production process, a physics-based system is difficult to model and has high computational costs; thus, its accuracy and reliability cannot be effectively guaranteed. With the development of information technology and improvements in management standards, many industrial enterprises are now equipped with enterprise resource planning, manufacturing execution, and supervisory control and data acquisition systems. As a result, huge quantities of industrial data are being accumulated, covering production plans, device/network statuses, scheduling schemes, and performance indices, among others. Meanwhile, artificial intelligence, especially data-driven machine learning approaches, has made great progress in recent decades. Bearing this in mind, this book examines the construction of intelligent prediction models based on real data.

Data-based prediction methods require only real data instead of complex modeling of production mechanisms to build non-linear relationships between inputs and outputs. In general, the feature selection is performed first to reasonably determine the most relevant variables for the input alternatives. Next, using the significant indicators as the output, the prediction model is created using machine learning methods, such as artificial neural networks and support vector machines. In addition, because the values of the hyper-parameters greatly affect the prediction performance, it is necessary to deploy online optimization, in which minimization of the forecasting error is typically regarded as the objective function. This book uses the decision-making problem of energy systems in a process industry as a basis. A series of data-based prediction models are presented in Chap. 3 to 6 of this book, with a class of specific optimization methods provided in Chap. 7.

This book introduces a variety of classifications for categorizing prediction models. Depending on whether other variables are introduced, these models can be classified as time series-based or factors-based. The former typically exposes underlying variation patterns by learning from the historic values of the variable itself, whereas the latter aims to establish a relationship between the output and other possible related variables. Furthermore, data-based prediction models can be classified according to their different demands or production statuses, such as long-term or short-term for the requirements of horizon predictions, single-output or multi-output for the number of indicators, and single-model or multi-model collaboration for the number of utilized models; all of these model types will be well illustrated in this book. It should also be noted that all models in this book are constructed using continuous variables rather than discrete variables.

Given the importance of decision-making and the typicality of the related techniques, the case studies in this book are all conducted using an energy system of the steel industry. To provide a comprehensive discussion, we also present a two-stage predictive scheduling method in the last chapter of this book. This method will not only be beneficial for increasing production efficiency, but it could also lead to a breakthrough for intelligent vehicles in the overall management and optimization of the industry.

Dalian, China Jun Zhao
Dalian, China Wei Wang
Qingdao, China Chunyang Sheng

Audience and Goal of This Book

This book is intended for engineers who apply prediction techniques to industrial problems, as well as scholars and Ph.D. candidates who are interested in the topics of this book to increase their knowledge of data-based prediction approaches. Although several machine learning techniques are well described in this book, some details have been omitted due to the limited space.

All of the authors have been immersed in research on data-based prediction for industrial applications for nearly a decade. This book aims to summarize their numerous works, especially in the steel industry. The goal of this book is to introduce intelligent methods and their real application. We hope this book will enlighten our readers and encourage them to apply our methods in other practical fields. Although most of the cases use the characteristics of energy systems in the steel industry, we have attempted to formulate prediction problems in a common way so that a general solution can be obtained. With the rapid development of artificial intelligence technologies and newly emerging machine learning methods, data-based prediction has great potential to acquire valuable results and achieve intelligent manufacturing and scheduling in industry.

Acknowledgements

This book was supported by the National Key Research and Development Program of China under Grant 2017YFA0700300 and the National Natural Sciences Foundation of China under Grants 61533005, 61522304 and U1560102.

We would like to express our deep gratitude to many people for their inspiration and assistance with this book. In particular, we are grateful to Dr. Zhongyang Han for his contribution to the writing of Chap. 6. We also give thanks to Dr. Long Chen for providing the original drafts of Chaps. 7 and 8. We thank Dr. Hongqi Zhang and Dr. Tianyu Wang for writing drafts of Chap. 2 and Chap. 9, respectively. We thank Dr. Ying Liu, Dr. Zheng Lv, and Dr. Linqing Wang for offering their valuable comments and suggestions on the whole manuscript. In addition, many thanks are due to a number of graduate students, including Feng Jin, Zhiming Lv, Qiang Wang, Wenlin Zhang, Yanwei Zhai, Yang Liu, Chun Qin, and Jian Sun, for their discussions on special issues.

Contents

Chapter 1
Introduction

Abstract This chapter gives an overall introduction to this book. First, we discuss the importance of the prediction for industrial process. Then, we divide the data-driven prediction methodology discussed in this book into a number of categories. Specifically, there are three categories, i.e., data feature-based methods, time scale-based ones, and prediction reliability-based ones. Besides, considering the characteristics of prediction modeling and industrial demands, this book introduces some commonly used prediction techniques, including the time series-based methods, the factor-based methods, the prediction intervals (PIs) construction methods, and the granular-based long-term prediction methods.

Forecast is an ancient topic in the world, which is probably born due to the human fears of the future. In ancient world, forecast is employed to predict the variety of environment and things in the environment. According to the prediction results, human can understand the environment in the future, and then makes some changes to suit the environment for a more comfortable life. A typical prediction activation is weather forecasting for ancient people since the variation of weather and climate has a great relationship on the people's daily life. Different from the accurate prediction nowadays, the weather forecasting in ancient time belongs to rough and long-term version. However, the ancient forecasts have made great achievements, which enrich astronomy and calendars. Taking the lunar ephemeris and 24 solar terms as examples, one can get a thorough understanding on how the moon and sun work, respectively. Meanwhile, one can also know the impact of other celestial body on the earth. Based on the researches on the ephemeris and the 24 solar terms, people can grasp the tides of a river and precipitation law along the river, which has a great meaning to the agrarian society. Although the ancient forecast is not as accurate as nowadays prediction, it still has great meanings and awes inspiring.

Nowadays, prediction and forecast techniques are more and more applied in many fields of science and research, such as weather, business, finance, traffic system, fault detection, and industry system. Nowadays, the commonly used prediction methods are different from the ancient ones. We still take the weather forecasting as an example. According to the moisture in the air, diurnal temperature,

J. Zhao et al., *Data-Driven Prediction for Industrial Processes and Their Applications*, Information Fusion and Data Science,
https://doi.org/10.1007/978-3-319-94051-9_1

wind magnitude, wind direct, and cloud layers, we can use some techniques, such as neural networks model [1], Gaussian process [2], or support vector machine [3], to understand the variation of the weather in the future 24 or 48 h. Although there are many different forecasting activations in a number of fields, and many different prediction methods in research domain, the aim of forecasting and prediction is constantly to improve the life quality and comfortability of peoples.

1.1 Why Prediction Is Required for Industrial Process

It is very important for typical industrial production to guarantee its safety, continuity, and product quality. Correspondingly, a large number of techniques related to faults detection, advanced control techniques and optimal scheduling, and decision-making have to be paid more attention in the corresponding researches. As for all the above-mentioned problems, the significant one that should be firstly solved is how to effectively guarantee the real-time ability of the control solution or the operations. Thus, forecasting or prediction concept could be premium strategy that should be introduced into the functions of industrial control, scheduling, and optimization [4–6].

With respect to fault diagnosis, aiming at analyzing and judge abnormal industrial conditions, the research categories of fault diagnosis include fault detection, fault type recognition, fault locating, and fault elimination. Fault prediction also named fault pre-estimation is capable of providing assistance for system recovery from the failure, which introduces the concept of prediction, namely when the system does not completely fall into fault states, the type and the location of fault can be predicted in advance [7]. As such, according to the predicted type and the locations, a series of effective operations can be adopted to correct and eliminate the fault.

In perspective of industrial control system, predictive control belongs to a class of computer control algorithm developed in recent years, which involves the steps of multi-step prediction, rolling optimization, and feedback correction [8]. It is often applied to complex industrial production process that is difficult to establish an accurate mathematical model. In such a framework, multi-step prediction is the most important part of predictive control, which aims at how to employ data-driven techniques to model a system with complex or unknown operational mechanism. If the model accuracy is high enough, the variation of the system states in the future can be effectively predicted. Based on the accurate prediction, the control measures should be implemented for the system in order to ensure safe and reliable production process when the control target changes from the expected range.

In addition, one can also consider the problems of industrial production planning and scheduling. When planning and decision-making is required for industrial plant, the planner in practice hopes to be able to master the important indicators of industrial system changes in advance as soon as possible. If so, forecasting or prediction can be introduced, which makes the operation of the system in the future effectively judged and make the enterprise decision-makers flexible to set the future production scheme [9]. Besides, with the help of prediction, if an industrial system in the future could run with a lower operating efficiency and even in an abnormal operating condition, an optimizing process could be provided to make the system return to normal or higher efficiency state.

1.2 Category of Data-Based Industrial Process Prediction

There are multiple classification strategies for data-based industrial process prediction techniques, in which three aspects are the most commonly considered, i.e., data characteristics, time scale of the data, and the amount of the predicting reliability.

1.2.1 Data Feature-Based Prediction

Based on the difference of the data feature, the prediction problems of industrial process can be divided into two categories. One is nonlinear time series-based prediction, in which the sample data usually involve high level noises and belong to the mode of one-dimension or multi-dimension time series data [10]. Another is factors-based one, which allows that the input data is the reasons of the output and the output is a class of result depending on the input data [11]. In general, time series-based prediction serves to a large amount of industrial demand. One can take the energy forecasting as an example. As for a large-scale industrial enterprise, the control and optimization problem of its energy system might require time series-based prediction. For instance, in a steel plant there are various energy media involving by-product gases, oxygen, power, hot steam, etc. It is necessary for such an industrial system to predict the related energy variables, such as gas consumption flow, oxygen consumption flow of steel-making, and blast furnace gas generation flow. Meanwhile, the factor-based prediction is similarly concentrated in the energy system of steel plant. Given that a number of energy variables have to be impacted by the corresponding factors, the factors-based prediction involves not only the common multi-input single-output (MISO) prediction model, but the multi-input multi-output prediction (MIMO) ones as well. One can also take another example that the pressure prediction of energy pipeline networks and the blast furnace gas tank level prediction are both the typical MISO version. While due to the fact that the converter gas system in steel plant is usually equipped by several subsystems, the converter gas tanks levels prediction belongs to a typical MIMO prediction.

1.2.2 Time Scale-Based Prediction

According to the differences of prediction time scale, the data-based industrial prediction can be divided into super-short-term prediction (usually less than

1 min), short-term prediction (less than 60 min), mid-long-term prediction (from 1 to 24 h) and long-term prediction (larger than 24 h). The difference of prediction time scale depends on the practical requirements of tasks, such as real-time control, production scheduling, and decision-making optimization. For instance, the super-short-term prediction is commonly used for process real-time control since the timeliness and control accuracy is the basic requirements for such a task [12]. And, the short-term versions such as short-term power load and by-product gas flow predictions serve as the foundation of industrial energy scheduling, which exhibits a large difference from the control problems since the objective of production scheduling could be related to a much longer time period, and its real-time requirement is relatively low compared to that of the process control function [13]. As such, the timeliness and accuracy of prediction can be appropriately reduced. While the mid-long-term prediction aims at assisting to optimize the efficiency of a certain system in a period of time or estimate the system states in advance, and the long-term prediction plays an important role on making macro decision [14].

1.2.3 Prediction Reliability-Based Prediction

In the industrial production domain, the reliability of the prediction result is strictly emphasized by decision-makers or real-time controller since the risky decision will be sometimes generated due to the lack of the prediction results reliability. Thus, in the perspective of whether considering the prediction reliability, one can also divide the industrial predictions tasks into two categories. The first one is the point-oriented prediction and another is the interval-based prediction, i.e., prediction intervals (PIs) construction. The point-oriented prediction, viewed as the commonly used forecasting mode, is performed by using a model to mine nonlinear mapping relationship hidden in the data. And, the model is generally established by a class of machine learning or data mining techniques. On the basis of the established model, the predicted output can be obtained when giving a specific input. As for the PI construction, which can be regarded as an extended version of the point-oriented prediction, PIs require to estimate the reliability of the prediction results by generally using a probabilistic-based approach. Since the prediction errors are often adopted to quantify the prediction reliability, the uncertainties caused by nonlinear prediction model and data noise are commonly considered for PIs construction [15]. The performance of PIs quantified by the PIs coverage probability (PICP) and the mean PIs width (MPIW) can be applied to evaluate the prediction reliability [15].

Although there are various partition criteria for industrial process prediction, one prediction task might often belong to a number of categories, for example, a short-term industrial time series prediction intervals construction. In the perspective of data feature, it belongs to a time series-based prediction task. With respect to time scale category, it is certainly a short-term prediction. And in terms of the reliability of prediction, it can be viewed as an interval-based prediction problem. Therefore, in this book, we do not present the industrial prediction specified as one classification mode, but introduce those belonging to the combined prediction cases, such as short-term time series prediction, short-term factor-based prediction, PIs for time series and factor-based problems, long-term prediction, etc.

1.3 Commonly Used Techniques for Industrial Prediction

Based on the above classification for industrial process prediction tasks, one notices that there exist a large number of prediction problems with multiple categories, which could be related to various modeling and computing techniques. In this book, we will introduce some methods for such problems according to the degree of practicability and difficulty. The main contents of this book are divided into four parts. (1) short-term industrial time series prediction; (2) factor-based short-term industrial prediction; (3) short-term industrial PIs construction; (4) granular-based long-term prediction. According to the above four classifications, we introduce some commonly used methods and techniques.

1.3.1 Time Series Prediction Methods

Time series prediction is the most commonly used technique applicable to the industrial process. There are various kinds of approaches dealing with such a task in literature. Linear regression is a well-known regression method aiming at learning the linear relationship between the variables hidden in time series data. While the nonlinear characteristic rooted in industrial time series could be hard to be captured by the generic linear regression model. The kernel-based Gaussian process regression model, capable of learning the nonlinearity of variables, has however been developed for the time series prediction [16]. In the Gaussian process framework, we dispense with the parametric model and instead define a prior probability distribution over functions directly. At first sight, it might seem difficult to work with a distribution over the uncountably infinite space of functions. However, as we shall see, for a finite training set we only need to consider the values of the function at the discrete set of input values corresponding to the training set and test set data points, and so in practice we can work in a finite space. As far as we know, Gaussian process has a comprehensive application in the field of prediction [17–19]. However, its merit is not limited to this. It can be combined with other models to achieve a much better performance of prediction. For instance, the one combined with the neural networks, such as Gaussian process-based recurrent neural network can be viewed as a typical example.

Artificial neural networks (ANNs) are computing systems inspired by the biological neural networks that constitute animal brains [20]. The original goal of an ANNs approach was to solve problems in the same way that a human brain does.

Over time, attention focused on matching specific mental abilities, leading to deviations from biology such as back propagation, or passing information in the reverse direction and adjusting the network to reflect that information. ANNs are used for time series prediction based on the fact that it is able to approximate any continuous nonlinear function at an arbitrary accuracy, provided the number of hidden neurons is sufficient. Thus, ANNs can be used to learn the nonlinear mapping function hidden in the time series. When using such models for some practical time series prediction problems, we will face two main drawbacks. The first one is that we cannot always reach the global optimization in a learning algorithm, sometimes falls in some local points. The second one is that the computational time of the training process is too long for real-time application in industry.

The kernel-based approach, typically the support vector machines (SVMs), can be regarded as another commonly used time series prediction methods, which are supervised learning models for classification and regression [21]. More formally, a support vector machine constructs a hyperplane or a set of hyperplanes in a high- or infinite-dimensional space. Intuitively, a good separation hyperplane has the largest distance to the nearest training data point of any class, since in general the larger the margin is, the lower the generalization error of the classifier becomes. As for application of SVMs, the significant problem is to choose the suitable kernel functions. Moreover, although the number of the parameters of the SVM is relatively small, it is still a difficult problem for parameters optimization.

All of the above methods have their own merits and drawbacks. For example, NNs generally have much stronger learning ability with a higher computational load, while SVMs are suitable for problems with a few samples and it exhibits a relatively lower computational load [22]. For different problems, we should select their suitable corresponding prediction methods. In most of the cases, according to the practical problems, the aforementioned models have to be improved to fit the exact tasks. Sometimes, two or more different models should be combined into a complex one for industrial application.

1.3.2 Factor-Based Prediction Methods

Similarly, a factor-based prediction is also considered as a regression task with input and output data sampled from the different sensors, which is different from the sampling form in the time series prediction. Feedforward neural networks (FNNs) are particularly predominant in the factor-based prediction field, while recurrent neural networks (RNNs) are commonly used for time series prediction since RNNs have an ability of short-term memory [23]. The iterative prediction usually aims at the time series prediction, in which we hope the output of the model can be remembered and used for the input data in the next step. As for the factor-based prediction, the predicted targets are determined by a series of input variables. There is a kind of causal relationship between the input and output variables. Since the input and output data come usually from various data sources, the iterative prediction mode is not suitable in a number of industrial process. As for the non-iterative prediction, FNNs are comparatively effective.

Except for the ANNs, the fuzzy modeling approaches are usually employed to make predictions, where TS fuzzy rule-based model that requires only a small number of rules exhibits high ability of describing complex and highly nonlinear relationship [24]. The number of rules is much smaller than that of the Ma-fuzzy model. The basic idea of the TS model is based on the fact that an arbitrary complex system is divided into some mutually inter-linked subsystems. Let regions, corresponding to individual subsystems, be determined in the state-space under consideration. The relationship between the inputs and output the subsystem in these regions can then be described as a simple functional dependence. Thus, the TS fuzzy model can be represented by a set of rules with each subsystem assigned to one rule, when the dependence is linear relationship.

Besides, SVMs model can also be used for the factor-based prediction. Differently, the nonlinear mapping relationship hidden in the factor-based prediction might be more complex. Thus, the multi-kernel learning least squares SVM (LSSVM) is generally employed for the factor-based prediction in some industrial task [25]. As for the multi-kernel learning, the construction of the kernel function or the selection of the suitable basic kernel is also an important problem for building the factor-based prediction model. In addition, factor-based prediction for the practical application is not limited to the single-output model. Sometimes, the predicted problems involve two or more outputs. For example, the converter gas of the Linz-Donawitz gas (LDG) system in steel industry usually contains two converters; meanwhile, there are two corresponding gasholders in the LDG system those have a coupling relationship. As such, the established prediction model has to involve two outputs. The multi-output LSSVM will also be introduced in this book to solve such a prediction problem with multi-outputs.

1.3.3 Methods for PIs Construction

Generally, the prediction intervals (PIs) can be constructed by using the prediction models combining with statistical techniques, such as the delta, the Bayesian, the mean-variance estimates (MVEs), and the bootstrap approach [15]. The existing techniques can be divided into two categories according to the number of sub-models, containing the single model-based method and the ensemble method. Firstly, only one single model is adopted to construct PIs, including the delta, the Bayesian, and the MVE-based method. The delta was firstly applied to construct PIs with the assumption that the variance of the noises was constant for all samples [26]. However, such assumption deviates from the real-world conditions. The Bayesian framework was also applied in the field of PIs construction [22]. But the accuracy of such method depended mainly on the prior knowledge when the number of training samples was small. Furthermore, calculating the Hessian matrix was a time-consuming process. The MVE method assumed that the nonlinear model could accurately estimate the target with the least computational load [27]. Because of the uncertainties in practice, the generalization capability of the prediction model was weak, which might lead to a large variability in the construction of PIs.

Secondly, an ensemble that consists of a number of local models is employed for PI construction. Moreover, the bootstrap-based ensemble is more stable than single model-based ones for practical prediction tasks [28]. However, the bootstrap method demands a high computing cost due to the complexity of the ensemble, which is troublesome for the application in predicting real-time problems. In addition, the performance of the ensemble depends on each individual. That is, if one individual network is biased, the entire model accuracy might be obviously deteriorated. As such, it is rather difficult to develop an effective learning method with consideration of the performance of each individual [28].

In view of the drawbacks of the predicted variance-based PIs construction, some other approaches pay direct attention to estimating the bounds of PIs without any assumption on data distribution. For example, a lower-upper bound estimation (LUBE) was proposed by designing a feedforward NN with two outputs for PIs construction [29]. An interval-weighted granular NNs were also reported in [30], whose connections were described by the interval-valued information granularity. For the granular NN, its exogenous inputs are generally numeric values, while its outputs are always interval-valued granularity. However, the above two approaches still faced an important challenge for PIs construction. The challenge is that the error accumulation of iterative prediction process should be solved when constructing the PIs. According to the interval-based algebraic operations, the iterative mode will be accompanied with serious error accumulation that makes the bounds of PIs diverge from the real observed targets.

All of the aforementioned PIs construction methods have their own shortcomings and cannot be directly applied for industrial prediction. In Chap. 5, we will introduce some improved version of the above techniques for practical industrial applications, including the predicted variance-based iterative PIs construction, the interval-valued model for non-iterative PIs construction methods, the PIs construction methods with the consideration of input uncertainties and the PIs construction methods for the practical application with the incomplete inputs.

1.3.4 Long-Term Prediction Intervals Methods

In many practical applications, we usually pay close attention to the prediction tendency, e.g., ascending or descending. Different from the short-term prediction as shown in previous subsections, the long-term prediction tendency is very useful corresponding to the manufacturing procedure in a practical scene. To make this realized, the information granulation and the fuzzy modeling techniques play important roles.

The role of information granulation is in the organization of detailed numerical data into some meaningful and operationally viable abstract knowledge

[31]. In particular, information granulation is aimed at splitting the problem into several manageable sub-problems for which we are in a position to produce effective solutions. The quality of such granulation (abstraction) is clearly related to the ability to retain the essence of the original data. There are many different granulation schemes [32]. In the case of temporal data, an essential characteristic is the temporal ordering of individual data. An intuitive approach to granulation of time series for forecasting purpose is to define a “temporal window” and to evaluate an appropriate granular representative within each segment of time series.

To construct abstract granules from segmented data, fuzzy clustering is typically employed [33]. On the basis of such information granules, we develop relationships between consecutive granules in order to forecast future values. A suitable choice of representation will greatly increase the ease and efficiency of time series processes. Applying the clustering method to a set of continuously non-overlapping windows of a signal is useful for grouping similar temporal patterns which are dispersed along the signal. Such a granular forecasting method is more efficient than the previous point-by-point forecasting method in dealing with the problems associated with high-dimensional pattern space in long-term forecasting and large amounts of data in long time series.

In Chap. 6, a set of granular computing (GrC)-based prediction models is presented to overcome the long-term prediction problems and the PIs construction techniques introduced in Chap. 5 are also considered for long-term prediction intervals task. Dynamic time warping (DTW) technique, fuzzy modeling, and some other algorithms are deployed for achieving such a task.

1.4 Summary

In Chap. 1, we investigated why forecasting or prediction was necessary for human beings in the course of the historical development. We compared and analyzed the similarity and difference between the ancient prediction mode and the modern prediction methods. Then, we introduced why the prediction task was required for the industrial process and also explained which industrial problems are studied without prediction. We can conclude that different kinds of industrial prediction problems are required to be solved for industrial application. In Sect. 1.2, we introduced many kinds of data-based industrial prediction methods in different scene of applications and described some pivotal problems that will be solved in this book. In Sect. 1.3, we introduced some commonly used techniques, which will be elaborately studied in this book, including the time series prediction methods, the methods for factor-based prediction, the methods for industrial PIs construction, and the granular-based methods for long-term prediction. In Chap. 2, we will introduce some data preprocessing techniques for the raw data, which is a necessary procedure before modeling so as to obtain an accurate model. From Chaps. 3 to 6, we will focus on the core of this book.

References

1. Base, L. T. (1995). *Neural networks for pattern recognition*. Oxford: Oxford University Press.
2. Rasmussen, C., & Williams, C. (2006). *Gaussian processes for machine learning*. Cambridge: MIT Press.
3. Vapnik, V. N. (1995). *The nature of statistical learning theory*. New York: Springer.
4. Ekkachai, K., & Nilkhamhang, I. (2016). Swing phase control of semi-active prosthetic knee using neural network predictive control with particle swarm optimization. *IEEE Transactions on Neural Systems and Rehabilitation Engineering, 24*(11), 1169.
5. Jian, L., & Gao, C. (2013). Binary coding SVMs for the multiclass problem of blast furnace system. *IEEE Transactions on Industrial Electronics, 60*(9), 3846–3856.
6. Zhang, Y., Teng, Y., & Zhang, Y. (2010). Complex process quality prediction using modified kernel partial least squares. *Chemical Engineering Science, 65*(6), 2153–2158.
7. Lakehal, A., & Tachi, F. (2017). Bayesian duval triangle method for fault prediction and assessment of oil immersed transformers. *Measurement and Control, 50*(4), 103–109.
8. Reese, B. M., & Collins, E. G., Jr. (2016). A graph search and neural network approach to adaptive nonlinear model predictive control. *Engineering Applications of Artificial Intelligence, 55*, 250–268.
9. Jiang, S. L., Liu, M., Lin, J. H., et al. (2016). A prediction-based online soft scheduling algorithm for the real-world steelmaking-continuous casting production. *Knowledge-Based Systems, 111*, 159–172.
10. Fu, T. C. (2011). A review on time series data mining. *Engineering Applications of Artificial Intelligence, 24*(1), 164–181.
11. Nurkkala, A., Pettersson, F., & Saxén, H. (2011). Nonlinear modeling method applied to prediction of hot metal silicon in the ironmaking blast furnace. *Industrial and Engineering Chemistry Research, 50*(15), 9236–9248.
12. Han, H. G., Zhang, L., Hou, Y., et al. (2016). Nonlinear model predictive control based on a self-organizing recurrent neural network. *IEEE Transactions on Neural Networks & Learning Systems, 27*(2), 402.
13. Costa, A., Crespo, A., Navarro, J., et al. (2008). A review on the young history of the wind power short-term prediction. *Renewable and Sustainable Energy Reviews, 12*(6), 1725–1744.
14. Wang, W., Pedrycz, W., & Liu, X. (2015). Time series long-term forecasting model based on information granules and fuzzy clustering. *Engineering Applications of Artificial Intelligence, 41(C*, 17–24.
15. Khosravi, A., Nahavandi, S., Creighton, D., et al. (2011a). Comprehensive review of neural network-based prediction intervals and new advances. *IEEE Transactions on Neural Networks, 22*(9), 1341–1356.
16. Brahim-Belhouari, S., & Bermak, A. (2004). Gaussian process for nonstationary time series prediction. *Computational Statistics & Data Analysis, 47*(4), 705–712.
17. Aye, S. A., & Heyns, P. S. (2017). An integrated Gaussian process regression for prediction of remaining useful life of slow speed bearings based on acoustic emission. *Mechanical Systems & Signal Processing, 84*, 485–498.
18. Keprate, A., Ratnayake, R. M. C., & Sankararaman, S. (2017). Adaptive Gaussian process regression as an alternative to FEM for prediction of stress intensity factor to assess fatigue degradation in offshore pipeline. *International Journal of Pressure Vessels & Piping, 153*, 45–58.
19. Wang, F., Su, J., & Wang, Z. (2018). Prediction of subsidence of buildings as a result of earthquakes by Gaussian process regression. *Chemistry & Technology of Fuels & Oils, 99*, 363–373.
20. Jani, D. B., Mishra, M., & Sahoo, P. K. (2017). Application of artificial neural network for predicting performance of solid desiccant cooling systems – A review. *Renewable & Sustainable Energy Reviews, 80*, 352–366.

21. Sujay, R. N., & Deka, P. C. (2014). Support vector machine applications in the field of hydrology: A review. *Applied Soft Computing Journal, 19*(6), 372–386.
22. Robert, C. (2012). *Machine learning, a probabilistic perspective*. Cambridge: MIT Press.
23. Jacobsson, H. (2005). Rule extraction from recurrent neural networks: A taxonomy and review. *Neural Computation, 17*(6), 1223–1263.
24. Beyhan, S. (2017). Affine TS fuzzy model-based estimation and control of Hindmarsh-Rose neuronal model. *IEEE Transactions on Systems Man & Cybernetics Systems, 47*(8), 2342–2350.
25. Li, Y., Zhang, W., Xiong, Q., et al. (2017). A rolling bearing fault diagnosis strategy based on improved multiscale permutation entropy and least squares SVM. *Journal of Mechanical Science & Technology, 31*(6), 2711–2722.
26. Wild, C. J., & Seber, G. A. F. (1989). *Nonlinear regression*. New York: Wiley.
27. Nix, D. A., & Weigend, A. S. (1994). Estimating the mean and variance of the target probability distribution. In *Proceedings of IEEE International Conference on Neural Networks* (Vol. 1, pp. 55–60), Orlando, FL, 1994.
28. Pan, L., & Politis, D. N. (2016). Bootstrap prediction intervals for Markov processes. *Computational Statistics & Data Analysis, 100*, 467–494.
29. Khosravi, A., Nahavandi, S., Creighton, D., et al. (2011b). Lower upper bound estimation method for construction of neural network-based prediction intervals. *IEEE Transactions on Neural Networks, 22*(3), 337–346.
30. Sheng, C., Zhao, J., & Wang, W. (2017). Map-reduce framework-based non-iterative granular echo state network for prediction intervals construction. *Neurocomputing, 222*, 116–126.
31. Skowron, A., & Stepaniuk, J. (2001). Information granules: Towards foundations of granular computing. *International Journal of Intelligent Systems, 16*(1), 57–85.
32. Pedrycz, W. (2005). *Knowledge-based clustering: From data to information granules*. Chichester: Wiley-Interscience.
33. Lu, W., Chen, X., Pedrycz, W., et al. (2015). Using interval information granules to improve forecasting in fuzzy time series. *International Journal of Approximate Reasoning, 57*(1), 1–18.

Chapter 2
Data Preprocessing Techniques

Abstract It is hard for raw industrial data accumulated by commonly implemented supervisory control and data acquisition (SCADA) system on-site to be directly employed to construct a prediction model, given that such data are always mixed with high level noise, missing points, and outliers due to the possible real-time database malfunction, data transformation, or maintenance. Thereby, the data preprocessing techniques have to be implemented, which usually contain anomaly data detection, data imputation, and data de-noising techniques. As for the issue of outliers, in this chapter, we introduce the anomaly detection methods based on fuzzy C means (FCM), *K*-nearest-neighbor (KNN), and dynamic time warping (DTW) algorithms. To tackle the missing data points problem, a series of data imputation methods are also described. After introducing the generic regression filling and expectation maximum methods, we supplement a varied window similarity measure method, the segmented shape-representation-based method, and the non-equal-length granules correlation method for industrial data imputation. With respect to the high level noise embodied in raw data, we then give an introduction to the well-known empirical mode decomposition (EMD) method. To verify the effectiveness of these methods, this chapter also provides a number of industrial case studies.

2.1 Introduction

With the rapid development of information technology, most of the industrial enterprises have accumulated huge amounts of data related to the manufacturing processes (often several gigabytes or more) via SCADA system [1, 2]. Data mining techniques enable to discover the implicit and extremely useful relationship between data patterns [3, 4]. However, these data are usually corrupted by industrial high level noise, and they often contain missing points and outliers, due to the poor real-time database malfunction, improper data management and maintenance, and data input errors, etc. In addition, the data exchange between enterprises might also cause the data inconsistencies. Therefore, it is very important to preprocess these industrial data before modeling or employing some data mining approaches.

J. Zhao et al., *Data-Driven Prediction for Industrial Processes and Their Applications*, Information Fusion and Data Science,
https://doi.org/10.1007/978-3-319-94051-9_2

Currently, there are a number of data preprocessing techniques including data cleaning, data integration, data reduction, data imputation, and data transformation. Data cleaning is to remove noise and correct inconsistencies involved in raw industrial data. Data integration merges those coming from multiple sources into a coherent data store, such as the data warehouse technique. Data reduction removes some samples or eliminates the irrelevant features by aggregating, eliminating redundant features, or clustering. Data transformation (e.g., normalization) is to scale the data to fall into a certain range [5]. These techniques can enhance the accuracy and efficiency of data mining algorithms, and they can be used together in a single modeling task [6, 7]. For example, data cleaning usually involves transformations to correct wrong data, such as by transforming all entries for a date field to a common format.

In this chapter, we will introduce these preprocessing techniques for industrial data and discuss three issues applied in industrial cases in details. The first issue in this chapter is the anomaly detection of data, which also named as outlier detection aims to identify items, events, or observations that do not conform to an expected pattern in a dataset [8, 9]. We employ an industrial by-product energy example generated in steel plant in order to explain the practical problem and its solving approach. The flow of by-product gas is an essential secondary energy in the steel industry, such as blast furnace gas (BFG) and coke oven gas (COG). Figure 2.1

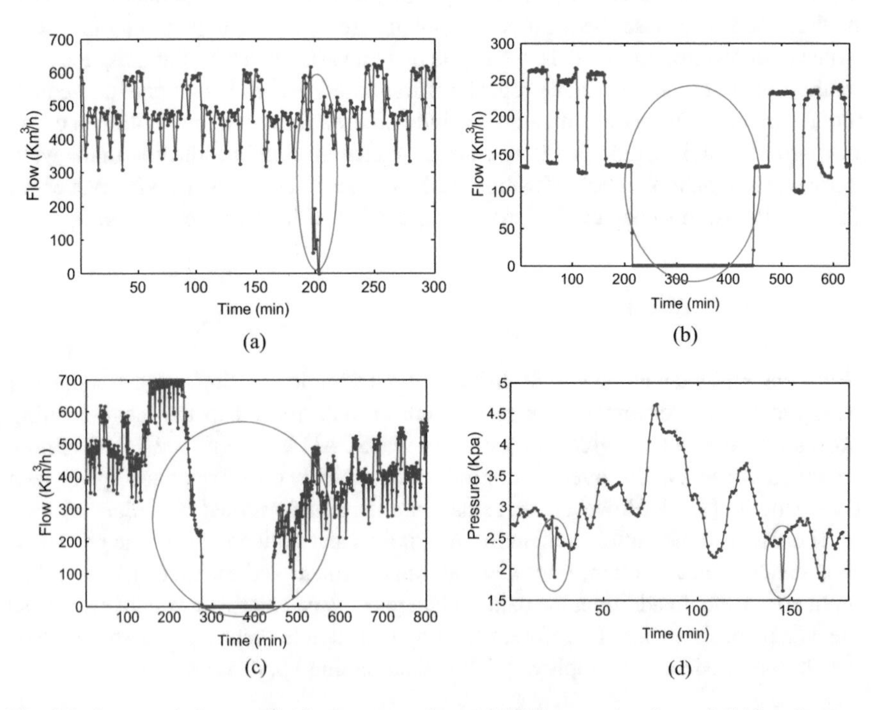

Fig. 2.1 Some typical anomalies in the energy data. (**a**) BFG generation flow of #1 blast furnace. (**b**) BFG consumption of hot blast stoves. (**c**) BFG generation flow of #2 blast furnace. (**d**) Main pipeline pressure of COG system

illustrates some anomalous situations occurred in the samples of BFG and COG, where the trend anomaly is presented in Fig. 2.1a–c, and the anomaly coming from the irregular data with an amplitude deviation from its neighborhood is regarded as the contextual anomaly or deviants, see Fig. 2.1d. As for the anomaly detection of data, the adaptive fuzzy C means (FCM) and KNN-AFCM algorithm will be introduced in this chapter [10].

The second issue in this chapter is data imputation for raw missing points. In statistics, missing data, or values, occur when no data value is stored for the variable [11, 12]. Since the industrial process is usually rather complicated and its data acquisition process might be frequently affected by the unexpected operational factors, the data-missing phenomenon usually occurs, which might lead to failure of model establishment or inaccurate information discovery [13]. In this chapter, the segmented shape-representation method and the non-equal-length granules correlation coefficient method will be detailed, which are mainly based on the sliding-window technology and the estimation of distribution algorithm.

The third issue is data de-noising, which is a process of removing noise from an original time series data [14, 15]. The cause of noise may be hardware failures, programming errors, or other interference when data is transformed. The empirical mode decomposition (EMD) technique for de-noising will be introduced in this section, which decomposes the original signals of interest into a number of oscillatory functions including the noise function.

This chapter focuses on the aforementioned problems in the field of industrial time series preprocessing. It also gives the application of data preprocessing with a specific industrial environment. Finally, in the last subsection we give a discussion about the following direction of future research.

2.2 Anomaly Data Detection

To detect outliers in a time series data, data clustering methods is usually adopted for such a task. Clustering a time series mainly considers two aspects, the selection of clustering algorithm and the selection of distance measure. In this chapter, we will introduce the commonly used clustering algorithms such as the *K*-nearest-neighbor (KNN), the FCM method, and their improved versions, i.e., the adaptive FCM (AFCM) and KNN-AFCM. Based on these clustering algorithms, we introduce two anomaly detection methods, including the trend anomaly detection based on AFCM, and the deviants detection based on KNN-AFCM.

2.2.1 K-*Nearest-Neighbor*

The *K*-nearest-neighbor (KNN) classification is one of the most simple and commonly used methods, and it is also one of the first choices for a classification task when there is little prior knowledge about the distribution of the data [16].

The Euclidean distance between a test sample and a specified training sample is usually used in a KNN classifier, which is defined by,

$$d(\mathbf{x}_i, \mathbf{x}_l) = \sqrt{(x_{i1} - x_{l1})^2 + (x_{i2} - x_{l2})^2 + \cdots + (x_{iM} - x_{lM})^2} \tag{2.1}$$

where $\mathbf{x}_i = (x_{i1}, x_{i2}, \ldots, x_{iM})^T$ and $\mathbf{x}_l = (x_{l1}, x_{l2}, \ldots, x_{lM})^T$ are two samples with M features, and the total number of input samples and features are n and M, respectively.

Thus, the training procedure of a KNN model is listed as follows:

Algorithm 2.1: Training the KNN Model

***Step 1**: Calculate the Euclidean distance between the test sample data and each training sample.*

***Step 2**: Sort these distances of the training samples calculated in Step 1 in ascending order.*

***Step 3**: Select the first K points of which their distances are the smallest among the training samples.*

***Step 4**: Compute the frequency of each class label appeared in these K points.*

***Step 5**: Treat the label with highest frequency as the predictive label of the test sample.*

2.2.2 *Fuzzy C Means*

Clustering analysis is an unsupervised learning technique, which aims to make the massive data be clustered into different clusters. The similarity between samples in one cluster becomes as small as possible so that samples assigned to different clusters have large dissimilarity. The fuzzy set was conceived as a result of an attempt to come to grips with the problem of pattern recognition in the context of imprecisely defined categories. In such cases, the belonging of a sample to a class is a matter of degree, instead of the question of whether or not a group of samples form a cluster. Among various fuzzy clustering algorithms, the FCM algorithm is one of the most widely used ones [17], which can determine the degree of membership of each sample to all the class centers by optimizing its objective function [18, 19]. In a FCM, the membership function of an element in the dataset reflects the membership degree of an element belonging to a fuzzy set, of which the value of its range is in [0, 1]. The closer its value approaches 1, the higher the degree of this element belonging to the fuzzy set is.

The details of a FCM are listed as below. Let J_{FCM} represent the objective function of FCM to be minimized, which is defined by

$$
\begin{aligned}
J_{\mathrm{FCM}} &= \sum_{j=1}^{n}\sum_{i=1}^{c} u_{ij}^{m} d_{ij}^{2} \\
\text{s.t.} \sum_{j=1}^{n} u_{ij} > 0, &\sum_{i=1}^{c} u_{ij} = 1, u_{ij} \in [0, 1]
\end{aligned}
\tag{2.2}
$$

where u_{ij} denotes the degree of fuzzy membership belonging to the ith class for the jth sample, and d_{ij} is a distance measure between the sample $\mathbf{x}_j$ and the clustering center $\mathbf{v}_i$. This distance measure can be the Euclidean distance defined by (2.1), as well as the Markov distance and so on. m denotes the fuzzy coefficients with its value being set in the range [1.5, 2.5], which is used to adjust the validity of the FCM partition [20]. n and c denotes the number of samples and cluster centers, respectively. Then, we give the update equation of the degree matrix of membership $\mathbf{U} = \{u_{ij}\}_{c\times n}$ and the clustering centers $\{\mathbf{v}_i\}_{i=1}^{c}$, respectively.

$$
u_{ij} = \frac{d_{ij}^{-\frac{2}{m-1}}}{\sum_{i=1}^{c} d_{ij}^{-\frac{2}{m-1}}}
\tag{2.3}
$$

$$
\mathbf{v}_i = \frac{\sum_{j=1}^{n} u_{ij}^{m} \mathbf{x}_j}{\sum_{j=1}^{n} u_{ij}^{m}}
\tag{2.4}
$$

Thus, the FCM procedure can be stepped as follows:

Algorithm 2.2: The FCM Algorithm

Model inputs: *the dataset* $\mathbf{X} = [\mathbf{x}_1, \mathbf{x}_2, \ldots, \mathbf{x}_i \ldots, \mathbf{x}_n]^{\mathrm{T}}$ *with each row representing a sample, the number of clustering centers c, the threshold ε for stopping the iteration, and the count variables k of iteration.*

Model outputs: *the clustering centers* $\{\mathbf{v}_i\}_{i=1}^{c}$, *the degree matrix of membership* $\mathbf{U} = \{u_{ij}\}_{c\times n}$.

Step 1: *Select the initial clustering centers* $\mathbf{v}_1^{(0)}, \mathbf{v}_2^{(0)}, \ldots, \mathbf{v}_c^{(0)}$, *and initialize the degree matrix* $\mathbf{U}^{(0)}$ *with its value selected randomly in the unit range. Let k=0.*

Step 2: *Update the degree matrix **U** by (2.3).*

Step 3: *Update the clustering centers* $\{\mathbf{v}_i\}_{i=1}^{c}$ *by (2.4).*

Step 4: *Calculate the objective function by (2.2), and let* $k = k + 1$.

Step 5: *If* $|J_{\mathrm{FCM}}{}^{k+1} - J_{\mathrm{FCM}}{}^{k}| \le \varepsilon$, *return to Step 2; otherwise, stop this iteration procedure and give the outputs.*

Finally, the adscription of all samples can be determined by the obtained degree matrix **U**, i.e., when $u_{jk} = \max_{1\le i\le c} \{u_{ik}\}$, the sample $\mathbf{x}_k$ belongs to the jth class.

2.2.3 Adaptive Fuzzy C Means

As for industrial applications, the data accumulated by the SCADA system often contain anomalistic points. For example, as for a typical industrial system presented in this chapter, the by-product gas system in steel manufacturing process, the deviant data points often occurred in the tendency of the regular production cadence. To deal with this anomaly mentioned above, this section employs an adaptive FCM (AFCM) method for anomaly detection, which transforms the detection process into a clustering problem on the basis of the characteristics of the industrial data.

The AFCM is based on the generic FCM, which introduces an adaptive vector and an index serving as the fitness in the objective function. And, the AFCM takes the differences among the data samples into account through the adaptive vector and the adaptive index, which control the clustering process together, targeting to deal with outliers for the quality of clustering.

Next, we will introduce the AFCM algorithm in details. J_{AFCM} denotes the objective function of AFCM, which is defined by

$$\begin{gathered} J_{\text{AFCM}} = \sum_{j=1}^{n} w_j^p \sum_{i=1}^{c} u_{ij}^m d_{ij}^2 \\ \text{s.t.} \sum_{j=1}^{n} u_{ij} > 0, \sum_{i=1}^{c} u_{ij} = 1, \prod_{j=1}^{n} w_j = 1 \end{gathered} \tag{2.5}$$

where w_j denotes the adaptive degree of the jth sample $\mathbf{x}_j$, which measures the influence of $\mathbf{x}_j$ on the objective function, and it is updated in the iterative process. $p(p \neq 0)$ is the adaptive exponent used to adjust the value of the adaptive degree. The constraint $\prod_{j=1}^{n} w_j = 1$ indicates the intrinsic correlation of the samples with respect to the objective function. And, u_{ij} denotes the degree of the fuzzy membership of the adaptive model, and m is the fuzzy coefficient. Then, we have the update equation of u_{ij}, $\mathbf{v}_i$, and w_j.

$$u_{ij} = \frac{d_{ij}^{-\frac{2}{m-1}}}{\sum_{i=1}^{c} d_{ij}^{-\frac{2}{m-1}}} \tag{2.6}$$

$$\mathbf{v}_i = \frac{\sum_{j=1}^{n} w_j u_{ij}^m \mathbf{x}_j}{\sum_{j=1}^{n} w_j u_{ij}^m} \tag{2.7}$$

$$w_j = \left[\frac{\left[\prod_{j=1}^{n} \left(\sum_{i=1}^{c} u_{ij}^m d_{ij}^2 \right) \right]^{1/n}}{\sum_{i=1}^{c} u_{ij}^m d_{ij}^2} \right]^{1/p} \tag{2.8}$$

In the next section, we will introduce a trend anomaly detection method based on the AFCM.

2.2.4 Trend Anomaly Detection Based on AFCM and DTW

As for the pseudo-periodic industrial data, considering that the data trend reflects the related similar production operation in manufacturing process, a DTW-based sequence stretching method is presented here to transform the similarity of the sequences with unequal length into the corresponding Euclidean distance metric, which serves to calculate the similarity degree between the sequences. Then, the AFCM is employed to find the anomaly data. We firstly describe the procedure of the DTW.

Considering two sequences $\mathbf{s} = s_1, s_2, \ldots, s_m$ and $\mathbf{t} = t_1, t_2, \ldots, t_n$, the distance matrix can be described by

$$\mathbf{D}_{m \times n} = \begin{bmatrix} d(s_1, t_1) & d(s_1, t_2) & \ldots & d(s_1, t_n) \\ d(s_2, t_1) & d(s_2, t_2) & \ldots & d(s_2, t_n) \\ \ldots & \ldots & \ldots & \ldots \\ d(s_m, t_1) & d(s_m, t_2) & \ldots & d(s_m, t_n) \end{bmatrix}.$$

One can use a grid to describe the matrix, shown in Fig. 2.2, where the crossing points represent the elements of the matrix. In this figure, the solid black line indicates the warping path, on which the crossing point can be denoted as p_k and its amount is K. To minimize the warping distance $\mathrm{DTW}(\mathbf{s}, \mathbf{t}) = \frac{1}{K} \sum_{j=1}^{n} \sum_{i=1}^{m} d(s_i, t_j)$, where the point $(s_i, t_j) \in \{p_1, \ldots, p_k, \ldots, p_K\}$, p_{k+1} can be determined when the set $\{d(s_{i+1}, t_j), d(s_i, t_{j+1}), d(s_{i+1}, t_{j+1})\}$ has its minimum. In addition, the warping path must be within the warping window, presented as Fig. 2.2.

We then give the definition of the central sequence as follows. Assuming that there are n sequences $\mathbf{x}_1, \mathbf{x}_2, \ldots, \mathbf{x}_n$, one can define $\mathrm{Sum}D_i = \sum_{j=1, j \neq i}^{n} \mathrm{DTW}(\mathbf{x}_i, \mathbf{x}_j)$ as the sum of DTW distances between $\mathbf{x}_i (i = 1, 2, \ldots, n)$ and the other $n - 1$ sequences. As such, we designate $\mathbf{x}_k$ as the central sequence when the value of $\mathrm{Sum}D_k$ is the minimum among the set $\{\mathrm{Sum}D_1, \mathrm{Sum}D_2, \ldots, \mathrm{Sum}D_n\}$. Based on the principle of DTW, the defined central sequence is used as the standard one to realize the

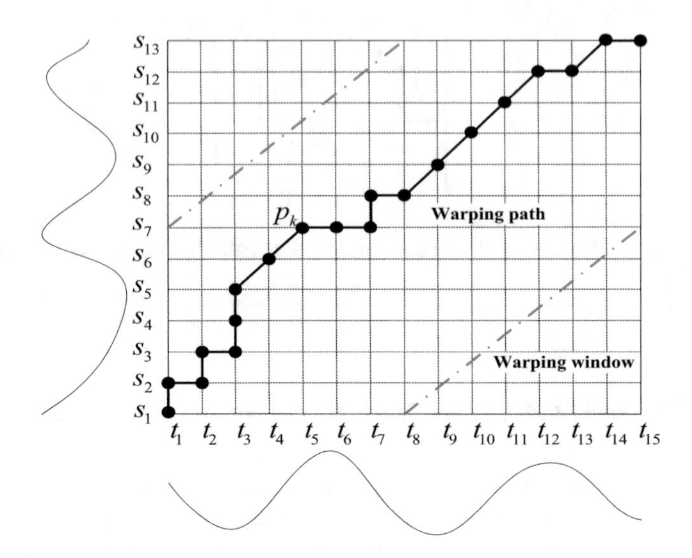

Fig. 2.2 The illustrative chart for DTW

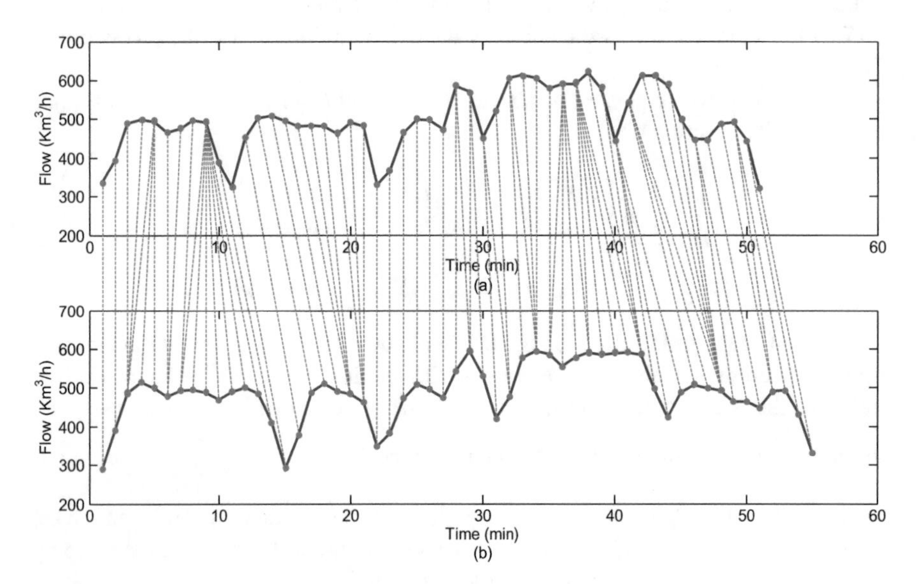

Fig. 2.3 Two similar sequences and their corresponding relationship on DTW path. (**a**) The original sequence *a*. (**b**) The original sequence *b*

comparison to the others. Supposing that two similar sequences **a** and **b**, one can assume **a** to be the central sequence and $\mathbf{b}'$ to be the stretched sequence from **b**. Then, the DTW path can be obtained, and the corresponding relationship on the path can be shown as Fig. 2.3, where Fig. 2.3a and b indicate the original central sequence and a generic one, respectively. In detail, if the *i*th point of **a** corresponds to the

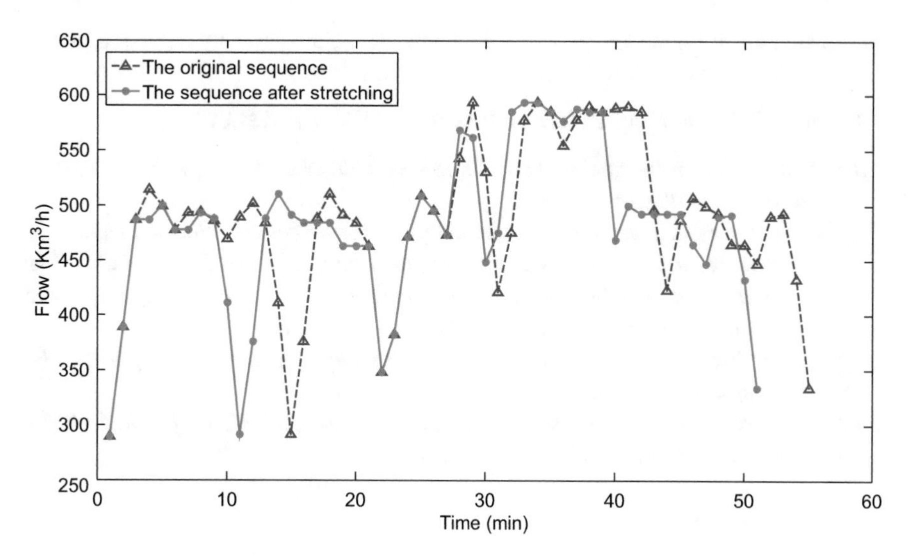

Fig. 2.4 The comparison of the original sequences and that after stretching

jth point of **b**, then the relationship can be represented as $a_i \rightarrow b_j$. In such a way, the DTW-based sequence stretching reads as follows:

1. If $a_i \rightarrow (b_j, b_{j+1}, \ldots, b_r)$, then $b_i' = \text{Avg}(b_j, b_{j+1}, \ldots, b_r)$;
2. If $(a_i, a_{i+1}, \ldots, a_r) \rightarrow b_j$, then $b_i' = b_j, b_{i+1}' = b_j, \ldots, b_r' = b_j$.

where the operator Avg denotes the average value of the multiple data points. It is clear that the stretched sequence $\mathbf{b}'$ is of equal length with **a**. Thus, by using the proposed stretching, the one-to-many or many-to-one relationship of the similar sequences can be transformed into the one-to-one relationship with equal length. For an illustrative presentation, Fig. 2.4 gives the original sequence **b** and the corresponding $\mathbf{b}'$ after stretching by using the standard one **a**, in which one can apparently observe that the stretched sequence can maintain the trend and amplitude characteristics of the original one without the loss of the sequence similarity.

Based on the DTW sequence stretching, the obtained sequence can be used to realize the AFCM clustering, in which the adaptive degree, w_j, can be calculated. In [10], the anomaly data can be determined by using an anomaly index $O_{\text{AFCM}}(\mathbf{x}_j) = \min\left\{w_j^p\right\}/w_j^p$, which means that the data point that makes the criterion equal to 1 can be detected as the anomaly data. However, such index can hardly reflect the practical anomaly situations because there always exists a data point that makes the index equal to 1. Therefore, we define a modified anomaly index in this section, $O_{\text{AFCM}}(\mathbf{x}_j) = 1/w_j^p$, by which the anomaly threshold can be reasonably designated according to the practical demands. Thus, the anomaly detection procedure for the pseudo-periodic data can be listed as follows:

Algorithm 2.3: Anomaly Detection Based on AFCM and DTW

Step 1*: Divide the entire sample dataset into n sequences* $\mathbf{x}_1, \mathbf{x}_2, \ldots, \mathbf{x}_n$ *by estimating its periodical feature.*
Step 2*: Get the central sequence* $\mathbf{x}_k$ *according to the proposed definition rule.*
Step 3*: Set the central sequence as the standard one, and stretch all of the other* $n - 1$ *sequences to construct the sequences with identical length, i.e., these sequences are with the same dimensionality.*
Step 4*: For higher detection accuracy, the sequences with same length can be divided into a series of smaller sequences.*
Step 5*: Cluster the obtained sequences by using the AFCM, and calculate the anomaly index to detect the anomaly data.*

2.2.5 Deviants Detection Based on KNN-AFCM

As for the irregular data points obtained in an industrial system, it can be discovered that the anomaly are mostly local deviants, as shown in Fig. 2.1d. Considering that there is often a large deviation between the deviant and its neighbors, here we present a method called KNN-AFCM, in which the correlative information of the neighboring points is introduced. The objective function for this clustering algorithm is given by

$$\begin{gathered} J_{\text{KNN-AFCM}} = \sum_{j=1}^{n} s_j w_j^p \sum_{i=1}^{c} u_{ij}^m \left\| \mathbf{x}_j - \mathbf{v}_i \right\|^2 + \beta \sum_{j=1}^{n} \sum_{i=1}^{c} s_j u_{ij}^m \left\| \overline{\mathbf{x}'}_j - \mathbf{v}_i \right\|^2 \\ \text{s.t.} \sum_{j=1}^{n} u_{ij} > 0, \sum_{i=1}^{c} u_{ij} = 1, \prod_{j=1}^{n} w_j = 1 \end{gathered} \tag{2.9}$$

where the first term in the right hand side of (2.9) aims to estimate the deviation degree between the point $\mathbf{x}_j$ and its neighbors; the second term affects the clustering result of the point $\mathbf{x}_j$ by its neighbors' value; $\overline{\mathbf{x}'}_j$ denotes the mean value of the neighbors of $\mathbf{x}_j$; and β denotes the coefficient of the mean value of the neighbors. In addition, $\bar{D} = \left(D_1 + \ldots D_j \ldots + D_{\text{n}}\right)/n$, where $D_j = \left|\mathbf{x}_j - \overline{\mathbf{x}'}\right|_j$. One can designate s_j to be the weight coefficient corresponding to $\mathbf{x}_j$, i.e., $s_j = a^{D_j/\bar{D}}\,(a \geq 1)$. Especially, when $a = 1$, the impact of the weight coefficient on the point $\mathbf{x}_j$ can be neglected. It is obvious that the points closer to the maximum or the minimum of a sequence will get a larger s_j. As such, the deviants that deviate the neighborhood trend can be effectively detected.

To minimize the constraint objective function defined by (2.9), one can use the Lagrangian extreme method. Considering the constraints, this optimization objective function is written as follows:

$$\begin{aligned} J_{\text{KNN-AFCM},\mu_1,\mu_2} &= \sum_{j=1}^{n} s_j w_j^p \sum_{i=1}^{c} u_{ij}^m \left\|\mathbf{x}_j - \mathbf{v}_i\right\|^2 + \beta \sum_{j=1}^{n}\sum_{i=1}^{c} s_j u_{ij}^m \left\|\overline{\mathbf{x}'}_j - \mathbf{v}_i\right\|^2 \\ &+ \mu_1\left(\sum_{i=1}^{c} u_{ij} - 1\right) + \mu_2\left(\prod_{j=1}^{n} w_j - 1\right) \end{aligned} \tag{2.10}$$

The derivatives of (2.10) with respective to u_{ij} and w_j are

$$\begin{cases} \dfrac{\partial J_{\text{KNN-AFCM},\mu_1,\mu_2}}{\partial u_{ij}} = m \times s_j w_j^p u_{ij}^{m-1}\left\|\mathbf{x}_j - \mathbf{v}_i\right\|^2 + m\beta s_j u_{ij}^{m-1}\left\|\overline{\mathbf{x}'}_j - \mathbf{v}_i\right\|^2 + \mu_1 \\ \dfrac{\partial J_{\text{KNN-AFCM},\mu_1,\mu_2}}{\partial w_j} = p s_j w_j^{p-1}\sum_{i=1}^{c} u_{ij}^m \left\|\mathbf{x}_j - \mathbf{v}_i\right\|^2 + \mu_2\left(\prod_{k=1,k\neq j}^{n} w_k\right) \end{cases} \tag{2.11}$$

Setting this to zero and solving for {μ$_1$ and μ_2 gives

$$\mu_1 = -m \times s_j w_j^p u_{ij}^{m-1}\left\|\mathbf{x}_j - \mathbf{v}_i\right\|^2 - m \times \beta s_j u_{ij}^{m-1}\left\|\overline{\mathbf{x}'}_j - \mathbf{v}_i\right\|^2 \tag{2.12}$$

$$\mu_2 = -\frac{p \times s_j w_j^{p-1}\sum\limits_{i=1}^{c} u_{ij}^m \left\|\mathbf{x}_j - \mathbf{v}_i\right\|^2}{\prod\limits_{k=1,k\neq j}^{n} w_k} \tag{2.13}$$

From (2.12) we can get the estimation of u_{ij}

$$u_{ij} = \left(\mu_1/(-m \times s_j)\right)^{1/m-1}\left(1/\left(w_j^p\left\|\mathbf{x}_j - \mathbf{v}_i\right\|^2 + \beta\left\|\overline{\mathbf{x}'}_j - \mathbf{v}_i\right\|^2\right)\right)^{1/m-1} \tag{2.14}$$

Summing the degrees of the fuzzy membership of the point $\mathbf{x}_j$ with respect to all the cluster centers gives

$$\sum_{i=1}^{c} u_{ij} = \left(\mu_1/(-m \times s_j)\right)^{1/m-1}\sum_{i=1}^{c}\left(1/\left(w_j^p\left\|\mathbf{x}_j - \mathbf{v}_i\right\|^2 + \beta\left\|\overline{\mathbf{x}'}_j - \mathbf{v}_i\right\|^2\right)\right)^{1/m-1} \tag{2.15}$$

Since the left hand side of (2.15) is equal to 1, one can get

$$\left(\mu_1/(-m \times s_j)\right)^{1/m-1} = \frac{1}{\sum\limits_{i=1}^{c}\left(1/\left(w_j^p\left\|\mathbf{x}_j - \mathbf{v}_i\right\|^2 + \beta\left\|\overline{\mathbf{x}'}_j - \mathbf{v}_i\right\|^2\right)\right)^{1/m-1}} \tag{2.16}$$

According to (2.14) and (2.16), the iterative rule of u_{ij} can be written as

$$u_{ij} = \frac{1}{\sum_{k=1}^{c} \left(\frac{w_j^p \left\| \mathbf{x}_j - \mathbf{v}_i \right\|^2 + \beta \left\| \overline{\mathbf{x}'}_j - \mathbf{v}_i \right\|^2}{w_j^p \left\| \mathbf{x}_j - \mathbf{v}_k \right\|^2 + \beta \left\| \overline{\mathbf{x}'}_j - \mathbf{v}_k \right\|^2} \right)^{\frac{1}{m-1}}} \tag{2.17}$$

Form (2.13), we can get

$$\mu_2 \left(\prod_{k=1,\, k \neq j}^{n} w_k \right) w_j = \left(-p \times s_j w_j^{p-1} \sum_{i=1}^{c} u_{ij}^m \left\| \mathbf{x}_j - \mathbf{v}_i \right\|^2 \right) w_j \tag{2.18}$$

With the resistance $\prod_{j=1}^{n} w_j = 1$, μ_2 can be rewritten as

$$\mu_2 = -p \times s_j w_j^{p-1} \sum_{i=1}^{c} u_{ij}^m \left\| \mathbf{x}_j - \mathbf{v}_i \right\|^2 \tag{2.19}$$

Besides,

$$w_j = \left(-\mu_2 / p \times s_j \sum_{i=1}^{c} u_{ij}^m \left\| \mathbf{x}_j - \mathbf{v}_i \right\|^2 \right)^{1/p} \tag{2.20}$$

From (2.19), the scalar μ_2 can be expressed as a function of the data samples $\{\mathbf{x}_j\}_{j=1}^{n}$, i.e.,

$$\mu_2 = (-p) \left(\prod_{j=1}^{n} s_j \right)^{\frac{1}{n}} \left(\prod_{j=1}^{n} \sum_{i=1}^{c} u_{ij}^m \left\| \mathbf{x}_j - \mathbf{v}_i \right\|^2 \right)^{\frac{1}{n}} \tag{2.21}$$

Replacing it into (2.20), then

$$w_j = \left(\left(\prod_{j=1}^{n} s_j \right)^{\frac{1}{n}} \Big/ s_j \right)^{\frac{1}{p}} \left(\frac{\left(\prod_{j=1}^{n} \sum_{i=1}^{c} u_{ij}^m \left\| \mathbf{x}_j - \mathbf{v}_i \right\|^2 \right)^{\frac{1}{n}}}{\sum_{i=1}^{c} u_{ij}^m \left\| \mathbf{x}_j - \mathbf{v}_i \right\|^2} \right)^{\frac{1}{p}} \tag{2.22}$$

Thus, one can simplify (2.22) as

$$w_j = \left(\frac{a}{s_j}\right)^{\frac{1}{p}} \left(\frac{\left(\prod_{j=1}^{n} \sum_{i=1}^{c} u_{ij}^m \left\| \mathbf{x}_j - \mathbf{v}_i \right\|^2 \right)^{\frac{1}{n}}}{\sum_{i=1}^{c} u_{ij}^m \left\| \mathbf{x}_j - \mathbf{v}_i \right\|^2} \right)^{\frac{1}{p}} \tag{2.23}$$

which is the self-adaptability iteration rule of w_j.

The partial derivative of the objective function with respective to $\mathbf{v}_i$ is given by

$$\frac{\partial J_{\text{KNN-AFCM},\mu_1,\mu_2}}{\partial \mathbf{v}_i} = 2\sum_{j=1}^{n} s_j w_j^p u_{ij}^m \left(\mathbf{x}_j - \mathbf{v}_i\right) + 2\beta \sum_{j=1}^{n} s_j u_{ij}^m \left(\overline{\mathbf{x}'}_j - \mathbf{v}_i\right) \tag{2.24}$$

Setting this partial derivative to be 0, one has an expression of $\mathbf{v}_i$,

$$\mathbf{v}_i = \frac{\sum_{j=1}^{n} \left(w_j s_j^p u_{ij}^m \mathbf{x}_j + \beta s_j^p u_{ij}^m \overline{\mathbf{x}'}_j\right)}{\sum_{j=1}^{n} \left(w_j s_j^p u_{ij}^m + \beta s_j^p u_{ij}^m\right)} \tag{2.25}$$

One can also obtain w_j^p from (2.23)

$$w_j^p = \frac{a}{s_j} \frac{\left(\prod_{j=1}^{n} \sum_{i=1}^{c} u_{ij}^m \left\| \mathbf{x}_j - \mathbf{v}_i \right\|^2 \right)^{\frac{1}{n}}}{\sum_{i=1}^{c} u_{ij}^m \left\| \mathbf{x}_j - \mathbf{v}_i \right\|^2} \tag{2.26}$$

It's obvious that the right hand side of (2.26) is a quantity, which does not contain p, so that w_j^p is integrated. Therefore, we replace w_j by w_j^p. At this point, the value of p indirectly affects the results of the cluster centers. Thus, the iteration rule of the ith cluster center $\mathbf{v}_i$ should be

$$\mathbf{v}_i = \frac{\sum_{j=1}^{n} \left(w_j s_j u_{ij}^m \mathbf{x}_j + \beta s_j u_{ij}^m \overline{\mathbf{x}'}_j\right)}{\sum_{j=1}^{n} \left(w_j s_j u_{ij}^m + \beta s_j u_{ij}^m\right)} \tag{2.27}$$

There is a large value of Euclidean distance between the deviant point and its clustering prototype, which means that the deviant point and its neighbors could not be located into the same cluster region. From (2.17), it can be observed that how a cluster the point $\mathbf{x}_j$ belongs to can be affected by not only its own value but also the one of its neighbors. If β is large enough, the clustering centers will be largely determined by its neighbors. As such, once $\mathbf{x}_j$ is a deviant point, the value of $\left\|\overline{\mathbf{x}'}_j - \mathbf{v}_i\right\|^2$ and $\|\mathbf{x}_j - \mathbf{v}_i\|^2$ will decrease during the learning procedure. Based on

(2.10) and (2.20), the value of w_j^p will adaptively decrease along with the minimization of the objective function (2.10). Thus, the larger the deviation between a point and its neighbors is, the smaller the value of its corresponding w_j^p is. Therefore, the anomaly index can be used to identify the local deviants.

2.2.6 Case Study

In this section, we report an industrial application of the AFCM and the KNN-AFCM methods for the anomaly detection of a blast furnace gas (BFG) system in a steel plant. Some periods of the pipeline pressure data of the BFG generation are shown in Fig. 2.5a, from which one can see that there are three deviant points marked by red circles. The results of KNN-AFCM, the SVM-based method and the AFCM for anomaly detection are illustrated as Fig. 2.5b, c, in which the last two methods perform relatively poor results for the deviant detection. In the perspective of the training time, Table 2.1 lists the comparative results by adopting these methods when the values of c changes. It is obvious from this table that the computing efficiency of KNN-AFCM is much higher than those of the SVM-based method. Although the computing by the KNN-AFCM is somewhat slower than those of the generic AFCM, the efficiency can be acceptable for the real application.

In addition, to illustrate the impacts of the weight coefficients s_j in KNN-AFCM on the detection results, one can randomly employ 300 data points of the pipeline

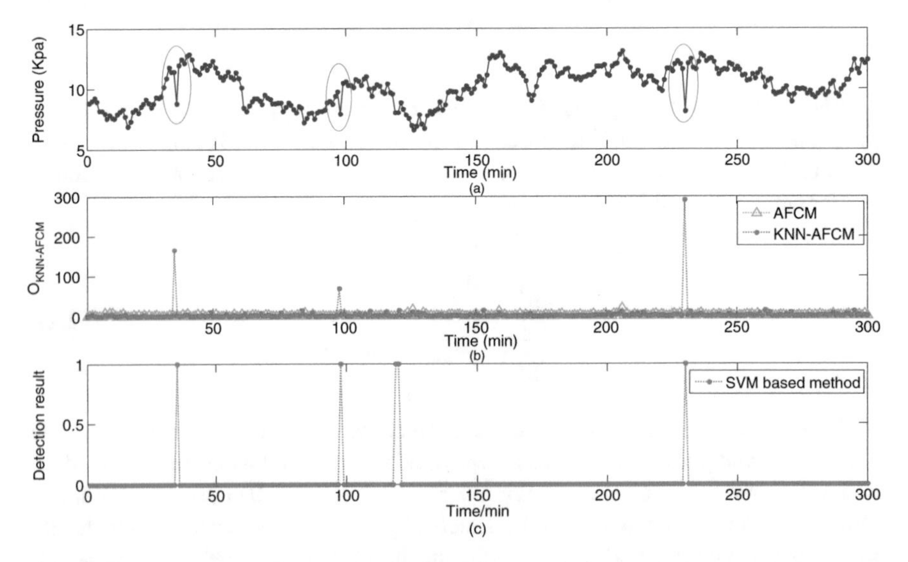

Fig. 2.5 The comparison of the deviant points detection by using the methods. (**a**) pipeline pressure of #5 blast furnace. (**b**) The value of the anomaly index. (**c**) Detection results by using the SVM based method

Table 2.1 Comparison of the training time of different methods

The values of c	KNN-AFCM (s)	AFCM (s)	SVM-based method (s)
2	1.110	1.172	14.516
3	1.282	0.844	13.937
4	2.047	0.781	13.766
5	2.750	1.047	13.859
6	2.765	1.875	14.031
7	2.953	1.938	14.328
8	2.859	2.235	14.078
9	3.734	2.344	14.125
10	4.297	2.594	13.781
11	4.718	2.516	14.328
12	4.422	2.719	14.297
13	4.766	2.266	13.859

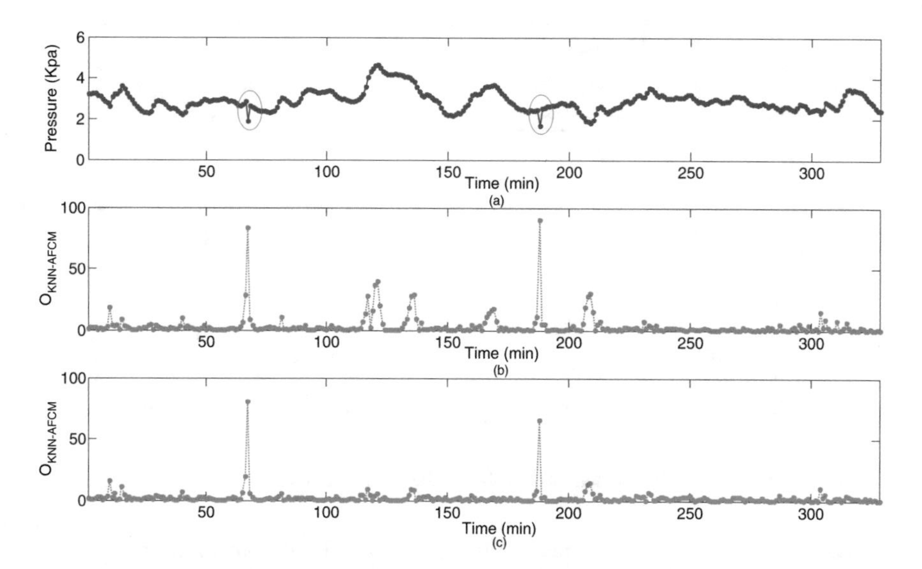

Fig. 2.6 Comparison of anomaly detection results when a varies. (**a**) Main pipeline pressure of COG system. (**b**) The values of anomaly index when $a = 1$. (**c**) The values of anomaly index when $a = 1.5$

pressure of the coke oven gas (COG) system, shown in Fig. 2.6a. And, Fig. 2.6b shows the detection results when $a = 1$ and $s_j = 1.0$, where the 67th point and the 188th point are detected as the deviants and their anomaly indexes are calculated as 83.78 and 90.10, respectively. However, once the anomaly threshed is unreasonably set, it is highly probable that the local maximum or the minimum (the 121th point and the 209th point) could be falsely detected as the deviants. For the KNN-AFCM method, if the value of a is 1.5, the value of the anomaly indexes of the local extreme

Table 2.2 The values of the clustering prototypes when a varies

The values of a	The clustering prototypes							
1.0	2.2562	2.4597	2.6686	2.8853	3.0933	3.3395	3.6500	4.2566
1.5	2.1679	2.4375	2.6744	2.9663	3.2880	3.5506	4.1490	4.4411

Table 2.3 The values of anomaly index of the three deviants in Fig. 2.3a when β varies

β	0.01	0.03	0.05	0.1	1	5	10	20	50	100
35th point	0.09	109.18	130.68	125.71	159.19	182.88	166.12	142.36	149.63	153.46
98th point	1.65	0.76	28.02	61.71	72.46	53.43	68.40	70.45	81.55	86.40
230th point	1.08	195.57	195.29	248.92	203.15	263.78	290.32	283.52	325.42	345.04

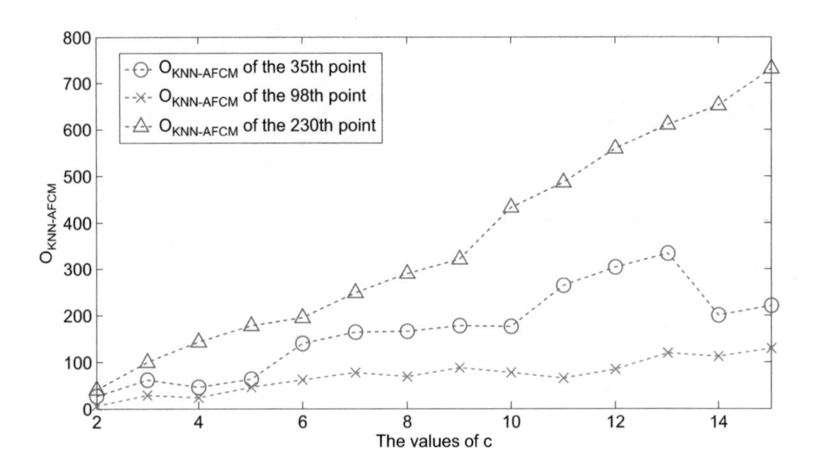

Fig. 2.7 The change of the anomaly index along with the increase of c

points can be effectively decreased due to the introduction of s_j, see Fig. 2.6c. Besides, Table 2.2 lists the value of the clustering prototypes when a equals 1.0 and 1.5, respectively. It is apparent that the values have a wider range by introducing the weight coefficient s_j, which can effectively avoid the misidentification when the normal point is the extreme ones.

The impacts of the parameters β and c on the detection results are also discussed here, and the results are listed in Table 2.3 for the sequence in Fig. 2.5a. One can see that if β is larger than 0.1, all the detection are reliable. Generally, the value of c makes a great impact on the effectiveness of the KNN-AFCM, and a larger c should be taken when a fine-granular division of the sequence is required. Figure 2.7 shows the trend of anomaly index when c increases. It can be observed that the KNN-AFCM algorithm is increasingly sensitive to the deviants along with the increase of the value of c. Meanwhile, the computing cost also increases. As such, one can empirically select the number of clusters in the range of [6, 10].

2.3 Data Imputation

2.3.1 Data-Missing Mechanism

The mechanism of missing data depends on whether or not the missing value of variable is related to the true value of the variable. The analysis of mechanism of missing data is important because the data imputation methods depend on the characteristics of dependencies of these missing mechanisms. Little and Rubin [21] made a comprehensive theoretical analysis of the missing data mechanism, which divides the mechanism into three categories, the missing complete at random (MCAR), the missing at random (MAR), and the not missing at random (NMAR).

One can designate $M = (M_{ij})$ denotes the missing matrix of the full dataset $Y = (Y_{ij})$, in which the missing mechanism is represented as $f(M|Y, \varphi)$. Such a missing mechanism shows the conditional distribution of M given Y, where φ denotes the unknown parameters, Y_{obs} is assumed as the observed part of Y, and Y_{mis} is assumed as the missing part. All the three missing mechanisms can be expressed as follows:

1. Missing complete at random:

$$f(M|Y,\varphi) = f(M,\varphi) \tag{2.28}$$

MCAR doesn't depend on the true value nor on the missing data. The data accumulation process is independent of other variables, and the missing points are fully random events.

2. Missing at random:

$$f(M|Y,\varphi) = f(M|Y_{\text{obs}},\varphi) \tag{2.29}$$

MAR depends on Y_{obs} but have nothing to do with Y_{mis}.

3. Not missing at random:

$$f(M|Y,\varphi) = f(M|Y_{\text{mis}},\varphi) \tag{2.30}$$

NMAR depends on observed and missing part of the dataset Y_{obs} and Y_{mis}.

However, the mechanism of time series deletion is different from the above situation. Let $\mathbf{y} = [y_1, y_2, \ldots, y_n]^{\text{T}}$ be a univariate random time series, in which some points are missing, where y_i is the ith sample of this sequence. $\mathbf{m} = [m_1,$

$m_2 \ldots\ldots, m_n]^{\mathrm{T}}$ denotes the vector of logical values corresponding to the elements in $\mathbf{y}$. If y_i is the missing point m_i=1; otherwise, m_i=0. The distributions over y_i and m_i are independent, so that

$$f(Y, M|\theta, \varphi) = f(Y|\theta)f(M|Y, \varphi) = \prod_{i=1}^{n} f(y_i|\theta) \prod_{i=1}^{n} f(m_i|y_i, \varphi) \tag{2.31}$$

in which $f(y_i|\theta)$ denotes the density function of y_i with the parameter θ, $f(m_i|y_i, \varphi)$ is the binomial distribution density function over m_i. $\Pr(m_i = 1|y_i, \varphi)$ is the missing probability over y_i. If the missing sample is independent to Y ($\Pr(m_i = 1|y_i, \varphi) = \varphi$), one can infer that the missing mechanism is the MCAR.

2.3.2 Regression Filling Method

Regression filling (RF) method uses a regression model to fill the missing points contained in data, in which the missing points are imputed by its predicted values [22]. The RF algorithm uses the auxiliary variable $X_k (k = 1, 2, \ldots, K)$ to establish the relationship with the objective function Z. First, the regression model is constructed by using the observed samples, then the values of the missing points are predicted by such a model. Assuming that the missing points is a linear function of the other variables, the prediction model is given by

$$Z_i = \beta_0 + \sum_{k=1}^{K} \beta_k X_{ki} + e_i \tag{2.32}$$

where β_k denotes the regression coefficients, and e_i is the corresponding Gaussian noise. This linear model can be estimated by a least square method.

2.3.3 Expectation Maximum

Expectation maximization (EM) method is a simple iterative probabilistic algorithm, which can deal with missing data [23, 24]. From an initial value of the missing data, their posterior distribution is calculated, then one can maximize the complete data log likelihood with respect to the missing values. By iterating the above procedure, the final results of missing points are obtained. Specifically, it is assumed that the dataset $\mathbf{Z} = (\mathbf{X}, \mathbf{Y})$ consists of the observed data $\mathbf{X}$ and the unobserved data $\mathbf{Y}$, with $\mathbf{X}$ and $\mathbf{Z} = (\mathbf{X}, \mathbf{Y})$ called incomplete dataset and complete one, respectively. $P(\mathbf{X}, \mathbf{Y}|\Theta)$ represents the joint probability density distribution of the complete $\mathbf{Z}$. The maximum likelihood estimation of this dataset can be obtained by integrating out $\mathbf{Y}$.

$$\mathrm{L}(\Theta;\mathbf{X}) = \log p(\mathbf{X}|\Theta) = \log \int p(\mathbf{X},\mathbf{Y}|\Theta)d\mathbf{Y} \tag{2.33}$$

where Θ denotes the parameters of the model. The EM method is divided into two steps: the expectation step and the maximization step. The log likelihood function of incomplete data is denoted as

$$\mathrm{Lc}(\Theta;\mathbf{Z}) = \log p(\mathbf{X},\mathbf{Y}|\Theta) \tag{2.34}$$

These two steps are as follows:

E step: calculate the posterior distribution over $\mathbf{Y}$ conditioned on old Θ

$$Q\left(\mathbf{Y},\Theta|\mathbf{Y}^{\mathrm{old}},\Theta^{\mathrm{old}}\right) = E\left\{\mathrm{Lc}(\Theta;\mathbf{Z})|\mathbf{X};\mathbf{Y}^{\mathrm{old}},\Theta^{\mathrm{old}}\right\} \tag{2.35}$$

M step: get new $\mathbf{Y}$ and Θ by maximizing $Q(\mathbf{Y},\Theta|\,\mathbf{Y}^{\mathrm{old}},\Theta^{\mathrm{old}})$ with respect to Θ.

First, the initial values of the missing points and the model parameters are selected randomly, then perform the E step and M step. By iterating over these two steps, the value of the likelihood of the model increases gradually and converges to a maximum at last. However, the EM is vulnerable to the initial values of the missing points and the parameters; therefore, their initial values have to be carefully chosen so as to avoid the local optima. Besides, the EM usually takes too much training time because of its iteration learning process.

2.3.4 *Varied Window Similarity Measure*

The varied window similarity measure (VWSM) method uses the sliding-window technique to find a sequence that matches the target one, and then matches the missing values according to this sequence [25]. The calculation steps of the VWSM are listed as follows:

Step 1: Find a comparative sequence V_2 that is similar to the target one V_1. $\delta = \cos(V_1, V_2)$ denotes the similarity coefficients of the two sequences, in range of [−1,1]. When $\delta < 0$, the target V_1 is opposite to the comparative sequence so that such sequence is abandoned. The smaller δ is, the smaller the similarity is.

Step 2: Perform missing data imputation. By using the above computation scheme for the similarity, the target sequence is assumed as V_τ, whose mostly matching sequence is V_q. $\mathbf{v}_\tau^{(m)}$ denotes the missing sequence of V_τ at the mth time. The value of V_q at the mth time is $\mathbf{v}_q^{(m)}$. Therefore, $\mathbf{v}_\tau^{(m)}$ can be obtained by $\mathbf{v}_q^{(m-1)}$ and $\mathbf{v}_q^{(m+1)}$. As shown in Fig. 2.8, the left division Q_l is

$$Q_l = \mathbf{v}_q^{(m-1)} - \mathbf{v}_\tau^{(m-1)} \tag{2.36}$$

The right division Q_l is given by

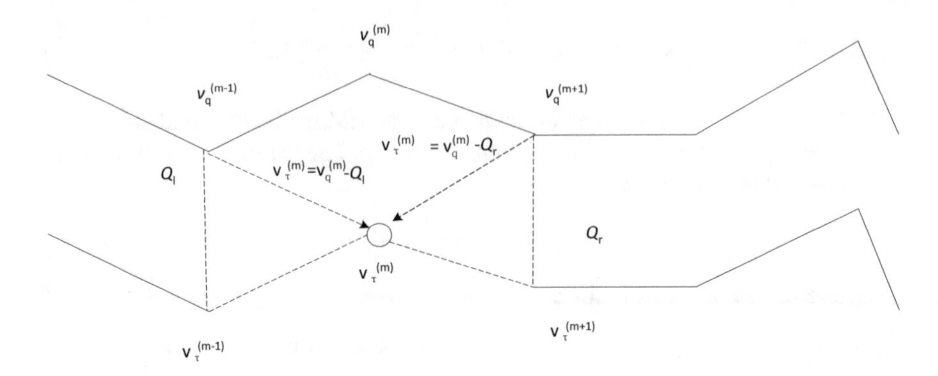

Fig. 2.8 The sliding-window imputation

$$Q_r = \mathbf{v}_q{}^{(m+1)} - \mathbf{v}_\tau{}^{(m+1)} \tag{2.37}$$

Therefore, the estimation of $\widehat{\mathbf{v}}_\tau{}^{(m)}$ is obtained by the left and right division

$$2\widehat{\mathbf{v}}_\tau{}^{(m)} = 2\mathbf{v}_q{}^{(m)} - Q_l - Q_r \tag{2.38}$$

and

$$\widehat{\mathbf{v}}_\tau{}^{(m)} = \mathbf{v}_q{}^{(m)} - 0.5\left(\mathbf{v}_q{}^{(m-1)} - \mathbf{v}_\tau{}^{(m-1)}\right) - 0.5\left(\mathbf{v}_q{}^{(m+1)} - \mathbf{v}_\tau{}^{(m+1)}\right) \tag{2.39}$$

Although such a VWSM is fairly easy to be implemented, it is difficult to find similar sequences by using simple cosine function because of the great characteristic of fluctuation in industrial data, which is usually with larger errors in filling process by only one similar sequence.

2.3.5 Segmented Shape-Representation Based Method

In this section, we present a segmented shape-representation based data imputation method. First, we introduce a key-sliding-window technique for sequence segmentation.

Key-Sliding-Window for Sequence Segmentation

As for a time series $S = (s_1, s_2, \ldots\ldots, s_k)$, it can be expressed as

$$S = \begin{cases} f_1\left(t, {w_1}^1\right) + e_1(t) \\ \quad \cdots\cdots \\ f_k(t, w_k) + e_k(t) \end{cases} \quad (2.40)$$

where the kth division of this sequence is denoted as f_k, and $e_k(t)$ denotes the white noise of the kth sequence. The mean square division is defined by

$$L = \sum_{i=1}^{i=k} l_i = \sum_{i=1}^{i=k} \sum_{j=0}^{j=m} \left(x_{a_{i-1}+j} - f_i\left(x_{a_{i-1}+j}, w_i\right)\right)^2 \quad (2.41)$$

where the number of samples in the ith segment is denoted as m, and the $(i-1)$th segmentation point is written as a_{i-1}. The division is the best when (2.41) reaches the minimal value [26].

Here, we present a sliding-window method for sequence segmentation. This method firstly set a piecewise maximum error $e_{\max}$ aiming to find the segmentation points, which does not set the number of sequence division k in advance. In the following, we present its details. The first point of the split time series is the sliding-window starting point. Then, slide the window backward and increase the number of sliding-window. For the first point of the split time series, the sliding-window begins to slide backward and the sliding-window number increases. If the accumulative error the sliding-window is greater than the predefined error limit $e_{\max}$, this point is used as a break one while the following point is a new starting point for the sliding-window. If the error is less than $e_{\max}$, the slide window continues to slide to the right, containing the next point and sequentially traversing all the points of the time series, looking for the sequence of segmentation points.

However, since the sliding-window method is time-consuming when partitioning all of the sequence in a traversed form and the partition point is ambiguous when two points are adjacent, this section presents a key-sliding-window method for non-equal-length segmentation to tackle the above problems, which combines the key point segmentation and the sliding-window method. Defining an area δ as shown in Fig. 2.9, the key points are the internal points in a sequence with the maximum or the minimum value compared to all the points in the area δ, which can reflect the change of main trend of the series. If the number of points between two adjacent key points is larger than 3, the steadiness of these data should be further verified. Based on the raw segmentation of the sequence by all the key points, this section performs a further partition through verifying the steadiness of data between the key points by using the sliding-window method. This procedure is illustrated as follows:

As shown in Fig. 2.9, s_i and s_j are the two adjacent key points in sequence S, which contain more points needed to be further divided by the sliding-window. s_k is an arbitrarily point between these two continuous points. The vertical distance between s_k and the fitting line is denoted as d_k. And, the fitting maximum error is denoted as e_{max}. If $\max(d_k) > e_{max}$, s_k is the segmentation point of the sliding-window.

Next, we will introduce the representation of the sequence segmentation.

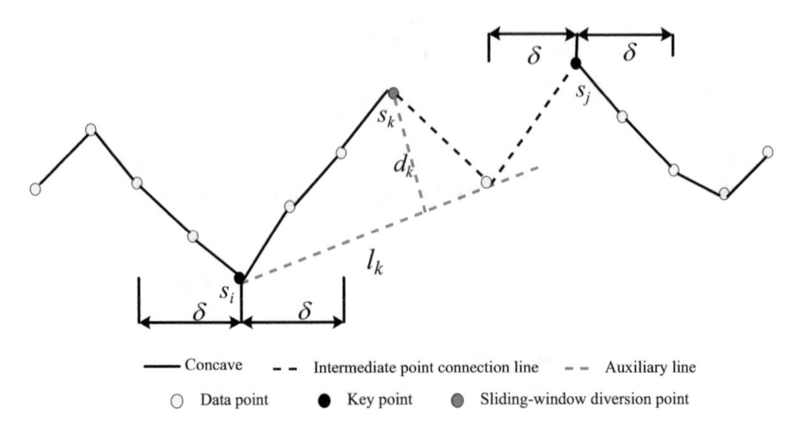

Fig. 2.9 Key-sliding-window segment method

Representation of Sequence Segmentation

The sequence S is divided into n segments by the key-sliding-window segmentation. The sequence is represented as (2.42), where s_{kl}, s_{kr}, and l_k denote the left start point, the right termination point, and the number of segments in the kth segment, respectively.

$$S = \{(s_{1l}, s_{1r}, l_1), \cdots, (s_{kl}, s_{kr}, l_k), \cdots, (s_{nl}, s_{nr}, l_n)\} \tag{2.42}$$

The bit operation method is a well-known representation method for segments [27]. The mean value of kth segment can be represented by $\frac{u_k = \sum_{i=l*k}^{i=l*(k+1)} s_i^{\ i}}{l}$, and $\frac{u = \sum_{i=1}^{n} s_i^{\ i}}{n}$ is the mean value of the sequence. The bit coding mode (BCM) [27] of S_k is represented by

$$s^{\mathrm{tr}}{}_k = \begin{cases} 1, u_k >= u \\ 0, u_k < u \end{cases} \tag{2.43}$$

The distance between sequence S_{i} and S_{j} is defined as

$$d_{ij} = \frac{\sum_{k=1}^{k=n} (s^{\mathrm{tr}}{}_i)_k \mathrm{xor}\left(s^{\mathrm{tr}}{}_j\right)_k}{n} \tag{2.44}$$

where $(s^{\mathrm{tr}}{}_i)_k\mathrm{xor}(s^{\mathrm{tr}}{}_j)_k$ is the XOR operation between the kth segments in S_i and S_j. From (2.44), the distance represents the similarity of two sequences. The larger this distance is, more similar the trends of these two sequences are. The higher the number of segments is, the higher the similarity of the two sequences is.

Since the bit operation method only makes use of the amplitude levels to measure the similarity of two sequences, which may lose information, this section represents a segment by using three terms including the amplitude levels, the trend and fluctuation levels of a sequence. The mean value of the sequence is represented as

$$\mathbf{u} = (u_1, \ldots\ldots, u_k, \ldots\ldots u_n) \tag{2.45}$$

where u_k denotes the mean value of the kth segment. The trend is denoted by s^{tr}, i.e.,

$$\mathbf{s}^{\mathrm{tr}} = ({s^{\mathrm{tr}}}_1, \ldots\ldots, {s^{\mathrm{tr}}}_k, \ldots\ldots, {s^{\mathrm{tr}}}_n) \tag{2.46}$$

If the trend of the kth segment is ascending, i.e., $s_{kr} > s_{kl}$, ${s^{\mathrm{tr}}}_k = 1$; otherwise, ${s^{\mathrm{tr}}}_k = 0$. The standard deviation of the sequence is denoted by $\boldsymbol{\sigma}$, i.e.,

$$\boldsymbol{\sigma} = (\sigma_1, \ldots\ldots, \sigma_k, \ldots\ldots, \sigma_n) \tag{2.47}$$

where σ_k is the standard deviation of the kth segment. Besides, for the sequence S, we define a deviation vector, which represents the degree of the deviation of the points from the mean values of the sequence, of the form

$$\mathbf{q} = (q_1, \ldots\ldots, q_k, \ldots\ldots, q_n) \tag{2.48}$$

where $q_k = \frac{u_k}{u}$ is the degree of deviation of the kth segment.

For two sequences S_i and S_j, the similarity coefficient of their kth segments is defined by

$$\rho_k = \left| q_{ik} + a\left(({s^{tr}}_i)_k xor({s^{tr}}_j)_k\right) q_{jk} \right| \frac{\max(\sigma_{ik}, \sigma_{jk}) * l_k}{\max(\sigma_{ik}, \sigma_{jk}) + b\min(\sigma_{ik}, \sigma_{jk})} \tag{2.49}$$

where a and b denote measures of how the amplitude levels and the fluctuation levels determine the similarity of the two sequences, respectively. Then, the similarity coefficient between S_i and S_j is given by

$$Q(i, j) = \sum_{k=1}^{n} \rho_k \tag{2.50}$$

Using this similarity measure, one can perform the data imputation based on the Gaussian process echo state network which will be introduced in Chap. 3. In the following, we introduce the whole procedure of data imputation for industrial time series.

Procedure of Data Imputation Based on Segmented Shape-Representation

Assuming that the incomplete sequence S contains two parts, i.e., S_{obs} and S_{abs} (the observed part and the missing part). Based on the aforementioned similarity measure method, we treat S_{obs} as the target sequence, then q sequences consisting a similarity set Θ are obtained. The sequences in Θ are treated as the training set of the Gaussian process echo state network, and the values of points in S_{abs} are evaluated by this trained model. Thus, the data imputation is implemented. Furthermore, the procedure of this data imputation is illustrated in Fig. 2.10.

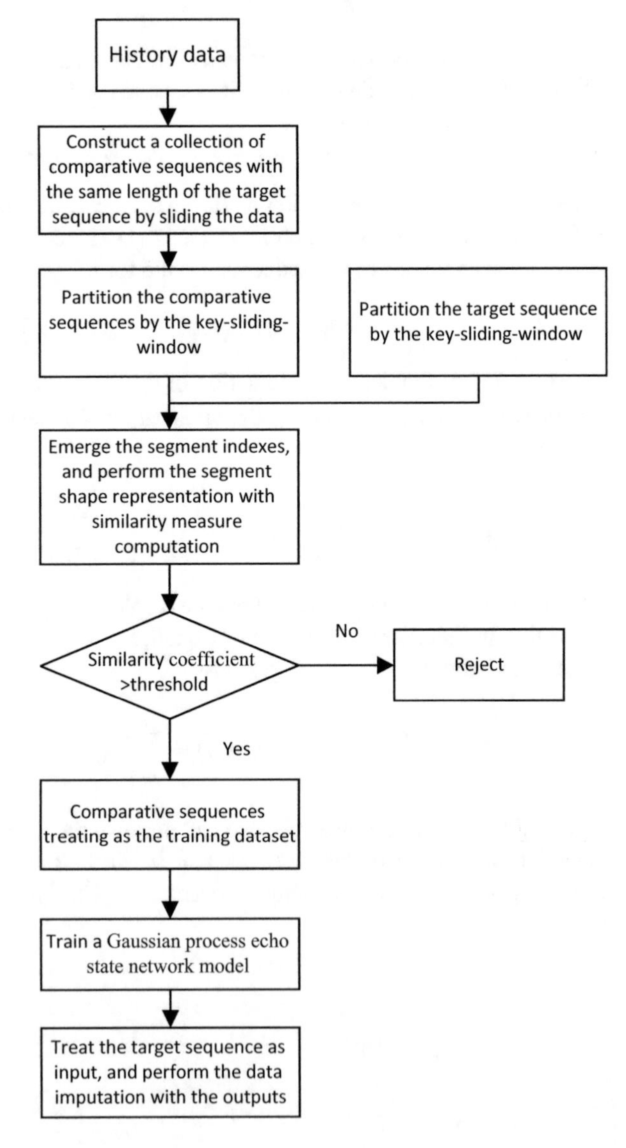

Fig. 2.10 The flow chart for the procedure of the data imputation based on the segmented shape-representation

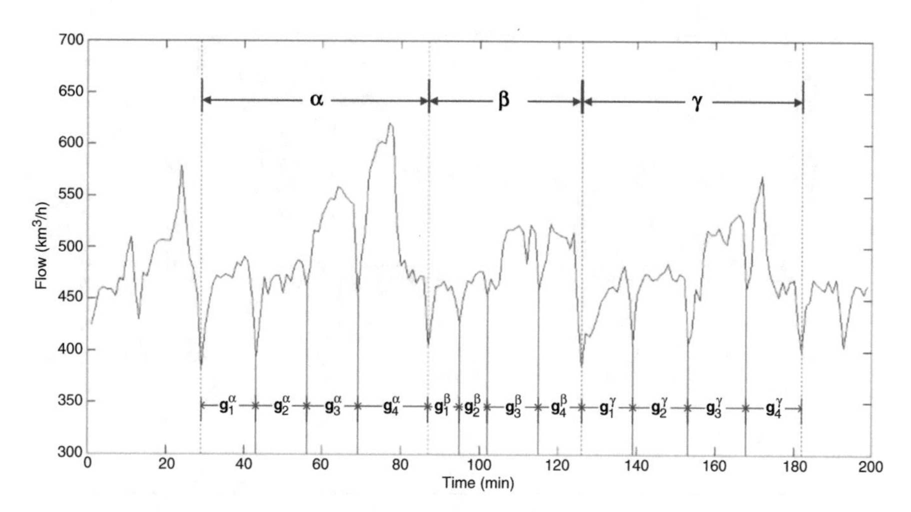

Fig. 2.11 Non-equal-length granules of the flow of BFG generation data

2.3.6 Non-equal-Length Granules Correlation

Industrial time series usually exhibits quasi-periodic features, and the non-equal-length sequences are often collected such that the existing sliding-window methods are not appropriate. Moreover, considering various period of industrial production links, this section provides a non-equal-length granules correlation coefficient (NGCC) method for data imputation [28].

Taking the sequence of the BFG generation flow as an example, one can designate each manufacturing link (production rhythm) as an information block (see $\boldsymbol{\alpha}$, $\boldsymbol{\beta}$, and $\boldsymbol{\gamma}$ in Fig. 2.11), and each block consists of a number of granules g_i^j, which represents the production features of a blast furnace. From Fig. 2.11, the granules g_i^j generally have different lengths. Thus, in order to analyze the degree of correlation of two identical manufacturing processes, their matching relationship should be obtained. Taking $\boldsymbol{\alpha}$ and $\boldsymbol{\beta}$ in Fig. 2.11 as example, considering the features of each granule, the best match-ups of these two sequences is presented as the red lines in Fig. 2.12, which is similar to Fig. 2.3. Then, the degree of correlation of these two sequences can be calculated.

Assuming these two information sequences $\boldsymbol{\alpha} = \{b_1^{\boldsymbol{\alpha}}, b_2^{\boldsymbol{\alpha}}, \ldots, b_m^{\boldsymbol{\alpha}}\}$ and $\boldsymbol{\beta} = \left\{b_1^{\boldsymbol{\beta}}, b_2^{\boldsymbol{\beta}}, \ldots, b_n^{\boldsymbol{\beta}}\right\}$, one can define $s_k = \left(b_i^{\boldsymbol{\alpha}}, b_j^{\boldsymbol{\beta}}\right)$ to describe the matching relationship between the points in these two sequences, which means the kth mapping that involves the points $b_i^{\boldsymbol{\alpha}}$ and $b_j^{\boldsymbol{\beta}}$. Defining $\{s_1, s_2, \ldots, s_t\}$ as the entire matching relations between $\boldsymbol{\alpha}$ and $\boldsymbol{\beta}$, this section converts $\boldsymbol{\alpha}$ and $\boldsymbol{\beta}$ into two new blocks with the same length, denoted by $\mathbf{b}^{\boldsymbol{\alpha}} = \{s_1^{\boldsymbol{\alpha}}, s_2^{\boldsymbol{\alpha}}, \ldots, s_t^{\boldsymbol{\alpha}}\}$ and $\mathbf{b}^{\boldsymbol{\beta}} = \left\{s_1^{\boldsymbol{\beta}}, s_2^{\boldsymbol{\beta}}, \ldots, s_t^{\boldsymbol{\beta}}\right\}$. And the quantity NGCC($\boldsymbol{\alpha}, \boldsymbol{\beta}$) is defined as follows:

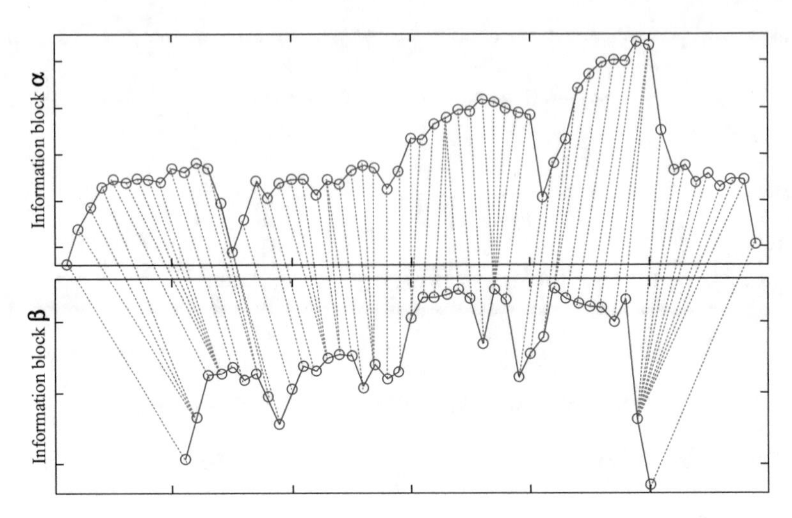

Fig. 2.12 The best mapping relationship of two sequences with non-equal-length information granule

$$\mathrm{NGCC}(\boldsymbol{\alpha},\boldsymbol{\beta}) = \mathrm{MAX}\left\{ \frac{\sum_{i=1}^{t}\left(s_i^{\boldsymbol{\alpha}} - \overline{\mathbf{b}^{\boldsymbol{\alpha}}}\right)\left(s_i^{\boldsymbol{\beta}} - \overline{\mathbf{b}^{\boldsymbol{\beta}}}\right)}{\sqrt{\sum_{i=1}^{t}\left(s_i^{\boldsymbol{\alpha}} - \overline{\mathbf{b}^{\boldsymbol{\alpha}}}\right)^2 \cdot \sum_{i=1}^{t}\left(s_i^{\boldsymbol{\beta}} - \overline{\mathbf{b}^{\boldsymbol{\beta}}}\right)^2}} \right\} \tag{2.51}$$

$$s_1 = \left(b_1^{\boldsymbol{\alpha}}, b_1^{\boldsymbol{\beta}}\right), s_t = \left(b_m^{\boldsymbol{\alpha}}, b_n^{\boldsymbol{\beta}}\right) \tag{2.52}$$

$$0 \le i - i' \le 1, 0 \le j - j' \le 1 \text{ if } s_k = \left(b_i^{\boldsymbol{\alpha}}, b_j^{\boldsymbol{\beta}}\right) \text{ and } s_{k-1} = \left(b_{i'}^{\boldsymbol{\alpha}}, b_{j'}^{\boldsymbol{\beta}}\right) \tag{2.53}$$

$$\forall s_k = \left(b_i^{\boldsymbol{\alpha}}, b_j^{\boldsymbol{\beta}}\right), |i - j| \le w_{\text{Length}} \tag{2.54}$$

where $\overline{\mathbf{b}^{\boldsymbol{\alpha}}}$ and $\overline{\mathbf{b}^{\boldsymbol{\beta}}}$ denote the expectation of $\mathbf{b}^{\boldsymbol{\alpha}}$ and $\mathbf{b}^{\boldsymbol{\beta}}$, respectively. The value of $\mathrm{NGCC}(\boldsymbol{\alpha},\boldsymbol{\beta})$ is the maximum correlation measure among all the possible matching relationships between $\boldsymbol{\alpha}$ and $\boldsymbol{\beta}$. And, the constraints are listed in the formulas (2.52)–(2.54). Since the blocks $\boldsymbol{\alpha}$ and $\boldsymbol{\beta}$ are time series, all of the points must be included in the mappings while remaining the time stamp in each block, see the constraint (2.53). Moreover, in a practical application, the constraint (2.54) depicts the time delay of the two points in a mapping, which must be within a reasonable range, and w_{Length} is the largest time delay. As such, the NGCC method can quantify the correlation of non-equal-length granules, and it can be addressed by solving the discrete optimization problem formulated by (2.51)–(2.54).

Calculation for NGCC

As for solving the NGCC, on one hand, it is impossible to calculate the value of NGCC if only knowing one mapping (a certain s_i) of two sequences; on the other hand, for two sequences, one cannot determine the number of the mappings in advance. The estimation of distribution algorithm (EDA) becomes an optimization task, which determines the evolutionary direction by building a probabilistic model of the solutions. Enlightened by the idea of the EDA, one can construct a probabilistic model of the mapping relations and designs the learning process for NGCC. Here, the probabilistic model of the solution space is represented by a matrix **P**

$$\mathbf{P} = \begin{bmatrix} \mathbf{p}^*_{(1,1)} & \mathbf{p}^*_{(1,2)} & \cdots & \mathbf{p}^*_{(1,n-1)} & \mathbf{p}^*_{(1,n)} \\ \mathbf{p}^*_{(2,1)} & & & \mathbf{p}^*_{(2,n-1)} & \mathbf{p}^*_{(2,n)} \\ \vdots & & \ddots & \vdots & \vdots \\ \mathbf{p}^*_{(m-1,1)} & \mathbf{p}^*_{(m-1,2)} & \cdots & \mathbf{p}^*_{(m-1,n-1)} & \mathbf{p}^*_{(m-1,n)} \\ \mathbf{p}^*_{(m,1)} & \mathbf{p}^*_{(m,2)} & \cdots & \mathbf{p}^*_{(m,n-1)} & -- \end{bmatrix} \tag{2.55}$$

where $\mathbf{p}^*_{(i,j)}$ is a probability distribution determining the mapping relationship next to the one of $\left(b_i^{\alpha}, b_j^{\beta}\right)$. Based on the constraints (2.53) and (2.54), each element of **P** is expressed by a set of conditional probabilities, i.e.,

$$\mathbf{p}^*_{(i,j)} = \begin{cases} \left[p_{(i,j)(i,j+1)} \quad p_{(i,j)(i+1,j)} \quad p_{(i,j)(i+1,j+1)}\right] & \text{if } i < m \text{ and } j < n \text{ and } |i-j| < w_{\text{Length}} \\ \left[p_{(i,j)(i,j+1)} \quad p_{(i,j)(i+1,j+1)}\right] & \text{if } i < m \text{ and } j < n \text{ and } i-j = w_{\text{Length}} \\ \left[p_{(i,j)(i+1,j)} \quad p_{(i,j)(i+1,j+1)}\right] & \text{if } i < m \text{ and } j < n \text{ and } j-i = w_{\text{Length}} \\ p_{(m,j)(m,j+1)} & \text{if } i = m \text{ and } j < n \\ p_{(i,n)(i+1,n)} & \text{if } i < m \text{ and } j = n \end{cases} \tag{2.56}$$

where $p_{(a_1,a_2)(a_3,a_4)}$ can be calculated by (2.57), and the sum of the elements of $\mathbf{p}^*_{(i,j)}$ is equal to 1.

$$p_{(a_1,a_2)(a_3,a_4)} = p\left(s_{k+1} = \left(b_{a_3}^{\alpha}, b_{a_4}^{\beta}\right) \middle| s_k = \left(b_{a_1}^{\alpha}, b_{a_2}^{\beta}\right)\right) \tag{2.57}$$

Let $\mathbf{S}' = \{\mathbf{s}_1, \mathbf{s}_2, \ldots, \mathbf{s}_l\}$ denotes the optimal solutions, which maximizes (2.57). During the iterative process, **P** is updated by (2.58)–(2.60). Here, one can define a local optimal probability matrix $\mathbf{P}'$

$$\mathbf{P}' = \begin{bmatrix} \mathbf{p}'^*_{(1,1)} & \mathbf{p}'^*_{(1,2)} & \cdots & \mathbf{p}'^*_{(1,n-1)} & \mathbf{p}'^*_{(1,n)} \\ \mathbf{p}'^*_{(2,1)} & & & \mathbf{p}'^*_{(2,n-1)} & \mathbf{p}'^*_{(2,n)} \\ \vdots & & \ddots & \vdots & \vdots \\ \mathbf{p}'^*_{(m-1,1)} & \mathbf{p}'^*_{(m-1,2)} & \cdots & \mathbf{p}'^*_{(m-1,n-1)} & \mathbf{p}'^*_{(m-1,n)} \\ \mathbf{p}'^*_{(m,1)} & \mathbf{p}'^*_{(m,2)} & \cdots & \mathbf{p}'^*_{(m,n-1)} & -- \end{bmatrix} \tag{2.58}$$

where $\mathbf{p}'^*_{(i,j)}$ represents the probability distribution of the solutions set $\mathbf{S}'$. Defining $s_k = \left(b_i^\alpha, b_j^\beta\right)$, if s_k is not in $\mathbf{S}'$, then $\mathbf{p}'^*_{s_k} = \mathbf{p}^*_{s_k}$; otherwise, the probability will be calculated by (2.59), and smoothed by (2.60)

$$p''_{s_k,s} = \frac{\text{num}(\mathbf{s}_l | s_k \in \mathbf{S}' \& s \in \mathbf{S}')}{\text{num}(\mathbf{s}_l | s_k \in \mathbf{S}')} \tag{2.59}$$

$$p'_{s_k,s_{k+1}} = \sum_s \exp(-\|s_{k+1}, s\|) \cdot p''_{s_k,s} \tag{2.60}$$

where the operator num(·) aims at counting the amount of the individuals that meet the conditions, $p''_{s_k,s}$ denotes the probability of the mapping s in the solution $\mathbf{S}'$, and the symbol $\|\|$ denotes the Euclidean distance of two mapping relations. Then, the probability matrix $\mathbf{P}$ can be updated by

$$\mathbf{P} = \eta\mathbf{P}' + (1-\eta)\mathbf{P} \tag{2.61}$$

where η is a factor for the new result that is a positive number of its value smaller than 1. The probability matrix of the solutions is evolved by repeating the aforementioned steps, until the model converged or reached the maximum of iteration. The learning procedure of the algorithm is as follows:

Algorithm 2.4: Calculation for NGCC

Step 1: *Initialize the probability matrix* **P** *using the uniform distribution of the unit range.*

Step 2: *Draw a set of solutions from* **P**.

Step 3: *Calculate the value of the objective function (2.51) for each solution, and find the set of better individuals* $\mathbf{S}'$.

Step 4: *Update the matrix* **P** *by (2.58–2.61).*

Step 5: *Return to Step 2 until* **P** *is converged or reached the maximum number of iteration.*

Step 6: *One can obtain the final value of the objective function of the optimal solution.*

NGCC-Based Correlation Analysis

The searching process for NGCC can be listed here, where θ is the correlation evaluation parameter, and the outputs of the algorithm involve the correlation NGCC $(\mathbf{x}, \mathbf{y})$ and the corresponding mapping relations $\mathbf{S}'$.

Algorithm 2.5: NGCC-Based Correlation Analysis

1. ***for*** $i = 1$ *to* $l - n + 1$ ***do***
2. $y_1 \leftarrow t_i$
3. *Initialize* $\mathbf{P}$
4. $\{NGCC(\mathbf{x}, \mathbf{y}), \mathbf{S}'\} \leftarrow$ *Calculation for the NGCC*
5. ***if*** $NGCC(\mathbf{x}, \mathbf{y}) \geq \theta$ ***then***
6. *break*
7. ***end if***
8. ***end for***
9. ***return*** $\{NGCC(\mathbf{x}, \mathbf{y}), \mathbf{S}'\}$

Correlation-Based Data Imputation

Let $\mathbf{t} = t_1, t_2, t_3, \ldots, t_l$ represent the observed data, and the sequence to be imputed is denoted as $\mathbf{x} = x_1, x_2, \ldots, x_m$. Here, the NGCC is employed to find the sequence $\mathbf{y} = y_1, y_2, \ldots, y_n$, which has the highest correlation with $\mathbf{x}$. Since the length of $\mathbf{y}$ might not be equal to that of $\mathbf{x}$, the mapping relations between them also need to be determined. The probability model of the solutions space is determined by

$$\mathbf{P} = \begin{bmatrix} \mathbf{p}^*_{(1,1)} & \mathbf{p}^*_{(1,2)} & \cdots & \mathbf{p}^*_{(1,m+w_{\text{Length}})} \\ \mathbf{p}^*_{(2,1)} & & & \mathbf{p}^*_{(2,m+w_{\text{Length}})} \\ \vdots & & \ddots & \vdots \\ \mathbf{p}^*_{(m-1,1)} & \mathbf{p}^*_{(m-1,2)} & \cdots & \mathbf{p}^*_{(m-1,m+w_{\text{Length}})} \end{bmatrix} \tag{2.62}$$

where $\mathbf{p}^*_{(i,j)}$ represents the probability distributions of the mapping (x_i, y_j). And,

$$\mathbf{p}^*_{(i,j)} = \begin{cases} \left[p_{(i,j)(i,j+1)} \quad p_{(i,j)(i+1,j)} \quad p_{(i,j)(i+1,j+1)}\right] & \text{if } |i-j| < w_{\text{Length}} \\ \left[p_{(i,j)(i,j+1)} \quad p_{(i,j)(i+1,j+1)}\right] & \text{if } i-j = w_{\text{Length}} \\ \left[p_{(i,j)(i+1,j)} \quad p_{(i,j)(i+1,j+1)}\right] & \text{if } j-i = w_{\text{Length}} \end{cases} \tag{2.63}$$

where $p_{(i,j)(i,j+1)}$ is the probability of mapping relation (x_i, y_{j+1}) when the mapping (x_i, y_j) is determined.

The sample $\mathbf{y}$ could be converted to the same length as the imputed sequence $\mathbf{x}$, Here, one can define a mapping relation $x_i \rightarrow y_j$ to represent the ith point of

$\mathbf{x}$ corresponding to the j-th point of $\mathbf{y}$. As such, a new formed sequence can be formulated by $\mathbf{y}' = y'_1, y'_2, \ldots, y'_m$. Then, one has

(a) If $x_i \to y_j, y_{j+1}, \ldots, y_r$, then $y'_i = \mathrm{Avg}(y_j, y_{j+1}, \ldots, y_r)$.
(b) If $x_i, x_{i+1}, \ldots, x_r \to y_j$, then $y'_i = y'_{i+1} = \ldots = y'_r = y_j$.

The imputation equations can be expressed as (2.64) for there are strong linear correlations in $\mathbf{x}$ and $\mathbf{y}'$.

$$\begin{bmatrix} \mathbf{x}_{\mathrm{obs}} \\ \mathbf{x}_{\mathrm{abs}} \end{bmatrix} = \begin{bmatrix} \mathbf{y}'_{\mathrm{obs}} & 1 \\ \mathbf{y}'_{\mathrm{abs}} & 1 \end{bmatrix} \cdot \begin{bmatrix} \alpha \\ \beta \end{bmatrix} \tag{2.64}$$

where $\mathbf{x}_{\mathrm{obs}}$ and $\mathbf{x}_{\mathrm{abs}}$ are the observed value and the absence one of $\mathbf{x}$, and the corresponding values of $\mathbf{y}'$ are $\mathbf{y}'_{\mathrm{obs}}$ and $\mathbf{y}'_{\mathrm{abs}}$, respectively. α and β are the coefficient and the deviation of the linear relationship, respectively. Thus, the missing value $\mathbf{x}_{\mathrm{abs}}$ can be determined by (2.65).

$$\mathbf{x}_{\mathrm{abs}} = \begin{bmatrix} \mathbf{y}'_{\mathrm{abs}} & 1 \end{bmatrix} \cdot \begin{bmatrix} \mathbf{y}'_{\mathrm{obs}} & 1 \end{bmatrix}^{-1} \cdot \mathbf{x}_{\mathrm{obs}} \tag{2.65}$$

The calculating steps of the proposed imputation method are as follows:

Algorithm 2.6: Correlation-Based Data Imputation

Step 1: *Select a long period of continuous historical data* $\mathbf{t} = t_1, t_2, t_3, \ldots$ *from the database, and determine the length of the production cycle m and the largest time delay* w_{Length} *according to the production features.*

Step 2: *Divide the original data associated with missing points into a number of data segments by the production cycle m, and select the segment with the least missing points as the imputed sequence* $\mathbf{x}$.

Step 3: *Search for the sample* $\mathbf{y}$ *and the corresponding mapping relations* $\mathbf{S}'$ *by using the NGCC-based correlation analysis.*

Step 4: *Reconstruct* $\mathbf{y}$ *into* $\mathbf{y}'$ *based on the mapping relations* $\mathbf{S}'$ *and the converting rules to make it equal length to* $\mathbf{x}$.

Step 5: *Impute the missing values by (2.65) according to the correlation between* $\mathbf{x}$ *and* $\mathbf{y}'$.

Step 6: *If there are no missing points in the new sequence, then stop the imputation; otherwise, go back to Step 2.*

2.3.7 Case Study

To verify the performance of the NGCC method, in this section, we perform the experiments of the correlation analysis using the flow of BFG dataset under a number of missing ratios. A period of the BFG data collected in 2016 in a steel plant serves as the experimental samples, of which the length of each sequence is 50.

The missing ratios of each sequence are set from 10% to 40%. The results of correlation analysis under these missing ratios are shown in Fig. 2.10.

As shown in Fig. 2.13a, the most relevant sequence $\mathbf{y}$, which has the maximum of correlation with respect to the sequence $\mathbf{x}$, can be searched from the historical sequences by using the NGCC method. In order to obtain the optimal sequence, one has to compare $\mathbf{x}$ one by one. $\mathbf{y}^{'}$ is constructed from $\mathbf{y}$ by the length transmission process, which is a description of the trend of $\mathbf{x}$. The NGCC method can also work even if there are missing points in $\mathbf{x}$, of which the experimental results are represented in Fig. 2.13b–e.

To quantify the performance of the NGCC method, the neighboring mean method, the spline interpolation method, and the KNN method are treated as the comparative approaches. One can use the evaluation criterions of root mean square error (RMSE) and mean absolute percentage error (MAPE) as the indexes of the imputation accuracy

$$\mathrm{RMSE} = \sqrt{\frac{1}{n}\sum_{i=1}^{n}\left(\widehat{y}_i - y_i\right)^2} \tag{2.66}$$

$$\mathrm{MAPE} = \frac{1}{n}\sum_{i=1}^{n}\frac{\left|\widehat{y}_i - y_i\right|}{y_i} \tag{2.67}$$

where n is the amount of missing points, $\widehat{y}_i$ is the imputed value, and y_i is the real value. In order to demonstrate the reliability of the imputation method, these experiments are conducted 50 times for each dataset under these missing ratios, and the missing points are randomly set for each time. $\mathrm{MAPE}_{\mathrm{Mean}}$ and $\mathrm{RMSE}_{\mathrm{Mean}}$ denote the averaged value of the MAPEs and RMSEs, which can be adopted to measure the imputation accuracy. The smaller $\mathrm{MAPE}_{\mathrm{Mean}}$ and $\mathrm{RMSE}_{\mathrm{Mean}}$ are, the better the imputation performance is. $\mathrm{MAPE}_{\mathrm{SD}}$ and $\mathrm{RMSE}_{\mathrm{SD}}$ are the standard deviation of the MAPEs and RMSEs. In industrial practice, the acquired original industrial data often exhibits the characteristics of large fluctuation with high level noises, such as the flow of BFG generation data. Considering the production rhythm of the blast furnace, the length of each incomplete sequence is set to be 50. To show the performance of the NGCC method, the imputation results (errors) of these methods are listed in Table 2.4, where the NGCC exhibits the highest accuracy under the missing ratios. Figure 2.14 illustrates the imputation results of the data with missing ratio 40%. It is apparent that the mean imputation method exhibits good performance when there are few consecutive missing points. However, for the multiple consecutive missing values, the data feature could not be well recovered, see the results in a-zone of Fig. 2.14. The spline interpolation used the spline function to learn the characteristics of the data, which could reflect the dynamic features. But, due to the large fluctuations and high level noises, the results might be largely deviated from the truth due to the overfitting of the spline function, see the results in the b-zone and d-zone of Fig. 2.14. The results of the KNN method might bring out some deviations of data tendency because of the uncertain periodicity of a

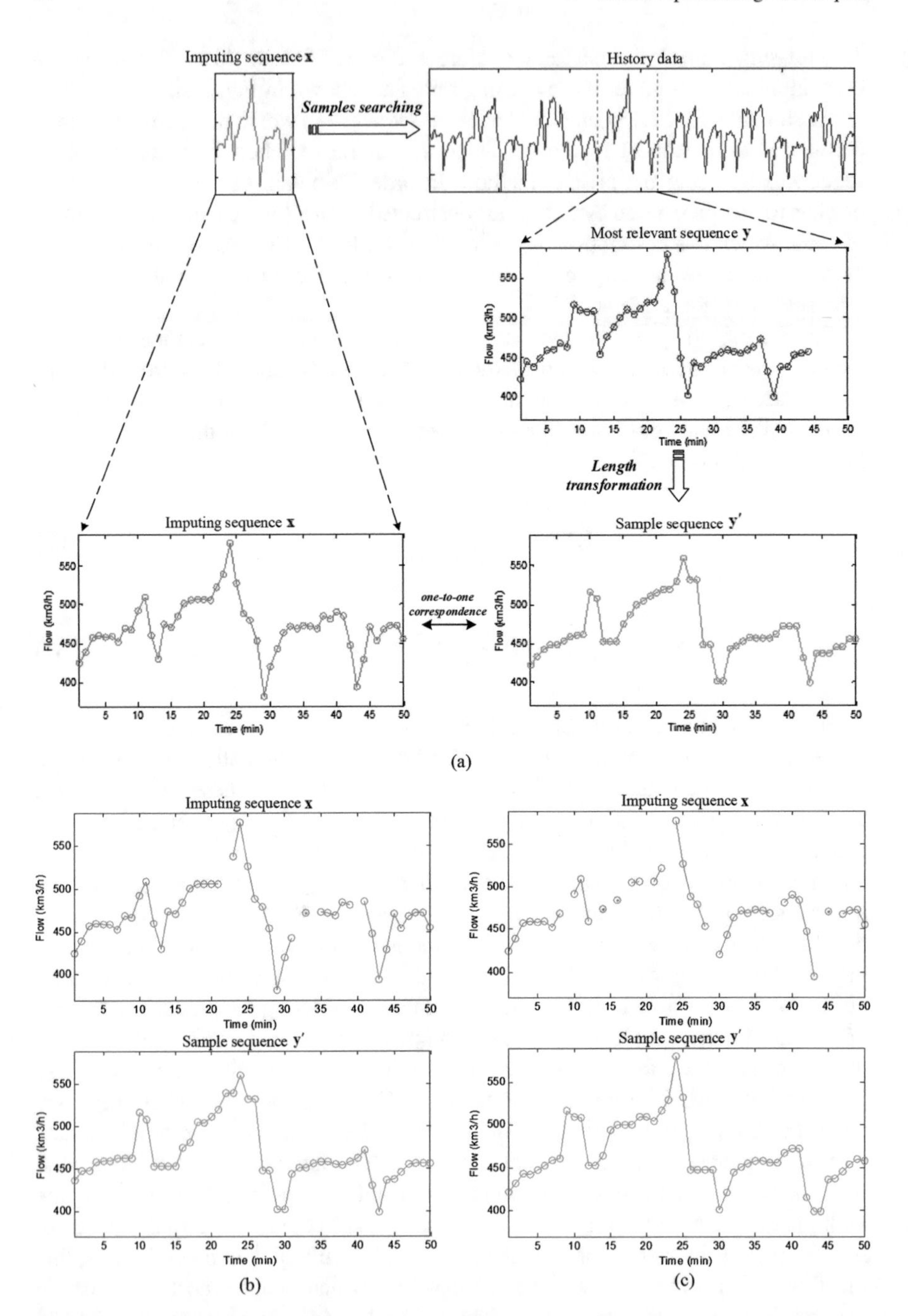

Fig. 2.13 Results of the correlation analysis under different missing ratios. (**a**) The process of correlation analysis and the analysis results of no missing points. (**b**) Analysis results of **x** with missing ratio 10%, where NGCC(**x**, **y**) = 0.9013. (**c**) Analysis results of **x** with missing ratio 20%, where NGCC(**x**, **y**) = 0.9390. (**d**) Analysis results of **x** with missing ratio 30%, where NGCC(**x**, **y**) = 0.8833. (**e**) Analysis results of **x** with missing ratio 40%, where NGCC(**x**, **y**) = 0.9428

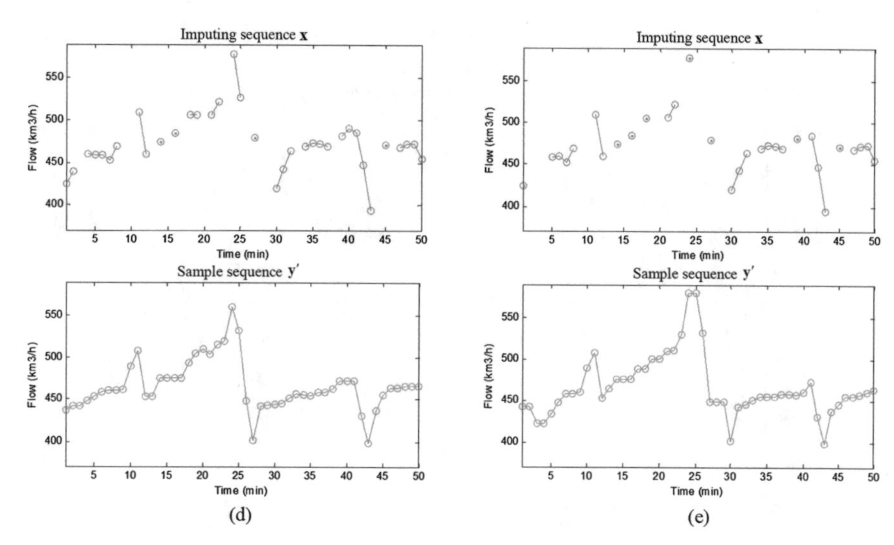

Fig. 2.13 (continued)

Table 2.4 Imputed errors of different methods for the flow of BFG data

	Missing ratio (%)	$MAPE_{Mean}$	$MAPE_{SD}$	$RMSE_{Mean}$	$RMSE_{SD}$
Mean method	10	4.0737	4.3948	23.0606	20.4363
	20	2.1368	2.3234	16.3392	11.2443
	30	3.1056	2.5454	19.6332	13.4543
	40	3.7513	3.6934	24.6896	17.2843
Interpolation method	10	4.1307	4.3533	23.1844	19.3533
	20	2.9615	2.1353	21.2155	10.2433
	30	5.0913	3.2565	30.1697	19.7557
	40	5.4542	4.7868	30.8046	20.6546
KNN method	10	5.6894	1.5685	32.5425	5.2353
	20	4.9752	1.1846	31.2106	6.6546
	30	4.7649	1.6496	29.6940	6.9443
	40	5.1302	2.4646	30.7083	8.3656
NGCC method	10	1.9450	1.1464	13.4572	5.3643
	20	1.4743	1.2759	11.3465	5.2464
	30	1.8353	1.5897	12.8745	6.3577
	40	2.3230	1.9058	14.8297	6.8395

manufacturing procedure, see the results in the c-zone and d-zone. In contrast, the NGCC method enables to learn the features of the industrial data and is capable of getting the more fitted samples, which could greatly improve the imputation accuracy.

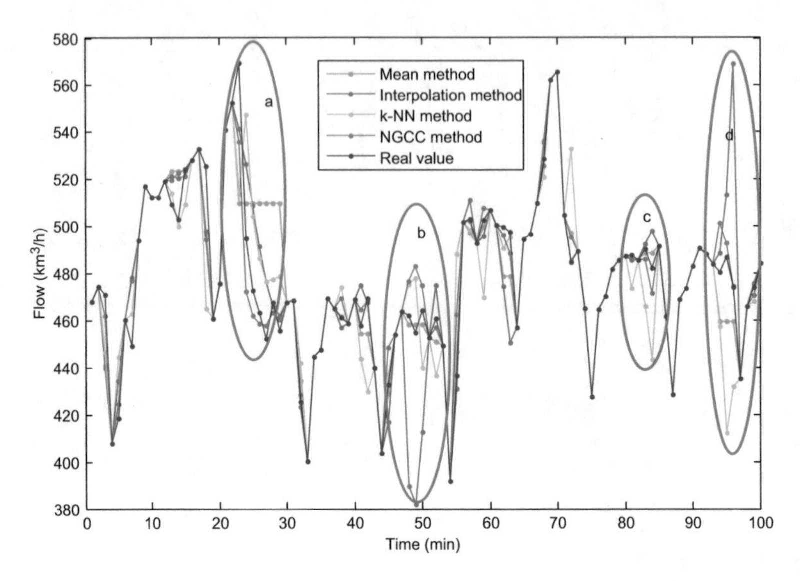

Fig. 2.14 Imputation results of the flow of BFG generation data with missing ratio 40%

2.4 Data De-noising Techniques

The industrial data accumulated by the SCADA system is often corrupted by high level noise, which could impact the quality of the prediction model. Therefore, it is very necessary to perform de-noising methods for removing noise before data-driven modeling. Currently, there are various de-noising techniques, e.g., the sliding averaging, the empirical mode decomposition (EMD) method, etc., in which the EMD is one of the mostly used de-noising method for the industrial data.

2.4.1 Empirical Mode Decomposition

For an industrial time series, the EMD method can decompose it into a number of components with trends [29], and these components form a complete and nearly orthogonal basis for the original signal. In addition, they can also be described by intrinsic mode functions (IMF). Specifically, an IMF can be any function with same number of extrema and zero crossings, whose envelopes are symmetric with respect to zero, and different IMFs have different local characteristics of time scales of the signal. Since the first IMF usually carries the most oscillating (high-frequency) components, it can be rejected to remove high-frequency components (e.g., random noise). EMD-based smoothing algorithms have been widely used in seismic data processing, where high-quality seismic records are highly demanded. Since the decomposition is based on the local characteristic of time scale of the data, it can be applied to nonlinear and nonstationary processes.

The procedure of extracting IMFs is called as sifting. We use the sifting process to perform de-noising for the noisy time series.

Algorithm 2.7: Extracting IMFs for Data De-noising

Step 1*: Extract the local maxima and minima of the signal $X(t)$. The upper envelope and the lower one of the signal are fitted by the three spline interpolation functions, and the mean value of the two envelopes is denoted by m_1. The difference between $X(t)$ and m_1 is $h_1(h_1 = X(t) - m_1)$. If h_1 is not an IMF, the decomposition process will be repeated for further decomposition. The result of the kth step is $h_{1k}(h_{1(k-1)} - m_{1k} = h_{1k})$, in which the first IMF is defined as $c_1(h_{1k} = c_1)$.*

Step 2*: The decomposition process is repeated by treating r_1 as the decomposition data that $r_1 = X(t) - c_1$. The loop is stopped when the remaining component is less than a predetermined value or a monotonic function. Finally, the signal is decomposed into n IMFs with the remaining component r_n so that $X(t) = \sum_{i=1}^{n} c_i + r_n$.*

Step 3*: The n IMF components are used to perform de-noising by using an adaptive threshold $T_i(T_i = \frac{(i-1)^2\sigma_i}{n^2}, 1 \le i \le k)$, where σ_i denotes the standard deviation of the ith scale signal.*

Step 4*: Using these envelopes, one can calculate the standard deviation σ_i^j between the ith signal and its neighborhood. If $\sigma_i^j > T_i$, the low-pass filter is used for de-noising.*

Step 5*: The de-noised IMF and the remaining $n - k$ components are reconstructed so as to recover the time series signal.*

2.4.2 *Case Study*

In order to verify the effectiveness of the above de-noising method, we employ the data of the BFG generation also from a steel plant as the practical example. A segment of BFG generation flow data containing 500 points is selected randomly for the experiments (sampling frequency is 1 min). As shown in Fig. 2.15, the data flow fluctuates randomly over time.

The corresponding IMFs produced by the EMD are illustrated in Fig. 2.16, where a total number of 15 IMFs are obtained and the last one is the remaining component. Using the low-pass filter on the decomposition of the former three IMF, the time series after reconstruction is shown in Fig. 2.17, from which the data is smoother than the original one.

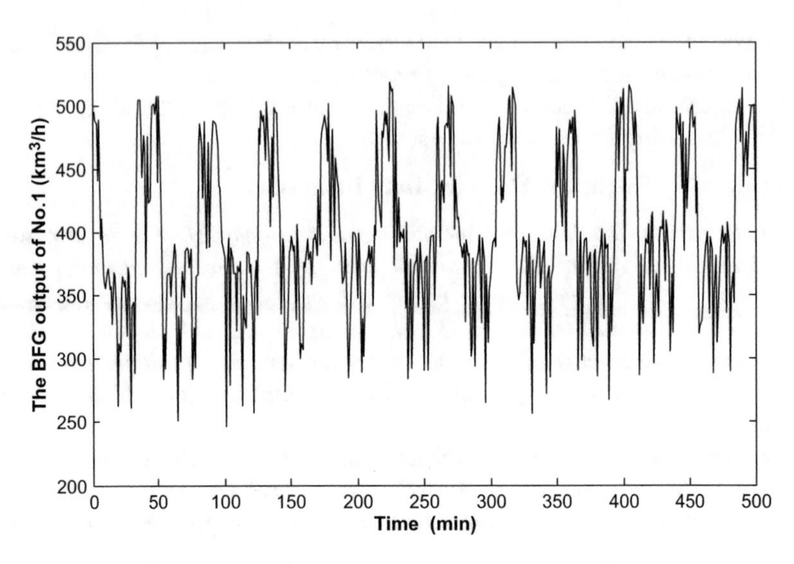

Fig. 2.15 A segment of BFG for blast furnace in a period

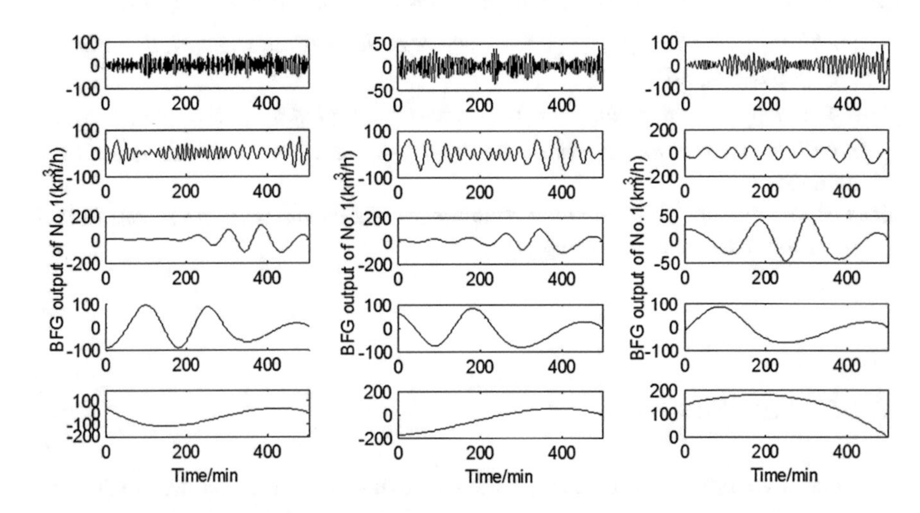

Fig. 2.16 Results of the EMD for the original data presented in Fig. 2.15

In addition, we test the smoothed samples and its corresponding original samples for time series prediction by using the echo state network (ESN) model [30, 31]. The input and output training dataset is constructed by a phase space reconstruction. The number of neurons in the reservoir of the ESN model is set to 100, the sparse proportion is 1%, the number of training samples is 500, and the spectral radius of the connection weight matrix **W** is 0.75. We treat the tanh() function as the internal activation function, and the linear activation function is used for outputs. The

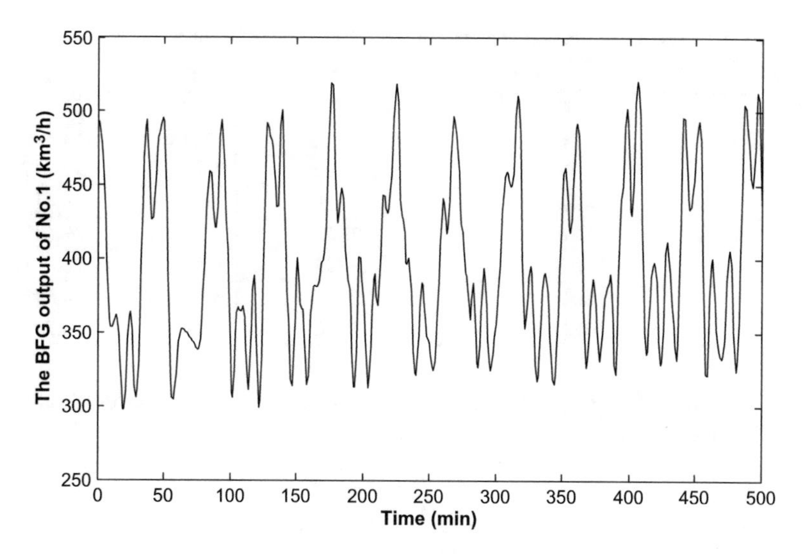

Fig. 2.17 The time series after de-noising by the EMD

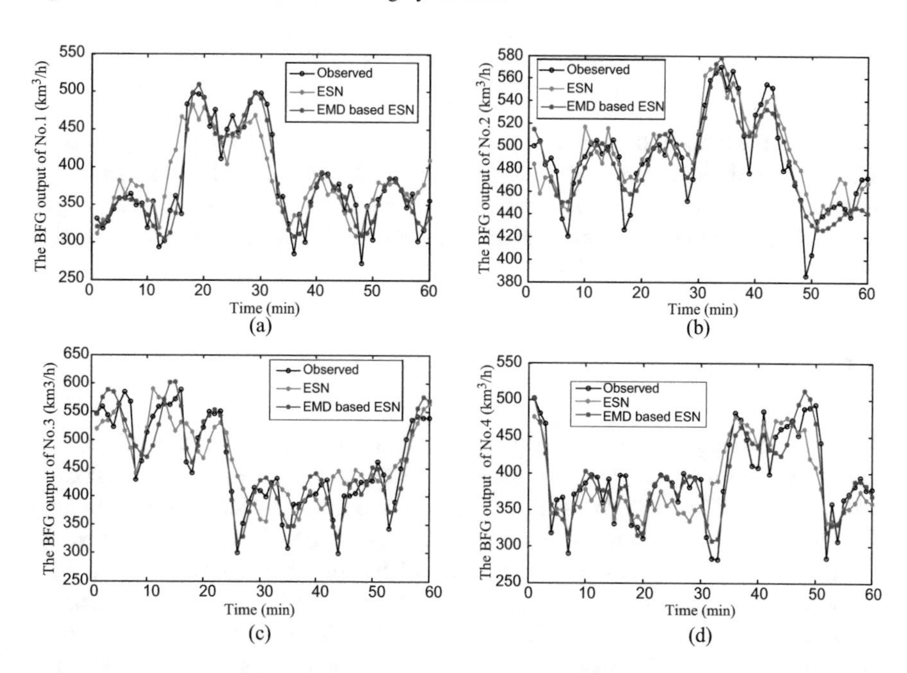

Fig. 2.18 Prediction results of BFG data (**a**) #1 BFG, (**b**) #2 BFG, (**c**) #3 BFG, (**d**) #4 BFG

prediction length is 60 min by one step iterative form. The prediction results are shown in Fig. 2.18a, the red line represents the actual values of data, and the blue line is the forecasted values using the smoothed samples. We conducted the same methods in predicting the same period of #2~#4 BFG and the results are shown in Fig. 2.18b–d, respectively. The normalized RMSE (NRMSE), the mean square error

Table 2.5 Comparison of prediction results of BFG data using different methods

Method	Predicted object	NRMSE	MSE	MAPE(%)
ESN	1BF	0.66796	628.4878	5.4726
	2BF	0.48902	433.156	4.4059
	3BF	0.65348	994.7694	7.8275
	4BF	0.58587	1057.936	6.4139
EMD-based ESN	1BF	0.3679	306.2257	3.6496
	2BF	0.30329	268.3751	3.0289
	3BF	0.35487	508.7191	4.6961
	4BF	0.30375	456.4997	4.3717

(MSE), and the MAPE defined by (2.67) are used as the measures for judging the prediction performance.

$$\text{NRMSE} = \sqrt{\frac{1}{T\|y_d\|^2}\sum_{t=1}^{T}(y(t) - y_d(t))^2} \tag{2.68}$$

$$\text{MSE} = \frac{1}{T}\sum_{t=1}^{T}(y(t) - y_d(t))^2 \tag{2.69}$$

where T is the number of predicted points, $y(t)$ is the predicted value, and $y_d(t)$ is the actual value of the samples. The accuracies of each method are listed in Table 2.5. From the experimental results, it can be obtained that using the smoothed samples the ESN model obtains higher prediction accuracy than that of the original samples.

2.5 Discussion

To improve the quality of the original data is a prerequisite for data mining and modeling, due to overcoming anomalies, noises, and missing points acquired in industrial data. Data preprocessing is a complex process, which contains anomaly detection, data imputation, and de-noising. At present, there is no generally effective framework or scheme for data preprocessing. The application of data mining algorithm in data preprocessing needs to be strengthened. Besides, when the amount of the industrial data is very huge, high computational load becomes a typical shortcoming of data preprocessing technique. For preprocessing massive data, it is necessary to consider appropriate human participation, which could prevent the occurrence of wrong pretreatment. For different industrial datasets, one has to select a suitable data preprocessing method so as to make the results more reasonable. Therefore, the preprocessing may be a recurring process, which can be performed multiple times until that a suitable method is found.

The data preprocessing described in this chapter is only a matter of data detection and imputation. The further processing is required when data with higher quality is needed. There is a practical demand for data preprocessing algorithms that can be used to make data mining more efficiently. Although numerous preprocessing approaches have been developed, the research of data preprocessing remains an active area due to the huge amount of inconsistent or dirty industrial data.

References

1. Keogh, E. (2005). Recent advances in mining time series data. knowledge discovery in databases: Pkdd 2005. *European Conference on Principles and Practice of Knowledge Discovery in Databases* (p. 6), Porto, Portugal, October 3–7, 2005, Proceedings. DBLP.
2. Adamo, J. M. (2001). *Data mining for association rules and sequential patterns*. Berlin: Springer.
3. Pyle, D. (1999). *Data preparation for data mining* (pp. 375–381). San Francisco: Morgan Kaufmann.
4. Kotsiantis, S. B., Kanellopoulos, D., & Pintelas, P. E. (2006). Data preprocessing for supervised leaning. *International Journal of Computer Science, 1*(2), 111–117.
5. Alpaydin, E. (2014). *Introduction to machine learning*. Cambridge: MIT press.
6. Fayyad, U., Piatetsky-Shapiro, G., & Smyth, P. (1996). From data mining to knowledge discovery in databases. *AI Magazine, 17*(3), 37.
7. Gama, J. (2010). *Knowledge discovery from data streams*. London: CRC Press.
8. Liu, H., & Motoda, H. (1998). *Feature extraction, construction and selection: a data mining perspective*. Boston: Kluwer Academic Publishers.
9. Chen, M., & Chen, L. (2008). An information granulation based data mining approach for classifying imbalanced data. *Information Sciences, 178*, 3214–3227.
10. Zhao, J., Liu, K., Wang, W., et al. (2014). Adaptive fuzzy clustering based anomaly data detection in energy system of steel industry. *Information Sciences, 259*, 335–345.
11. Akouemo, H. N., & Povinelli, R. J. (2014). Time series outlier detection and imputation. *PES General Meeting | Conference & Exposition* (pp. 1–5), 2014 IEEE. IEEE.
12. Aydilek, I. B., & Arslan, A. (2013). A hybrid method for imputation of missing values using optimized fuzzy c-means with support vector regression and a genetic algorithm. *Information Sciences, 233*, 25–35.
13. Fu, T. C. (2011). A review on time series data mining. *Engineering Applications of Artificial Intelligence, 24*(1), 164–181.
14. Jaeger, H., & Haas, H. (2004). Harnessing nonlinearity: predicting chaotic systems and saving energy in wireless communication. *Science, 304*(5667), 78–80.
15. Eftekhar, A., Toumazou, C., & Drakakis, E. M. (2013). Empirical mode decomposition: Real-time implementation and applications. *Journal of Signal Processing Systems, 73*(1), 43–58.
16. Monard, M. C. (2002). A study of K-nearest neighbour as an imputation method. *DBLP* (pp. 251–260).
17. Steinbach, M., Karypis, G., & Kumar, V. (2000, August 20–23). A comparison of document clustering techniques. In: *Proceedings of the Sixth ACM SIGKDD International Conference on Knowledge Discovery and DataMining* (pp. 174–181). Boston, MA, USA.
18. Bezdek, J. C. (1981). *Pattern recognition with fuzzy objective function algorithms*. New York: Plenum Press.
19. Pal, N. R., & Bezdek, J. C. (2002). On cluster validity for the fuzzy c-means model. *IEEE Transactions on Fuzzy Systems, 3*(3), 370–379.

20. Chiang, J. H., & Hao, P. Y. (2003). A new kernel-based fuzzy clustering approach: support vector clustering with cell growing. *IEEE Transactions on Fuzzy Systems, 11*(4), 518–527.
21. Little, R. J. A., & Rubin, D. B. (2002). *Statistical analysis with missing data* (pp. 87–88).
22. Gelman, A., & Hill, J. (2007). *Data analysis using regression and multilevel/hierarchical models*. London: Cambridge University press.
23. Dempster, A. P., Laird, N. M., & Rubin, D. B. Maximum likelihood from incomplete data via the EM algorithm. *Journal of the Royal Statistical Society: Series B: Methodological, 1977*, 1–38.
24. Rancourt, E., Särndal, C. E., & Lee, H. (1994). Estimation of the variance in the presence of nearest neighbor imputation. *Proceedings of the Section on Survey Research Methods* (pp. 888–893).
25. Buschman, T. J., & Miller, E. K. (2007). Top-down versus bottom-up control of attention in the prefrontal and posterior parietal cortices. *Science, 315*(5820), 1860–1862.
26. Navalpakkam, V., & Itti, L. (2006). An integrated model of top-down and bottom-up attention for optimal object detection. *Computer Society Conference on IEEE, 2*, 2049–2056.
27. Lu, K. F., Lin, S. K., & Qiao, J. Z. (2008). FSMBO: fast time series similarity matching based on bit operation. *Proceedings of the 9th International Conference for Young Computer Scientists*.
28. Lv, Z., Zhao, J., Liu, Y., et al. (2016). Data imputation for gas flow data in steel industry based on non-equal-length granules correlation coefficient. *Information Sciences, 367*, 311–323.
29. Rilling, G., Flandrin, P., & Goncalves, P. (2003). On empirical mode decomposition and its algorithms. *IEEE-EURASIP Workshop on Nonlinear Signal and Image Processing* (vol. 3, pp. 8–11). IEEER, Grado, Italy.
30. Kountouriotis, P. A., Obradovic, D., Goh, S. L., & Mandic, D. P. (2005). Multi-step forecasting using echo state networks. In: *Proceedings of International Conference on Computer as a Tool* (pp. 1574–1577). Belgrade, IEEE.
31. Shi, Z. W., & Han, M. (2007). Ridge regression learning in ESN for chaotic time series prediction. *Control and Decision, 22*(3), 258–267.

Chapter 3
Industrial Time Series Prediction

Abstract Time series prediction is a significant way for forecasting the variables involved in industrial process, which usually identifies the latent rules hidden behind the time series data of the variables by means of auto-regression. In this chapter we introduce the phase space reconstruction technique, which aims to construct the training dataset for modeling, and then a series of data-driven machine learning methods are provided for time series prediction, where some well-known artificial neural networks (ANNs) models are introduced, and a dual estimation-based echo state network (ESN) model is particularly proposed to simultaneously estimate the uncertainties of the output weights and the internal states by using a nonlinear Kalman-filter and a linear one for noisy industrial time series. In addition, the kernel based methods, including Gaussian processes (GP) model and support vector machine (SVM) model, are also presented in this chapter. Specifically, an improved GP-based ESN model is proposed for time series prediction, in which the output weights in ESN modeled by using GP avoids the ill-conditioned phenomenon associated with the generic ESN version. A number of case studies related to industrial energy system are provided to validate the performance of these methods.

3.1 Introduction

In Chap. 1, various categories of data-driven industrial process prediction (DDIPP) are introduced, which can be divided into industrial time series prediction and factors-based prediction due to the features of industrial data. With respect to whether the uncertainties of prediction model are considered, the DDIPP can be divided into point-oriented prediction and interval-based one. According to the difference of the time scale for prediction, such a prediction task can also be divided into super-short-term prediction, short-term one, mid-long-term one, and long-term one. From Chaps. 3 to 6, we will introduce the industrial time series prediction, factors-based industrial prediction, prediction intervals with data uncertainty, and granular computing-based long-term prediction, respectively. In this chapter, the industrial time series prediction is firstly introduced.

J. Zhao et al., *Data-Driven Prediction for Industrial Processes and Their Applications*, Information Fusion and Data Science,
https://doi.org/10.1007/978-3-319-94051-9_3

Time series prediction is a kind of auto-regression mode based on historical data. In brief, it aims to find and identify the developing and evolutionary latent laws behind the historical data, based on which one can make predictions in the future. However, since industrial data are often suffered from the issues of high level noise, abnormal or missing points, the raw time series is inappropriate to build a data-driven model directly. In Chap. 2, a number of preprocessing techniques for industrial data are introduced. With the pretreatment for the raw data, the industrial time series will be more suitable for the data model construction.

The first concern in this chapter is how to quantify the time delay and the embedding dimensionality of a data-driven prediction model by using a period of industrial time series. The commonly used technique is the phase space reconstruction, of which its related approaches will be introduced in Sect. 3.2. The core content of this chapter is provided from Sects. 3.3 to 3.6, in which a series of typical models for industrial time series prediction are illustrated, including linear regression, neural networks (NN), support vector machine (SVM), and Gaussian process (GP). In Sect. 3.3, this book will firstly present the linear regression model that is suitable for the regression task with lower complexity. The advantages of linear regression lie in the simple model structure and the demand of low computational load. However, its learning ability is relatively limited when facing with industrial data. It is generally combined with the kernel learning approach and the Bayesian-based one for more complex industrial problems in order to obtain a much stronger learning ability. In Sect. 3.4, the GP-based regression methods are exhibited, whose principle assumes a prior Gaussian distribution over the latent functions more suitable to learn a nonlinear relationship. Moreover, it could be combined with a NN-based model for more complex regression task. In Sect. 3.5, NNs-based models for time series prediction are reported, where the reservoir computing network (RNN) is one of the most commonly used neural networks-based approach. In this section, we firstly study the extended Kalman-filtering (EKF)-based Elman network for time series prediction, and then a specific RNN, named echo state network (ESN) is introduced. In the rest of this section, some improved forms of ESN for industrial time series prediction will also be proposed, including the SVD-based ESN, ESNs with leaky integrator neurons, and dual estimation-based ESN. Section 3.6 will give the SVM-based regression, in which the basic concept of SVM is firstly provided. Although the SVM was originally proposed for solving a classification problem, it has been proved that it also has a stronger capability for regression problem, especially for industrial time series prediction. The advantage of SVM is that it can be adopted for the problems with fewer training samples and with much smaller parameters than that of NN. In order to deeply understand SVM for time series prediction, this chapter also introduces the LSSVM, the Gamma test-based LSSVM, the sample selection-based LSSVM, and the Bayesian treatment for LSSVM.

3.2 Phase Space Reconstruction

Phase space reconstruction can fully trace out the underlying orbit of a deterministic chaotic system, which provides a simplified, multi-dimensional representation of a univariate nonlinear time series. Phase space can be constructed by two classes of methods, the derivative reconstruction and the coordinates delay reconstruction. While, in practice, there is usually absence of the prior information on nonlinear time series, and numerical differentiation is very sensitive to the errors; therefore, the derivative reconstruction can be hardly performed. In contrast, the coordinates delay reconstruction approach is a sound way for a time series.

For a scalar time series $\{x(1), x(2), \ldots, x(i), \ldots, x(N)\}$, the dynamics can be fully embedded in m-dimensional phase space, where the components of each state vector $\mathbf{s}_i(m)$ are defined through the delay coordinates,

$$\mathbf{s}_i(m) = (x(i), x(i+\tau), x(i+2\tau), \ldots, x(i+(m-1)\tau)) \tag{3.1}$$

where $i = 1, 2, \ldots, N-(m-1)\tau/\Delta t$, τ is the delay time that is the average length of memory of the chaotic system, m is the embedded dimensionality that can be considered the minimal amount of state variables required to describe the system, and Δt represents the sampling interval. Phase space reconstruction in a dimension m allows to interpret the underlying dynamics in the form of an m-dimensional map $f_T(\cdot)$ of the form,

$$\mathbf{s}_{j+T}(m) = f_T(\mathbf{s}_j(m)) \tag{3.2}$$

where $\mathbf{s}_{j+T}(m)$ is the vector denoting the state of the system at time $j + T$ (the future state), and T refers to the lead time.

The determination of the two parameters, τ and m, is very significant to the phase space reconstruction technique, which extremely impacts the identification of hidden dynamics. According to the Takens theorem [1], the delay time τ and the embedded dimensionality m can be fixed to any value for an infinite length and noise-free univariate time series. However, the practical time series cannot satisfy these two conditions on length and noise, i.e., τ and m must be picked out carefully. In literature, there are some ways to optimize them, and in the following the commonly used methods will be introduced.

3.2.1 *Determination of Embedding Dimensionality*

There are many methods for determining the embedding dimensionality, including the false nearest-neighbor algorithm (FNN) [2], and the Cao method [3], which are both popular approaches.

False Nearest-Neighbor Method (FNN)

FNN can be considered as one of the most effective ways for optimizing the embedding dimensionality. The basic thought of this method is to discover the FNNs of the state vector $\mathbf{s}_i(m)$ by discerning the change of the distance between $\mathbf{s}_i(m)$ and its nearest neighbors when the embedding dimensionality is raised from m to $m + 1$. The geometric structure of the chaotic system can be uncovered if there is no FNNs for $\mathbf{s}_i(m)$, with m being the optimal embedding dimensionality. This method is specified below.

Assuming that $\mathbf{s}_{n(i,m)}(m)$ is the nearest neighbor of $\mathbf{s}_i(m)$ with embedding dimensionality m, i.e.,

$$\|\mathbf{s}_{n(i,m)}(m) - \mathbf{s}_i(m)\| = \min_{j=1,\ldots,N,j\neq i} \|\mathbf{s}_j(m) - \mathbf{s}_i(m)\| \tag{3.3}$$

where $\|\cdot\|$ denotes the Euclidean metric, and $n(i,m)$ $(1 \leq n(i,m) \leq N - m\tau)$ is an integer such that $\mathbf{s}_{n(i,m)}(m)$ is the nearest neighbor of $\mathbf{s}_i(m)$ in the m-dimensional phase space in the sense of distance $\|\cdot\|$. For a univariate time series, along with the increase of the embedding dimensionality, if $\|\mathbf{s}_{n(i,m)}(m + 1) - \mathbf{s}_i(m + 1)\|$ is much larger than $\|\mathbf{s}_{n(i,m)}(m) - \mathbf{s}_i(m)\|$, the nearest neighbor $\mathbf{s}_{n(i,m)}(m + 1)$ is treated to be false. Specifically, let the below inequality hold

$$\alpha(i,m) = \frac{\|\mathbf{s}_{n(i,m)}(m+1) - \mathbf{s}_i(m+1)\|}{\|\mathbf{s}_{n(i,m)}(m) - \mathbf{s}_i(m)\|} \geq R_{\mathrm{T}} \tag{3.4}$$

where $\mathbf{s}_{n(i,m)}(m + 1)$ is a FNN of $\mathbf{s}_i(m + 1)$, and R_{T} is the threshold value that can be selected in the range of [10, 50].

With the embedding dimensionality m starting from one, the percentage of the FNN is computed for each m. When the percentage does not enlarge with the increase of m or it is smaller than a small value (e.g., 5%), the optimal m will be detected, which can be considered as the fact that a geometric structure of the chaotic system is uncovered completely.

Cao Method

Cao method [3] is an improvement version of the FNN. From (3.4), one can see that the threshold value should be determined by the derivative of the underlying signal. Therefore, for different phase points i, $\alpha(i,m)$ should have different threshold values R_{T} at least in principle. Furthermore, various time series data may have different threshold values. These imply that it is very difficult and even impossible to give an appropriate and reasonable threshold value, which is independent of the dimensionality d and each trajectory's point, as well as the studied time series.

Therefore, the Cao method gives an improvement over the FNN method to tackle the above problem. Considering a time series $\{x(1), x(2), \ldots, x(i), \ldots, x(N)\}$, the mean value of $\alpha(i, m)$ is given by

$$E(m) = \frac{1}{n - m\tau} \sum_{i=1}^{n-m\tau} \alpha(i, m) \tag{3.5}$$

$E(m)$ is dependent only on the dimensionality m and the lag τ. To investigate its variation from m to $m + 1$, one can define the quantity

$$E_1(m) = \frac{E(m+1)}{E(m)} \tag{3.6}$$

It can be found that $E_1(m)$ stops changing when m is greater than some value m_0 if the time series comes from an attractor. Then, $m_0 + 1$ is the minimal dimensionality one looks for.

3.2.2 *Determination of Delay Time*

The purpose of time delay embedding is to unfold the projection back to a multivariate state-space that is representative of the original system. If the time delay is too small, the reconstructed attractor is compressed along the identity line, and this is called *redundancy*. If the time delay is too large, the attractor dynamics may become causally disconnected, and this is called *irrelevance*. Therefore, choosing an appropriate delay time is very crucial to reconstruct original dynamic system.

Autocorrelation Function Method

The autocorrelation function method is a very full-blown one for computing the time delay. For a discrete-time series, one can provide its autocorrelation function, i.e.,

$$R(\tau) = \frac{1}{N} \sum_{i=1}^{N-\tau} x(i)x(i+\tau) \tag{3.7}$$

The curve of $R(\tau)$ can be depicted with the increase of τ, and when the value of $R(\tau)$ declines to the $1 - e^{-1}$ times of the initial value $R(0)$, i.e., $R(\tau^*) = (1 - e^{-1})R(0)$, this time delay τ^* can be regarded as the optimal one.

Although the autocorrelation function method is a simple and effective method for computing the time delay, it can only provide the linear autocorrelation of a time series. By this means, there is no correlation between $x(i)$ and $x(i + \tau)$, while there may exist high correlation between $x(i)$ and $x(i + 2\tau)$. Besides, the coefficient $1 - e^{-1}$ may be inaccurate in such a way.

Mutual Information Method

The autocorrelation function is only concerned about the linear relation between time series, which ignores the nonlinear dynamics of chaos system. To tackle this limitation, Fraser and Swinney [4] proposed a mutual information method for revealing the nonlinear performance of the system.

Assuming that there are two discrete-time series, $\{g(1), g(2), \ldots, g(n)\}$ and $\{q(1), q(2), \ldots, q(m)\}$ for variables G and Q, respectively, the information entropies of these two time series are formulated by

$$H(G) = -\sum_{i=1}^{n} P_g(g_i)\log_2 P_g(g_i) \tag{3.8}$$

$$H(Q) = -\sum_{i=1}^{m} P_q(q_j)\log_2 P_q(q_j) \tag{3.9}$$

where $P_g(g_i)$ and $P_q(q_j)$ are the probability distribution over G and Q, respectively. The mutual information of G and Q is defined by

$$I(Q, G) = H(Q) + H(G) - H(Q, G) \tag{3.10}$$

where the joint entropy over G and Q is $H(Q, G) = -\sum_{i=1}^{n}\sum_{j=1}^{m} P_{Q,G}(q_j, g_i)\log_2 P_{Q,G}(q_j, g_i)$.

We define $[G, Q] = [x(t), x(t + \tau)]$, where G denotes time series $x(t)$, and Q is the time series $x(t + \tau)$ with time delay τ. Therefore, $I(Q, G)$ is a function of τ, written as $I(\tau)$ which measures the uncertainty of $x(t + \tau)$ given the system $x(t)$, with $I(\tau) = 0$ denoting that there is no correlation between $x(t)$ and $x(t + \tau)$. Moreover, the minimum of $I(\tau)$ represents the degree of irrelevance between $x(t)$ and $x(t + \tau)$. The optimal time delay can be set to the first minimum of $I(\tau)$. The key point in the mutual information method is computing the distributions $P_{Q,G}(q_j, g_i)$, $P_g(g_i)$, and $P_q(q_j)$.

3.2.3 *Simultaneous Determination of Embedding Dimensionality and Delay Time*

There is another idea that the embedding dimensionality m and the time delay τ should be interrelated because the practical time series has finite length and is usually corrupted by industrial noise. That is to say, the relation of m and τ has an impact on the time window $t_w = (m - 1)\tau$, and for a specific time series, fixing t_w, the impropriate matching of m and τ will cause the fact that the structure of the reconstructed phase space is no longer equivalent to the original space. Thereby, the methods for computing m and τ simultaneously were born, e.g., the C-C method [5].

The C-C method assumed that the embedding dimensionality and the time delay is correlated. Assuming that τ_s denotes the sampling time of the time series, $\tau_d = t\tau_s$ is the delay time, $\tau_w = (m - 1)\tau_d$ denotes the delay window, τ_p is the mean orbital period ($\tau_w \geq \tau_p$), $\tau(\tau = t)$ denotes the time delay, N is the size of the dataset, and $M = N - (m - 1)\tau$. The correlation dimension is widely used in many fields for the characterization of strange attractors. The correlation integral for the embedded time series is as the following function:

$$C(m, N, r, t) = \frac{2}{M(M-1)} \sum_{1 \leq i \leq j \leq M} \Theta\big(r - \|\mathbf{s}_i(m) - \mathbf{s}_j(m)\|\big), \quad r > 0 \tag{3.11}$$

where $\Theta(a) = 0$, if $a < 0$; $\Theta(a) = 1$, if $a \geq 0$, t is the index lag, and $C(m, N, r, t)$ is a cumulative distribution function, which represents the probability that the distance between arbitrary two points in the state-space is no greater than r.

Defining the statistic $S(m, N, r, t)$ as

$$S(m, N, r, t) = C(m, N, r, t) - C^m(1, N, r, t) \tag{3.12}$$

and since the length of the time series is finite, the partition strategy is employed to compute $S(m, N, r, t)$. For a time series $\{x(i)\}$, $i = 1, 2, \ldots, N$, one can decompose it to a number of uncorrelated partitions with size t, i.e.,

$$\begin{aligned} \mathbf{x}^1 &= \{x(1), x(t+1), x(2t+1), \ldots\} \\ \mathbf{x}^2 &= \{x(2), x(t+2), x(2t+2), \ldots\} \\ &\cdots\cdots \\ \mathbf{x}^t &= \{x(t), x(2t), x(3t), \ldots\} \end{aligned} \tag{3.13}$$

Thus, the computation formula of $S(m, N, r, t)$ is given by

$$S(m, N, r, t) = \frac{1}{t} \sum_{s=1}^{t} \left[C_s\left(m, \frac{N}{t}, r, t\right) - C_s^m\left(1, \frac{N}{t}, r, t\right) \right] \tag{3.14}$$

and when $N \to \infty$, then

$$S(m, r, t) = \frac{1}{t} \sum_{s=1}^{t} [C_s(m, r, t) - C_s^m(1, r, t)], \; m = 2, 3, \ldots \tag{3.15}$$

For a fixed m and t, $S(m, r, t)$ will be identically equal to 0 for all r if the data are independent identical distribution (i.i.d.) and $N \to \infty$. However, the size of the real datasets is finite, and the data may be serially correlated. Thus, one will have $S(m, r, t) \neq 0$, and the locally optimal times may be either the zero crossings of $S(m, r, t)$ or the times at which $S(m, r, t)$ shows the least variation with r since it indicates a nearly uniform distribution of points. Hence, one can select several representative values r_j and define

$$\Delta S(m,t) = \max\{S(m,r_j,t)\} - \min\{S(m,r_j,t)\} \tag{3.16}$$

which measures the variation of $S(m, r, t)$ with r_j. The locally optimal times t are then the zero crossings of $S(m, r, t)$ and the minima of $\Delta S(m, t)$. The zero crossings of $S(m, r, t)$ should be nearly the same for all m and r, and the minima of v should be nearly the same for all m; otherwise, the times t is not locally optimal for $S(m, r, t)$. Thus, the delay time τ_d corresponds to the first one of these locally optimal times.

The proper estimation of m, N, and r can be found according to the BDS statistic [5]. In [6], such an investigation was carried out using time series generated from a variety of asymptotic distributions. Time series with three sample sizes, $N = 100$, 500, and 1000 were generated by Monte Carlo simulations from six asymptotic distributions: a standard normal distribution, a student-t distribution with three degrees of freedom, a double exponential distribution, a chi-square distribution with four degrees of freedom, a uniform distribution, and a bimodal mixture of normal distributions. These studies led to the conclusion that m should be between 2 and 5 and r should be between $\sigma/2$ and 2σ. In addition, the asymptotic distributions were well approximated by finite time series when $N > 500$.

Here, according to the above statistical conclusion, one can define $N = 3000$, $m = 2, 3, 4, 5$, $r_i = i\sigma/2$, $i = 1, 2, 3, 4$, and σ denotes the standard deviation of the data. Then,

$$\bar{S}(t) = \frac{1}{16}\sum_{m=2}^{5}\sum_{j=1}^{4} S(m,r_j,t) \tag{3.17}$$

$$\Delta\bar{S}(t) = \frac{1}{4}\sum_{m=2}^{5}\Delta S(m,t) \tag{3.18}$$

$$S_{\mathrm{cor}}(t) = \Delta\bar{S}(t) + |\bar{S}(t)| \tag{3.19}$$

The optimal delay time τ_d can be found by searching the first zero crossings of $\bar{S}(t)$ or the first local minima of $\Delta\bar{S}(t)$. The optimal time is the index lag t, for which $\bar{S}(t)$ and $\Delta\bar{S}(t)$ are both closest to 0. If one assigns the equal importance to the two quantities, then one may simply look for the minimum of the quantity $S_{\mathrm{cor}}(t)$, and this optimal time gives the delay time window $\tau_w = t\tau_s$.

3.3 Linear Models for Regression

3.3.1 Basic Linear Regression

Linear regression is one typical model of supervised learning, whose goal is to learn the hidden mapping relationship $\mathbf{w}$ between the target variables $\{t_n\}$ and the given input $\{\mathbf{x}_n\}$ with its dimensionality m. With the mapping parameters $\mathbf{w}$, one can make predictions for the value of the target t^* given a set of new inputs $\mathbf{x}^*$.

Here, the linear regression is introduced in details. Assuming that the input dataset is $\mathbf{X} = [\mathbf{x}_1, \mathbf{x}_2, \ldots, \mathbf{x}_N]^{\mathrm{T}}$ with its input dimensionality m, the output dataset is $\mathbf{t} = [t_1, t_2, \ldots, t_N]^{\mathrm{T}}$, and the hidden mapping relationship between $\mathbf{x}_i$ and t_i is written as $\mathbf{w}$, the simplest linear model for regression can be described by a linear combination of the input variables.

$$y(\mathbf{x}, \mathbf{w}) = \mathbf{w}^{\mathrm{T}} \cdot \mathbf{x} = w_0 + w_1 x_1 + \cdots + w_m x_m \tag{3.20}$$

where $\mathbf{x} = (x_0, x_1, x_2, \ldots, x_m)^{\mathrm{T}}$ and $x_0 = 1$. The parameter $\mathbf{w}$ is denoted by $\mathbf{w} = (w_0, w_1, \ldots, w_m)^{\mathrm{T}}$ with dimensionality $m + 1$.

However, the above linear model is only suitable for solving very simple problems. As for regression tasks in complex industrial application, a kernel-based linear regression method will be introduced, which is much stronger that involves a linear combination of fixed nonlinear basis functions of the form

$$y(\mathbf{x}, \mathbf{w}) = w_0 + \sum_{j=1}^{m} w_j \phi_j(\mathbf{x}) \tag{3.21}$$

where $\phi_j(\mathbf{x})$ are known as basic functions. If $\phi_0(\mathbf{x}) = 1$, one can simplify the formula (3.21) as

$$y(\mathbf{x}, \mathbf{w}) = \sum_{j=0}^{m} w_j \phi_j(\mathbf{x}) = \mathbf{w}^{\mathrm{T}} \boldsymbol{\phi}(\mathbf{x}) \tag{3.22}$$

where $\boldsymbol{\phi} = (\phi_0, \ldots, \phi_m)^{\mathrm{T}}$. There are many optional choices for the basic functions while the exponent function is the most commonly used one, that is

$$\phi_j(x) = \exp\left\{ -\frac{(x - \mu_j)^2}{2\sigma^2} \right\} \tag{3.23}$$

Due to the limited learning abilities of linear regression and kernel-based one, combining them with some other nonlinear techniques, such as TS fuzzy model and neural networks, could be a good attempt. Taking the ESN as an example, the mathematical formula of ESN contains two parts, the update equation of the neurons states in the dynamic reservoir and the output equation [7]. If the activation functions of the neurons in the dynamic reservoir are set as a kind of nonlinear function, then the ones of the output neurons can be linear function to reduce the complexity of the model. Since the introduction of nonlinear mapping in the update process of states, the whole model exhibits a stronger nonlinear learning ability even though the output equation is linear. The details will be illustrated in the following sections about the NNs-based model.

As for the linear regression, it is also valuable to discuss whether there exists a solution of the mapping parameters $\mathbf{w}$. The least squares technique is usually employed for the solution of linear regression to deal with the non-full rank problems of coefficient matrix. The typical kernel-based regression model can be written by $y_n = \mathbf{w}^{\mathrm{T}}\boldsymbol{\phi}(\mathbf{x}_n)$ with inputs $\mathbf{x}_n$ and output t_n. For the specific training samples $\{\mathbf{x}_n, t_n\}_{n=1}^{N}$, a system of linear equations can be obtained, i.e.,

$$\sum_{n=1}^{N} t_n = \mathbf{w}^{\mathrm{T}} \sum_{n=1}^{N} \boldsymbol{\phi}(\mathbf{x}_n) \tag{3.24}$$

If the dimensionality of parameters $\mathbf{w}$ is M, the column dimensionality of the coefficient matrix of (3.24) is also equal to M. Since the number of the training data samples is N, the dimensionality of the coefficient matrix is $N \times M$. To make the coefficient matrix to be a full-rank one, the two sides of the equality sign in (3.24) should multiply $\boldsymbol{\phi}^{\mathrm{T}}(\mathbf{x}_n)$. And then a new system of linear equals is obtained

$$\sum_{n=1}^{N} t_n \boldsymbol{\phi}^{\mathrm{T}}(\mathbf{x}_n) = \mathbf{w}^{\mathrm{T}} \sum_{n=1}^{N} \boldsymbol{\phi}(\mathbf{x}_n)\boldsymbol{\phi}^{\mathrm{T}}(\mathbf{x}_n) \tag{3.25}$$

To simplify this description, (3.25) can be rewritten as

$$\mathbf{t}\boldsymbol{\Phi}^{\mathrm{T}} = \mathbf{w}^{\mathrm{T}}\boldsymbol{\Phi}\boldsymbol{\Phi}^{\mathrm{T}} \tag{3.26}$$

where $\boldsymbol{\Phi}$ is the coefficient matrix with its dimension $N \times M$ of the form

$$\boldsymbol{\Phi} = \begin{pmatrix} \phi_0(\mathbf{x}_1) & \phi_1(\mathbf{x}_1) & \cdots & \phi_{M-1}(\mathbf{x}_1) \\ \phi_0(\mathbf{x}_2) & \phi_1(\mathbf{x}_2) & \cdots & \phi_{M-1}(\mathbf{x}_2) \\ \vdots & \vdots & \ddots & \vdots \\ \phi_0(\mathbf{x}_N) & \phi_1(\mathbf{x}_N) & \cdots & \phi_{M-1}(\mathbf{x}_N) \end{pmatrix} \tag{3.27}$$

According to (3.26), the description of the optimal parameters $\mathbf{w}_{\mathrm{ML}}$ is given by

$$\mathbf{w}_{\mathrm{ML}} = \left(\Phi^{\mathrm{T}}\Phi\right)^{-1}\Phi^{\mathrm{T}}\mathbf{t} \tag{3.28}$$

Although $\Phi^{\mathrm{T}}\Phi$ is a square matrix, it might not be a full-rank matrix. The existence of the pseudo-inverse matrix $(\Phi^{\mathrm{T}}\Phi)^{-1}$ must be discussed, which goes against to the industrial application.

To sum up, there are two drawbacks of the linear regression when directly used for industrial prediction. The first is that the noises are not effectively considered into modeling. Another is that the least squares method often produces abnormal solutions.

3.3.2 *Probabilistic Linear Regression*

In this subsection, we consider a probabilistic linear regression model for prediction, such as Bayesian linear regression, which can infer the posterior distribution over the mapping relationship $\mathbf{w}$. With the trained probabilistic model, the distribution of output can be inferenced given a new input. Bayesian technique can not only guarantee the existence of the solution of linear regression, but also consider the uncertainty caused by data noises, which lays the groundwork for the following research on prediction intervals. Considering noisy industrial data, the mapping between the input and output is described as

$$t = y(\mathbf{x}, \mathbf{w}) + \gamma \tag{3.29}$$

where γ denotes the Gaussian white noise. Here, a prior probability distribution is assigned to the model parameters $\mathbf{w}$. Generally, this prior distribution is given by a Gaussian distribution of the form $p(\mathbf{w}) = \mathcal{N}(\mathbf{w}|\mathbf{m}_0, \mathbf{S}_0)$ with mean $\mathbf{m}_0$ and covariance $\mathbf{S}_0$. To simplify the treatment, one considers a form of the Gaussian prior, which is a zero mean isotropic Gaussian distribution governed by a single precision hyperparameter α, i.e.,

$$p(\mathbf{w}|\alpha) = \mathcal{N}\left(\mathbf{w}|0, \alpha^{-1}\mathbf{I}\right) \tag{3.30}$$

To obtain the predictive distribution, the posterior distribution of the weights, which is proportional to the product of the likelihood function and the prior according to the Bayes' rule, can be computed. If there is sufficient quantity of data, due to the choice of a conjugate Gaussian prior distribution, the posterior will also be Gaussian in the form

$$p(\mathbf{w}|D) = \mathcal{N}(\mathbf{w}|\mathbf{m}_N, \mathbf{S}_N) \tag{3.31}$$

We evaluate this distribution according to the prior and data-based likelihood function, where this posterior is composed at least into two parts, including the prior and the likelihood information, i.e.,

$$p(\mathbf{w}|D, \alpha, \beta) \propto \exp\left(-\frac{\beta}{2}\sum_{n=1}^{N}\{t_n - \mathbf{w}^{\mathrm{T}}\boldsymbol{\phi}(\mathbf{x}_n)\}^2 - \frac{\alpha}{2}\mathbf{w}^{\mathrm{T}}\mathbf{w}\right) \tag{3.32}$$

The log of the posterior distribution is given by the sum of the log likelihood and the log of the prior and, as a function of $\mathbf{w}$,

$$\ln p(\mathbf{w}|D, \alpha, \beta) = -\frac{\beta}{2}\sum_{n=1}^{N}\{t_n - \mathbf{w}^{\mathrm{T}}\phi(\mathbf{x}_n)\}^2 - \frac{\alpha}{2}\mathbf{w}^{\mathrm{T}}\mathbf{w} + \mathrm{const} \tag{3.33}$$

The maximization of this posterior distribution with respect to $\mathbf{w}$ is therefore equivalent to the minimization of the sum-of-squares error function with the addition of a quadratic regularization term. Therefore, the maximum a posteriori (MAP) estimation for $\mathbf{w}$ could be calculated. Thus, the predictive distribution is defined by

$$p(t|D,\alpha,\beta) = \int p(t|\mathbf{w},\beta)p(\mathbf{w}|D,\alpha,\beta)d\mathbf{w} \tag{3.34}$$

where D denotes the training dataset. The conditional distribution $p(t|\mathbf{w},\beta)$ of the target variable is the likelihood function, which describes the difference between the observed targets and the prediction results based on the parameters $\mathbf{w}$. It can be seen that (3.34) involves the convolution of two Gaussian distributions, making use of the existing results from [8], and then the predictive distribution

$$p(t|D,\alpha,\beta) = \mathcal{N}\left(t|\mathbf{m}_N^{\mathrm{T}},\boldsymbol{\phi},(\mathbf{x}),\sigma_N^2,(\mathbf{x})\right) \tag{3.35}$$

where the variance $\sigma_N^2(\mathbf{x})$ is given by

$$\sigma_N^2(\mathbf{x}) = \frac{1}{\beta} + \boldsymbol{\phi}^{\mathrm{T}}(\mathbf{x})\mathbf{S}_N\boldsymbol{\phi}(\mathbf{x}) \tag{3.36}$$

where the first term in the right hand side of (3.36) represents the noise on the data, whereas the second term reflects the uncertainty associated with $\mathbf{w}$. Because of the noise process and the distribution over $\mathbf{w}$ are independent Gaussians, their variances are additive.

3.4 Gaussian Process-Based Prediction

3.4.1 Kernel-Based Regression

The intrinsic of kernel-based regression method is similar to the linear regression model. Let us consider the mapping between the input and output, given by

$$t_n = f(\mathbf{x}_n,\mathbf{w}) + \gamma_n \tag{3.37}$$

where γ_n denotes a Gaussian white noise whose value is chosen independently for each observation t_n. And, $\mathbf{w}$ is the unknown parameters. The simplest assumption on f is a linear function, i.e., $f(\mathbf{x}_n) = \mathbf{a}^{\mathrm{T}}\mathbf{x}_n + b$. However, it is obvious that the above assumption cannot satisfy industrial prediction problem. In practice, f is generally a nonlinear function. Thus, consider a specific mapping $\phi : \mathbb{R}^N \mapsto \mathcal{H}$. Without loss of generality, $\mathcal{H}$ is usually assumed to be reproducing kernel Hilbert space, which can be viewed as an inner product space. Thus, assume

$$f(\mathbf{x}_n) = \langle \mathbf{w}, \boldsymbol{\phi}(\mathbf{x}_n)\rangle + b, \quad \mathbf{w}\in\mathcal{H} \tag{3.38}$$

and usually, (3.38) can be written as

$$y = f(\mathbf{x}_n,\mathbf{w}) = \mathbf{w}^{\mathrm{T}}\boldsymbol{\phi}(\mathbf{x}_n) \tag{3.39}$$

For evaluating the parameter $\mathbf{w}$, the regularized sum-of-squares error for the optimization of $\mathbf{w}$ can be defined as

$$J(\mathbf{w}) = \frac{1}{2}\sum_{n=1}^{N}\{t_n - \mathbf{w}^{\mathrm{T}}\boldsymbol{\phi}(\mathbf{x}_n)\}^2 + \frac{\lambda}{2}\mathbf{w}^{\mathrm{T}}\mathbf{w} \quad (3.40)$$

where $\lambda \geq 0$. Since the function $J(\mathbf{w})$ is convex, one can obtain the optimal value $\mathbf{w}_{\mathrm{MP}}$ for $\mathbf{w}$ when the derivative of $J(\mathbf{w})$ with respect to $\mathbf{w}$ equals 0.

$$\frac{\partial J(\mathbf{w})}{\partial \mathbf{w}} = \beta\sum_{n=1}^{N}\{\mathbf{w}^{\mathrm{T}}\boldsymbol{\phi}(\mathbf{x}_n) - t_n\}\boldsymbol{\phi}(\mathbf{x}_n) + \mathbf{w} = 0 \quad (3.41)$$

From (3.41), one can see that the solution for $\mathbf{w}$ takes the form of a linear combination of the vectors $\boldsymbol{\phi}(\mathbf{x}_n)$, with coefficients that are functions of $\mathbf{w}$.

$$\mathbf{w}_{\mathrm{MP}} = -\beta\sum_{n=1}^{N}\{\mathbf{w}^{\mathrm{T}}\boldsymbol{\phi}(\mathbf{x}_n) - t_n\}\boldsymbol{\phi}(\mathbf{x}_n) = \sum_{n=1}^{N} a_n\boldsymbol{\phi}(\mathbf{x}_n) = \Phi^{\mathrm{T}}\mathbf{a} \quad (3.42)$$

where Φ is the design matrix, whose nth row is given by $\boldsymbol{\phi}(\mathbf{x}_n)^{\mathrm{T}}$. Here, the vector $\mathbf{a} = (a_1, \ldots, a_N)^{\mathrm{T}}$ with the definition,

$$a_n = -\frac{1}{\lambda}\{\mathbf{w}^{\mathrm{T}}\boldsymbol{\phi}(\mathbf{x}_n) - t_n\} \quad (3.43)$$

Instead of dealing with $\mathbf{w}$, the least squares algorithm in terms of the parameter vector $\mathbf{a}$ is reformulated, giving rise to a dual representation. Substituting $\mathbf{w} = \Phi^{\mathrm{T}}\mathbf{a}$ into $J(\mathbf{w})$, one has

$$J(\mathbf{a}) = \frac{1}{2}\mathbf{a}^{\mathrm{T}}\mathbf{K}\mathbf{K}\mathbf{a} - \mathbf{a}^{\mathrm{T}}\mathbf{K}\mathbf{t} + \frac{1}{2}\mathbf{t}^{\mathrm{T}}\mathbf{t} + \frac{\lambda}{2}\mathbf{a}^{\mathrm{T}}\mathbf{K}\mathbf{a} \quad (3.44)$$

where $\mathbf{K} = \Phi\Phi^{\mathrm{T}}$is an $N \times N$ symmetric matrix with its elements $K_{nm} = \boldsymbol{\phi}(\mathbf{x}_n)^{\mathrm{T}}$ $\boldsymbol{\phi}(\mathbf{x}_m) = k(\mathbf{x}_n, \mathbf{x}_m)$.

Setting the gradient of $J(\mathbf{a})$ with respect to $\mathbf{a}$ to zero, one has

$$\mathbf{a} = (\mathbf{K} + \lambda\mathbf{I}_N)^{-1}\mathbf{t} \quad (3.45)$$

Besides, substituting this back into the linear regression model, the following prediction for a new input $\mathbf{x}$.

$$y(\mathbf{x}) = \mathbf{w}^{\mathrm{T}}\phi(\mathbf{x}) = \mathbf{a}^{\mathrm{T}}\Phi\boldsymbol{\phi}(\mathbf{x}) = \mathbf{k}^{\mathrm{T}}(\mathbf{x})(\mathbf{K} + \lambda\mathbf{I}_N)^{-1}\mathbf{t} \quad (3.46)$$

where we define the vector $\mathbf{k}(\mathbf{x})$ with elements $k_n(\mathbf{x}) = k(\mathbf{x}_n, \mathbf{x})$. Thus, one sees that the dual formulation allows the solution to a least squares problem to be expressed entirely in terms of the kernel function $k(\mathbf{x}, \mathbf{x}')$. One can therefore work directly in terms of kernels and avoid the explicit introduction of the feature vector $\boldsymbol{\phi}(\mathbf{x})$, which allows us implicitly to use feature spaces of high, even infinite, dimensionality.

3.4.2 Gaussian Process for Prediction

In the GP viewpoint, one dispenses with the parametric model and instead defines a prior probability distribution over functions directly. At first sight, it might seem difficult to work with a distribution over the uncountably infinite space of functions. However, as can be seen, for a finite training set we only need to consider the values of the function at the discrete set of input values $\mathbf{x}_n$ corresponding to the training set and test set data points. In practice, this can work in a finite space. Here, the basic principle of GP-based prediction model will be introduced.

Considering the noise processes γ_n in (3.37) as a Gaussian prior distribution with covariance β^{-1}, one has

$$p(t_n|y_n, \mathbf{x}_n) = \mathcal{N}\left(t_n|y_n, \beta^{-1}\right) \tag{3.47}$$

where $y_n = \mathbf{w}^{\mathrm{T}}\phi(\mathbf{x}_n)$ and β is the precision of noise. Because the noise is independent for each data point, the joint distribution of the target values $\mathbf{t} = (t_1, t_2, \ldots, t_N)$ conditioned on the values of $\mathbf{y} = (y_1, y_2, \ldots, y_N)$ and $\mathbf{X} = (\mathbf{x}_1, \mathbf{x}_2, \ldots, \mathbf{x}_N)$ is given by an isotropic Gaussian of the form

$$p(\mathbf{t}|\mathbf{y}) = \mathcal{N}\left(\mathbf{t}|\mathbf{y}, \beta^{-1}\mathbf{I}_N\right) \tag{3.48}$$

where $\mathbf{I}_N$ denotes the $N \times N$ unit matrix.

From the definition of a GP, the marginal distribution $p(\mathbf{y})$ is given by a Gaussian, whose mean is zero vector and whose covariance is defined by a Gram matrix $\mathbf{K}$, i.e.,

$$p(\mathbf{y}|\mathbf{X}) = \mathcal{N}(\mathbf{y}|\mathbf{0}, \mathbf{K}) \tag{3.49}$$

with $\mathbf{K}$ being the covariance matrix, i.e.,

$$\mathbf{K} = \mathrm{cov}(\mathbf{y}) = \mathbb{E}(\mathbf{y}\mathbf{y}^{\mathrm{T}}) = \Phi\mathbb{E}(\mathbf{w}\mathbf{w}^{\mathrm{T}})\Phi^{\mathrm{T}} \tag{3.50}$$

The kernel function that forms $\mathbf{K}$ is typically chosen to express the property. And, the smaller the distance of any two points $\mathbf{x}_n$ and $\mathbf{x}_m$ is, the nearer their corresponding values $y(\mathbf{x}_n)$ and $y(\mathbf{x}_m)$ are. If the prior distribution over $\mathbf{w}$ is assumed to be a normal Gaussian distribution $\mathcal{N}(\mathbf{0}, \mathbf{I})$, then $\mathbf{K} = \Phi\Phi^{\mathrm{T}}$, which is an $N \times N$ symmetric matrix with elements $K_{nm} = \langle\phi(\mathbf{x}_n), \phi(\mathbf{x}_m)\rangle = k(\mathbf{x}_n, \mathbf{x}_m)$. To find the marginal distribution $p(\mathbf{t})$, conditioned on the input values $\mathbf{x}_1, \ldots, \mathbf{x}_N$, one need to integrate over $\mathbf{y}$. This can be done by making use of the results from the last section for the linear Gaussian model. Using the Bayes' rule, one can see that the marginal distribution of $\mathbf{t}$ is given by

$$p(\mathbf{t}|\mathbf{X}) = \int p(\mathbf{t}|\mathbf{y})p(\mathbf{y}|\mathbf{X})d\mathbf{y} = \mathcal{N}\left(\mathbf{t}|\mathbf{0}, \mathbf{K} + \beta^{-1}\mathbf{I}_N\right) \tag{3.51}$$

As such, the GP was employed to build a probabilistic model for the data points. The goal in regression, however, is to make predictions of the target variables for new inputs, given a set of training data. Let us suppose that $\mathbf{t} = (t_1, t_2, \ldots, t_N)$, corresponding to the input values $\mathbf{X} = (\mathbf{x}_1, \mathbf{x}_2, \ldots, \mathbf{x}_N)$, comprise the observed training set, and our goal is to predict the target variable t_{N+1} for a new input $\mathbf{x}_{N+1}$. It requires that we evaluate the predictive distribution $p(t_{N+1}|\mathbf{t})$. Note that this distribution is conditioned also on the variables $\mathbf{x}_1$, $\mathbf{x}_2$, …, $\mathbf{x}_N$ and $\mathbf{x}_{N+1}$. However, to keep the notation simple we will not show these conditioning variables explicitly.

To compute the predictive distribution $p(t_{N+1}|\mathbf{t})$, one can begin by writing down the joint distribution $p(t_{N+1}, \mathbf{t})$. From (3.51), the joint distribution over $\mathbf{t}$, t_{N+1} is given by

$$\begin{pmatrix} \mathbf{t} \\ t_{N+1} \end{pmatrix} \sim \mathcal{N}\left(0, \begin{bmatrix} \mathbf{K} + \beta^{-1}\mathbf{I}_N & \mathbf{k}(\mathbf{X}, \mathbf{x}_{N+1}) \\ \mathbf{k}(\mathbf{x}_{N+1}, \mathbf{X}) & k(\mathbf{x}_{N+1}, \mathbf{x}_{N+1}) \end{bmatrix}\right) \tag{3.52}$$

with $\mathbf{k}(\mathbf{X}, \mathbf{x}_{N+1}) = \mathbf{k}(\mathbf{x}_{N+1}, \mathbf{X}) = [k(\mathbf{x}_{N+1}, \mathbf{x}_1), \ldots, k(\mathbf{x}_{N+1}, \mathbf{x}_N)]$. According to the properties of the Gaussian distribution, the conditional distribution $p(t_{N+1}|\mathbf{t})$ is a Gaussian distribution with its mean and covariance,

$$\begin{aligned} m(\mathbf{x}_{N+1}) &= \mathbf{k}(\mathbf{x}_{N+1}, \mathbf{X})\left(\mathbf{K} + \beta^{-1}\mathbf{I}_N\right)^{-1}\mathbf{t} \\ \sigma^2(\mathbf{x}_{N+1}) &= k(\mathbf{x}_{N+1}, \mathbf{x}_{N+1}) - \mathbf{k}(\mathbf{x}_{N+1}, \mathbf{X})\left(\mathbf{K} + \beta^{-1}\mathbf{I}_N\right)^{-1}\mathbf{k}(\mathbf{X}, \mathbf{x}_{N+1}) \end{aligned} \tag{3.53}$$

which is the key result that defines GP-based prediction. Since $\mathbf{k}$ is a vector of function for the test input $\mathbf{x}_{N+1}$, the predictive distribution is a Gaussian, whose mean and variance both depend on $\mathbf{x}_{N+1}$.

The GP regression model is widely used in various field of prediction. However, its merit is not limited to this point. It can be combined with other models to achieve much better performance for prediction. For instance, the GP can be adopted to combine with the neural networks.

3.4.3 Gaussian Process-Based ESN

In this subsection, a GP is used to substitute the generic linear regression of the output of ESN to form the learning model [9]. Combining with a GP, the advantage of short-term memory of dynamical reservoir (DR) in ESN is retained. Then, the recursive relation of the GPESN reads as,

$$\begin{aligned} \mathbf{x}(k+1) &= \tanh(\mathbf{W}^{\text{in}}\mathbf{u}(k+1) + \mathbf{W}\mathbf{x}(k)) \\ y(\mathbf{x}; \mathbf{w}) &= \boldsymbol{\phi}(\mathbf{x}(k))^{\mathrm{T}}\mathbf{w} \end{aligned} \tag{3.54}$$

where $\boldsymbol{\phi}(\cdot)$ is the basis kernel function, and $\mathbf{w}$ is the weights vector of the transformed linear model that is assigned to a Gaussian prior distribution with mean $\mathbf{0}$ and covariance matrix $\mathbf{Z}$. Assuming $f(\mathbf{x}) = y(\mathbf{x}; \mathbf{w}) = \boldsymbol{\phi}(\mathbf{x})^{\mathrm{T}}\mathbf{w}$ and the covariance function $k(\mathbf{x}, \mathbf{x}')$, the statistical feature of the GP is determined by the mean and the covariance function

$$E[f(\mathbf{x})] = \boldsymbol{\phi}(\mathbf{x})^{\mathrm{T}} E(\mathbf{w}) = 0 \tag{3.55}$$

$$\begin{aligned} k(\mathbf{x}, \mathbf{x}') &= E[\left(f(\mathbf{x}) - m(\mathbf{x})\right)\left(f(\mathbf{x}') - m(\mathbf{x}')\right)] \\ &= \boldsymbol{\phi}(\mathbf{x})^{\mathrm{T}} E[\mathbf{w}\mathbf{w}^{T}]\boldsymbol{\phi}(\mathbf{x}') = \boldsymbol{\phi}(\mathbf{x})^{\mathrm{T}} \Sigma \boldsymbol{\phi}(\mathbf{x}') \end{aligned} \tag{3.56}$$

Therefore, $f(\mathbf{x}) \sim \mathbb{N}(0, \boldsymbol{\phi}(\mathbf{x})^{\mathrm{T}} \sum \boldsymbol{\phi}(\mathbf{x}'))$.

Assuming that the latent function y is corrupted by noise, $t = f(\mathbf{x}) + \varepsilon$, where ε is an independent Gaussian white noise with mean 0 and covariance σ_n^2. Then, the prior distribution of y is denoted as $y \sim \mathbb{N}\left(0, \mathbf{K} + \sigma_n^2 \mathbf{I}\right)$, where $\mathbf{K}$ is a positive definite symmetric matrix and each item in $\mathbf{K}$ can be described as $k(\mathbf{x}_p, \mathbf{x}_q)$, expressing the correlation between the states $\mathbf{x}_p$ and $\mathbf{x}_q$. Therefore, the joint Gaussian prior distribution of the training output $\mathbf{t}$ of the sample data and the testing output f_* is given by

$$\begin{bmatrix} \mathbf{t} \\ f_* \end{bmatrix} \sim \mathbb{N}\left(0, \begin{bmatrix} \mathbf{K} + \sigma_n^2 \mathbf{I} & \mathbf{k}(\mathbf{X}, \mathbf{x}_*) \\ \mathbf{k}(\mathbf{X}, \mathbf{x}_*) & k(\mathbf{x}_*, \mathbf{x}_*) \end{bmatrix}\right) \tag{3.57}$$

Typically, one can select the Gaussian kernel covariance function

$$k\left(\mathbf{x}_p, \mathbf{x}_q\right) = \sigma_f^2 \exp\left[-\frac{1}{2l^2}\left(\mathbf{x}_p - \mathbf{x}_q\right)^2\right] \tag{3.58}$$

where σ_f^2 is the coefficient, and l defines the characteristic length scales of the covariance. Then, the covariance of two observed outputs is formulated as,

$$\operatorname{cov}\left(y_p, y_q\right) = k\left(\mathbf{x}_p, \mathbf{x}_q\right) + \sigma_n^2 \delta_{pq} = \sigma_f^2 \exp\left[-\frac{1}{2l^2}\left(\mathbf{x}_p - \mathbf{x}_q\right)^2\right] + \sigma_n^2 \delta_{pq} \tag{3.59}$$

where δ_{pq} denotes a Kronecker delta that is one if $p = q$ and zero otherwise. Based on the Bayes' theorem, one has the conditional probability of $p(f_* | \mathbf{x}_*, \mathbf{X}, \mathbf{t}) = \int p(f_* | \mathbf{x}_*, \mathbf{w}) p(\mathbf{w} | \mathbf{X}, \mathbf{t}) d\mathbf{w}$. Then,

$$(f_* | \mathbf{x}_*, \mathbf{X}, \mathbf{t}) \sim \mathbb{N}\left(\sigma_n^{-2} \boldsymbol{\phi}(\mathbf{x}_*)^{\mathrm{T}} \mathbf{A}^{-1}, \boldsymbol{\phi}, (\mathbf{X}), \mathbf{t}, , \boldsymbol{\phi}(\mathbf{x}_*)^{\mathrm{T}} \mathbf{A}^{-1}, \boldsymbol{\phi}, (\mathbf{x}_*)\right) \tag{3.60}$$

where

$$\mathbf{A} = \sigma_n^{-2} \boldsymbol{\phi}(\mathbf{X}) \boldsymbol{\phi}(\mathbf{X})^{\mathrm{T}} + \boldsymbol{\Sigma}^{-1} \tag{3.61}$$

For the detailed derivations, one can refer to [8]. Combined with (3.56), the conditional probability distribution reads as

$$(f_* | \mathbf{x}_*, \mathbf{X}, \mathbf{t}) \sim \mathbb{N}(E[f_* | \mathbf{x}_*, \mathbf{X}, \mathbf{t}], \operatorname{cov}(f_*)) \tag{3.62}$$

where

$$E[f_*|\mathbf{x}_*, \mathbf{X}, \mathbf{t}] = \mathbf{k}(\mathbf{x}_*, \mathbf{X})\left[\mathbf{K}(\mathbf{X}, \mathbf{X}) + \sigma_n^2\mathbf{I}\right]^{-1} - \mathbf{t} \tag{3.63}$$

and

$$\mathrm{cov}(f_*) = k(\mathbf{x}_*, \mathbf{x}_*) - k(\mathbf{x}_*, \mathbf{x}_*)\left[\mathbf{K} + \sigma_n^2\mathbf{I}\right]^{-1}\mathbf{k}(\mathbf{x}_*, \mathbf{X}) \tag{3.64}$$

Therefore, the mean of the predictive distribution is given by

$$y_* = \mathbf{k}(\mathbf{x}_*, \mathbf{X})\left[\mathbf{K} + \sigma_n^2\mathbf{I}\right]^{-1}\mathbf{t} \tag{3.65}$$

where $\mathbf{x}$ denotes the dataset of sample input, $\mathbf{t}$ denotes the sample output, $\mathbf{x}_*$ denotes the test input, and $\mathbf{k}(\mathbf{x}_*, \mathbf{X})$ stands for the covariance function matrix between $\mathbf{x}_*$ and $\mathbf{x}_1$, $\mathbf{x}_2$, ..., $\mathbf{x}_N$, i.e., $\mathbf{k}(\mathbf{x}_*, \mathbf{X}) = [k(\mathbf{x}_*, \mathbf{x}_1), k(\mathbf{x}_*, \mathbf{x}_2), \ldots, k(\mathbf{x}_*, \mathbf{x}_N)]$. The predicted variance of the output is computed by (3.64).

Since we typically have only rather limited and vague information about the properties exhibited by the parameters, it may not be easy to specify all aspects of the covariance function with confidence in practical applications. As mentioned in [8], we refer to the parameters of the covariance function as hyper-parameters to emphasize that they are parameters of a non-parametric model. To turn the GPESN model into powerful practical tools for the studied industrial problem, the model selection issue must be addressed along with the modeling process. Let us note that the hyper-parameters presented in (3.64) and (3.65), such as σ_f, l, and σ_n, should be determined in advance. In general, the parameters σ_f^2 describe the typical squared amplitude of deviation from the mean, and l denotes the typical length scale, over which the function varies. These hyper-parameters directly affect the prediction results. The generic cross-validation combined with grid search had been widely used to optimize the parameters for data-driven learning model (see [10] for details). However, the optimal parameters exhibit a close relationship with the grid partition. If the partition granularity is rather coarse, the optimum might be easily missed. On the contrary, when the partition granularity is too fine, the computational cost of traversing in allowable period. In addition, the dimension of l equals the number of neurons present in the DR that is usually larger than 100. Thus, it is practically impossible to determine the optimal hyper-parameters by grid search.

In this subsection, we will introduce how to use the Bayes' theorem based to maximize the marginal likelihood function method in order to select these hyper-parameters and optimize them by the conjugate gradient method. Here, we just discuss the basic principle and the detailed process of optimization will be explained in Chap. 7. Under the assumption of the GP, the priori is Gaussian, i.e., $f \mid \mathbf{X}\tilde{}\mathcal{N}(0, \mathbf{K})$, then the logarithm of marginal likelihood function [8] reads as

$$\log p(f|\mathbf{X}) = -\frac{1}{2}f^{\mathrm{T}}\mathbf{K}^{-1}f - \frac{1}{2}\log|\mathbf{K}| - \frac{n}{2}\log 2\pi \tag{3.66}$$

Then, one can obtain the logarithm of marginal likelihood function directly by observing that

$$y \sim \mathcal{N}\left(0, \mathbf{K} + \sigma^2 \mathbf{I}\right) \tag{3.67}$$

and defining the set of hyper-parameters $\boldsymbol{\theta} = [\sigma_f, l, \sigma_n]$

$$\log p(y|\mathbf{X}, \boldsymbol{\theta}) = -\frac{1}{2} y^{\mathrm{T}} \mathbf{K}_y^{-1} - \frac{1}{2} \log\left|\mathbf{K}_y\right| - \frac{n}{2} \log 2\pi \tag{3.68}$$

with

$$\mathbf{K}_y = \mathbf{K} + \sigma^2 \mathbf{I} \tag{3.69}$$

It is clearly that log$p$ is the marginal likelihood of $\boldsymbol{\theta}$. To determine the optimal $\boldsymbol{\theta}$, one can resort to maximizing the marginal likelihood. Taking the partial derivatives of marginal likelihood with respect to the hyper-parameters gives

$$\begin{aligned} \frac{\partial}{\partial \boldsymbol{\theta}} \log p(y|\mathbf{X}, \boldsymbol{\theta}) &= \frac{1}{2} y^{\mathrm{T}} \mathbf{K}_y^{-1} \frac{\partial \mathbf{K}_y}{\partial \boldsymbol{\theta}} \mathbf{K}_y^{-1} y - \frac{1}{2} \mathrm{tr}\left(\mathbf{K}_y^{-1} \frac{\partial \mathbf{K}_y}{\partial \boldsymbol{\theta}}\right) \\ &= \frac{1}{2} \mathrm{tr}\left(\left(\boldsymbol{\alpha}\boldsymbol{\alpha}^{\mathrm{T}} \mathbf{K}_y^{-1}\right) \frac{\partial \mathbf{K}_y}{\partial \boldsymbol{\theta}}\right) \end{aligned} \tag{3.70}$$

where $\boldsymbol{\alpha} = \mathbf{K}_y^{-1} y$, and tr$(\cdot)$ denotes the trace of a square matrix. Accordingly, the optimal hyper-parameters can be estimated.

The underlying algorithm can be described as a sequence of the following steps:

Algorithm 3.1: Learning a GPESN Model

Step 1: Obtain the time series sample dataset, the length of samples, and initiate the embedded dimension m and the fundamental delay time z.

Step 2: Initiate the reservoir of ESN, and the parameters $\mathbf{W}^{\mathrm{in}}$, $\mathbf{W}$, *and* $\mathbf{W}^{\mathrm{back}}$.

Step 3: Train the hybrid neural network using the sample data, and update its states by (3.54).

Step 4: Set the initial time, gather the states of ESN $\mathbf{x}(k)$, *and the supervision output $y(k)$, to form a series of "state-output" pairs for the GP.*

Step 5: Initiate the selection of the hyper-parameters of GP and train the GP by using the activated "state-output" patterns.

Step 6: Determine the optimal hyper-parameters by maximizing the marginal likelihood function.

Step 7: Get a period of industrial time series before the prediction, complete data preprocessing, and start prediction.

Step 8: Calculate the states value of DR and produce the prediction results given by (3.64) and (3.65).

3.4.4 Case Study

Steel industry is usually accompanied with high energy consumption and environmental pollution, in which the by-product gas is regarded as the one of the useful energy resources. It is very meaningful to predict the gas flow for reasonable secondary energy scheduling. However, since the historic acquired data for the SCADA are always mixed with noise, the existing predictions with any indication of the accuracy are less reliable for industrial application. With the progress of energy resources crisis in the world, the scientific by-product energy utilization plays a crucial role on the production cost and the energy efficiency. Therefore, the accurate gas flow prediction will greatly contribute to the reasonable energy scheduling. In practice, the workers usually raise more concerns on the estimated accuracy on the energy flow, with which the real-time decision-making operations are implemented. Thus, it is very significant to increase the prediction accuracy of the energy flows based on the real-time requirement of industrial application.

In this case study, we consider a coke oven gas (COG) system of a steel plant, and Fig. 3.1 depicts the structure of the COG system. Six coke ovens are viewed as the gas producing units that can supply 300 km^3 COG per hour (on average) into the pipeline network, whose calorific power is about 18,000 kJ/Nm^3. The transportation system usually consists of pipelines, mixing stations, and pressure stations. And, the consumption users primarily comprise two categories, which are the general users such as hot blast stoves, hot rolling plant, cold rolling plant, chemical products recovery (CPR), and the adjustable users such as low pressure boiler and internal electrical power plant. The remainder gas could be stored into the gasholders, one of which is practically with the capacity of 150 km^3 and another is 300 km^3. The gas consumption of the adjustable users can be adjusted under some constraints in order to meet the balance requirement of the whole COG system. In addition, there is a

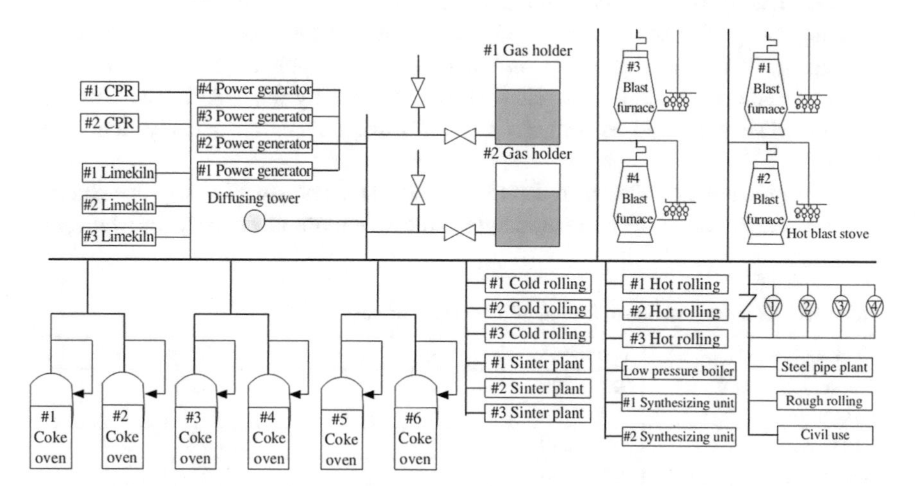

Fig. 3.1 Structure of the COG system in steel industry

diffusing tower in the COG system, which emits the redundant gas by burning for the purpose of maintaining the system security while the temporarily surplus gas exceeds the capacity of gasholder level.

In this subsection, to verify the performance of the GPESN, the prediction with the use of real-time data from a steel plant is performed. Based on the modeling of the historical data, the GPESN can predict the gas flow of the future 60 min. To indicate the effectiveness of the GPESN, we randomly selected three of the units in the COG system of the steel plant, which are the consumption amount of #1,2 coke oven, that of the #1 blast furnace, and that of #3 hot rolling plant as the simulation examples.

Here, 1000 continuous historical data in the energy data center was adopted to engage in the processes of learning and prediction. The sampling frequency is 1 min, i.e., In the GPESN, 1000 pairs of input-output dataset are used to activate the DR, and then the first 100 samples are discarded to get rid of the initial transient process of learning. In this way, the GP is learnt by using the remaining 900 pairs of patterns to mine the relationship between the states of network and the output. With a long-term observation for the dynamics of gas units in the plant, the embedded dimensionality of the model is set to be ten empirically. For the appropriate parameters setting of ESNs, 100 times independent experiments using the different period of practical data are conducted for each unit, and the sound values are listed as Table 3.1.

To quantify the validity of the GPESN, some comparative analysis is carried out by using other learning methods based on the same dataset, including a neural network trained by general back propagation (BP) algorithm, SVM, ESN, and the GPESN. To make the comparison fair, the experiment-based manner is used to select the relatively good settings of each method, where we design the number of layers of BP is three and the number of neurons in hidden layer is ten. Considering the training efficiency, in SVM the sensitive coefficient is 0.01, and the similar parameters settings of GPESN are given to the generic ESN. The testing experiments are done by using five groups of different 1000-min continuous data from this plant collected in August, 2010, each of which is independently completed for ten times. The average prediction results comprehensively evaluated by the MAPE, the NRMSE, and the MSE are listed as Table 3.2 (these three indices are already defined in Chap. 2).

From the statistics shown in Table 3.2 and the graphs in Fig. 3.2, the GPESN provides the lowest errors compared to the others, in terms of not only the average of

Table 3.1 Experimental determination of parameters of the GPESN

Prediction units	Size of reservoir	Sparseness of W (%)	Spectral radius of W
Consumption of #1–2 coke oven	200×200	2	0.85
#1 Blast furnace	200×200	3	0.85
#3 Hot rolling plant	100×100	3	0.80

Table 3.2 Average prediction error statistics obtained by using four prediction methods

Prediction units	Method	NRMSE	MSE	MAPE (%)
Inside consumption of #1–2 coke oven	BP	0.3190	0.0256	1.4663
	SVM	0.2131	0.0114	0.9298
	ESN	0.1528	0.0059	0.6575
	GPESN	0.1263	0.0040	0.5706
#1 Blast furnace	BP	0.2918	0.2347	7.2210
	SVM	0.2578	0.1832	4.2311
	ESN	0.2312	0.1473	3.7905
	GPESN	0.0705	0.0137	1.5456
#3 Hot rolling plant	BP	0.3606	0.5749	4.1663
	SVM	0.3006	0.3996	3.6590
	ESN	0.2711	0.3249	3.0021
	GPESN	0.2116	0.1981	2.3607

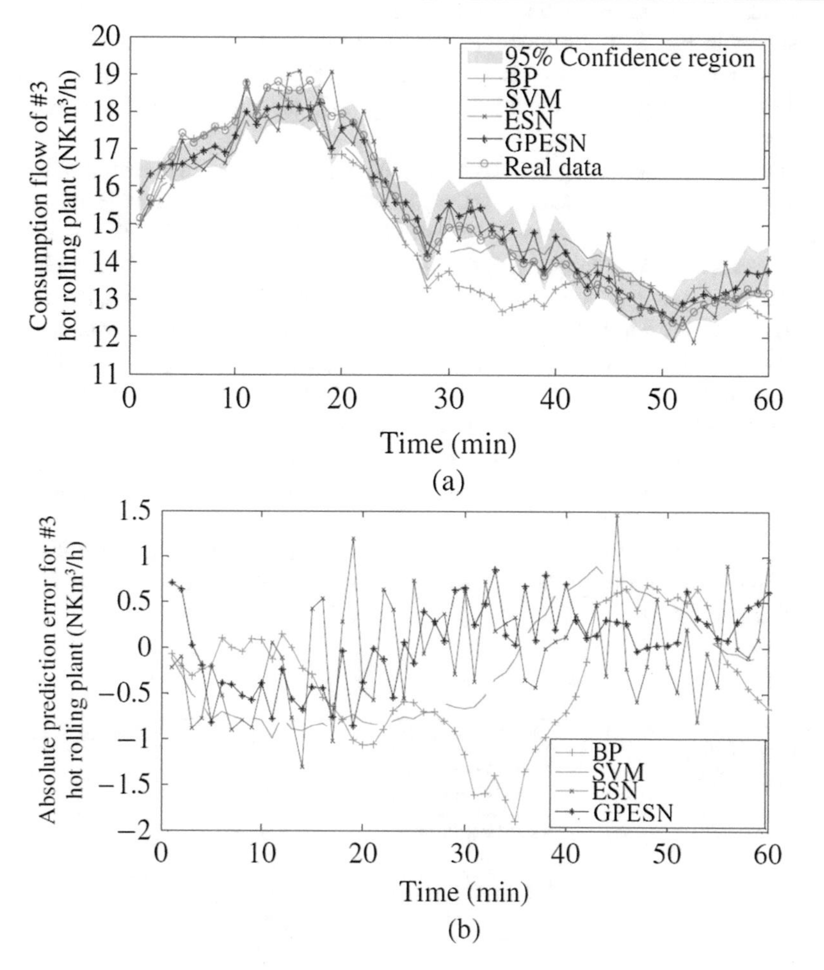

Fig. 3.2 Prediction results of the consumption flow of #3 hot rolling plant (a) Predictive tendency comparison (b) Curves of absolute errors

prediction errors (NRMSE and MSE) but also the relative percentage of the errors (MAPE). As such, these statistics reflect the average prediction accuracy; in addition, one can compare the error values of fitted time series that differ in level. It is analyzed that the BP belongs to a static neural network, and since the core of the network is to realize the learning process by gradient descent, it tends to bring about the local optimum and usually a large prediction error. ESN has a dynamic reservoir that can memorize the system dynamics in a short term, so the results for the time series prediction, such as the consumptions of #1–2 coke oven, #1 blast furnace, and #3 hot rolling plant, are statistically better than those produced by the SVM. Such practical results are completely consistent with the theoretical analysis. With respect to the GPESN, the results of the time series prediction are better in comparison to the others obtained by means of the three-error evaluation criterion.

Here, we randomly select the gas consumption flow of #3 hot rolling to show the tendency comparison and 95% confidence region calculated by the GPESN in Fig. 3.2a. Their corresponding absolute errors by these learning methods are also illustrated in Fig. 3.2b. Furthermore, 95% confidence region illustrated by the gray zone in Fig. 3.2a are also visualized, which helps us to identify the variation tendency of the predicted units. From Fig. 3.2, the GPESN performs the best among all the four methods.

In addition, when using the classical ESN, the output weight sometimes exceeds the order of 1012, which leads to an unforeseen or exceptional result. Here, we also verify the difference between the generic ESN and the GPESN with this respect. Another set of 1500 continuous data points from this plant is chosen to complete the experiments, where the first 1000 data are employed to learn the model, and the rest of the dataset aims to test the model. Here, assuming the two methods with identical DR scale and the parameters, the independent experiments are conducted 200 times by ESN and the GPESN. Then, the failure intensities are listed in Table 3.3. Note that there is no failure prediction at all when using the GPESN, while some large output weights occur if using the ESN. Therefore, one would conclude that the GPESN can completely avoid the over-fitting prediction generated by the industrial data that is mixed by noise.

All in all, the GP-based ESN exhibits the best prediction abilities for the COG real-time flow or the gasholder level compared to the other machine learning approaches.

Table 3.3 comparison of prediction failure times by ESN and GPESN

Prediction units	Method	Number of failures
Inside consumption of #1–2 coke oven	ESN	3
	GPESN	0
#1 Blast furnace	ESN	0
	GPESN	0
#3 Hot rolling plant	ESN	4
	GPESN	0

3.5 Artificial Neural Networks-Based Prediction

Artificial neural networks (ANNs) are computing systems inspired by biological neural networks that constitute animal brains. Such systems learn to do tasks by considering examples, generally without task-specific programming. An ANN is based on a collection of connected units called artificial neurons. Each connection between neurons can transmit a signal to another neuron. The receiving neuron can process the signal and then signal downstream neurons connected to it. Neurons may have state, generally represented by real numbers, typically between 0 and 1. Neurons and synapses may also have a weight that varies as learning proceeds, which can increase or decrease the strength of the signal that it sends downstream. Furthermore, they may have a threshold such that only if the aggregate signal is below (or above) that level, it will be sent to the downstream neurons. Typically, neurons are organized in layers, and different layers may perform different kinds of transformations on their inputs. Signals travel from the first (input layer), to the last (output layer), possibly after traversing the layers multiple times.

The original goal of the neural network approach was to solve problems in the same way that a human brain would. Over time, attentions focused on matching specific mental abilities, leading to deviations from biology such as back propagation, or passing information in the reverse direction and adjusting the network to reflect that information.

Recently, NNs have been used on a variety of tasks, including computer vision, speech recognition, machine translation, social network filtering, playing board and video games, medical diagnosis, and in many other domains. In this chapter, the application of neural networks in industrial process prediction will be mainly introduced, where industrial data usually contains noise and is accompanied with missing points and abnormal points that could be coped with some data preprocessing techniques. Thus, most of the attentions are paid to the study of using the RNNs to model with noisy data.

3.5.1 RNNs for Regression

Elman network is a local RNN that mainly consists of four layers, the input, the hidden, the context, and the output layer, whose structure is shown in Fig. 3.3. Such a network possesses massive parallel connections not only between the hidden and the output, but also between the hidden and the input as well as the context units [11]. The output of hidden layer goes forward to the output and goes back to the context layer [12, 13]. The self-connections of the context nodes make it sensitive to historical input and qualified for the efficiency in dynamic system modeling [14]. Meanwhile, the neurons layout of Elman network is regularly layered and organized, where the signals can simultaneously arrive at the neurons inputs; therefore, the internal states can be simultaneously calculated. In other words, if

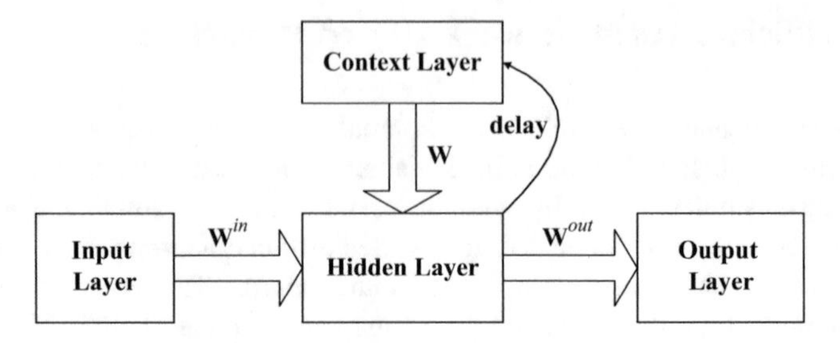

Fig. 3.3 The structure of an Elman network

one views each neuron as a processing unit, then the whole network can be regarded as a large distributed system, which means that such network structure makes it particularly applicable for parallelized modeling [12].

If the numbers of input units, hidden units, context units, and output units are K, N, T, and L, respectively, and this network subjects to the constraint that $N = T$. The state equations are formulated as

$$\begin{aligned} \mathbf{x}(k+1) &= f\left(\mathbf{W}^{\text{in}}\mathbf{u}(k+1) + \mathbf{W}\mathbf{x}(k)\right) \\ \mathbf{y}(k+1) &= f^{\text{out}}(\mathbf{W}^{\text{out}}\mathbf{x}(k+1) + \mathbf{r}_k) \end{aligned} \tag{3.71}$$

where $\mathbf{W}^{\text{in}}$, $\mathbf{W}$, and $\mathbf{W}^{\text{out}}$ are the weights of input, internal states and output, respectively, and their dimensionalities are $N \times K$, $N \times N$, and $L \times N$, respectively. $\mathbf{u}(k)$,$\mathbf{x}(k)$, and $\mathbf{y}(k)$ denote the input, the internal states, and the output at time point k, respectively, i.e., $\mathbf{u}(k) = [u_1(k), \ldots, u_K(k)]^{\text{T}}$, $\mathbf{x}(k) = [x_1(k), \ldots, x_N(k)]^{\text{T}}$, and $\mathbf{y}(k) = [y_1(k), \ldots, y_L(k)]^{\text{T}}$. f is the activation of internal neurons, e.g., a hyperbolic tangent function, f^{out} is the output activation, and $\mathbf{r}_k$ is the Gaussian white noise.

The training of a RNN aims to calculate the optimal connected weights of the network. Since the states update process and the output process described in (3.71) are interrupted by the noise, the EKF (it will be introduced in Chap. 7 in details) that transformed the training process into a nonlinear filtering problem will be one of the most effective methods to obtain the optimal weights. By linearizing the mean and covariance of the filter, EKF estimates a process by using a form of feedback control [15]. The time update equations obtain the priori estimates of the weight vector $\widehat{\mathbf{w}}(k|k-1)$ and the error covariance matrix $\mathbf{P}(k|k-1)$.

$$\widehat{\mathbf{w}}(k|k-1) = \widehat{\mathbf{w}}(k-1) \tag{3.72}$$

$$\mathbf{P}(k|k-1) = \mathbf{F}(k-1)\mathbf{P}(k-1)\mathbf{F}^{\text{T}}(k-1) + \mathbf{Q}(k-1) \tag{3.73}$$

where the state transition matrix $\mathbf{F}(k) = \mathbf{I}$ and the dimensionalities of $\mathbf{F}$ and $\mathbf{P}$ are both $G \times G$, and that of $\widehat{\mathbf{w}}$ is G. $\mathbf{Q}(k-1)$ with $G \times G$ is the covariance of the process noise $\mathbf{q}_k$. And the measurement update equations read as

$$\mathbf{K}(k) = \mathbf{P}(k|k-1)\mathbf{H}(k)^{\mathrm{T}}\left[\mathbf{H}(k)\mathbf{P}(k|k-1)\mathbf{H}(k)^{\mathrm{T}} + \mathbf{R}(k)\right]^{-1} \quad (3.74)$$

$$\widehat{\mathbf{w}}(k|k) = \widehat{\mathbf{w}}(k|k-1) + \mathbf{K}(k)[\mathbf{y}(k) - \widehat{\mathbf{y}}(k)] \quad (3.75)$$

$$\mathbf{P}(k|k) = \mathbf{P}(k|k-1) - \mathbf{K}(k)\mathbf{H}(k)\mathbf{P}(k|k-1) \quad (3.76)$$

where $\mathbf{y}(k)$ is the expected output with length L; $\mathbf{K}(k)$ is the gain with $G \times L$; $\mathbf{P}(k|k)$ is the error covariance; $\mathbf{H}(k)$ is the measurement matrix with $L \times G$; and $\mathbf{R}(k)$ with $L \times L$ is the covariance of the measurement noise $\mathbf{r}_k$. Since the computation of the Jacobian matrix is rather difficult in EKF-based modeling, a truncated back propagation through time (BPTT) to estimate $\mathbf{H}(k)$ was reported in [16]. However, that method had to transform the RNN into a feedforward network.

Next, we introduce a simplified and direct calculation for the Jacobian matrix. First, without loss of generality, one can formulate the nonlinear discrete-time equations of the network as

$$\mathbf{w}(k+1) = \mathbf{w}(k) + \mathbf{q}_k \quad (3.77)$$

and

$$\mathbf{y}(k+1) = f^{\mathrm{out}}\left(\mathbf{W}^{\mathrm{out}} f\left(\mathbf{W}^{\mathrm{in}}\mathbf{u}(k+1) + \mathbf{W}\mathbf{x}(k)\right) + \mathbf{r}_k\right) \quad (3.78)$$

where $\mathbf{y}(k+1)$ is the expected output at time $k+1$. $\mathbf{w}(k+1)$ is the weight of the entire network at time $k+1$; $\mathbf{u}(k+1)$ is the external input; $\mathbf{x}(k)$ is the internal states at k; and both $\mathbf{q}_k$ and $\mathbf{r}_k$ denote Gaussian white noises. To apply EKF to the network training, we define the weight vector $\mathbf{w}$ with dimensionality $G \times 1$ and $G = N \times K + N \times N + L \times N$ for combining all the connection weights ($\mathbf{W}^{\mathrm{in}}, \mathbf{W}, \mathbf{W}^{\mathrm{out}}$) of the network. Then, one can obtain

$$\mathbf{w} = \begin{pmatrix} \mathrm{Res}\left(\mathbf{W}^{\mathrm{in}}\right) \\ \mathrm{Res}(\mathbf{W}) \\ \mathrm{Res}\left(\mathbf{W}^{\mathrm{out}}\right) \end{pmatrix} \quad (3.79)$$

where the designed operator $\mathrm{Res}(\mathbf{A}) = \left(\mathbf{a}_1^{\mathrm{T}}\ \mathbf{a}_2^{\mathrm{T}} \cdots\ \mathbf{a}_N^{\mathrm{T}}\right)^{\mathrm{T}}$, where $\mathbf{A} = (\mathbf{a}_1\ \mathbf{a}_2 \cdots\ \mathbf{a}_N)^{\mathrm{T}}$ and $\mathbf{a}_i = (\mathrm{a}_{i1}\ \mathrm{a}_{i2} \cdots \mathrm{a}_{iK})^{\mathrm{T}}$. This operator aims to transform the weight matrix into a column vector. Here, all the weights are regarded as the system state required to be estimated, thus the network modeling aims to obtain the optimal $\mathbf{w}$ by minimizing the mean square error between the expected output and the predicted one.

To obtain the exact expression of $\mathbf{H}$, some definitions should be firstly described. One can assume that $\left(\mathbf{w}_i^{\mathrm{out}}\right)^{\mathrm{T}}$ denotes the ith row of $\mathbf{W}^{\mathrm{out}}$, where $i = 1, 2, \ldots, L$. $\mathbf{I}$ is an identity matrix with dimensionality $L \times L$, and $\mathbf{i}_i$ denotes the ith column of $\mathbf{I}$. The defined operator $\otimes$ aims to obtain the multiplication of the corresponding elements of two vectors with identical dimensionalities. Thus, the Jacobian matrix can be formulated as the partial derivatives of the output with respect to the weight

$$\mathbf{H} = \frac{\partial \mathbf{y}}{\partial \mathbf{w}^{\mathrm{T}}} = \begin{pmatrix} \frac{\partial y_1}{\partial \mathbf{w}^{\mathrm{T}}} \\ \frac{\partial y_2}{\partial \mathbf{w}^{\mathrm{T}}} \\ \vdots \\ \frac{\partial y_L}{\partial \mathbf{w}^{\mathrm{T}}} \end{pmatrix} = \begin{pmatrix} \frac{\partial y_1}{\partial w_1} & \frac{\partial y_1}{\partial w_2} & \cdots & \frac{\partial y_1}{\partial w_G} \\ \frac{\partial y_2}{\partial w_1} & \frac{\partial y_2}{\partial w_2} & \cdots & \frac{\partial y_2}{\partial w_G} \\ \vdots & \vdots & \ddots & \vdots \\ \frac{\partial y_L}{\partial w_1} & \frac{\partial y_L}{\partial w_2} & \cdots & \frac{\partial y_L}{\partial w_G} \end{pmatrix} \tag{3.80}$$

where $\partial y_i/\partial \mathbf{w}^{\mathrm{T}}$ denotes the partial derivatives of the ith output with respect to the weights, and $\partial y_i/\partial w_j$ ($j = 1, 2, \ldots, G$) denotes the partial derivatives of the ith output with respect to the jth weight.

From (3.79), the computation of **H** can be divided into three parts. For Res($\mathbf{W}^{\mathrm{in}}$), Res(**W**) and Res($\mathbf{W}^{\mathrm{out}}$), their partial derivatives are denoted as $\mathbf{H}_1$, $\mathbf{H}_2$, and $\mathbf{H}_3$, respectively. Their dimensionalities are $L \times (N \times K)$, $L \times (N \times N)$, and $L \times (L \times N)$, respectively. We detail the three parts for computing **H** in the following.

Scenario 1: At the input layer,

$$\mathbf{H}_1 = \begin{pmatrix} \mathbf{h}_{1,1}^{\mathrm{T}} \\ \mathbf{h}_{1,2}^{\mathrm{T}} \\ \vdots \\ \mathbf{h}_{1,L}^{\mathrm{T}} \end{pmatrix} \tag{3.81}$$

with $\mathbf{h}_{1,\,i} = \mathrm{Res}(\partial y_i/\partial \mathbf{W}^{\mathrm{in}})$, and

$$\begin{aligned} \frac{\partial y_i}{\partial \mathbf{W}^{\mathrm{in}}} &= \frac{\partial \left\{ f^{\mathrm{out}} \left[\left(\mathbf{w}_i^{\mathrm{out}}\right)^{\mathrm{T}} f\left(\mathbf{W}^{\mathrm{in}}\mathbf{u}(k+1) + \mathbf{W}\mathbf{x}(k)\right) + \mathbf{r}_{k_i} \right] \right\}}{\partial \mathbf{W}^{\mathrm{in}}} \\ &= \left[\mathbf{w}_i^{\mathrm{out}} \otimes f'\left(\mathbf{W}^{\mathrm{in}}\mathbf{u}(k+1) + \mathbf{W}\mathbf{x}(k)\right)\right]\mathbf{u}^{\mathrm{T}}(k+1) \end{aligned} \tag{3.82}$$

Scenario 2: At the hidden layer,

$$\mathbf{H}_2 = \begin{pmatrix} \mathbf{h}_{2,1}^{\mathrm{T}} \\ \mathbf{h}_{2,2}^{\mathrm{T}} \\ \vdots \\ \mathbf{h}_{2,L}^{\mathrm{T}} \end{pmatrix} \tag{3.83}$$

with $\mathbf{h}_{2,\,i} = \mathrm{Res}(\partial y_i/\partial \mathbf{W})$, and

$$\frac{\partial y_i}{\partial \mathbf{W}} = \left[\mathbf{w}_i^{\mathrm{out}} \bigotimes f'\left(\mathbf{W}^{\mathrm{in}}\mathbf{u}(k+1) + \mathbf{W}\mathbf{x}(k)\right)\right]\mathbf{x}^{\mathrm{T}}(k) \tag{3.84}$$

Scenario 3: At the output layer

$$\mathbf{H}_3 = \begin{pmatrix} \mathbf{h}_{3,1}^{\mathrm{T}} \\ \mathbf{h}_{3,2}^{\mathrm{T}} \\ \vdots \\ \mathbf{h}_{3,L}^{\mathrm{T}} \end{pmatrix} \tag{3.85}$$

with $\mathbf{h}_{3,\,i} = \mathrm{Res}(\partial y_i/\partial \mathbf{W}^{\mathrm{out}})$, and

$$\frac{\partial y_i}{\partial \mathbf{W}} = \mathbf{i}_i\left(\mathbf{W}^{\mathrm{in}}\mathbf{u}(k+1) + \mathbf{W}\mathbf{x}(k)\right)^{\mathrm{T}} \tag{3.86}$$

Then, one can obtain $\mathbf{H} = [\mathbf{H}_1, \mathbf{H}_2, \mathbf{H}_3]$. The modeling process of the Elman network with EKF can be listed as follows. Although RNN are commonly used for the industrial prediction, its training process is generally complex with high computational load. In Chap. 8, a GPU-based parallel computing for the computation of the Jacobian matrix will be introduced.

Thus, the modeling process of the Elman network with EKF can be listed as follows:

Algorithm 3.2: Modeling the Elman Network with EKF

Step 1: Initialize the network parameters, including the weight vector $\mathbf{w}$*, the internal state vector* $\mathbf{x}$*, the embedded dimensionality* K*, and the number of internal neurons* N*.*

Step 2: Determine the number of sample data M*, and the training set* $(\mathbf{u}^{\mathrm{T}}(k), \mathbf{d}^{\mathrm{T}}(k))$*,* $(k = 1, 2, 3, \ldots, M)$*, where* $\mathbf{u}(k)$ *is the kth input sample,* $\mathbf{u}(k) = [u_1(k), u_2(k), \ldots, u_K(k)]^{\mathrm{T}}$*; and* $\mathbf{d}(k)$ *is the kth output sample,* $\mathbf{d}(k) = [d_1(k), d_2(k), \ldots, d_L(k)]^{\mathrm{T}}$*.*

Step 3: Set the initial values of $\mathbf{P}$*,* $\mathbf{Q}$*, and* $\mathbf{R}$ *of the EKF. The values of* $\mathbf{Q}$ *and* $\mathbf{R}$ *are invariable in iteration, and the value of* $\mathbf{P}$ *is updated by (3.76).*

Step 4: Obtain the priori estimates $\widehat{\mathbf{w}}(k|k-1)$ *and* $\mathbf{P}(k|k-1)$ *at k by (3.72) and (3.73).*

Step 5: Obtain the weights $\widehat{\mathbf{w}}(k|k)$ *and the error covariance* $\mathbf{P}(k|k)$ *at k by (3.74)–(3.76).*

Step 6: Repeat Step 4–5 until obtaining an optimal $\mathbf{w}$ *that minimizes the MSE between the expected output and the predicted one. Then, calculate the optimal* $\mathbf{W}^{\mathrm{in}}$*,* $\mathbf{W}$*, and* $\mathbf{W}^{\mathrm{out}}$ *of trained network.*

3.5.2 ESN for Regression

ESN consists of an input layer, a dynamical reservoir (DR), and an output layer, whose structure is illustrated in Fig. 3.4. The DR comprises many neurons with sparse connectivity each other, exhibiting the ability of short-term memory. ESN

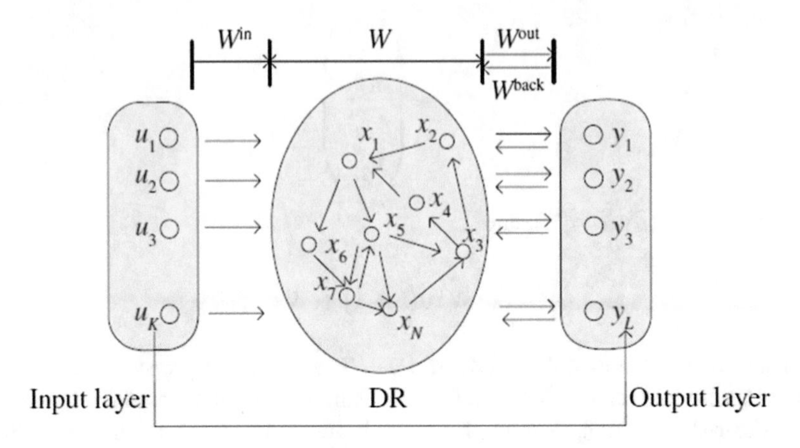

Fig. 3.4 The structure of an ESN

shows excellent performance for time series prediction [17]. The recursive relation governing ESN is described as follows:

$$\begin{aligned} \mathbf{x}(k+1) &= f(\mathbf{W}^{\text{in}}\mathbf{u}(k) + \mathbf{W}\mathbf{x}(k) + \mathbf{W}^{\text{back}}\mathbf{y}(k)) \\ \mathbf{y}(k+1) &= f^{\text{out}}\left(\mathbf{W}^{\text{out}}[\mathbf{x}(k+1); \mathbf{y}(\mathbf{k})]\right) \end{aligned} \tag{3.87}$$

where the number of input units is equal to K, the number of neurons in the DR is N, and the size of the output layer is equal to L. The input and the observed output are, respectively, $\mathbf{u}(k) = [u_1(k), \ldots, u_K(k)]^{\mathrm{T}}$ and $\mathbf{y}(k) = [y_1(k), \ldots, y_L(k)]^{\mathrm{T}}$. Here, $\mathbf{W}^{\text{in}}$ is the connection matrix describing the relationships between the elements located in the input and the DR. $\mathbf{W}$ is the weight matrix of the neurons in DR. To provide memorization capability, $\mathbf{W}$ should be a sparse matrix whose connectivity level is 1–5% and the value of spectral radius should be less than 1. $\mathbf{W}^{\text{back}}$ denotes the feedback weight matrix representing connections between the output layer and the DR. Then, the states of DR $\mathbf{x}(k)$ can be calculated by (3.87) based on the random initialized $\mathbf{W}^{\text{in}}$, $\mathbf{W}$, and $\mathbf{W}^{\text{back}}$.

The key of training an ESN lies in the calculation of the output weight matrix $\mathbf{W}^{\text{out}}$. The function f^{out} is often chosen to be the identity function. However, it usually suffers from numerical problems, such as the ill-conditioned matrix or output weight with very large value that might occur when the data are corrupted by noise. Through a great number of industrial simulation experiments using the practical data, it is observed that the reservoir has an eigenvalue spread in the order of 10^{12} or even larger. This is typically accompanied by very large output weights that lead to relatively poor prediction results. Jaeger [18] also claimed such phenomenon made the online learning algorithm more difficult, and the ESN with very large output weights led to degraded generation performance. It can be analyzed that the traditional ESN determines the output weights $\mathbf{W}^{\text{out}}$ by the generic linear regression, which means a linear system concerning the state matrix of DR, $\mathbf{M}$ and the supervision output signals $\mathbf{T}$, need to be solved

$$\mathbf{M}(\mathbf{W}^{\text{out}})^{\text{T}} = \mathbf{T} \tag{3.88}$$

where $\mathbf{M} = [(\mathbf{u}(k_0), (\mathbf{x}(k_0))^{\text{T}}, \ldots, (\mathbf{u}(k_n), (\mathbf{x}(k_n))^{\text{T}}]$ and $\mathbf{T} = [(f^{\text{out}} y(k_0)), \ldots, (f^{\text{out}} y(k_n))]^{\text{T}}$. Then, $(\mathbf{W}^{\text{out}})^{\text{T}} = \mathbf{M}^{-1}\mathbf{T}$ and the mean least square error for training $\mathbf{W}^{\text{out}}$ is given by

$$E = \sigma^2 \sum_{i=1}^{t} \frac{1}{\lambda_i} \tag{3.89}$$

where λ_is denote the eigenvalues of $\mathbf{M}^{\text{T}}\mathbf{M}$. Once the original time series contains high level noise or other abnormal cases (such as anomaly), $\mathbf{M}^{\text{T}}\mathbf{M}$ will be almost singular, which means that at least one of eigenvalues approaches zero. To overcome such drawback, [19] proposed to add some noise to the dynamic reservoir of ESN during the training process. Here, the noise intensity was generally determined in an ad hoc way that suffers from the lack of reliability. In [20], the ridge regression was used in modeling the reservoir state-space to improve the performance of ESN. However, these methods can work well only if the model was ill conditioned. Furthermore, it is difficult to determine the values of the regularization coefficients. If we adopt cross-validation to select these coefficients, then such method became very time-consuming, which has very limited value in practical application.

3.5.3 SVD-Based ESN for Industrial Prediction

In this subsection, the SVD method for estimating the parameters of ESN model is introduced. From the above subsection, one can see that the ill-conditioned solutions are raised from some minimum eigenvalues (almost zero) since these eigenvalues can lead to a very large mean squares error [21]. Thus, the SVD method is employed here to solve the eigenvalues of $\mathbf{M}$ in (3.90), and some minimal eigenvalues are abandoned when computing the pseudo-inverse matrix of $\mathbf{M}$.

According to the principle of SVD [22], there always exist two matrices $\mathbf{U}_M$ and $\mathbf{V}_M$ which satisfies the following equation.

$$\mathbf{U}_M^{\text{T}}\mathbf{M}\mathbf{V}_M = \text{diag}\{\sigma_1, \sigma_2, \ldots, \sigma_p\} \tag{3.90}$$

where $\mathbf{U}_M \in \mathbb{R}^{(T-T_0+1)\times(T-T_0+1)}$and $\mathbf{V}_M \in \mathbb{R}^{(K+N)\times(K+N)}$, $p = \min\{(T - T_0 + 1), (K + N)\}$, $\sigma_1 \geq \sigma_2 \geq \cdots \geq \sigma_p \geq 0$, and σ_i is the singular value of matrix $\mathbf{M}$. Then, the singular values decomposition of the matrix $\mathbf{M}$ can be written as

$$\mathbf{M} = \mathbf{U}_M \mathbf{\Sigma}^{-1} \mathbf{V}_M^{\text{T}} \tag{3.91}$$

with $\mathbf{\Sigma} = \text{diag}\{\sigma_1, \ldots, \sigma_r, 0, \ldots, 0\}$. Then, the generalized inverse of $\mathbf{M}$ equals

$$\mathbf{M}^{\dagger} = \mathbf{V}_M \boldsymbol{\Sigma}^{-1} \mathbf{U}_M^{\mathrm{T}} \tag{3.92}$$

with

$$\boldsymbol{\Sigma} = \begin{bmatrix} \boldsymbol{\Sigma}_r^{-1} & 0 \\ 0 & 0 \end{bmatrix}_{(K+N)(T-T_0+1)} \tag{3.93}$$

$$\boldsymbol{\Sigma}_r^{-1} = \mathrm{diag}\{\sigma_1^{-1}, \ldots, \sigma_r^{-1}\} \tag{3.94}$$

As for the output weights, one can use the least squares technique to compute

$$\left(\widehat{\mathbf{W}}^{\mathrm{out}}\right)^{\mathrm{T}} = \mathbf{M}^{\dagger}\mathbf{T} = \mathbf{V}_M \boldsymbol{\Sigma}_M^{-1} \mathbf{U}_M^{\mathrm{T}} \mathbf{T} \tag{3.95}$$

Then, the output weights of the ESN model can be calculated as

$$\widehat{\mathbf{W}}^{\mathrm{out}} = \left(\sum_{j=1}^{r} \left(\frac{u_j^{\mathrm{T}} \mathbf{T}}{\sigma_j}\right) v_j\right)^{\mathrm{T}} \tag{3.96}$$

Besides, one has

$$w_k = \sum_{j=1}^{r} \frac{v_{kj}}{\sigma_j} \sum_{i=1}^{T-T_0+1} u_{ij} T_i, \quad k = 1, \ldots, K+N \tag{3.97}$$

According to (3.97), a very small eigenvalue can lead to an ill-conditioned solution when using the least square technique. Therefore, one can set a parameter δ to select the effective eigenvalues and abandon these very small eigenvalues. $\sigma_1 \geq \cdots \geq \sigma_r \geq \delta \geq \sigma_{r+1} \geq \cdots \geq \sigma_{T-T_0+1}$. Generally, the value of δ could be set as $\delta = \mu \|M\|_{\infty}$, where μ is the computing accuracy.

To guarantee the stability of the SVD-based ESN, one has to set suitable spectral radium for the DR. With the introduction of the SVD, the output weights are unique and have globally optimal point. In addition, the SVD-based training technique is very simple and needs not to calculate the partial derivatives of the nonlinear model, which eliminates the difficulties of training process.

3.5.4 ESNs with Leaky Integrator Neurons

Although the parameters of the ESN including $\mathbf{W}^{\mathrm{in}}$, $\mathbf{W}$, and $\mathbf{W}^{\mathrm{back}}$ may be initialized randomly in the phase of the network design, one can find that as for the applications in industry, the dynamics of ESN calls for quite different requirements, which mainly depends not only on the value of the spectral radius and the sparsity level of $\mathbf{W}$, but also the inputs of the network. Certainly, the expert experience can be employed to decide the values of these parameters, but this might result in a low quality of the resulting prediction. A systematic method to select both input variables and the level of connectivity available through a feedforward neural network with a single hidden layer was presented in [23]. A successful industrial application of this method was reported in [24]. In this section, we consider an improved version of the ESN to be used for prediction of various BFG units, and add a compensation signal to serve as a certain input to eliminate the impact coming from input noise. The corresponding recurrent expression reads as follows:

$$\mathbf{x}(n+1) = f\left[\mathbf{W}^{\text{in}}(\mathbf{u}(n+1)+c) + \alpha\mathbf{W}\mathbf{x}(n) + \beta\mathbf{W}^{\text{back}}y(\text{n})\right] \tag{3.98}$$

where c is the input compensatory signal, α and β are proportionality coefficients of the DR states and the output feedback, respectively. These parameters are optimized online during the design of the model. To ensure the effectiveness of the learning procedure, a step-variable least mean square (LMS) algorithm is adopted to optimize the parameters of the network. Here, the error $\varepsilon(n)$ and the squared error $E(n)$ are described by

$$\varepsilon(n) = d(n) - y(n), \quad E(n) = \|d(n) - y(n)\|^2 \tag{3.99}$$

If considering $g(\cdot)$ to be a linear activation function, the optimization process can be realized as

$$o(n+1) = o(n) + s(n+1)\varepsilon(n)\frac{\partial\varepsilon(n)}{\partial o} \tag{3.100}$$

$$s(n+1) = ps(n) + qE(n) \tag{3.101}$$

$$\frac{\partial\varepsilon(n)}{\partial o} = -\mathbf{W}^{\text{out}}(n)\left[\frac{\partial x(n)}{\partial o}; \mathbf{0}^{\text{T}}\right] \tag{3.102}$$

Equation (3.100) serves as an update rule for the three parameters of the network, where o contains c, α, and β. The update for these variables is realized by means of (3.101), where $s(n)$ denotes the length of the nth step. Here, p and q denote the corresponding coefficients, respectively. During the initialization, the step length is comparably large, which promotes a rapid, yet quite rough search to move to the most promising region of the search space. In the sequel, this length is gradually decreased. (3.102) offers the computational details regarding $\partial\varepsilon(n)/\partial o$, where $\mathbf{0} = (0, 0, \ldots, 0)$. Combining (3.98) and (3.87) gives

$$\frac{\partial x(n)}{\partial c} = \dot{f}(X_n) \cdot \left[\mathbf{W}^{\text{in}} + \alpha W\frac{\partial x(n-1)}{\partial c} + \beta\mathbf{W}^{\text{back}}W^{\text{out}}\left[\frac{\partial x(n-1)}{\partial c}; \mathbf{0}^{\text{T}}\right]\right] \tag{3.103}$$

$$\frac{\partial x(n)}{\partial \alpha} = \dot{f}(X_n) \cdot \left[\mathbf{W}x(n-1) + \alpha \mathbf{W} \frac{\partial x(n-1)}{\partial \alpha} + \beta \mathbf{W}^{\text{back}} \mathbf{W}^{\text{out}} \left[\frac{\partial x(n-1)}{\partial \alpha}; \mathbf{0}^{\text{T}} \right] \right] \tag{3.104}$$

$$\frac{\partial x(n)}{\partial \beta} = \dot{f}(X_n) \cdot \left[\alpha \mathbf{W} \frac{\partial x(n-1)}{\partial \beta} + \mathbf{W}^{\text{back}} y(\text{n}-1) + \beta \mathbf{W}^{\text{back}} \mathbf{W}^{\text{out}} \left[\frac{\partial x(n-1)}{\partial \beta}; \mathbf{0}^{\text{T}} \right] \right] \tag{3.105}$$

where (3.103)–(3.105) form the essence of the iterative optimization for $\partial x(n)/\partial o$. In this section, assuming that $\partial x(0)/\partial o = 0$, the activation function $f(\cdot)$ is specified as $\tanh(\cdot)$. In this case, the corresponding derivative reads as

$$f'(x) = \frac{4}{2 + \exp(2x) + \exp(-2x)} \tag{3.106}$$

Using this improved version of the ESN with leaky integrator neurons, the prediction is realized in the following fashion.

Algorithm 3.3: Modeling the ESN with Leaky Integrator Neurons

Step 1: Initiate the parameters of the ESN, the connections $\mathbf{W}^{\text{in}}$, $\mathbf{W}$, *and* $\mathbf{W}^{\text{back}}$, *set up an initial state of DR, (x(0)), as well as the values of the parameters c,* α, *and* β.

Step 2: Input a training sample to update the DR using (3.98).

Step 3: Determine the states of DR, the sample input vector and the output vector, and check whether the requirement on the specified number of samples for calculating $\mathbf{W}^{\text{out}}$*has been satisfied. If so, calculate* $\mathbf{W}^{\text{out}}$*using (3.87); otherwise, go back to Step 2.*

Step 4: Input a sample data, calculate the output using (3.87) and (3.98), and obtain the value of error $\varepsilon(n)$.

Step 5: Calculate the gradient with respect to each parameter for the optimization.

Step 6: Calculate the current step length of the LMS by (3.101) and update the parameter value with the use of (3.100) and $\mathbf{W}^{\text{out}}$.

Step 7: Check the convergence of the learning process. If the convergence has been achieved, go to Step 8; otherwise, go to Step 4.

Step 8: The training has been completed. The prediction can be made using (3.87) and (3.98).

3.5.5 Dual Estimation-Based ESN

When using the ESN to establish a regression model for noisy time series, the output uncertainty is considered in the above model. In general, the independent Gaussian noise sequences reflecting the difference between the observation and the expected output are introduced into the output formula. However, as for the noisy time series, the internal states are still uncertain. In this subsection, we consider an ESN model, whose internal states and outputs are combined with the additive noises in the form

$$\begin{aligned} \mathbf{x}_k &= f\left(\mathbf{W}^{\text{in}}\mathbf{u}_k + \mathbf{W}\mathbf{x}_{k-1}\right) + \nu_{k-1} \\ y_k &= \mathbf{W}^{\text{out}} \cdot [\mathbf{u}_k, \mathbf{x}_k] + n_k \end{aligned} \tag{3.107a}$$

where y_k is a scalar quantity showing that the network is a single-output model. $\nu_{k-1} \in \mathbb{R}^{N \times 1}$ and n_k are independent white Gaussian noise sequence with mean zero and covariance $\mathbf{R}^{\nu}$ and σ_n^2, respectively. In (3.107a), ν_{k-1} reflects the uncertainty of internal states, and n_k reflects the output uncertainty.

Since the output weights are unknowns as well as the uncertain internal states, we should consider another separate state space formulation for the underlying output weights which is written as

$$\begin{aligned} \mathbf{W}_k^{\text{out}} &= \mathbf{W}_k^{\text{out}} + \mathbf{q}_{k-1} \\ y_k &= \mathbf{W}_k^{\text{out}}\left[\mathbf{u}_k, \widehat{\mathbf{x}}_k\right] + n_k \end{aligned} \tag{3.107b}$$

Since both the internal states and the output weights are unknown, it is a tough task to train the aforementioned model by using traditional learning methods, such as linear regression or recursive least squares method. The dual estimation works by alternating between the model estimation and the process state estimation, whose main advantage lies in that the model can be estimated with the unknown process state [25]. Making full use of this advantage, a nonlinear/linear dual estimation is adopted to estimate the internal states and the output weights of the established ESN. Figure 3.5 describes the brief structure of the dual estimation model. It is regarded in nature as an optimization algorithm that recursively determines the internal state $\widehat{\mathbf{x}}_k$

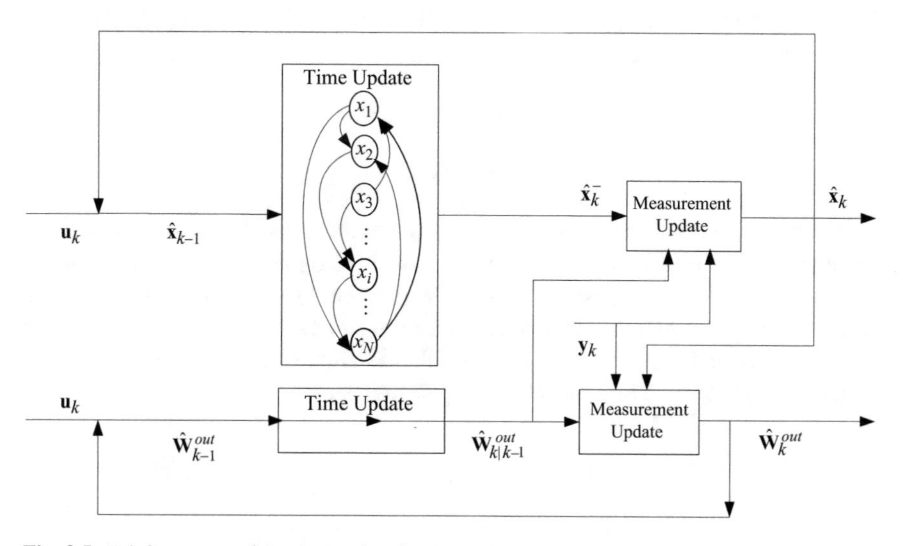

Fig. 3.5 Brief structure of the dual estimation-based ESN

and output weights $\widehat{\mathbf{W}}_k^{\text{out}}$ via minimizing the cost function, a joint binary function of $\widehat{\mathbf{x}}_k$ and $\widehat{\mathbf{W}}_k^{\text{out}}$ given by

$$J\left(\mathbf{x}_k, \mathbf{W}_k^{\text{out}}\right) = \sum_k^n \left\{ \left(\sigma_n^2\right)^{-1} \left(y_k - \mathbf{W}_k^{\text{out}} \cdot \left[\mathbf{u}_k, \mathbf{x}_k\right]\right)^{\mathrm{T}} \left(y_k - \mathbf{W}_k^{\text{out}} \cdot \left[\mathbf{u}_k, \mathbf{x}_k\right]\right) \right. \\ \left. + \left(\mathbf{x}_k - \widehat{\mathbf{x}}_{k|k-1}\right)\left(\mathbf{R}^{\nu}\right)\left(\mathbf{x}_k - \widehat{\mathbf{x}}_{k|k-1}\right) \right\} \tag{3.108}$$

with $\widehat{\mathbf{x}}_{k|k-1} = f\left(\widehat{\mathbf{x}}_{k|k-1}, \mathbf{u}_k, \mathbf{W}^{\text{in}}, \mathbf{W}\right)$. It is obvious that the priori estimation of internal states is the function of input weight $\mathbf{W}^{\text{in}}$ and internal weight $\mathbf{W}$ that are determined in the process of estimation. To minimize the cost function with respect to $\widehat{\mathbf{x}}_k$ and $\widehat{\mathbf{W}}_k^{\text{out}}$, the partial derivative of (3.108) are calculated as

$$\frac{\partial J\left(\mathbf{x}_k, \mathbf{W}_k^{\text{out}}\right)}{\partial \mathbf{W}_k^{\text{out}}} = \sum_{k=1}^n \left\{ 2\left(\sigma_n^2\right)^{-1} \left(y_k - \mathbf{W}_k^{\text{out}}\left[\mathbf{u}_k, \mathbf{x}_k\right]\right) \right\} \cdot \left(-\left[\mathbf{u}_k, \mathbf{x}_k\right]\right) \tag{3.109}$$

$$\frac{\partial J\left(\mathbf{x}_k, \mathbf{W}_k^{\text{out}}\right)}{\partial \mathbf{x}_k} = \sum_{k=1}^n \left\{ 2\left(\sigma_n^2\right)^{-1} \left(y_k - \mathbf{W}_k^{\text{out}}\left[\mathbf{u}_k, \mathbf{x}_k\right]\right) \cdot \mathbf{W}_k^{\text{out},\mathbf{x}} + \mathbf{Q}(\mathbf{x}_k) \right\} \tag{3.110}$$

$$\mathbf{Q}(\mathbf{x}_k) = \left(\mathbf{R}^{\nu}\right)^{-1}\left(\mathbf{x}_k - \widehat{\mathbf{x}}_{k|k-1}\right) + \left(\left(\mathbf{R}^{\nu}\right)^{-1}\right)\left(\mathbf{x}_k - \widehat{\mathbf{x}}_{k|k-1}\right) \tag{3.111}$$

where $\mathbf{W}_k^{\text{out},\mathbf{x}}$ is a partitioned matrix of $\mathbf{W}_k^{\text{out}}$. Since the matrix $\mathbf{R}^{\nu}$ is diagonal, the inverse matrix of $\mathbf{R}^{\nu}$ is equal to the transpose of the inverse of $\mathbf{R}^{\nu}$, i.e., $(\mathbf{R}^{\nu})^{-1} = ((\mathbf{R}^{\nu})^{-1})^{\mathrm{T}}$. Then, (3.111) can be described as

$$\mathbf{Q}(\mathbf{x}_k) = 2\left(\mathbf{R}^{\nu}\right)^{-1}\left(\mathbf{x}_k - \widehat{\mathbf{x}}_{k|k-1}\right) \tag{3.112}$$

If the minimum of $J\left(\mathbf{x}_k, \mathbf{W}_k^{\text{out}}\right)$ exists, (3.109) and (3.111) should be equal to zero. Considering the constraint $[\mathbf{u}_k, \mathbf{x}_k] \neq 0$, one has

$$y_k - \mathbf{W}_k^{\text{out}}\left[\mathbf{u}_k, \mathbf{x}_k\right] = 0 \tag{3.113}$$

$$\mathbf{x}_k = \widehat{\mathbf{x}}_{k|k-1} \tag{3.114}$$

These two formulas show that the estimation of the internal states is unique if the optimum of (3.108) exists, and the uniqueness of the output weights $\widehat{\mathbf{W}}_k^{\text{out}}$ depends on the fact that $[\mathbf{u}_k, \mathbf{x}_k]$ is positive definite. In addition, (3.113) also explains that the linear regression is an inadequate method for the parameters determination of ESN due to its ill-condition; thus, the Kalman-filter is employed to estimate the output weights, which will be introduced in Chap. 7 in details.

3.5.6 Case Study

This subsection presents some experimental cases to verify the performance of these aforementioned NN-based time series prediction methods. The experimental data are the historical data of the by-product gas in steel industry accumulated by the SCADA system.

This case study considers the BFG system of a steel plant, and it is a complex multi-input and multi-output system, which involves blast furnaces, transportation networks, and a series of gas users, see Fig. 3.6. Four blast furnaces viewed as the generation units can supply into the transportation network on average 1.8 million BFG per hour. The transportation system includes pipelines, mixing stations, and pressure stations; and the gas users primarily comprise coking oven, hot rolling plant, cold rolling plant, chemical products recovery (CPR), low pressure boiler (LPB), synthesizing unit (SU), and power generator. Since the hot blast stoves expend quite a quantity of BFG, and be continuously switched, the generated gas flow into transportation will frequently fluctuate. In addition, there are often abnormal situations in the production such as blast reduction, equipment parameters changes, etc. To maintain the system stability, the operators have to timely judge the trend of tank level according to the generation and consumption amount of the BFG, and subsequently adjust some users' demands. Therefore, the prediction of these consumption and generation gas flows can provide guidance of energy scheduling.

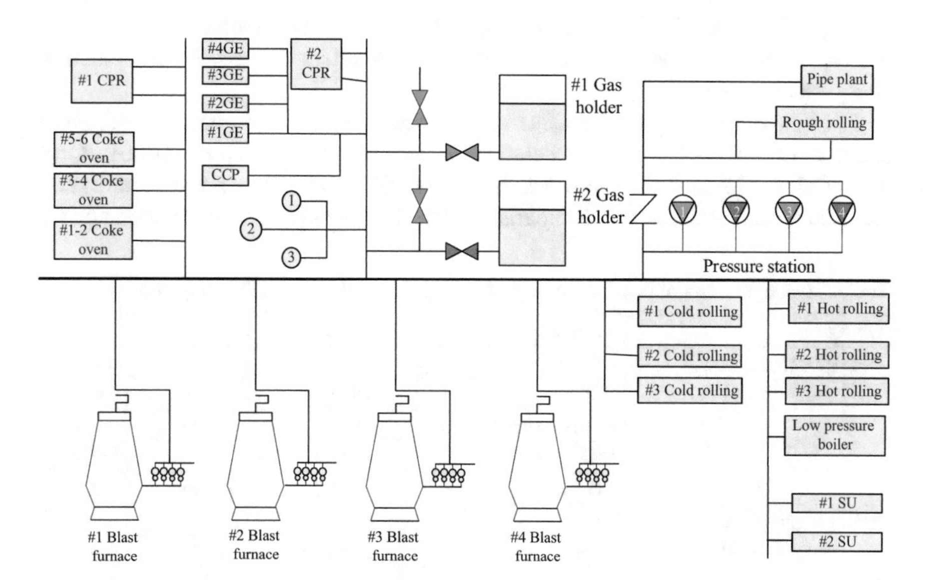

Fig. 3.6 A brief pipeline structure of BFG system

Extended Kalman-Filter-Based Elman Network

In this case, the performance of the extended Kalman-filter-based Elman network (EKFEN) for industrial application is evaluated. Specifically, one can apply it to an energy system prediction of steel industry. The real-world gas flows coming from a steel plant of China, ranged between April 1 and April 8, 2012 are employed. Without loss of generality, the data of the BFG flows are randomly chosen as the studied units, including the BFG consumption on coke oven and COG consumption on blast furnace. Many comparative experiments are conducted to clarify the approach effectiveness on the accuracy and the computing cost.

The BFG generation flow of #1 blast furnace and BFG consumption flow of #1–2 coke oven are randomly selected to conduct the experiments. Based on the experimental approach presented previously, the initial parameters are set as $\mathbf{P}(0|0) = 1000\mathbf{I}$, $\mathbf{Q}(0|0) = 0.0001\mathbf{I}$, and $\mathbf{R}(0|0) = 100\mathbf{I}$. For a fair competition, the optimal parameters in each modeling are also determined by the experimental method. Here, 1000 continuous data points are selected from the online database as the training sample, and the next 60 data points as the testing data. The statistic prediction results are reported in Table 3.4, and one of the random selected performances including the predicted values and the errors are illustrated in Figs. 3.7 and 3.8.

From Figs. 3.7 and 3.8, it can be clearly seen that the EKFEN presents the highest prediction accuracy compared to the others. In the statistics perspective, Table 3.4 lists the multiple quantitative evaluation indices for the prediction accuracy, which can comprehensively indicate the advantages of the EKFEN. As for the industrial data-based experiments, the accuracies of the EKFEN are obviously higher than that by the EKF with BPTT. By analyzing Table 3.4, the accuracies of the EKFEN present great improvements, which will be an important concern for the practical energy scheduling.

For a further validation, the COG consumption on #1 blast furnace is selected as another instance. The initial parameters are set as $\mathbf{P}(0|0) = 1000\mathbf{I}$, $\mathbf{Q}(0|0) = 0.0001\mathbf{I}$, and $\mathbf{R}(0|0) = 100\mathbf{I}$. Similarly, 50 times independent experiments are also conducted for the comparison. The comparative results that include a

Table 3.4 Comparison of the statistic prediction results by using five different methods

Object	Method	MAPE	RMSE	NRMSE
BFG generation flow of #1 blast furnace	BP	7.729	50.484	0.012
	ESN	8.203	52.860	0.013
	Kernel learning-based	9.428	63.148	0.016
	EKF with BPTT	6.504	41.856	0.010
	EKFEN	5.797	39.344	0.010
BFG consumption flow of #1–2 coke oven	BP	1.1307	2.0251	0.0018
	ESN	0.6730	1.2420	0.0011
	Kernel learning-based	0.6003	1.0482	0.0009
	EKF with BPTT	0.6262	1.1701	0.0010
	EKFEN	0.5373	0.9821	0.0008

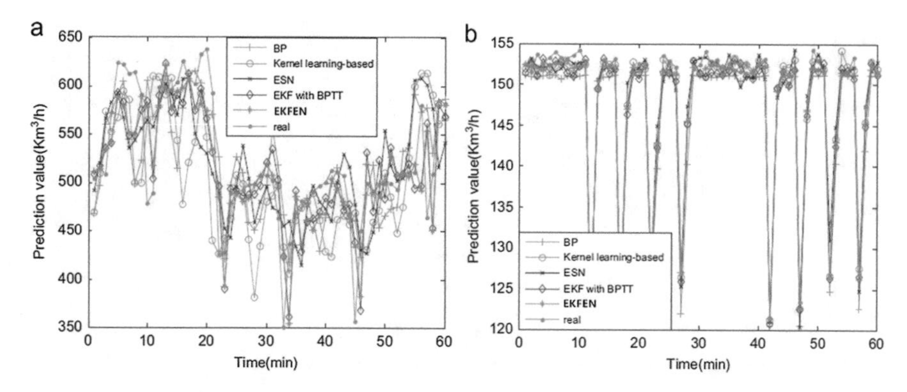

Fig. 3.7 Comparisons of the prediction results produced by the five methods. (**a**) BFG generation flow of #1 blast furnace and (**b**) BFG consumption flow of #1,2 coke oven

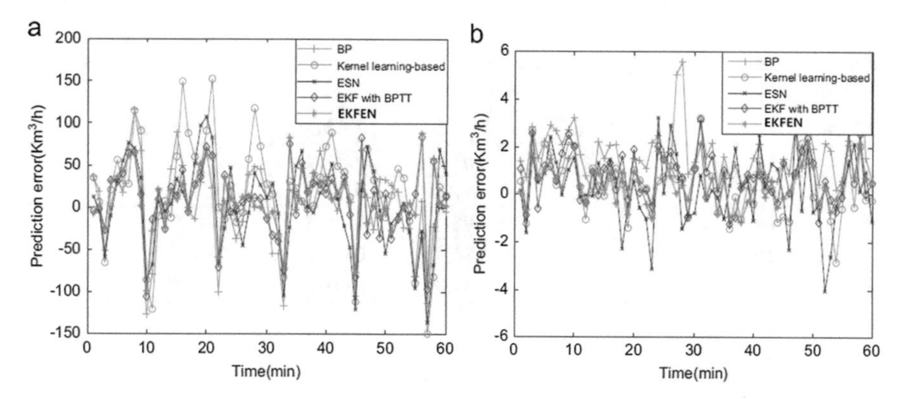

Fig. 3.8 Comparisons of the obsolute prediction errors produced by the five methods. (**a**) BFG generation flow of #1 blast furnace and (**b**) BFG consumption flow of #1,2 coke oven

segment of the gas flow are presented in Fig. 3.9, and the average statistic results are reported in Table 3.5.

It is apparently that from Fig. 3.9a, b the performance of the EKFEN is also the best, and the statistic results of the multiple experiments also draw conclusion. From this table, the errors of the EKFEN are lower than those by the others as for the overall evaluation indices. The BP and the ESN might generate the prediction failure in the experiments: while, the EKFEN can reach the stable prediction results. With long-term application in this steel plant, the EKFEN exhibits a good stability, which guarantees the feasibility of the EKFEN prediction for the industrial application. Overall, compared to other methods, the EKFEN exhibits the highest prediction accuracies and the applicable stability, which greatly results in the scientific energy scheduling guidance for steel enterprise.

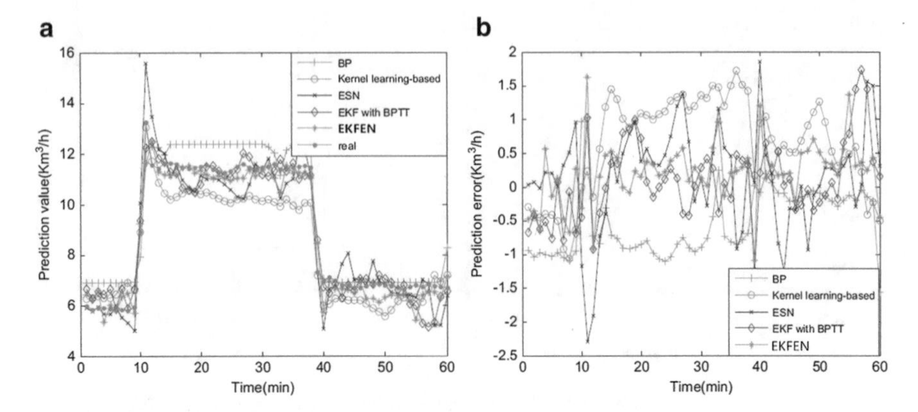

Fig. 3.9 Comparison of prediction for COG consumption flow on #1 blast furnace by the five methods (**a**) prediction results (**b**) absolute errors

Table 3.5 Comparisons of prediction error and computing time for COG consumption on #1 blast furnace

Method	MAPE	RMSE	NRMSE	Computational time(s)
BP	6.6715	0.6707	0.0094	8.827
ESN	5.9436	0.5778	0.0081	0.580
Kernel learning-based	7.4398	0.9511	0.0133	1.967
EKF with BPTT	5.3572	0.5638	0.0078	28.641
EKFEN	5.1094	0.5188	0.0072	23.797

SVD-Based ESN for Industrial Prediction

To verify the performance of the SVD-based ESN (SVD-ESN) for industrial prediction, this method can predict the generation flow of the BFG in steel plant. To improve the quality of the industrial data, the EMD technique introduced in Chap. 2 is adopted here for de-noising. In this case, the prediction is completed based on the de-noising data. The dimensionality of the dynamic reservoir of the ESN is set as 100, the degree of sparse connection of the internal weights is set as 1%, the spectral radius of the matrix of the internal weights is set as 0.75, and tanh() is the activation function of the neurons in the dynamic reservoir. The number of the training samples is 500 and the predicted series is 60 points after these 500 points. The predicted object is the generation of four blast furnaces, and the prediction results are shown in Fig. 3.10. From Fig. 3.10, one can see that the SVD-ESN can predict the gas generation of these four blast furnace accurately.

To further evaluate the performance of the SVD-ESN, a set of comparative experiments are conducted in this chapter, including the back propagation (BP) network, the radial basis function (RBF) network and the generic ESN. The predicted results are shown in Fig. 3.11, from which it can be seen that the SVD-ESN provides a much better performance. For a more intuitionistic comparison, three-

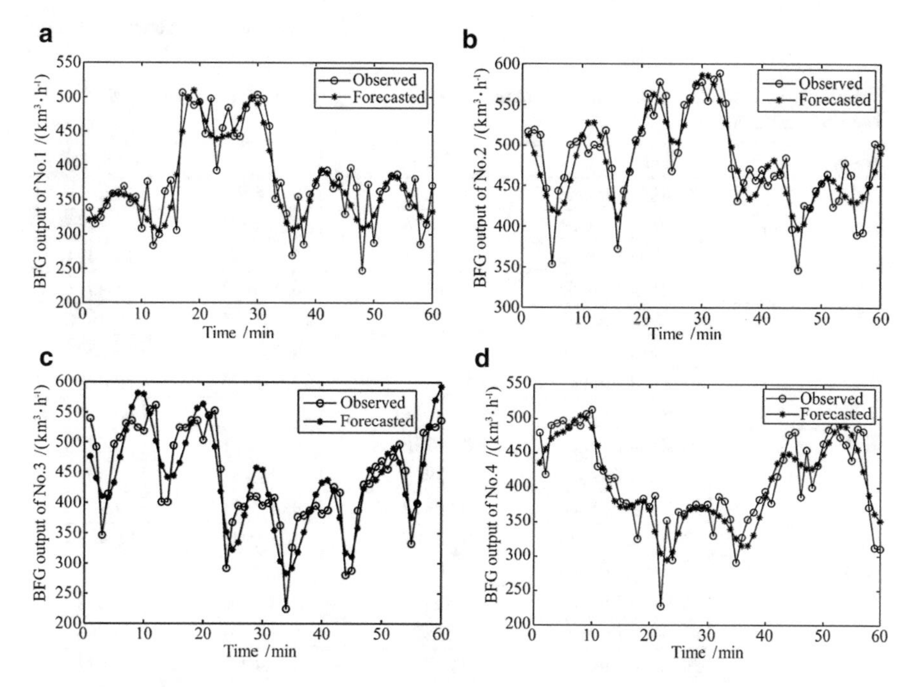

Fig. 3.10 Prediction results of the generation of four blast furnaces (**a**) #1, (**b**) #2, (**c**) #3, and (**d**) #4

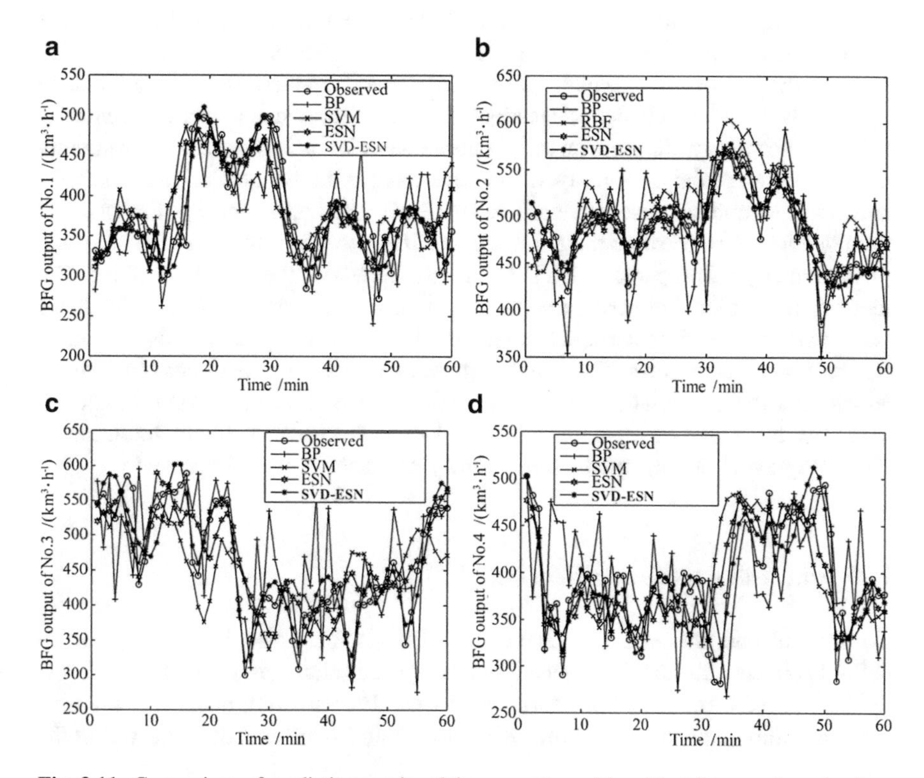

Fig. 3.11 Comparison of prediction results of the generation of four blast furnaces by using four different methods (**a**) #1, (**b**) #2, (**c**) #3, and (**d**) #4

Table 3.6 Comparative results of the generation prediction of gas flow of 4 blast furnaces by uisng four different methods

Method	Object	NRMSE	MSE	MAPE (%)
BP	#1 Blast furnace	0.8953	1575.6205	11.0069
	#2 Blast furnace	0.73196	1420.4001	6.0137
	#3 Blast furnace	1.42595	2660.4778	14.7471
	#4 Blast furnace	1.1239	1828.3902	13.43
RBF	#1 Blast furnace	0.84915	1147.1764	6.4906
	#2 Blast furnace	0.65034	981.1209	5.6527
	#3 Blast furnace	1.1327	2221.1976	12.0461
	#4 Blast furnace	0.98015	2571.2525	10.1721
ESN	#1 Blast furnace	0.66796	628.4878	5.4726
	#2 Blast furnace	0.48902	433.156	4.4059
	#3 Blast furnace	0.65348	994.7694	7.8275
	#4 Blast furnace	0.58587	1057.936	6.4139
SVD-ESN	#1 Blast furnace	0.3679	306.2257	3.6496
	#2 Blast furnace	0.30329	268.3751	3.0289
	#3 Blast furnace	0.35487	508.7191	4.6961
	#4 Blast furnace	0.30375	456.4997	4.3717

error evaluation indexes involving MSE, NRMSE, and MAPE are employed here, and the comparative results are shown in Table 3.6. From Table 3.6, the SVD-ESN exhibits the best performance among all these three indexes.

From the comparative results, both the BP network and the RBF network belong to static NNs, in which the connections between different neurons are generally fixed. In such a way, the performance of the network is easily affected by the number of the neurons of the hidden layer. Moreover, as for the BP network and the RBF network, the gradient descent algorithm that easily falls into a local optimizaiton is usually adopted to learning the parameters of the network, which will lead to a big prediction error. Compared to the BP and RBF networks, the ESN has a short-term memory to save the information coming from the teacher signals. Thus, ESN has a better performance for time series prediciton. However, the generic ESN is suitable for the data without noise. Thus, in this subsection, a SVD-based improved ESN is adopted for the industrial time series prediction. Meanwhile, the EMD is employed to remove the noise hidden in the industrial data. In this case, the ill-condition can be effectively avoid and the prediciton accuracy can be effectively improved.

ESN with Leaky Integrator Neurons

To demonstrate the effectiveness of the ESN with leaky integrator neurons (ESN-LIN), the data of BFG generation or consumption flows acquired on May, 2009 obtained from a steel plant was employed. The sampling interval of the time series is 1 min. In practice, the flow variations within 1 h (60-min data) can reflect the

Table 3.7 Optimal normalized ESN parameters of the typical BFG units

BFG generator or user	c	α	β
Generation amount of #1 blast furnace	−0.0519	0.1934	0.0015
Consumption flow of #1–2 coke oven	0.0158	0.7504	0.0012
Consumption flow of #1 hot rolling plant	0.0634	0.8327	0.0147
Consumption flow of #1 cold rolling plant	−0.0915	0.8584	0.2884

corresponding the dynamics characteristics of these production units. Thus, the input dimensionality of the improved ESN is set to be 60. The inputs used in this prediction problem are denoted as $u(t) = [z(t-60), z(t-59), \ldots, z(t-1)]$, where $z(1)$, $z(2)$, …, $z(n)$ is the time series describing the value of flow amounts. The one-dimensional output is $z(t)$. Several typical BFG units, including the BFG generator #1 blast furnace, the users #1–2 coke oven, #1 hot rolling plant, and #1 cold rolling plant, are taken as representative examples since they usually exhibit a significant impact on the holder level. To set up a sound trade-off between the dynamic memorization capacity of the ESN and the real-time requirements of the application, the number of nodes in the DR is set to 200 whereas the connectivity of is set up to be 2%. In the adaptive LMS, the values of the parameters are selected by running some preliminary set of experiments. As a result, the value of is set to 0.95; is set to 0.0005, and the initial step length is equal to 0.05. Although the variable-step LMS is relatively computationally demanding, the prediction of all of BFG generators or users takes about 1–2 min. Considering that the sampling interval of the time series is 1 min and the predicted time horizon is 60 min, the time requirements for online prediction can be fully satisfied.

One can randomly choose a series of consecutive 2000 data as the training data, and use the predictor built on their basis for the subsequent 60 min of data. The optimal normalized values of the parameters of the ESN obtained by the learning process are listed in Table 3.7. The prediction results are shown in Fig. 3.12. The maximal relative error is always less than 5% (on average), which can fully require the practical demands.

To further contrast the advantages of the ESN-LIN, one can consider four other prediction methods when carrying out the comparative analysis, which are: 1) manual estimation coming from scheduling workers (MAN); 2) radial basis function network (RBF) reported in [8]; 3) SVM coming from [8]; and 4) a "generic" ESN. The generic version of the ESN can be regarded as a special case of the ESN-LIN with fixed global parameters ($c = 0$ and $\alpha = \beta = 1$). Randomly choosing the flow data to carry out a comparative analysis, the results are reported in Fig. 3.13. Through a quick visual inspection, it can be concluded that the ESN-LIN results in the best prediction performance. The NRMSE and MAPE are employed to quantify the prediction quality, and the statistical analysis of the prediction results read as Table 3.8, which becomes apparent that the accuracy of the ESN-LIN is definitely higher than that reported when running the four other methods. More specifically, the manual estimation based on the workers experience gives only an approximate flow trend of the BFG units. As anticipated, its prediction quality is the worst. The

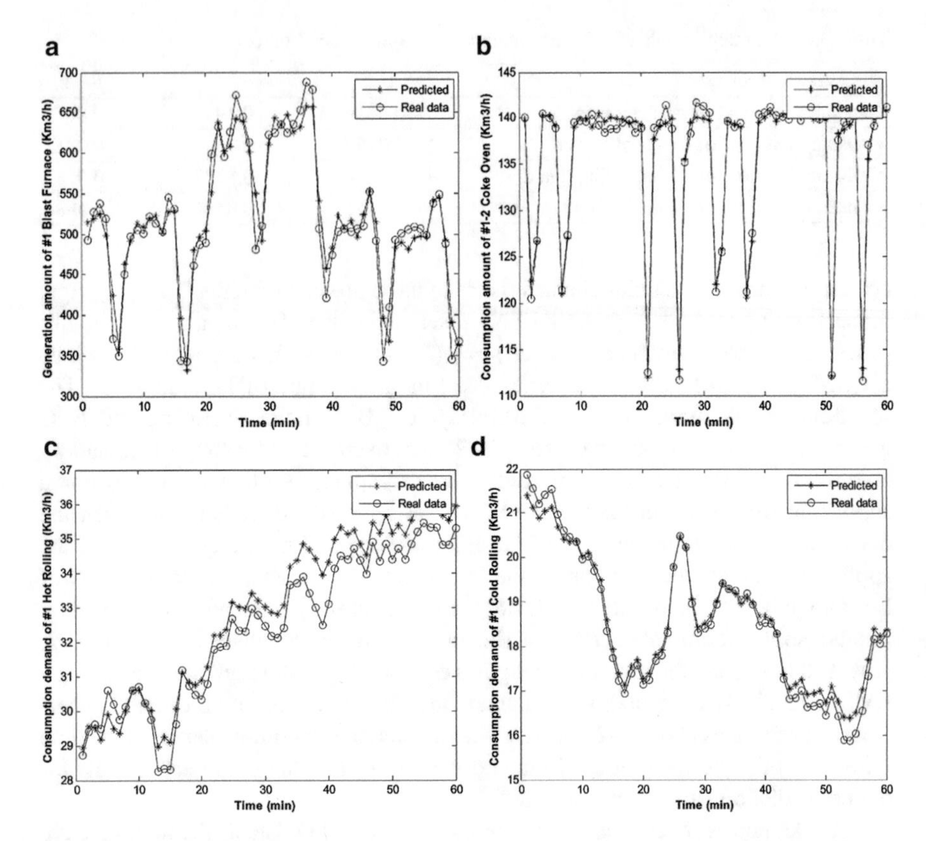

Fig. 3.12 Prediction results of the typical BFG generator or users (**a**) generation amount of #1 blast furnace, (**b**) consumption amount of #1–2 coke oven, (**c**) consumption amount of #1 hot rolling plant, (**d**) consumption amount of #1 cold rolling plant

RBF network, realizing a static input-output mapping, comes with a large prediction error. The performance of the generic ESN (although the network was able to avoid the local minima in its training process) is negatively influenced by existing noise. On the other hand, the ESN-LIN not only is less affected by disturbances, but also adapts to the dynamics of different BFG units. Altogether this leads to the best prediction results.

In order to study the impact of the global parameters on the performance of the network and further validate the effectiveness of the ESN-LIN, some units are also selected to complete the prediction as shown in Table 3.9. Here, 100 times independent experiments were repeated for each unit, and the results indicate that the ESN-LIN yields higher accuracy than those provided by the generic version of the ESN. As mentioned above, the ESN-LIN comes with better generalization capabilities. Note that the DR can effectively memorize the information obtained from training sample and afterwards minimize the training error via optimizing the output

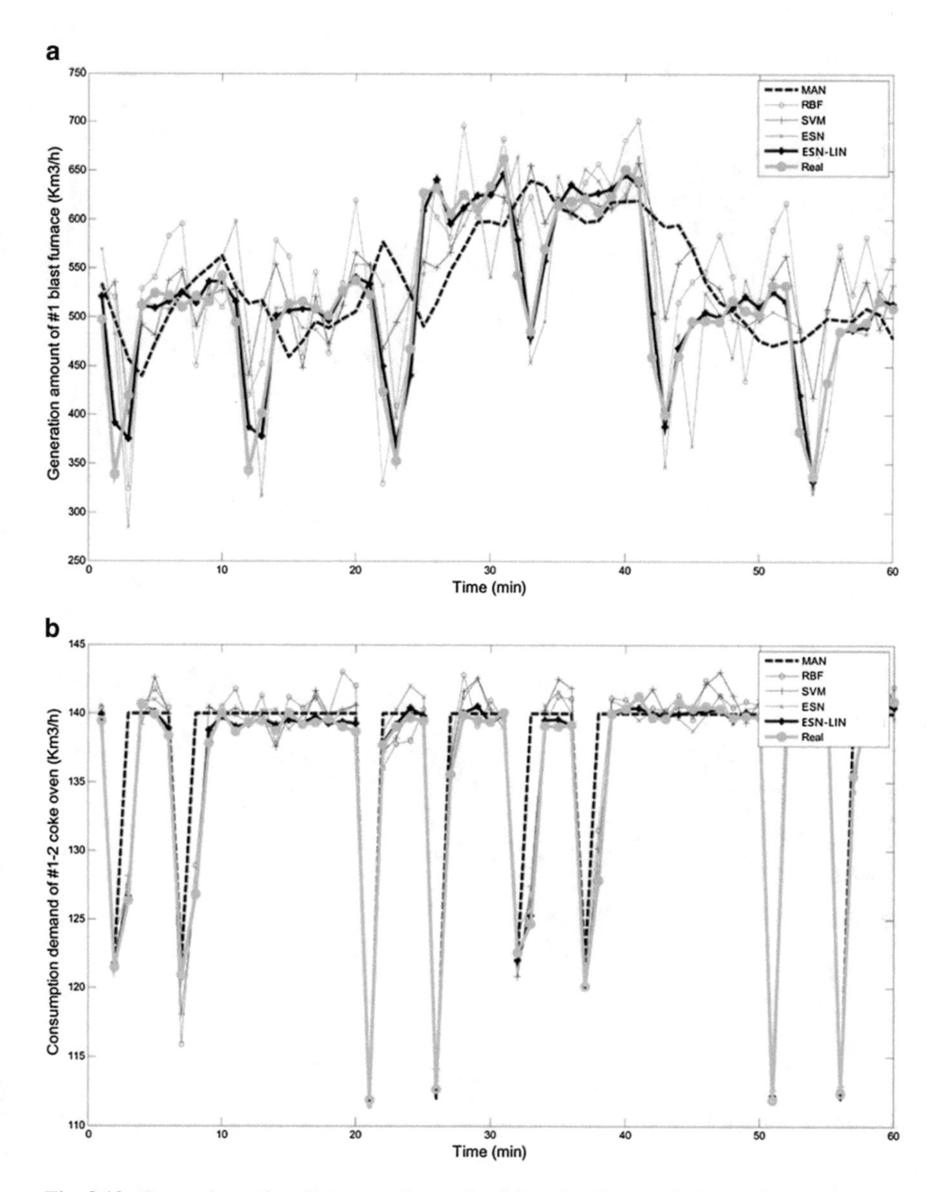

Fig. 3.13 Comparison of prediction results produced by using four methods; see the details in the text. (**a**) Generation amount of #1 blast furnace, (**b**) consumption demand of #1–2 coke oven, (**c**) consumption demand of #1 hot rolling, (**d**) consumption demand of #1 cold rolling

weights by linear regression. This way of training helps avoid local minima, which are quite common when designing neural networks. Second, the optimized global parameters of the ESN-LIN are suitable for different units, which make the ESN capable of dealing with the dynamics of the objects.

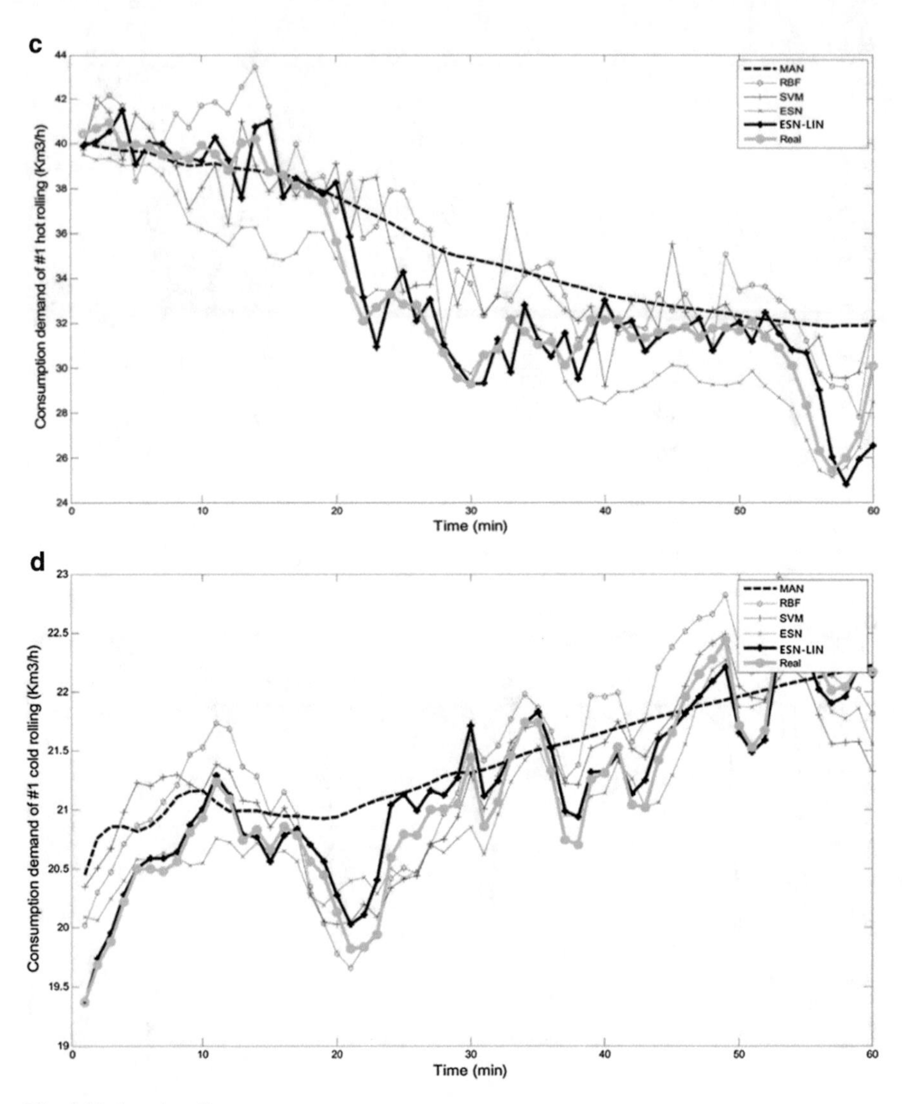

Fig. 3.13 (continued)

Dual Estimation-Based ESN

In this case study, the flow prediction for BFG is studied. The #1 blast furnace of a steel plant presented in Fig. 3.6 is chosen as the practical example to verify the effectivensss of the dual estimation-based ESN (DE-ESN). Considering the fact that the flow variation within 1 h can fundamentally reflect the dynamics feature of

Table 3.8 Results of prediction error: A comparative analysis

Prediction	Method	NRMSE	MAPE (%)
Generation amount of #1 blast furnace	MAN	0.9672	12.5912
	RBF	0.7694	10.5003
	SVM	0.7365	10.0914
	ESN	0.7150	8.9587
	ESN-LIN	0.2208	2.7905
Consumption demand of #1–2 coke oven	MAN	0.4426	1.2761
	RBF	0.2015	1.0140
	SVM	0.1773	0.8570
	ESN	0.1515	0.7553
	ESN-LIN	0.0467	0.2292
Consumption demand of #1 hot rolling	MAN	0.6279	6.7087
	RBF	0.5797	6.4917
	SVM	0.5621	5.8672
	ESN	0.4651	4.6961
	ESN-LIN	0.2828	2.8877
Consumption demand of #1 cold rolling	MAN	0.6614	2.0005
	RBF	0.5949	1.9569
	SVM	0.5347	1.6483
	ESN	0.3936	1.2598
	ESN-LIN	0.2145	0.6276

Table 3.9 Comparative analysis of predict quality: generic ESN and the ESN-LIN

System units	The number of training samples	The number of nodes in DR	NRMSE (on average)	
			Generic ESN	ESN-LIN
#1 Blast furnace	2000	200	0.7214	0.2237
#2 Blast furnace	2000	200	0.6976	0.2195
#3 Blast furnace	2000	200	0.7143	0.2108
#4 Blast furnace	2000	200	0.7229	0.2246
#1–2 Coke oven	1500	100	0.1521	0.0458
#3–4 Coke oven	1500	100	0.1528	0.0492
#5–6 Coke oven	1500	100	0.1631	0.0497
#1 Hot rolling plant	1000	100	0.4651	0.2828
#2 Hot rolling plant	1000	100	0.5106	0.3174
#3 Hot rolling plant	1000	100	0.4578	0.2904
#1 Cold rolling plant	1000	100	0.3921	0.2095
#2 Cold rolling plant	1000	100	0.3354	0.2109
#3 Cold rolling plant	1000	100	0.4127	0.2351

gas flow, the input dimensionality of the DE-ESN is set to 60. We randomly select the flow data of continuous 1060 min for the model training and testing, and use 30 epochs per run to train the constructed network. Namely, each epoch consists of 1000 samples and the length of each sample is 61. Thus, a set of samples was

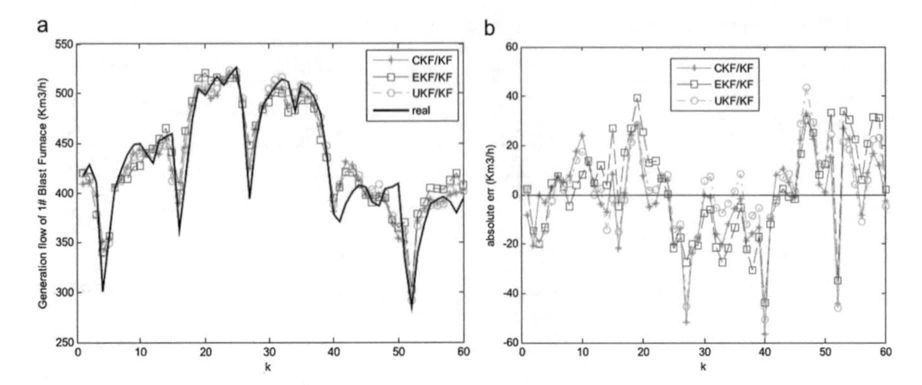

Fig. 3.14 Prediction results for the gas flow prediction by using the three dual estimation based ESN (**a**) Prediction curve and (**b**) prediction error

Table 3.10 The comparative analysis of prediction performance

Method	RMSE (km^3/h)	MAPE	Computational time (s)
EKF/KF-based ESN	30.7087	5.8729	38.91
UKF/KF-based ESN	29.2645	5.4194	260.91
CKF/KF-based ESN	28.3257	5.2379	80.85
KF-based ESN	35.1557	6.3798	21.56
Generic ESN	34.2332	6.3621	2.36

obtained, see $\{(\mathbf{u}_i, y_i)\}_{i=1}^{1000}$, where $y_i = u(i + \Delta)$, $\mathbf{u}_i = [u(i - (d_E - 1)\Delta), u(i - (d_E - 2)\Delta), \ldots, u(i)]$, $d_E = 60$, $\Delta = 1$.

A group of comparative prediction results for the gas flow randomly selected also in 2010, are shown in Fig. 3.14. From Fig. 3.14a, a 60-min prediction is presented by the three DE-ESNs and their absolute prediction errors are shown as Fig. 3.14b. Likewise, the average absolute error by the CKF/KF dual estimation one is the lowest. The corresponding quantified indexes for the prediction performance can be listed in Table 3.10.

Here, the experiments also involve the three methods, the DE-ESN, the ESN based on KF without state estimation and the generic ESN, whose comparative results are presented in Fig. 3.15, from which the DE-ESN exhibits the best performance on prediction accuracy, and the absolute error is meanwhile depicted in Fig. 3.15b. Table 3.10 gives the quantified statistics for the prediction results, where RMSE, MAPE, and computational time serve to evaluate the running quality. Likewise, the accuracy of the first three methods is definitely higher than that of other methods. On the comprehensive perspective of practical application, the CKF/KF-based improved ESN is a more effective method for the prediction problem of the industrial BFG generation flow.

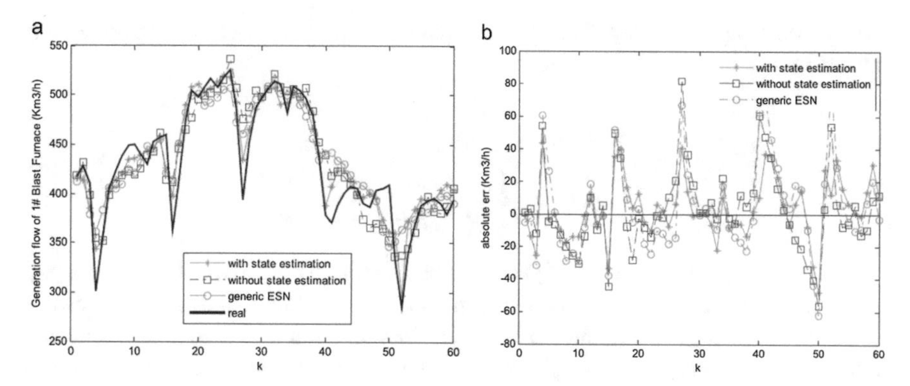

Fig. 3.15 Comparison of prediction results produced by three methods. (**a**) Prediction curve and (**b**) prediction error

3.6 Support Vector Machine-Based Prediction

The generic NNs for practical predictions often face two main drawbacks. The first is that its learning algorithms suffer from the local optimum problem and the over-fitting. The second is that its training has heavy computational load. In this case, the support vector machine (SVM) is proposed by Vapnik [26] for supervised learning. More formally, a SVM constructs a hyperplane or set of hyperplanes in a high- or infinite-dimensional space, which can be used for classification, regression, or other tasks. Intuitively, a good separation is achieved by the hyperplane, which has the largest distance to the nearest training data point of any class, since in general the larger the margin is, the lower the generalization error of the classifier is.

3.6.1 *Basic Concept of SVM*

One can suppose that the training data $\{(\mathbf{x}_1, y_1), \ldots, (\mathbf{x}_l, y_l)\} \subset \chi \times \mathbb{R}$, where χ denotes the space of the input patterns. These might be, for instance, an exchange rates for some currency measured at subsequent days together with corresponding econometric indicators. Our goal is to find a function $f(\mathbf{x})$ that has at most ε deviation from the obtained targets y_i for all the training data, and at the same time is as flat as possible. That is, the errors are tolerated if they are less than ε, but any deviation larger than ε will not be accepted.

One can begin by describing the case of linear functions f, taking the form

$$f(\mathbf{x}) = \langle \mathbf{w}, \mathbf{x} \rangle + b \quad \text{with} \quad \mathbf{w} \in \chi, b \in \mathbb{R} \tag{3.115}$$

where $\langle \cdot, \cdot \rangle$ denotes the dot product in χ. A small $\mathbf{w}$ in (3.115) should be found in order to make the model as flat as possible. One way to ensure this is to minimize the norm, i.e., $\|\mathbf{w}\|^2 = \langle \mathbf{w}, \mathbf{w} \rangle$. This problem can be written as a convex optimization problem,

$$\begin{aligned} &\text{minimize} \quad \frac{1}{2}\|\mathbf{w}\|^2 \\ &\text{subject to} \begin{cases} y_i - \langle \mathbf{w}, \mathbf{x}_i \rangle - b \le \varepsilon \\ \langle \mathbf{w}, \mathbf{x}_i \rangle + b - y_i \le \varepsilon \end{cases} \end{aligned} \tag{3.116}$$

The tacit assumption in (3.116) is that such a function f approximates all pairs ($\mathbf{x}_i$, y_i) with ε precision. However, one also might want to allow for some further errors sometimes. Analogously to the "soft margin" loss function [27] used in [28], one can introduce slack variables ξ_i, ξ_i^* to cope with the infeasible constraints of the optimization (3.116). Hence, the formulation stated in [26] is given by

$$\text{minimize} \quad \frac{1}{2}\|\mathbf{w}\|^2 + C\sum_{i=1}^{l}\left(\xi_i + \xi_i^*\right) \tag{3.117}$$

$$\text{subject to} \begin{cases} y_i - \langle \mathbf{w}, \mathbf{x}_i \rangle - b \le \varepsilon + \xi_i \\ \langle \mathbf{w}, \mathbf{x}_i \rangle + b - y_i \le \varepsilon + \xi_i^* \\ \xi_i, \xi_i^* \ge 0 \end{cases} \tag{3.118}$$

The constant $C > 0$ determines the trade-off between the flatness of f and the amount up to which deviations larger than ε are tolerated. This corresponds to the so-called ε - intensitive lost function $|\xi|_\varepsilon$ described by

$$|\xi|_\varepsilon := \begin{cases} 0 & if |\xi| < \varepsilon \\ |\xi| - \varepsilon & \text{otherwise} \end{cases} \tag{3.119}$$

Figure 3.16 depicts the situation graphically. Only the points outside the shaded region contribute to the cost insofar, as the deviations are penalized in a linear

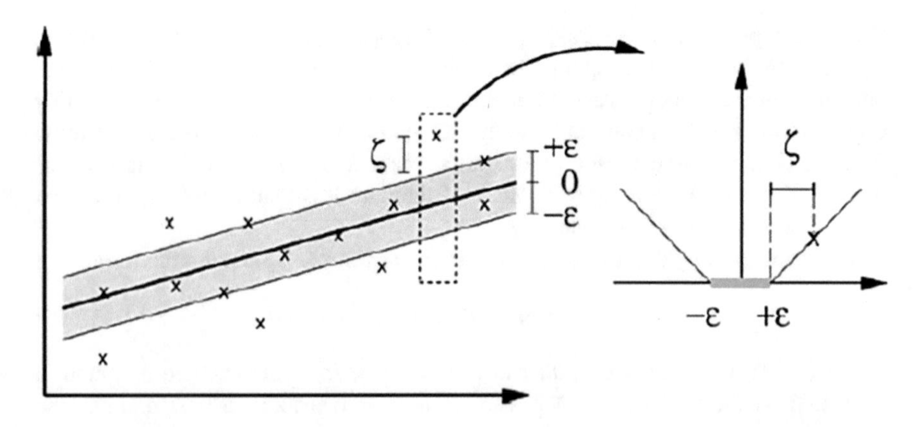

Fig. 3.16 The soft margin loss setting for a linear SVM [30]

fashion. It figures out that in most cases the optimization problem (3.117) can be solved more easily in its dual formulation. Moreover, the dual formulation provides the key for extending SV machine to nonlinear functions. Hence, one can use a standard dualization method utilizing Lagrange multipliers, as described in [29].

3.6.2 SVMs for Regression

Extending support vector machines to regression problems while preserving the property of sparseness, a regularized error function is minimized in simple linear regression,

$$\frac{1}{2}\sum_{n=1}^{N}\{y_n - t_n\}^2 + \frac{\lambda}{2}\|\mathbf{w}\|^2 \tag{3.120}$$

To obtain sparse solutions, the quadratic error function is replaced by an ε - intensitive error function [26], which gives zero error if the absolute difference between the prediction $y(\mathbf{x})$ and the target t is less than ε where $\varepsilon > 0$. A simple example of an ε - intensitive error function, having a linear cost associated with errors outside the insensitive region

$$E_\varepsilon(y(\mathbf{x}) - t) = \begin{cases} 0 & \text{if } |y(\mathbf{x}) - t| < \varepsilon \\ |y(\mathbf{x}) - t| - \varepsilon, & \text{otherwise} \end{cases} \tag{3.121}$$

Therefore, a regularized error function is minimized

$$C\sum_{n=1}^{N} E_\varepsilon(y(\mathbf{x}_n) - t_n) + \frac{1}{2}\|\mathbf{w}\|^2 \tag{3.122}$$

where $y(\mathbf{x}_n)$ is defined by (3.115). And, the inverse regularization parameter, denoted by C, appears in front of the error term.

As before, one can re-express the optimization problem by introducing slack variables. For each point $\mathbf{x}_n$, two slack variables are needed, including $\xi_n \geq 0$ and $\xi_n^* \geq 0$, where $\xi_n > 0$ corresponds to a point, for which $t_n > y(\mathbf{x}_n) + \varepsilon$, and $\xi_n^* > 0$ corresponds to one, for which $t_n < y(\mathbf{x}_n) - \varepsilon$.

The condition for a target point to lie inside the ε - *tube* is that $y(\mathbf{x}_n) - \varepsilon \leq t_n \leq y(\mathbf{x}_n) + \varepsilon$. The introduction of the slack variables allows points to lie outside the tube provided the slack variables are non-zero, and the corresponding conditions are

$$\begin{cases} t_n \leq y(\mathbf{x}_n) + \varepsilon + \xi_n \\ t_n \geq y(\mathbf{x}_n) - \varepsilon - \xi_n^* \end{cases} \tag{3.123}$$

The error function for support vector regression can then be written as

$$C\sum_{n=1}^{N}\left(\xi_n+\xi_n^*\right)+\frac{1}{2}\|\mathbf{w}\|^2 \tag{3.124}$$

which must be minimized subject to the constraints $\xi_n \geq 0$ and $\xi_n^* \geq 0$ as well as (3.123). This can be achieved by introducing Lagrange multipliers $a_n \geq 0$, $a_n^* \geq 0$, $\mu_n \geq 0$, and $\mu_n^* \geq 0$. Then, one has a Lagrangian function to be minimized,

$$\begin{aligned} L = C\sum_{n=1}^{N}\left(\xi_n+\xi_n^*\right)+\frac{1}{2}\|\mathbf{w}\|^2-\sum_{n=1}^{N}\left(\mu_n\xi_n+\mu_n^*\xi_n^*\right) \\ -\sum_{n=1}^{N}a_n(\varepsilon+\xi_n+y_n-t_n)-\sum_{n=1}^{N}a_n^*\left(\varepsilon+\xi_n^*-y_n+t_n\right) \end{aligned} \tag{3.125}$$

Substituting for $y(\mathbf{x})$ using (3.115) and setting the derivatives of the Lagrangian function with respect to $\mathbf{w}$, b, ξ_n, and ξ_n^* to zero, then

$$\frac{\partial L}{\partial \mathbf{w}}=0 \Rightarrow \mathbf{w}=\sum_{n=1}^{N}\left(a_n-a_n^*\right)\phi(\mathbf{x}_n) \tag{3.126}$$

$$\frac{\partial L}{\partial b}=0 \Rightarrow \sum_{n=1}^{N}\left(a_n-a_n^*\right)=0 \tag{3.127}$$

$$\frac{\partial L}{\partial \xi_n}=0 \Rightarrow a_n+\mu_n=C \tag{3.128}$$

$$\frac{\partial L}{\partial \xi_n^*}=0 \Rightarrow a_n^*+\mu_n^*=C \tag{3.129}$$

Using these results to eliminate the corresponding variables from the Lagrangian function, one can see that the dual problem involves maximizing

$$\begin{aligned} \tilde{L}\left(a,a^*\right) = -\frac{1}{2}\sum_{n=1}^{N}\sum_{m=1}^{N}\left(a_n-a_n^*\right)\left(a_m-a_m^*\right)k(\mathbf{x}_n,\mathbf{x}_m) \\ -\varepsilon\sum_{n=1}^{N}\left(a_n+a_n^*\right)+\sum_{n=1}^{N}\left(a_n-a_n^*\right)t_n \end{aligned} \tag{3.130}$$

with respect to $\{a_n\}$and $\left\{a_n^*\right\}$, where we introduced the kernel $k(\mathbf{x},\mathbf{x}')=\phi(\mathbf{x})^{\mathrm{T}}\phi(\mathbf{x}')$. Again, this is a constrained maximization and to find the constraints it is noticeable that $a_n \geq 0$ and $a_n^* \geq 0$ are both required. Also, $\mu_n \geq 0$ and $\mu_n^* \geq 0$ together with (3.128) and (3.129) require $a_n \leq C$, $a_n^* \leq C$, therefore one has the box constraints

$$0 \leq a_n \leq C \tag{3.131}$$

$$0 \leq a_n^* \leq C \tag{3.132}$$

together with the condition (3.127).

Substituting (3.127) into (3.115), the prediction for new inputs can be made in the form

$$y(\mathbf{x}) = \sum_{n=1}^{N} (a_n - a_n^*) k(\mathbf{x}, \mathbf{x}_n) + b \tag{3.133}$$

which is again expressed in terms of the kernel function.

The corresponding KKT conditions, which state that at the solution the product of the dual variables and the constraints must vanish, are given by

$$a_n(\varepsilon + \xi_n + y_n - t_n) = 0 \tag{3.134}$$

$$a_n^* (\varepsilon + \xi_n^* - y_n + t_n) = 0 \tag{3.135}$$

$$(C - a_n)\xi_n = 0 \tag{3.136}$$

$$(C - a_n^*)\xi_n^* = 0 \tag{3.137}$$

from which one can obtain the useful results. First of all, a coefficient a_n can only be non-zero if $\varepsilon + \xi_n + y_n - t_n = 0$, which implies that the data point either lies on the upper boundary of the ε – *tube* ($\xi_n = 0$) or lies above the upper boundary ($\xi_n > 0$). Similarly, a non-zero value for a_n^* implies $\varepsilon + \xi_n^* - y_n + t_n = 0$ and such points must lie either on or below the lower boundary of the ε - *tube*.

Furthermore, the two constraints $\varepsilon + \xi_n + y_n - t_n = 0$ and $\varepsilon + \xi_n^* - y_n + t_n = 0$ are incompatible, since by adding them together note that ξ_n and ξ_n^* are nonnegative, while ε is strictly positive. And thus for each data point $\mathbf{x}_n$, either a_n or a_n^* must be zero.

The support vectors are those data points that contribute to predictions given by (3.133), in other words, either $a_n \neq 0$ or $a_n^* \neq 0$. These are points that lie on the boundary of the ε - *tube* or outside the tube. All points within the tube follows that $a_n = a_n^* = 0$. Therefore, one can obtain a sparse solution, and the only terms that must be evaluated in the predictive model (3.133) are those support vectors.

In addition, considering a data point $\{\mathbf{x}_n, y_n\}$ that satisfies $0 \leq a_n \leq C$, $\xi_n = 0$, and $\varepsilon + y_n - t_n = 0$, and using (3.115) and (3.126) and solving for b, one can obtain

$$\begin{aligned} b &= t_n - \varepsilon - \mathbf{w}^{\mathrm{T}} \phi(\mathbf{x}_n) \\ &= t_n - \varepsilon - \sum_{m=1}^{N} (a_m - a_m^*) k(\mathbf{x}_n, \mathbf{x}_m) \end{aligned} \tag{3.138}$$

An analogous result for b can also be obtained by considering a point, for which $0 \leq a_n^* \leq C$, and for robustness, one can average over all such estimates of b.

3.6.3 Least Square Support Vector Machine

A support vector regression model can be formulated by

$$y = \mathbf{w}^{\mathrm{T}}\varphi(\mathbf{x}) + b \tag{3.139}$$

where $\mathbf{x}$ is the input, y is its corresponding output; $\varphi(\cdot) : \mathbb{R}^{l} \to \mathbb{R}^{p}$ is the mapping from input space to a high-dimensional feature space; and $\mathbf{w} \in \mathbb{R}^{p}$, $b \in \mathbb{R}$, are the coefficients in high-dimensional feature space and the bias, respectively. The construction of a least square support vector machine (LSSVM) comes from solving the following constraint-based optimization problem [31],

$$\begin{aligned} &\min J\{\mathbf{w}, b, e\} = \frac{1}{2}\mathbf{w}^{\mathrm{T}}\mathbf{w} + \frac{\gamma}{2}\sum_{i=1}^{N} e_i^2 \\ &\text{s.t.} \quad y_i = \mathbf{w}^{\mathrm{T}}\varphi(\mathbf{x}_i) + b + e_i \end{aligned} \tag{3.140}$$

where γ is the regularization factor that establishes a sound trade-off between the smoothness of the model and its accuracy, and i denotes the ith sample. The total number of samples is N. $e_i \in \mathbb{R}$ is the fitting error of the ith sample. Similarly, with reference to the solution of SVM itself, a Lagrange multiplier α is introduced to make the optimization problem constraint free. Then, the resulting optimization problem can be expressed as follows:

$$L\{\mathbf{w}, b, e, \boldsymbol{\alpha}\} = \frac{1}{2}\mathbf{w}^{\mathrm{T}}\mathbf{w} + \frac{\gamma}{2}\sum_{i=1}^{N} e_i^2 - \sum_{i=1}^{N} \alpha_i\left(\mathbf{w}^{\mathrm{T}}\varphi(\mathbf{x}_i) + b + e_i - y_i\right) \tag{3.141}$$

The optimal solution is obtained through solving for the following constraints:

$$\begin{cases} \dfrac{\partial L}{\partial \mathbf{w}} = 0 \to \mathbf{w} = \displaystyle\sum_{i=0}^{N} \alpha_i \varphi(\mathbf{x}_i) \\ \dfrac{\partial L}{\partial b} = 0 \to \displaystyle\sum_{i=1}^{N} \alpha_i = 0 \\ \dfrac{\partial L}{\partial e_i} = 0 \to \alpha_i = \gamma e_i \quad i = 1, 2, \ldots, N \\ \dfrac{\partial L}{\partial \alpha_i} = 0 \to \mathbf{w}^{\mathrm{T}}\varphi(\mathbf{x}_i) + b + e_i - y = 0 \quad i = 1, 2, \ldots, N \end{cases} \tag{3.142}$$

Eliminating e and $\mathbf{w}$ from the above expressions, and using the Mercer condition [32], the system can be expressed as

$$\begin{bmatrix} 0 & \vec{1}^{\mathrm{T}} \\ \vec{1} & \mathbf{K} + \gamma^{-1}\mathbf{I} \end{bmatrix} \cdot \begin{bmatrix} b \\ \boldsymbol{\alpha} \end{bmatrix} = \begin{bmatrix} 0 \\ \mathbf{y} \end{bmatrix} \tag{3.143}$$

where $\mathbf{K}$ is the kernel matrix with $\mathbf{K}_{i,j} = k(x_i, x_j) = \varphi(x_i)^{\mathrm{T}}\varphi(x_j)$, and the Gaussian radial basis$k(x_i, x_j) = \exp\left(-\|x_i - x_j\|^2/\sigma\right)$ is usually adopted. $\mathbf{I}$ is an identity matrix, and $\vec{1} = [1, 1, \ldots, 1]^T$. $\mathbf{y}$ is the column vector involving the output sample. Solving (3.143), one can obtain $\boldsymbol{\alpha} = \mathbf{A}^{-1}(\mathbf{y} - b\vec{1})$, $b = (\vec{1}^T\mathbf{A}^{-1}\mathbf{y})/(\vec{1}^T\mathbf{A}^{-1}\vec{1})$, with $\mathbf{A} = \mathbf{K} + \gamma^{-1}\mathbf{I}$.

Thus, the regression model can be trained using the training samples. Subsequently, the regression-based prediction is realized in the form

$$y = \sum_{i=1}^{N} \alpha_i k(x, x_i) + b \tag{3.144}$$

In the regression model, as the width of Gaussian kernel and the regularization factor γ are fixed in advance, the over-fitting or poor fitting might occur when the dynamics of sample data changes. Thus, the model cannot adaptively capture the tendency of the sample data, which will reduce the prediction accuracy. To address this problem, an online LSSVM modeling based on the effective noise estimation for the real-time prediction of gas flow in the subsection will be presented in Chap. 7.

3.6.4 Sample Selection-Based Reduced SVM

The industrial training samples will extremely affect the performance of the LSSVM. If their numbers are too small, which may not contain all the features of the sample space, the model will exhibit a poor prediction ability. If a training dataset with very large size is chosen, it may also affect the prediction performance since some training samples may be redundant, requiring high computational load. Therefore, the reduced SVM is proposed to reduce the scale of the support vector, which is the prune part of the training samples. However, as for the generic reduced SVM, the training samples are pruned randomly, which may lose some valuable information. In this subsection, a pruning method based on the community detection of complex network will be introduced, which can reserve the samples with typical characteristics. Based on the novel pruning method, the generalization ability and the computational complexity of the prediction model will be improved obviously since the training samples can cover different operational conditions and with a minimal scale.

The community detection method takes the module maximum as the optimization objective to obtain the optimal community division and to further select the valuable training samples, which can avoid the influence coming from the randomness. Considering an original training dataset from the industrial field,

$$S = \{[\mathbf{x}_i(t), y_i(t-1)], \quad y_i(t) | i = 0, 1, \ldots, N\} \tag{3.145}$$

where $\mathbf{x}_i(t)$ and $y_i(t-1)$ serves as two input samples, and $y_i(t)$ is the output sample.

In this book, the Euclidean distance is adopted to calculate the connected relationship of a complex network. The formula of the Euclidean distance is written as

$$m_{ij} = \sqrt{\sum_{i \neq j} \left(s_{ik} - s_{jk}\right)^2} \tag{3.146}$$

As such, one can compute the distance m_{ij} between two samples $[\mathbf{x}_i(t), y_i(t-1)]$, $y_i(t)$ and $[\mathbf{x}_j(t), y_j(t-1)]$, $y_j(t)$. Thus, the distance matrix $\mathbf{M}$ for any two samples are

$$\mathbf{M} = \begin{bmatrix} m_{11} & m_{12} & \cdots & m_{1N} \\ m_{21} & m_{22} & \cdots & m_{2N} \\ \vdots & \vdots & \ddots & \vdots \\ m_{N1} & m_{N2} & \cdots & m_{NN} \end{bmatrix} \tag{3.147}$$

with $m_{ij} = m_{ji}$ and $m_{ii} = 0$.

To obtain an adjacency matrix $\mathbf{M}'$, a threshold parameter R is set here. All the elements in the matrix $\mathbf{M}$ will be compared with R. If the values of some elements are lower than that of R, the two related samples are viewed as adjacent ones; otherwise, it will be viewed as non-contiguous.

$$m'_{ij} = \begin{cases} 1, & \text{if node } i \text{ and } j \text{ are connected} \\ 0, & \text{else} \end{cases} \tag{3.148}$$

Then, the adjacency matrix $\mathbf{M}'$ reads as

$$\mathbf{M}' = \begin{bmatrix} m'_{11} & m'_{12} & \cdots & m'_{1N} \\ m'_{21} & m'_{22} & \cdots & m'_{2N} \\ \vdots & \vdots & \ddots & \vdots \\ m'_{N1} & m'_{N2} & \cdots & m'_{NN} \end{bmatrix} \tag{3.149}$$

To explain how to use the community detection of complex networks to select valuable samples, the community detection of complex networks is firstly introduced. A complex network is a graph (network) with non-trivial topological features, which do not occur in simple networks such as lattices or random graphs but often occur in graphs modeling of real systems. Most social, biological, and technological networks display substantial non-trivial topological features, with patterns of connection between their elements that are neither purely regular nor purely random. Such features include a heavy tail in the degree distribution, a high clustering coefficient, assortativity or disassortativity among vertices, community structure, and hierarchical structure.

A typical network is commonly composed by nodes or vertices and edges or links. The nodes denote the different elements of one system and the edges can describe the relationship among different nodes. If there is one relationship between two nodes, these two nodes are linked with an edge which can be directed or undirected. The degree of a vertex is the number of the edges connected to that vertex. For undirected networks, it can be computed as

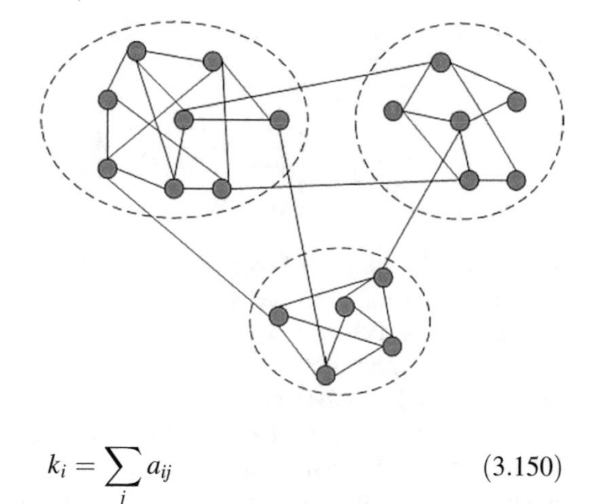

Fig. 3.17 Community structure of complex networks

$$k_i = \sum_j a_{ij} \tag{3.150}$$

where the value of a_{ij} denotes whether the two nodes i and j are connected. If i and j are connected, $a_{ij} = 1$; otherwise, $a_{ij} = 0$.

In the field of complex networks, community structures in a network are sets of nodes which can be easily grouped (potentially overlapping) such that each set of nodes is densely connected internally. As shown in Fig. 3.17, one can see a network with community structure. In this case there are three communities, denoted by the dashed circles, which has dense internal links, but between these communities there is only lower density of external links. To determine the optimal community division and the number of the community, the evaluation index named modularity is provided by [33]

$$Q = \sum_i \left(e_{ii} - a_i^2\right) \tag{3.151}$$

where e_{ii} denotes the ratio between the edges in the ith community and the edges of the whole network. $a_i = \sum_j e_{ij}$ denotes the ratio between the edges of the nodes in community i connected to other community and the edges of the whole network. The stronger the structure of the community is, the larger the value of Q will be. Here, a hierarchical clustering is employed to integrate different communities. Originally, each node can be viewed as one community, and two communities can be integrated in one step. The integration of the community should follow one rule that one should choose the integration resulting in the maximal increase of Q. The variety of the value of Q after the community integration can be described as

$$\Delta Q = e_{ij} + e_{ji} - 2a_i a_j = 2\left(e_{ij} - a_i a_j\right) \tag{3.152}$$

After n steps, when $\Delta Q < 0$, the value of Q can achieve the maximum. Then, the structure of the network is optimized. Since the original communities are described by the nodes, if assuming that i and j are connected by one edge, then $e_{ij} = 1/2m$;

otherwise, $e_{ij} = 0$. Meanwhile, $a_i = k_i/2m$. The original value of the matrix can be set as

$$\Delta Q = \begin{cases} \frac{1}{2m} - \frac{k_i k_j}{(2m)^2}, & \text{if } i \text{ is connected with } j \\ 0, & \text{otherwise} \end{cases} \tag{3.153}$$

where m is the number of edges of the whole network.

Based on the above principle about the community detection, the optimal community can be detected by computing the value of ΔQ. Using the community detection algorithm for the matrix $\mathbf{M}'$, the samples with the similar characteristics will be divided into the same community.

As for a data-driven modeling, the quality of the training samples has a great influence on the generalization ability and the prediction accuracy. When the similar training samples account for a large proportion in the sample population to other samples, the learning process of the prediction model will face the over-fitting due to the redundancy. After community detection, the samples located on the edge of the community and connected with the other community can be viewed as the redundancy samples. While, the samples in one community that do not connect with the samples in other community or has very sparse connection can be viewed as the samples with typical characteristics. To eliminate the redundancy of the training samples and effectively quantify the specification of the samples, an evaluation index named "joint binding degree" is defined as follows:

$$c_i = k_{i-\text{in}}/k_i \tag{3.154}$$

where $k_{i-\text{in}}$ is the number of the edges linked with node i in the community and k_i is the degree of node i. Generally, a large "joint binding degree" represents that the node belongs to the community with a higher probability. $c_i = 1$ means the node i belongs to this community completely while $c_i < 1$ means the node i is located on the edge of the community and probably has a similarity relation with the nodes in other community.

According to (3.154), the joint binding degree of each node can be calculated. Furthermore, one can sort the nodes in the community in descending order in accordance with the joint binding degrees. Finally, choosing the effective samples from different communities according to the value of the joint binding degree of the samples, the training dataset contains most of operational conditions in the industrial process, which enables to obviously improve generalization ability of the prediction model. The steps of implementation are listed as follows using the community detection of complex networks for the sample selection.

Algorithm 3.4: Sample Selection Using the Community Detection of Complex Networks

Step 1: Construct the training dataset by sampling from the industrial real-time database. Each training sample can be viewed as one node of the complex network.

Step 2: Compute the Euclidean distance m_{ij} of any two samples i and j in the training dataset according to (3.147). And then the distance matrix **M** *is obtained.*

Step 3: Choose a suitable threshold R. Compare each element m_{ij} of **M** *with R to obtain the adjacent matrix* $\mathbf{M}'$.

Step 4: Compute the value of ΔQ and use the community detection algorithm for the complex network represented by $\mathbf{M}'$.

Step 5: Based on the results of community detection, the joint binding degree of the nodes can be computed according to (3.154). Sort the nodes of one community in a descending order.

Step 6: Choose the effective samples according to the joint binding degrees to construct the training dataset.

Based on the above community detection, one new training dataset that belongs to the original dataset can be constructed, i.e., $\{(\mathbf{x}_i, d_i)\}_{i\in S} \subset \{(\mathbf{x}_i, d_i)\}_{i=1}^{N}$. Then, the parameters of the generic LSSVM can be rewritten as

$$\mathbf{w} = \sum\nolimits_{i\in S} \alpha_i k(\mathbf{x}_i, \cdot) \tag{3.155}$$

Then, the optimization objective can be described as

$$\min L(b, \boldsymbol{\alpha}) = \frac{1}{2}\boldsymbol{\alpha}^{\mathrm{T}}\mathbf{K}\boldsymbol{\alpha} + \frac{C}{2}\sum_{i=1}^{N}\left(d_i - \sum_{j\in S}\alpha_j\phi(\mathbf{x}_j)^{\mathrm{T}}\phi(\mathbf{x}_i) - b\right)^2 \tag{3.156}$$

with $\mathbf{K}_{ij} = k(\mathbf{x}_i, \mathbf{x}_j)$, $i, j \in S$.

To solve the optimization problem defined by (3.156), one can set

$$\frac{\partial L}{\partial b} = 0 \tag{3.157}$$

and

$$\frac{\partial L}{\partial \alpha_i} = 0 \tag{3.158}$$

According to (3.157) and (3.158), the linear equation can be derived as follows:

$$\left(\begin{bmatrix} 0 & \mathbf{0}^{\mathrm{T}} \\ \mathbf{0} & \mathbf{K}/C \end{bmatrix} + \mathbf{Z}\mathbf{Z}^{\mathrm{T}}\right)\begin{bmatrix} b \\ \alpha \end{bmatrix} = \mathbf{Zd} \tag{3.159}$$

where $\mathbf{Z} = \begin{bmatrix} \tilde{\mathbf{1}}^{\mathrm{T}} & \widehat{\mathbf{K}} \end{bmatrix}^{\mathrm{T}}$, and $\widehat{K}_{ij} = k(\mathbf{x}_i, \mathbf{x}_j), i\in S, j = 1, \ldots, N$.

Based on the above inference, the reduced regression function can be derived, in the form

$$f(\mathbf{x}) = \sum_{i\in S} \alpha_i k(\mathbf{x}_i, \mathbf{x}) + b \tag{3.160}$$

3.6.5 *Bayesian Treatment for LSSVM Regression*

The result of a generic LSSVM model is deterministic without considering the data uncertainty, which is inappropriate to deal with noisy industrial data. One could introduce the Bayesian-based approach into the modeling of LSSVM, which could be performed at two levels. To infer the primal weight space parameters **w** and b are at level 1, to infer the hyper-parameters μ and β are at level 2 [31].

Probabilistic Interpretation of LSSVM Regressor (Level 1): Predictive Mean and Error Bars

Calculation of Maximum Posterior

The model in primal weight space is $y(\mathbf{x}) = \mathbf{w}\phi(\mathbf{x}) + b$ with $\mathbf{w} \in \mathbb{R}^{n_h}$, $b \in \mathbb{R}$ and $\phi(\cdot)$ denotes a nonlinear mapping from the original input vector to a feature space. The training data set is denoted as $D = \{\mathbf{x}_k, y_k\}_{k=1}^{N}$. One can consider here a more general case of non-constant variance of the noise, which means that one can take the hyper-parameters $\boldsymbol{\beta} = [\beta_1, \cdots, \beta_N]$ instead of one single value β. In the primal weight space, an optimization problem should be solved.

$$\begin{aligned} &\min_{\mathbf{w}, b, e} J_p(\mathbf{w}, e) = \mu E_W + \sum_{k=1}^{N} \beta_k E_{D,k} \\ &\text{st.} \quad e_k = y_k - \left[\mathbf{w}^{\mathrm{T}}\phi(\mathbf{x}_k) + b\right], k = 1, 2, \ldots, N \end{aligned} \tag{3.161}$$

with

$$E_W = \frac{1}{2}\mathbf{w}^{\mathrm{T}}\mathbf{w} \tag{3.162}$$

and

$$E_{D,k} = \frac{1}{2}e_k^2 = \frac{1}{2}\left(y_k - \left[\mathbf{w}^{\mathrm{T}}\phi(\mathbf{x}_k) + b\right]\right)^2 \tag{3.163}$$

It can provide a Lagrangian

$$\mathcal{L}(\mathbf{w}, b, e; \alpha) = J_p(\mathbf{w}, e) - \sum_{k=1}^{N} \alpha_k \left(y_k - \left[\mathbf{w}^{\mathrm{T}}\phi(\mathbf{x}_k) + b\right] - e_k\right). \tag{3.164}$$

Then, one can get the following derivatives:

$$\begin{cases} \frac{\partial \mathcal{L}}{\partial \mathbf{w}} = 0 \rightarrow \mathbf{w} = \frac{1}{\mu}\sum_{k=1}^{N}\alpha_k \phi(\mathbf{x}_k) \\ \frac{\partial \mathcal{L}}{\partial b} = 0 \rightarrow \sum_{k=1}^{N}\alpha_k y_k = 0 \\ \frac{\partial \mathcal{L}}{\partial e_k} = 0 \rightarrow \alpha_k = \beta e_k, \qquad k = 1, \cdots, N \\ \frac{\partial \mathcal{L}}{\partial \alpha_k} = 0 \rightarrow y_k\left[\mathbf{w}^{\mathrm{T}}\phi(\mathbf{x}_k) + b\right] - 1 + e_k = 0, k = 1, \ldots, N \end{cases} \tag{3.165}$$

which gives the following dual problem for solving $\boldsymbol{\alpha}$, b:

$$\begin{bmatrix} 0 & 1_v^{\mathrm{T}} \\ 1_v^{\mathrm{T}} & \frac{1}{\mu}\Omega + D_\beta^{-1} \end{bmatrix}\begin{bmatrix} b \\ \hline \boldsymbol{\alpha} \end{bmatrix} = \begin{bmatrix} 0 \\ \hline \mathbf{y} \end{bmatrix} \tag{3.166}$$

with $D_\beta = \mathrm{diag}([\beta_1;\ldots;\beta_N])$, $y = [y_1;\ldots;y_N]$,$1_v = [1;\ldots;1]$, and $\Omega_{kl} = K(\mathbf{x}_k, \mathbf{x}_l) = \phi(\mathbf{x}_k)^T\phi(\mathbf{x}_l)$ for k, $l = 1, 2, \ldots, N$. In the dual space, the model becomes

$$y(\mathbf{x}) = \frac{1}{\mu}\sum_{k=1}^{N}\alpha_k K(\mathbf{x}, \mathbf{x}_k) + b \tag{3.167}$$

with support values α_k and bias term b.

The posterior distribution over $\mathbf{w}$ and b at the level 1 is given by

$$p(\mathbf{w}, b|D, \mu, \beta_{1,\ldots,N}, \mathcal{H}_\sigma) = \frac{p(D|\mathbf{w}, b, \mu, \beta_{1,\ldots,N}, \mathcal{H}_\sigma)}{p(D|, \mu, \beta_{1,\ldots,N}, \mathcal{H}_\sigma)}p(\mathbf{w}, b|\mu, \beta_{1,\ldots,N}, \mathcal{H}_\sigma) \tag{3.168}$$

where $\mathcal{H}_\sigma$ denotes the LSSVM model. The likelihood for regression $p(D|\mathbf{w}, b, \mu, \beta_{1,\ldots,N}, \mathcal{H}_\sigma)$ can be formulated by

$$p(D|\mathbf{w}, b, \mu, \beta_{1,\ldots,N}, \mathcal{H}_\sigma) = \prod_{k=1}^{N}\sqrt{\frac{1}{2\pi\beta_k^{-1}}}\exp\left(-\frac{1}{2}\frac{e_k^2}{\beta_k^{-1}}\right) \tag{3.169}$$

where $1/\beta_k$ is considered as the variance of the noise e_k. Then, the posterior becomes

$$\begin{aligned} & p(\mathbf{w} - \mathbf{w}_{\mathrm{MP}}, b - b_{\mathrm{MP}}|D, \mu, \beta_{1,\ldots,N}, \mathcal{H}_\alpha) \\ & = \frac{1}{\sqrt{(2\pi)^{(n_h+1)}\det\mathbf{Q}}}\exp\left(-\frac{1}{2}\mathbf{g}^{\mathrm{T}}\mathbf{Q}^{-1}\mathbf{g}\right) \end{aligned} \tag{3.170}$$

where $\mathbf{g} = [\mathbf{w} - \mathbf{w}_{\mathrm{MP}}; b - b_{\mathrm{MP}}]$ and $\mathbf{Q} = \mathrm{Cov}([\mathbf{w}; b], [\mathbf{w}; b])$ with Hessian matrix $\mathbf{H} = \mathbf{Q}^{-1}$. This covariance matrix is related to the Hessian of the quadratic cost function as follows:

$$\mathbf{Q}{=}\mathbf{H}^{-1} = \begin{bmatrix} (\mu\mathbf{I}+\beta\mathbf{G})^{-1} & -(\mu\mathbf{I}+\beta\mathbf{G})^{-1}\mathbf{H}_{12}\mathbf{H}_{22}^{-1} \\ -\mathbf{H}_{22}^{-1}\mathbf{H}_{12}^{T}(\mu\mathbf{I}+\beta\mathbf{G})^{-1} & \mathbf{H}_{22}^{-1}+\mathbf{H}_{22}^{-1}\mathbf{H}_{12}^{T}(\mu\mathbf{I}+\beta\mathbf{G})^{-1}\mathbf{H}_{12}\mathbf{H}_{22}^{-1} \end{bmatrix} \tag{3.171}$$

where

$$\begin{aligned} \mathbf{H}_{11} &= \mu\mathbf{I}+\boldsymbol{\gamma}D_{\beta}\boldsymbol{\gamma}^{\mathrm{T}} \\ \mathbf{H}_{12} &= \boldsymbol{\gamma}D_{\beta}\mathbf{1}_{v} \\ \mathbf{H}_{22} &= \textstyle\sum_{k=1}^{N}\beta_k{=:}s_{\beta} \end{aligned} \tag{3.172}$$

$\boldsymbol{\gamma} = [\phi(\mathbf{x}_1),\ldots,\phi(\mathbf{x}_N)], \mathbf{G} = \boldsymbol{\gamma}\mathbf{M}_c\boldsymbol{\gamma}^T$, and the centering matrix $\mathbf{M}_c = I-(1/N)\mathbf{1}_v\mathbf{1}_v^T$.

Moderated Output of LSSVM Regressor

The predictive mean of a point $\mathbf{x}$ and the error bar of this prediction are formulated by

$$\widehat{y}(\mathbf{x}) = \mathbf{w}_{\mathrm{MP}}^{\mathrm{T}}\phi(\mathbf{x})+b_{\mathrm{MP}} \tag{3.173}$$

where $\mathbf{w}_{\mathrm{MP}}^{\mathrm{T}}$ and b_{MP} satisfy the conditions for optimality related to (3.161), and α, b are the solution to the linear KKT system (3.166). The variance of the prediction is

$$\sigma_{\hat{y}}^2(\mathbf{x}) = 1/[\beta(\mathbf{x})]+\sigma_z^2(\mathbf{x}) \tag{3.174}$$

where

$$\begin{aligned} \sigma_z^2(\mathbf{x}) &= \varepsilon\left[(z-z_{\mathrm{MP}})^2\right] \\ &= \varepsilon\left[\left((\mathbf{w}^T\phi(\mathbf{x})+b)-\left(\mathbf{w}_{\mathrm{MP}}^{\mathrm{T}}\phi(\mathbf{x})+b_{\mathrm{MP}}\right)\right)^2\right] \\ &= \psi^{\mathrm{T}}(\mathbf{x})\mathbf{H}^{-1}\psi(\mathbf{x}) \end{aligned} \tag{3.175}$$

with $\psi(\mathbf{x}) = [\phi(\mathbf{x}); 1]$. The value $1/\beta(\mathbf{x})$ is the variance of the noise corresponding to input $\mathbf{x}$. The computation for $\sigma_z^2(\mathbf{x})$ can be obtained without explicit knowledge of the mapping $\phi(\cdot)$. Using matrix algebra and replacing inner products by the related kernel function, the expression for $\sigma_z^2(\mathbf{x})$ in the dual space is derived as [31]:

$$\begin{aligned} \sigma_z^2(\mathbf{x}) &= \boldsymbol{\theta}^{\mathrm{T}}(\mathbf{x})\mathbf{U}_G\mathbf{Q}_D\mathbf{U}_G^{\mathrm{T}}\boldsymbol{\theta}(\mathbf{x})+\frac{1}{\mu}K(\mathbf{x},\mathbf{x})-\frac{2}{s_{\beta}}\boldsymbol{\theta}^{\mathrm{T}}(\mathbf{x})\mathbf{U}_G\mathbf{Q}_D\mathbf{U}_G^{\mathrm{T}}\Omega D_{\beta}\mathbf{1}_v \\ &+\frac{2}{\mu s_{\beta}}\boldsymbol{\theta}^{\mathrm{T}}(\mathbf{x})D_{\beta}\mathbf{1}_v+\frac{1}{s_{\beta}^2}\mathbf{1}_v^{\mathrm{T}}D_{\beta}\Omega\mathbf{U}_G\mathbf{Q}_D\mathbf{U}_G^{\mathrm{T}}\Omega D_{\beta}\mathbf{1}_v+\frac{1}{\mu s_{\beta}^2}\mathbf{1}_v^{\mathrm{T}}D_{\beta}\Omega D_{\beta}\mathbf{1}_v \end{aligned} \tag{3.176}$$

where $\mathbf{Q}_D = (\mu\mathbf{I}+D_G)^{-1}-\mu^{-1}\mathbf{I}$, $\boldsymbol{\theta}(\mathbf{x}) = [K(\mathbf{x},\mathbf{x}_1);\cdots K(\mathbf{x},\mathbf{x}_N)]$, and $\mathbf{U}_G$, D_G are related to the eigenvalue decomposition

$$\left(D_\beta - \frac{1}{s_\beta} D_\beta \mathbf{1}_v \mathbf{1}_v^{\mathrm{T}} D_\beta\right) \Omega_{v_{G,i}} = \lambda_{G,i} v_{G,i}, \quad i = 1, \ldots, N_{\text{eff}} \le N - 1 \tag{3.177}$$

and $D_\beta = \text{diag}([\lambda_{G,1}; \ldots; \lambda_{G,N_{\text{eff}}}])$, $U_G = \left[(v_{G,1} \Omega v_{G,1})^{1/2} v_{G,1}; \ldots; (v_{G,\mathrm{N_{eff}}} \Omega v_{G,\mathrm{N_{eff}}})^{1/2} v_{G,\mathrm{N_{eff}}}\right]$.

In this case of a constant noise variance assumption, one assumes the constant value $1/\beta$.

Inference of Hyper-Parameters (Level 2)

We proceed at level 2 in this section. The posterior distribution over the hyper-parameters is

$$p(\mu, \beta_{1,\ldots,\mathrm{N}} | D, \mathcal{H}_\sigma) = \frac{p(D|\mu, \beta_{1,\ldots,\mathrm{N}}, \mathcal{H}_\sigma)}{p(D|\mathcal{H}_\sigma)} p(\mu, \beta_{1,\ldots,\mathrm{N}} | \mathcal{H}_\sigma) \tag{3.178}$$

Assuming a uniform prior distribution over $\log \mu$ and $\log \beta_{1,\ \ldots,\ N}$ and

$$p(\mu, \beta_{1,\ldots,\mathrm{N}} | \mathcal{H}_\sigma) = p(\mu|\mathcal{H}_\sigma) p(\beta_{1,\ldots,\mathrm{N}} | \mathcal{H}_\sigma) \tag{3.179}$$

Hence,

$$p(\mu, \beta_{1,\ldots,\mathrm{N}} | D, \mathcal{H}_\sigma) \propto p(D|\mu, \beta_{1,\ldots,\mathrm{N}}, \mathcal{H}_\sigma) \tag{3.180}$$

and

$$p(\mu, \beta_{1,\ldots,\mathrm{N}} | D, \mathcal{H}_\sigma) \propto \exp(-J_p(\mathbf{w}_{\mathrm{MP}}, b_{\mathrm{MP}})) \times \sqrt{\mu^{n_h} \prod_{k=1}^{N} \beta_k \sqrt{\det \mathbf{H}^{-1}}} \tag{3.181}$$

where

$$\begin{aligned} J_p(\mathbf{w}, b) &= J_p(\mathbf{w}_{\mathrm{MP}}, b_{\mathrm{MP}}) \\ &\quad + \frac{1}{2}\left([\mathbf{w}; b] - [\mathbf{w}_{\mathrm{MP}}; b_{\mathrm{MP}}]\right)^T \mathbf{H}\left([\mathbf{w}; b] - [\mathbf{w}_{\mathrm{MP}}; b_{\mathrm{MP}}]\right) \end{aligned} \tag{3.182}$$

and

$$J_p(\mathbf{w}_{\mathrm{MP}}, b_{\mathrm{MP}}) = \mu E_W(\mathbf{w}_{\mathrm{MP}}) + \beta E_D(\mathbf{w}_{\mathrm{MP}}, b_{\mathrm{MP}}). \tag{3.183}$$

By taking the negative logarithm of this expression, the following optimization problem with respect to the hyper-parameters is formulated:

$$\min_{\mu,\beta_{1,\dots,N}} J(\mu,\beta_{1,\dots,N}) = \mu E_W(\mathbf{w}_{\mathrm{MP}}) + \sum_{k=1}^{N}\beta_k E_{D,k}(\mathbf{w}_{\mathrm{MP}}, b_{\mathrm{MP}})$$
$$+\frac{1}{2}\sum_{i=1}^{N_{\mathrm{eff}}}\log(\mu+\lambda_{G,\mathrm{i}}) - \frac{N_{\mathrm{eff}}}{2}\log\mu - \frac{1}{2}\sum_{k=1}^{N}\log\beta_k + \frac{1}{2}\log\left(\sum_{k=1}^{N}\beta_k\right) \tag{3.184}$$

With the non-constant values of β_k, this becomes a nonlinear optimization in the case of many data points.

Inference of Kernel Parameters and Model Comparison

The inference at level 3 is similar to the classification case. Considering the case of a constant β, one has

$$P(D|\mathbf{H}_\sigma) \propto P(D|\mu_{\mathrm{MP}}, \beta_{\mathrm{MP}}, \mathcal{H}_\sigma)\frac{\sigma_{\mu|D}\sigma_{\beta|D}}{\sigma_\mu \sigma_\beta} \tag{3.185}$$

With the same assumptions as made in classification case, this leads to the following evidence function, according to which the models can be ranked.

$$P(D|\mathcal{H}_\sigma) \propto \sqrt{\frac{\mu_{\mathrm{MP}}^{N_{\mathrm{eff}}}\beta_{\mathrm{MP}}^{N-1}}{(d_{\mathrm{eff}}-1)(N-d_{\mathrm{eff}})\prod_{i=1}^{N_{\mathrm{eff}}}\left(\mu_{\mathrm{MP}}+\beta_{\mathrm{MP}}\lambda'_{G,i}\right)}} \tag{3.186}$$

where

$$\mathbf{M}c\Omega v_{G,i} = \lambda'_{G,i}v_{G,i} \quad \text{for} \quad i=1,\dots,N_{\mathrm{eff}} \le N-1 \tag{3.187}$$

and

$$d_{\mathrm{eff}} = 1 + \sum_{i=1}^{N_{\mathrm{eff}}}\beta\lambda'_{G,i}/\left(\mu+\beta\lambda'_{G,i}\right) \tag{3.188}$$

3.6.6 Case Study

For industrial application, here one can evaluate the performance of LSSVM model based on online parameter optimization with parallel computing [34], which will be introduced in Chap. 8 in details. This case study considers the Linz-Donawitz converter gas (LDG) system of a steel plant. The LDG system consists of LD converters for gas generation, pipeline networks for gas transportation, a series of gas users for consumption and gas holder for temporary storage. A brief structure of the LDG system of a steel plant is presented in Fig. 3.18, where 6 LD converters supply on average over 270,000 Nm^3 LDG per hour into the pipeline networks. First,

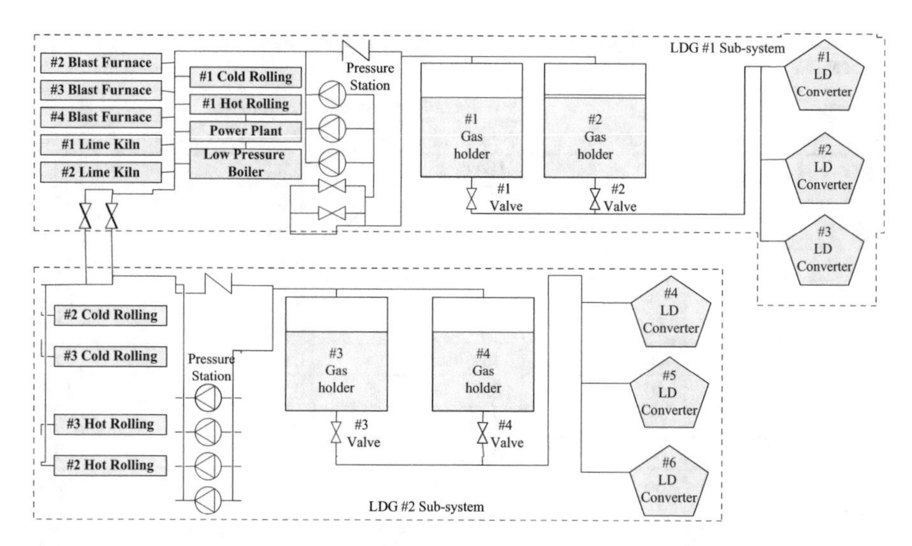

Fig. 3.18 Structural chart of LDG system used in steel industry

the generated LDG is temporarily stored in gas holders, each of which has the capacity of 160,000 m^3. The gas is then transported to other production units such as hot blast stove, hot rolling process, cold rolling plant, and low pressure boiler. If the generation amount of LDG in a period is larger than the consumption amount of the users, the gas holder might not be able to accumulate the surplus LDG. Thus, the useful gas energy has to be diffused into the environment for the safety reasons, which will seriously affect the environment. On the other hand, if the accumulative generation of LDG is less than the consumption demand, much more fossil energy such as coal or natural gas will have to be used as the alternative energy for sustaining the overall production. Since the operation status of the gas users frequently changes, the flow variations of the generation and the consumption fluctuate continuously. Therefore, the energy scheduling workers must monitor the flow of LDG in real-time and estimate the gas holder level for the scheduling and the system security. The manual estimation based on worker experience was still the main method being used in practice.

As for the practical online prediction problem, a comparative experiment is shown to analyze the efficiency of model learning between the serial LSSVM and the parallel one, where the optimized parameters have been determined in advance. One can randomly select the various objectives in LDG system with different number of sample data and input dimension to complete the experiments, whose results are illustrated in Table 3.11. From this table, the acceleration ratio gradually increases along with the increase of the number of sample data toward a certain input dimension. Apparently, the computing time growth of the parallel learning process is substantially slower than that of the serial one.

To further demonstrate an overall effectiveness of the parallel strategies for parameters optimization of the LS-SVM (those parallel strategies will be introduced in Chap. 8 in details), we consider 50 groups of energy data present in different

Table 3.11 Comparison of the training time using different number of samples and input dimension

Number of sample data	Input dimension	Computational time for training (s)		Acc
		Serial LS-SVM	Parallel LS-SVM	
100	2	0.035	0.004	8.9
200	2	0.204	0.01	20.4
450	2	2.25	0.051	44.0
800	2	12.48	0.205	60.9
1000	2	24.61	0.374	65.8
180	60	0.249	0.043	5.8
320	60	1.067	0.093	11.5
450	60	2.67	0.135	19.8
1000	60	25.98	0.555	46.8
180	80	0.283	0.054	5.2
320	80	1.15	0.107	10.7
450	80	2.85	0.161	17.7
1000	80	26.67	0.639	41.7
320	160	1.51	0.192	7.8
450	160	3.46	0.287	12.0
1000	160	29.61	0.902	32.8

Table 3.12 Results of comparison using the parallel online optimization and other three methods

Prediction objective	Method	Results		
		MAPE%	NRMSE	Training time (s)
LDG consumption of #2 blast furnace	BP-NN	10.59	0.1164	150.5
	SVM	7.35	0.0989	132.8
	LSSVM with fixed parameters	7.26	0.0686	1.1
	Parallel LS-SVM	3.49	0.0400	22.2
LDG consumption of #4 blast furnace	BP-NN	13.43	0.1674	46.4
	SVM	11.85	0.1524	24.4
	LSSVM with fixed parameters	11.84	0.1174	0.3
	Parallel LS-SVM	5.79	0.0898	6.4

periods to verify the prediction model. A 60-min variation of the objectives is predicted and compared when using the LSSVM with the fixed parameters, back propagation neural network (BP-NN), and SVM. One can take the NRMSE and MAPE to quantify the quality of prediction.

It can be seen from Table 3.12 that the BP-NN based on the principle of experience risk minimum gives the worst results, and its computational time is also the longest because the network weights determined by a gradient descent algorithm tend to slow down the learning process, which might also be trapped in possible local optimum. Although the prediction precision is evidently enhanced, the generic SVM solving a convex quadratic programming should require a quantity of computational time, which hardly meets the requirement of real-time prediction for the LDG system. The prediction based on LSSVM solves a series of linear equations to train the regression model. Although such a version with fixed parameters uses the least modeling time owing to getting rid of the validation process, it cannot adaptively adjust the kernel-based model according to the dynamics of gas flow. Therefore, its prediction error is obviously larger than that of the online optimized LSSVM. As for the computing time, in general practice, it is acceptable to predict the LDG flow online within 1 min, thus we propose the real-time optimized LSSVM for the industry application, in which the modeling process is carried out at designated time interval to meet the demand of online prediction. To clearly visualize the prediction, the results in Fig. 3.19 are illustrated. The parallel online LSSVM can

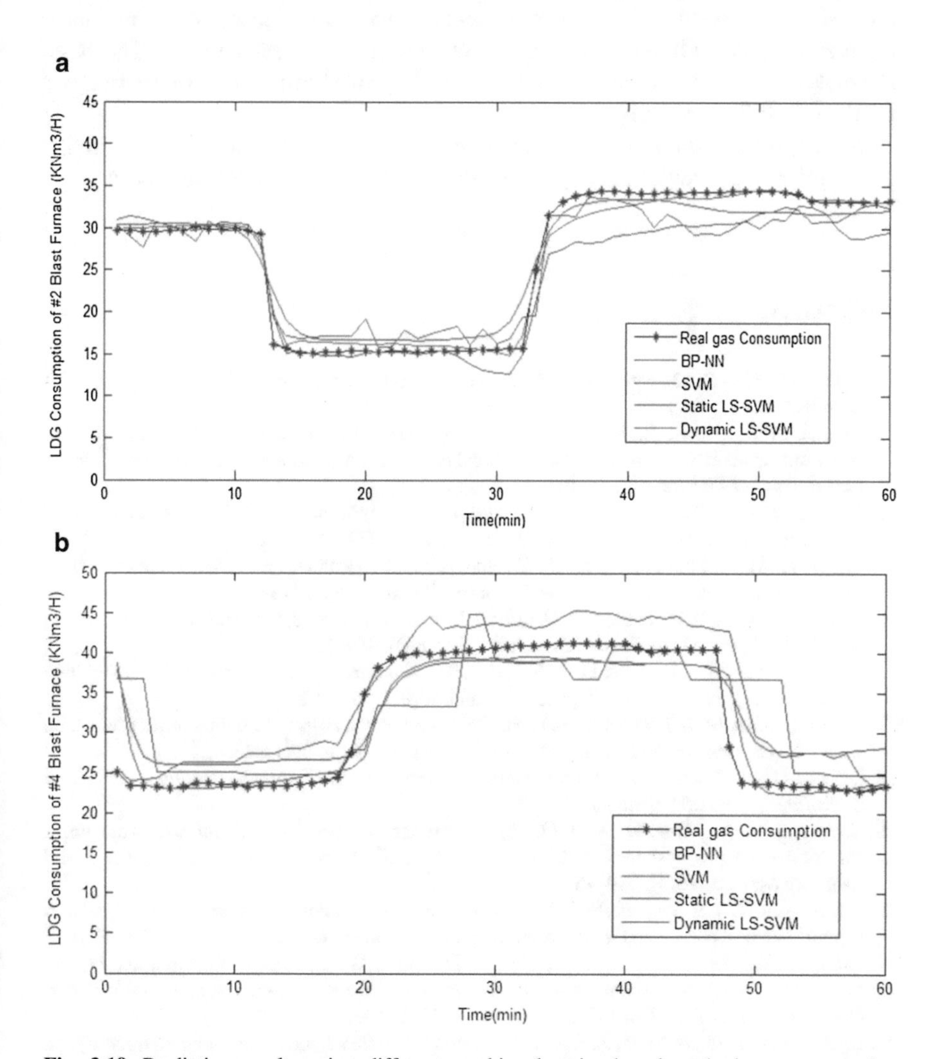

Fig. 3.19 Prediction results using different machine learning-based methods—a comparative analysis. (**a**) LDG consumption flow of #2 blast furnace. (**b**) LDG consumption flow of #4 blast furnace

well track the real energy variation for not only the gas holder level but also the LDG consumption amount of gas users.

3.7 Discussion

In this chapter, the issue of industrial time series prediction was paid more attention. First, the problem related to industrial time series prediction was discussed, which belongs to the methods of phase space reconstruction. Then, some commonly used methods for industrial time series prediction were introduced, including linear regression model, GP-based model, ANNs and SVM-based methods. The basic principles are introduced firstly and then some improved forms are further discussed for the industrial time series.

Although these time series prediction methods are based on the specified industrial applications, they can also provide references for other industrial scenes.

References

1. Takens, F. (1981). Detecting strange attractors in turbulence. *Lecture Notes in Math, 898*, 361–381.
2. Kennel, M. B., Brown, R., & Abarbanel, H. D. (1992). Determining embedding dimension for phase-space reconstruction using a geometrical construction. *Physical Review A Atomic Molecular & Optical Physics, 45*(6), 3403–3411.
3. Cao, L. (1997). Practical method for determining the minimum embedding dimension of a scalar time series. *Physica D-Nonlinear Phenomena, 110*(1-2), 43–50.
4. Fraser, A. M., & Swinney, H. L. (1986). Independent coordinates for strange attractors from mutual information. *Physical Review A General Physics, 33*(2), 1134.
5. Kim, H. S., Eykholt, R., & Salas, J. D. (1999). Nonlinear dynamics, delay times, and embedding windows. *Physica D Nonlinear Phenomena, 127*(1–2), 48–60.
6. Brock, W. A., Hsieh, D. A., & LeBaron, B. (1991). *Nonlinear dynamics, chaos, and instability: Statistical theory and economic evidence*. Cambridge: MIT Press.
7. Han, M., & Xu, M. (2018). Laplacian echo state network for multivariate time series prediction. *IEEE Transactions on Neural Networks and Learning System, 29*(1), 238–244.
8. Bishop, C. M. (2006). *Pattern recognition and machine learning (Information Science and Statistics)*. New York: Springer.
9. Zhao, J., Liu, Q., Wang, W., et al. (2012). Hybrid neural prediction and optimized adjustment for coke oven gas system in steel industry. *IEEE Transactions on Neural Networks and Learning Systems, 23*(3), 439–450.
10. An, S., Liu, W., & Venkatesh, S. (2007). Fast cross-validation algorithms for least squares support vector machine and kernel ridge regression. *Pattern Recognition, 40*(8), 2154–2162.
11. Hong, W. G., Feng, Q., Yan, C. L., Wen, L. D., & Lu, W. (2008). Identification and control nonlinear systems by a dissimilation particle swarm optimization-based Elman neural network. *Nonlinear Analysis Real World Applications, 9*, 1345–1360.
12. Li, X., Chen, Z. Q., & Yuan, Z. Z. (2000). Nonlinear stable adaptive control based upon Elman networks. *Applied Mathematics–A Journal of Chinese Universities, Series, B15*, 332–340.

13. Liou, C. Y., Huang, J. C., & Yang, W. C. (2008). Modeling word perception using the Elman network. *Neurocomputing, 71*, 3150–3157.
14. Köker, R. (2005). Reliability-based approach to the inverse kinematics solution of robots using Elman's networks. *Engineering Applications of Artificial Intelligence, 18*(6), 685–693.
15. Welch, G., & Bishop, G. (1995). *An introduction to the Kalman filter*, Technical Report TR 95-041. University of North Carolina, Department of Computer Science.
16. Jaeger, H. (2002). *Tutorial on training recurrent neural networks, covering BPTT, RTRL, EKF and "Echo State Network" approach, Technical Report GMD Report 159*. German National Research Center for Information Technology.
17. Farkaš, I., Bosák, R., & Gergeľ, P. (2016). Computational analysis of memory capacity in echo state networks. *Neural Networks, 83*, 109–120.
18. Jaeger, H., & Haas, H. (2004). Harnessing nonlinearity: Predicting chaotic systems and saving energy in wireless communication. *Science, 304*(5667), 78–80.
19. Jaeger, H. (2005). Reservoir riddles: Suggestions for echo state network research (pp. 1460–1462). In *Proceedings of the International Joint Conference on Neural Networks*.
20. Shi, Z. W., & Han, M. (2007). Support vector echo-state machine for chaotic time-series prediction. *IEEE Transactions on Neural Network, 18*(2), 359–372.
21. Liu, Y., Zhao, J., & Wang, W. (2009). Improved echo state network based on data-driven and its application in prediction of blast furnace gas output. *Acta Automatica Sinica., 35*, 731–738.
22. Golub, G. H., & van Loan, C. F. (1983). *Matrix computations*. Baltimore: The Johns Hopkins University Press.
23. Saxén, H., & Pettersson, F. (2005). A simple method for selection of inputs and structure of feedforward neural networks. *Computers & Chemical Engineering, 30*(6), 1038–1045.
24. Saxen, H., & Pettersson, F. (2007). Nonlinear prediction of the hot metal silicon content in the blast furnace. *Transactions of the Iron & Steel Institute of Japan, 47*(12), 1732–1737.
25. Wan, E. A., & Nelson, A. T. (2001). Dual extended Kalman filter methods. In S. Haykin (Ed.), *Kalman filtering and neural networks* (pp. 123–174). Chichester: Wiley.
26. Vapnik, V. (1995). *The nature of statistical learning theory*. New York: Springer.
27. Bennett, K. P., & Mangasarian, O. L. (1992). Robust linear programming discrimination of two linearly inseparable sets. *Optimization Methods and Software, 1*, 23–34.
28. Cortes, C., & Vapnik, V. (1995). Support vector networks. *Machine Learning, 20*, 273–297.
29. Fletcher, R. (1989). *Practical methods of optimization*. New York: Wiley.
30. Schölkopf, B., & Smola, A. J. (2002). *Learning with kernels*. Cambridge: MIT Press.
31. Gestel, V., Suykens, J. A. K., et al. (2001). Financial time series prediction using least squares support vector machines within the evidence framework. *IEEE Transactions on Neural Networks, 12*(4), 809–821.
32. Suykens, J., & Vandewalle, J. (1999). Least squares support vector machines classifiers. *Neural Processing Letters, 9*(3), 293–300.
33. Diykh, M., Li, Y., & Wen, P. (2017). Classify epileptic EEG signals using weighted complex networks based community structure detection. *Expert Systems with Applications, 90*(30), 87–100.
34. Zhao, J., Wang, W., Pedrycz, W., et al. (2012). Online parameter optimization-based prediction for converter gas system by parallel strategies. *IEEE Transactions on Control Systems Technology, 20*(3), 835–845.

Chapter 4
Factor-Based Industrial Process Prediction

Abstract This chapter gives the factors-based prediction methods for industrial processes. Different from the mentioned time series-based prediction, this kind of approaches construct a forecasting model by treating the process variables (not the output or target variables) called "factors" as the model inputs, rather than the auto-regression mode used in time series version. To select the factors from lots of candidates, this chapter firstly introduces some commonly used feature selection approaches such as gray correlation method and convolution-based methods. As for the single-output model, this chapter introduces NNs-based model, Takagi-Sugeno (T-S) fuzzy model, and SVM. In particular, a multi-kernel setting of a LSSVM model can perform better explanatary ability for learning a nonlinear model and fit the regression problem more effectively. Besides, this chapter also introduces a multi-output LSSVM model, which considers the single fitting error of each output and the combined error as well, and aims at the issues of multiple interactional outputs in industrial system. This chapter also provides some case studies on industrial energy system for performance verification.

4.1 Introduction

Both time series-based prediction and factors-based prediction belong to supervised learning, but they differ from the way of their training datasets construction. As for time series prediction, the samples of input and output are both drawn from same process variable, thus the law hidden in a single time series is to be discovered in this scene. In contrast, as for the factors-based version, the input and the output are usually collected from different process variables, which have different physical meanings, and its aim is to find a mapping between the target and its input variables called "factors." In such a way, the prediction can be performed given a new sample of input. Figure 4.1 shows a diagrammatic sketch of a factors-based prediction model, where $x_1, x_2, \ldots, x_m$ denotes the input variables and y is the target variable.

For instance, the industrial prediction of the by-product gasholder level in a steel system belongs to a class of factors-based task. The fluctuation of the value of the

J. Zhao et al., *Data-Driven Prediction for Industrial Processes and Their Applications*, Information Fusion and Data Science,
https://doi.org/10.1007/978-3-319-94051-9_4

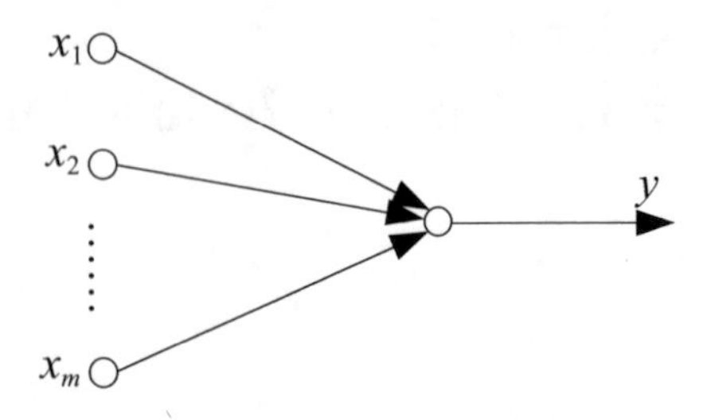

Fig. 4.1 Diagrammatic sketch of factors-based prediction model

gasholder level is caused by a comprehensive affection of the flow of gas of generation and consumption in the energy system. One can also pay attention to another example, the prediction of the pressure in the energy pipeline network, where the value of pressure is a function of the steam flow in different positions of the pipeline network. Based on the differences of specific industrial demands, factors-based industrial prediction can be divided into two types, the single-output prediction and multi-outputs one. The most commonly used techniques for the single-output version involve NNs-based model, TS fuzzy model, and SVR model. While as for the multi-outputs one, the multi-output NNs and multi-output LSSVM are usually employed.

In addition, there is another significant issue on how to determine the effective influential factors for the targets. Given a very complex industrial system, there are a large number of optional input variables, which might all be related to the predicted targets. However, there might be some redundant variables related to the target, having negative effects in prediction and increasing the computational load of modeling. Thus, we have to firstly identify the effective input variables by some means before modeling.

In this chapter, we firstly introduce some methods of the factors determination in Sect. 4.2, including the gray correlation, convolution-based technique, and Bayesian technique for automatic relevance determination (ARD). In Sect. 4.3, an introduction of three commonly used methods including NNs, TS fuzzy model, and multi-kernel learning for LSSVM, for factors-based single-output prediction are provided. In Sect. 4.4, we further study factors-based multi-output models, the NNs, and LSSVM models. Finally, we provide a summary about all the above contents and discuss the trends of research in the future.

4.2 Methods of Determining Factors

In general, the factors are determined on the basis of the knowledge of the experts, which is simple and sometimes effective. While it might also make mistakes due to the complexity and variation of the industrial environment. The data-based methods can select factors from the historical data so as to avoid the aforementioned drawbacks. In this section, we introduce three approaches, the gray correlation method, the convolution-based method, and the Bayesian technique of automatic relevance.

4.2.1 Gray Correlation

The gray correlation technique is often used to select the factors based on the accumulated data in an industrial prediction scene. One can also take the prediction of BFG gasholder level as an example to introduce an improved gray correlation method. There are many process factors producing or consuming the by-product gas in the industrial energy system, which may affect the variation of the value of gasholder level. If all these factors are considered as the inputs of the prediction model, the prediction accuracy might decrease due to the existence of the redundant variables. In this case, one can quantify the contribution of the factors to the holder level online and select part of them with high correlation. Additionally, due to the distribution of gas pipeline network, the change of BFG flow may be reflected by the variation of the holder level after time delay. To address such a problem, we intend to use a concept of gray correlation with time delay.

Let us recall that the fundamental idea of gray correlation is to quantify correlation by determining a proximity level of the trend of corresponding data [1, 2]. Gray correlation coefficient analysis is to determine whether variables are correlated to the predicted targets as well as to compute the quantified degree of their correlation. By calculating the characteristic curves and the degree of geometrical similarity between these curves, the main factors related to the target variable can be determined.

Supposing there is a set of time series,

$$\begin{aligned} &\left\{x_1^{(0)}(i)\right\}, \quad i = 1, 2, \ldots, N_1 \\ &\left\{x_2^{(0)}(i)\right\}, \quad i = 1, 2, \ldots, N_2 \\ &\quad\vdots \\ &\left\{x_m^{(0)}(i)\right\}, \quad i = 1, 2, \ldots, N_m \end{aligned} \tag{4.1}$$

where the natural numbers N_1, N_2, …, N_m could be different. Besides, there is a father time series $\left\{x_0^{(0)}(i)\right\}$, $i = 1, 2, \ldots, N_0$, and $\{x_k^{(0)}(i)\}$, $k = 1, 2, \cdots, m$ are all treated as its child one.

In the data preprocessing section, we firstly divide each time series by the mean value of the corresponding series. The correlation coefficient between father and child series is defined as

$$r_{0k}(i) = \frac{1}{N_0}\sum_{i=1}^{N_0}\xi_{0k}(i) \tag{4.2}$$

with

$$\xi_{0k}(i) = \frac{\min_k \min_i \left|x_0^{(0)}(i) - x_k^1(i)\right| + \rho \max_k \max_i \left|x_0^{(0)}(i) - x_k^1(i)\right|}{\left|x_0^{(0)}(i) - x_k^1(i)\right| + \rho \max_k \max_i \left|x_0^{(0)}(i) - x_k^1(i)\right|} \tag{4.3}$$

where $i = 1, 2, \ldots, N$; $k = 1, 2, \ldots, m$, ρ is the differentiation coefficient, which aims to diminish the ill-conditioned effect away from a big absolute error and to improve the prominence difference of the obtained correlation coefficient. The value of $\rho \in (0, 1)$ is usually assigned between 0.3 and 0.7.

For an industrial case, due to the large fluctuation and the high time-variability of the process variables present in the practical system, it is insufficient to determine the correlation only by the standardized increments of time series as this computing does not capture the effect of signal change rate. To alleviate this shortcoming, an improvement is introduced for the gray correlation. Given two pieces of time series, $X_1 = \{x_1(1), x_2(2), \ldots, x_n(n)\}$ and $X_2 = \{x_1(1), x_2(2), \ldots, x_n(n)\}$, the gray correlation for X_1 and $X_2$with delay d is defined as

$$\gamma(X_1, X_2, d) = \frac{1}{n-d-1}\cdot\sum_{k=2}^{n-d}\operatorname{sgn}(\Delta y_1(k)\cdot\Delta y_1(k+d))\cdot\mu(k,d)\cdot\nu(k,d) \tag{4.4}$$

where $\Delta y_i(k)$ is the standardized increment of signal i on time stamp k, $\Delta y_i(k) = x_i(k) - x_i(k-1)/D_i$; $(i = 1, 2;\ D_i = (1/(n-1))\sum_{k=2}^{n}|x_i(k) - x_i(k-1)|)$ $\mu(k, d)$ and $\nu(k,d)$ are the increment correlation coefficient and the change rate correlation coefficient, respectively.

The value of the correlation coefficient may be positive or negative depending on the sign of $\Delta y_1(k) \cdot \Delta y_2(k + d)$. When the two signals have identical increments and change rates in a certain time period, then the correlation degree equals 1.

Case Study

In this case study, we verify the effectiveness of the improved gray correlation method by considering the BFG system in steel industry which is depicted in Chap. 3, see Fig. 3.6. In the BFG system, although the gasholder can be treated as a buffer storage unit, its total capacity is hard enough to completely respond to the variation present in the BFG system. Besides, there are often abnormal conditions occurring in the production process such as blast reduction, user shutdown, or equipment fault. Such circumstance also leads to system imbalance. Therefore, it is very important to monitor and predict the value of the gasholder level based on the various possible impact factors. This case study performs the feature selection for

Table 4.1 The calculated results of gray correlation technique for the BFG system presented in Fig. 3.6

No.	Impact factors of system	#1 Gasholder level: Gray correlation degree	#1 Gasholder level: Time delay (min)	#2 Gasholder level: Gray correlation degree	#2 Gasholder level: Time delay (min)
1	#1 Blast furnace	*0.8283*	1	*0.8376*	1
2	#2 Blast furnace	*0.8492*	1	*0.8604*	1
3	#3 Blast furnace	*0.8531*	0	*0.8618*	1
4	#4 Blast furnace	*0.8624*	0	*0.8735*	0
5	#1 Gasholder level ($t-1$)	*0.8412*	–	–	–
6	#2 Gasholder level ($t-1$)	–	–	*0.8379*	–
7	#1–2 coke oven	*0.7439*	1	*0.7347*	1
8	#3–4 coke oven	*0.7845*	1	*0.7216*	1
9	#5–6 coke oven	*0.7232*	1	*0.7276*	1
10	#1 Power generator	*0.2159*	2	*0.2751*	2
11	#2 Power generator	*0.2467*	2	*0.2107*	2
12	#3 Power generator	*0.4109*	2	*0.4246*	1
13	#4 Power generator	*0.5726*	1	*0.5173*	1
14	CCPP	*0.3165*	1	*0.3741*	1
15	#1 Low pressure boiler	*0.3095*	1	*0.3747*	1
16	#1 Hot rolling plant	*0.2238*	2	*0.2434*	2
17	#2 Hot rolling plant	*0.2173*	2	*0.2109*	2
18	#3 Hot rolling plant	*0.2842*	2	*0.2847*	2
19	#1 Cold rolling plant	*0.1273*	2	*0.2111*	2
20	#2 Cold rolling plant	*0.1095*	2	*0.1221*	2
21	#3 Cold rolling plant	*0.0742*	2	*0.0841*	2
22	Steel pipe plant	*0.0903*	1	*0.0679*	1
23	Rough rolling	*0.0725*	1	*0.0902*	1
24	#2 Low pressure boiler	*0.0114*	1	*0.0104*	2
25	#1 Synthesizing unit	*0.0296*	2	*0.0178*	2
26	#2 Synthesizing unit	*0.0152*	2	*0.0157*	2
27	#1 Chemical products recovery	*0.0012*	2	*0.0018*	2
28	#2 Chemical products recovery	*0.0015*	2	*0.0015*	2

predicting the gasholder level by using the improved gray correlation method. We select a suitable threshold of the correlation level for all the BFG units and choose those with high correlation (viz. exceeding the threshold) to predict the gasholder level.

After removing noises, the correlation between BFG units and the holder level can be obtained by using the above gray correlation model with time delay. The calculated correlation values are listed in Table 4.1, where the time delay of each unit is determined based on production experience expressed by experts. It is apparent that the amount of the BFG generation has the highest correlation level with gasholders, and the units with a small amount of consumption or positioned at a long distance from the gasholder have limited influence on the holder level. One can pick the BFG generators or users whose correlation value is higher than 0.1, as the input for gasholder level. As seen from this table, the total number of the BFG units impacting on the gasholder, named as impact factors, is 28; while the number of selected units with high correlation is 20, which has been decreased by about 30% when the correlation model is used. Such operation largely reduces the complexity of the prediction model for the gasholder and makes the method more amenable to the online realization.

4.2.2 Convolution-Based Methods

In mathematics, convolution is a mathematical operation on two functions (generally f and g); it produces a third function that is typically viewed as a modified version of one of the original functions, treating the integral of the pointwise multiplication of the two functions as a function of the amount that one of the original functions is translated. Convolution is similar to cross-correlation, which is widely applied in various fields including probability, statistics, computer vision, natural language processing, image and signal processing, engineering, and differential equations. Convolution can be defined for functions of groups other than Euclidean space. For example, periodic functions, such as the discrete-time Fourier transform, can be defined on a circle and convolved by periodic convolution. A discrete convolution can be defined for functions on the set of integers. The generalizations of convolution have applications in the field of numerical analysis and numerical linear algebra, and in the design and implementation of finite impulse response filters in signal processing.

The convolution of f and g is written as $f * g$, using an asterisk or star. It is defined as the integral of the product of the two functions after one is reversed and shifted. As such, it is a particular kind of integral transform

$$(f * g)(t) \overset{\mathrm{def}}{=} \int_{-\infty}^{\infty} f(\tau)g(t-\tau)d\tau = \int_{-\infty}^{\infty} f(t-\tau)g(\tau)d\tau \tag{4.5}$$

where t does not need to represent the time domain. Thus, the convolution formula can be described as a weighted average of the function $f(\tau)$ at the moment t, where the weight is given by $g(-\tau)$ which is simply shifted by amount t. As t changes, the weighting function emphasizes different parts of the input function.

For function f and g supported on only $[0, \infty)$ (i.e., zeros for negative arguments), the integration limits can be truncated, resulting in

$$(f*g)(t) \stackrel{\text{def}}{=} \int_0^t f(\tau)g(t-\tau)d\tau \quad \text{for } f,g\colon [0,\infty) \to \mathbb{R} \tag{4.6}$$

In this case, the Laplace transform is more appropriate than the Fourier transform below and boundary terms become relevant. A primary convention as one often sees is given by

$$f(t)*g(t) \stackrel{\text{def}}{=} \underbrace{\int_{-\infty}^{\infty} f(\tau)g(t-\tau)d\tau}_{(f*g)(t)} \tag{4.7}$$

For instance, $f(t) * g(t - t_0)$ is equivalent to $(f * g)(t - t_0)$, but $f(t - t_0) * g(t - t_0)$ is in fact equivalent to $(f * g)(t - 2t_0)$.

Case Study

Also, we consider to apply the convolution-based methods to determine the input factors of a prediction model for the BFG system, in which each user or unit might have contribution to the variation of the gasholder level to some degree. That is, some users might have a direct contribution to its variation, while others might contribute minor influences. To ensure the efficiency and the accuracy of the prediction, we intend to find the users with direct contributions by using the convolution-based correlation.

In this subsection, the degree of correlation between two industrial time series is computed by the convolution formula. Theoretically, the larger the value of the convolution is, the higher the degree of their correlation gets. Since the variation of the holder level is simultaneously affected by the generation and consumption of the users, the time series with same length are synchronously sampled from the sensors of different users. Assuming that $\mathbf{t}(m)$ is the time series data of the gasholder level and $\mathbf{u}(m)$ is the corresponding one of a user, the degree of correlation between these two series can be calculated by

$$\mathbf{y}(k) = \sum_{i=1}^{m} \mathbf{t}(i)\mathbf{u}(k-i) \tag{4.8}$$

where m denotes the length of the target series, and the maximal value of the parameter k is $2m - 1$. The maximal value of $\mathbf{y}(k)$ denotes the correlation between the gasholder level and the gas flow. The larger the maximal value of $\mathbf{y}(k)$ is, the higher the degree of correlation obtains. In addition, the time delay between them is adopted to describe the lag level of the gas flow to the gasholder level, when the value of $\mathbf{y}(k)$ reaches the maximum.

We firstly choose the influential factors by the experts of domains. The users in the BFG system that have no influence on the gasholder should be removed, while the ones that might generate actions should be left. Then, we calculate the

Table 4.2 Results of convolution of blast furnace gasholder level and flow of users

No.	Users	Convolution	Time delay
1	1# Blast furnace	$1.0991*10^7$	1
2	2# Blast furnace	$9.9218*10^6$	0
3	3# Blast furnace	$9.0289*10^6$	0
4	4# Blast furnace	$9.1445*10^6$	0
5	1,2# coke oven	$3.0014*10^6$	0
6	3,4# coke oven	$3.3475*10^6$	0
7	5,6# coke oven	$3.4948*10^6$	0
8	1# synthesizing unit	$1.9541*10^5$	2
9	2# synthesizing unit	$1.5078*10^5$	3
10	1# cold rolling	$3.8935*10^5$	0
11	2# cold rolling	$1.1418*10^5$	0
12	5# cold rolling	$4.9540*10^5$	0
13	1800 cold rolling	$3.0955*10^5$	0
14	1# hot rolling	$5.7135*10^5$	0
15	2# hot rolling	$2.2221*10^5$	0
16	3# hot rolling	$2.2510*10^5$	0
17	1# power generator	$2.0103*10^6$	0
18	2# power generator	$5.8691*10^5$	0
19	3# power generator	$1.6624*10^6$	0
20	4# power generator	$1.4849*10^7$	0
21	CCP power plant	$7.2855*10^6$	0
22	Low pressure boiler	$1.4283*10^6$	0
23	Rough rolling	$3.4583*10^5$	0

convolution between the gasholder level and these selected users to determine the input factors for the prediction model.

To further understand the effectiveness of the convolution-based correlation technique for the BFG dataset, a series of experiments is conducted, and their experimental results are shown in Table 4.2. One can choose the users that have larger degree of relation to the gasholder level, such as the generation of the 1#~4# blast furnaces and the consumption of the 1#~6# coke ovens.

4.2.3 Bayesian Technique of Automatic Relevance

The implementation of Bayesian regularized ANNs assumes a single rate of weight decay for all the network weights, but the scaling properties of networks suggest that weights in different network layers should employ different regularization coefficients. By separating the weights into a number of classes [3, 4], developed a method for soft network pruning called automatic relevance determination (ARD). In ARD, the weights are divided into three classes, the one for each input (corresponding to the weights from the input to the hidden layer), the one for the hidden layer biases, and the one for each output (corresponding to the weights from the hidden layer to the output). In problems with many input variables, some of which may be irrelevant to the prediction of the output, the ARD method allows the network to "measure" the importance of each input by estimating their corresponding weights so as to effectively turn off those which are not relevant. This allows all variables, including those who have little impact on the output, to be included in the analysis without ill-effect, as the irrelevant variables have their weights reduced automatically. On the other hand, as for the problems with very large numbers of inputs, it may be more efficient to remove variables with large decay rates and train a new network on a reduced set of inputs, especially if the trained network is to be used to screen a very large virtual database. ARD has two main advantages over other pruning methods. The first one is firmly based on probability theory, and the second one lies in it carried out automatically.

We consider a dataset consisting of m examples with *n-dimensional* input vector $\mathbf{x}_\nu = [x_{\nu 1}, x_{\nu 2}, \ldots, x_{\nu n}]^{\mathrm{T}}$, which corresponds to target $y_\nu \in R$ (where ν are pattern indices). Next, the algorithms for regression will be introduced. We assume the following likelihood for the data [3]:

$$p\left(y|\mathbf{w}, \sigma^2\right) = \left(2\pi\sigma^2\right)^{-\frac{m}{2}} \exp\left(-\frac{1}{2\sigma^2}\|y - \mathbf{\Phi w}\|^2\right) \tag{4.9}$$

with $\mathbf{w} = (w_0, \ldots, w_n)$. The gene expression datasets which will be subsequently considered are generally linearly separable [5], thus, the matrix $\mathbf{\Phi}$ is given by

$$\mathbf{\Phi} = \begin{pmatrix} 1 & x_{11} & \ldots & x_{1n} \\ 1 & x_{21} & \ldots & x_{2n} \\ \ldots & \ldots & \ldots & \ldots \\ 1 & x_{m1} & \ldots & x_{mn} \end{pmatrix} \tag{4.10}$$

with the first column handling the bias w_0 in the hypothesis function

$$f(\mathbf{x}_\nu) = \sum_{j=1}^{n} w_j x_{\nu j} + w_0 \tag{4.11}$$

Pruning the ith column of $\mathbf{\Phi}$ aims to implement the feature selection. As for a relevance vector machine (RVM), one can assume a prior favoring sparse hypotheses [6], which realized sparsity in input features rather than sparsity in examples with linear model,

$$p(\mathbf{w}|\boldsymbol{\alpha}) = (2\pi)^{-\frac{n+1}{2}} \prod_{i=0}^{n} \alpha_i^{1/2} \exp\left(-\frac{\alpha_i w_i^2}{2}\right) \tag{4.12}$$

where $\alpha_0, \ldots, \alpha_n$ is a set of hyper-parameters that control the prior distribution over the weights. We then integrate out the weights from (4.9) and (4.12) to obtain the marginal likelihood:

$$p(y|\boldsymbol{\alpha}, \sigma^2) = (2\pi)^{-\frac{m}{2}}\left|\mathbf{B} + \boldsymbol{\Phi}\mathbf{A}^{-1}\boldsymbol{\Phi}^{\mathrm{T}}\right|^{-\frac{1}{2}}\exp\left[-\frac{1}{2}y^{\mathrm{T}}\left(\mathbf{B} + \boldsymbol{\Phi}\mathbf{A}^{-1}\boldsymbol{\Phi}^{\mathrm{T}}\right)^{-1}y\right] \tag{4.13}$$

where $\mathbf{A} =$ diag $(\alpha_0, \alpha_1, \ldots, \alpha_n)$, $\mathbf{B} = \sigma^{-2}\mathbf{I}_m$ and we have additionally used the Woodbury-Sherman-Morrison matrix identity,

$$\left(\mathbf{B} + \boldsymbol{\Phi}\mathbf{A}^{-1}\boldsymbol{\Phi}^{\mathrm{T}}\right)^{-1} = \mathbf{B} - \mathbf{B}\boldsymbol{\Phi}\boldsymbol{\Sigma}\boldsymbol{\Phi}^{\mathrm{T}}\mathbf{B} \tag{4.14}$$

with $\boldsymbol{\Sigma} = (\mathbf{A} + \boldsymbol{\Phi}^{\mathrm{T}}\mathbf{B}\boldsymbol{\Phi})^{-1}$. The posterior over the weights is formulated as

$$p(\mathbf{w}|y, \boldsymbol{\alpha}, \sigma^2) = (2\pi)^{-\frac{(n+1)}{2}}|\boldsymbol{\Sigma}|^{-\frac{1}{2}}\exp\left[-\frac{1}{2}(\mathbf{w} - \boldsymbol{\mu})^{\mathrm{T}}\boldsymbol{\Sigma}^{-1}(\mathbf{w} - \boldsymbol{\mu})\right] \tag{4.15}$$

with mean vector $\boldsymbol{\mu} = \boldsymbol{\Sigma}\boldsymbol{\Phi}^{\mathrm{T}}\mathbf{B}y$. The algorithms we now detail are based upon exploiting the *type-II maximum likelihood* principle. The objective is to maximize the marginal likelihood (4.13) with respect to the hyper-parameters in order to obtain point estimates of their values. These can then be substituted back into (4.15) in order to give an updated posterior distribution for the weights, which is typically summarized by its mean.

We can obtain an iterative update formula for $\boldsymbol{\alpha}$ by taking the natural log of (4.13) and differentiating this expression with respect to $\boldsymbol{\alpha}$. This differentiation is given by the following formula:

$$\frac{\partial \mathbf{M}^{-1}}{\partial \boldsymbol{\alpha}} = -\mathbf{M}^{-1}\left(\frac{\partial \mathbf{M}^{-1}}{\partial \boldsymbol{\alpha}}\right)\mathbf{M}^{-1} \tag{4.16}$$

where

$$\frac{\partial \ln |\mathbf{M}|}{\partial \boldsymbol{\alpha}} = \mathrm{Tr}\left(\mathbf{M}^{-1}\frac{\partial \mathbf{M}}{\partial \boldsymbol{\alpha}}\right) \tag{4.17}$$

and $\mathbf{M}$ is an arbitrary matrix. Then, the update formula for α_i at the optimum is derived as

$$\alpha_i = \gamma_i/\mu_i^2 \tag{4.18}$$

where $\gamma_i = 1 - \alpha_i\Sigma_{ii}$ and

$$\left(\sigma^2\right)^{\text{new}} = \frac{\|y - \boldsymbol{\Phi}\boldsymbol{\mu}\|^2}{\left(m - \sum_i \gamma_i\right)} \tag{4.19}$$

These formulas suggest an algorithmic approach for regression, in which we iteratively update $\boldsymbol{\alpha}$ and σ, and intermediate the quantities such as $\boldsymbol{\mu}$ and $\boldsymbol{\Sigma}$ [6]. In practice, during re-estimation, many of α_i approach to infinity. As for those α_i, the corresponding individual weight posteriors $p(w_i| y, \boldsymbol{\alpha}, \sigma^2)$ become infinitely peaked at zero, which implies that the corresponding ith column in $\boldsymbol{\Phi}$ can be pruned. During the execution of the algorithm with forward selection, such features were removed and in the simulations described below we used $\alpha_i > 10^{12}$ as the pruning criterion. The re-introduction of pruned features was also allowed if the gradient of the marginal likelihood, (4.13), with respect to α_i is less than a bound.

On the other hand, we have to here consider the adaptation for classification problem, where $y_\nu \in \{0, 1\}$. The linear model is generalized by applying the logistic sigmoid function $g(f) = 1/(1 + e^{-f})$ to $f(\cdot)$. The likelihood of the dataset is then written as [7],

$$p(y|\mathbf{w}) = \prod_{\nu=1}^{m} g(f(\mathbf{x}_\nu))^{y_\nu}[1 - g(f(\mathbf{x}_\nu))]^{1-y_\nu} \tag{4.20}$$

Owing to this modification, the weight posterior is no longer analytically obtainable, but for the given values of $\boldsymbol{\alpha}$ an effective approximation can be obtained by using a Gaussian centered at the maximum of $p(\mathbf{w}|\, y, \boldsymbol{\alpha})$. Finding this maximum is equivalent to a standard optimization of a regularized logistic model and different optimization methods can be applied to solve it [8].

The covariance $\boldsymbol{\Sigma} = (-\nabla\nabla \log p(y, \mathbf{w}|\, \boldsymbol{\alpha}))^{-1}$ of the approximating Gaussian is now equal to $(\boldsymbol{\Phi}^{\mathrm{T}}\mathbf{B}\boldsymbol{\Phi} + \mathbf{A})^{-1}$, where $\mathbf{B}$ is an $m \times m$ diagonal matrix with $\mathbf{B}_{\nu\nu} = g(f(\mathbf{x}_\nu))[1 - g(f(\mathbf{x}_\nu))]$. The hyper-parameters $\{\alpha_i\}$ are still updated as in (4.18).

Besides, since the covariance matrices between regression and classification models are similar, one can readily expand the regression algorithm to the task of classification as follows [8]. Let $y_\nu \in \{-1, +1\}$, $\mathbf{A} = \mathrm{diag}\,(\alpha_0, \alpha_1, \ldots, \alpha_n)$, and $\mathbf{B} = \mathrm{diag}\left(1/\sigma_1^2, \ldots, \sigma_\nu^2, \ldots, 1/\sigma_m^2\right)$. One can update the hyper-parameters $\{\alpha_i\}$ using (4.18). However, as for $\left\{\sigma_\nu^2\right\}$, we have the following formula to determine $\mathbf{B}$.

$$\left(\sigma_\nu^2\right)^{\text{new}} = \frac{1}{g(f(\mathbf{x}_\nu))[1 - g(f(\mathbf{x}_\nu))]} \tag{4.21}$$

4.3 Factor-Based Single-Output Model

4.3.1 Neural Networks-Based Model

Feedforward neural networks (FNNs) are commonly composed by the input layer, hidden layers, and the output layer, where the hidden layers can be only one or much more layers. Each neuron receives the outputs of all the neurons of the front layer and transforms its own output to the next. It is noticeable that there is no feedback at all existing in such a network. Figure 4.2 shows the schematic diagram of one feedforward neural network with only one hidden layer.

where the network only has the feedforward output. If the number of the inputs equals m and the number of the neurons of the hidden layer equals N, $w_{i,\,j}$ $(i = 1, 2, \ldots, m; j = 1, 2, \ldots, N)$ describes the connection between the neurons of

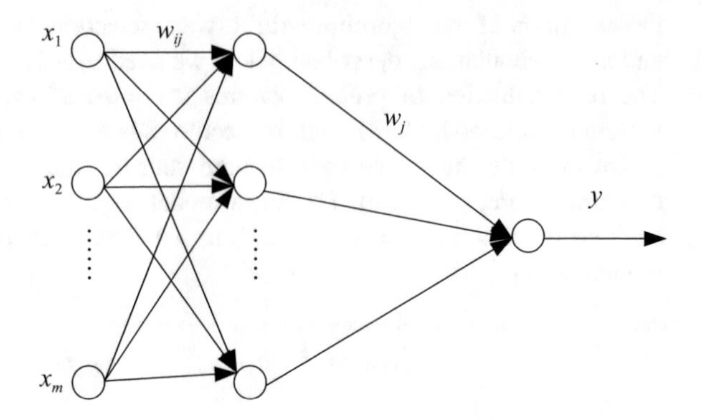

Fig. 4.2 The schematic diagram of one FNN with only one hidden layer

the input layer and the ones of the hidden layers. w_j describes the connection between the neurons of the hidden layer and the output neuron.

$$u_j(n) = \sum_{i=1}^{m} w_{i,j} x_i(n) \tag{4.22}$$

$$\nu_j(n) = f\left[u_j(n)\right] = f\left[\sum_{i=1}^{m} w_{i,j} x_i(n)\right] \tag{4.23}$$

$$o(n) = \sum_{i=1}^{N} w_j \nu_j(n) \tag{4.24}$$

$$y(n) = f^{\text{out}}[o(n)] = f^{\text{out}}\left[\sum_{i=1}^{N} w_j \nu_j(n)\right] \tag{4.25}$$

where f and f^{out} are the activation functions of the neurons of the hidden layer and the output one, respectively. $u_j(n)$ and $\nu_j(n)$ denote the input and output of the hidden layer, respectively. $o(n)$ is the input of the output neuron. The cost function of this neural network can be defined by

$$J(n) = e^2(n) = [d(n) - y(n)]^2 = \left\{d(n) - f^{\text{out}}\left[\sum_{i=1}^{N} w_j \nu_j(n)\right]\right\}^2 \tag{4.26}$$

where $d(n)$ is the observed target.

Therefore, a NN with m inputs, N hidden neurons and a single linear output unit defines a nonlinear parameterized mapping from an input $\mathbf{x}$ to an output y given by the following relationship:

$$y = y(\mathbf{x}; \mathbf{w}) = \sum_{j=0}^{h}\left[w_j f\left(\sum_{i=0}^{d} w_{ji} x_i\right)\right] \tag{4.27}$$

The parameters of the NN are given by the weights and the biases that connect the layers between them, which governs the nonlinear mapping between the inputs and the outputs, denoted by the weight vector **w**.

The parameters **w** can be estimated during the phase called training or learning, where the NN is trained by using a dataset of n input and output samples, pairs of the form $D = \{\mathbf{x}_i, t_i\}_{i=1}^{n}$. Generally, we choose the cost function (4.26) as the optimization objective to be minimized. However, as for the noisy industrial data, we usually use the Bayesian regularization technique to eliminate the negative impact caused by the noise hidden in the data.

Given that the regression can be undertaken by a NN with an output $y(\mathbf{x}^*, \mathbf{w})$ depending on a new input and a set of model weights **w**, the conditional distribution $p(t^* | \mathbf{x}^*, D)$ can be written as the integral over these parameters

$$p(t^* | \mathbf{x}^*, D) = \int p(t^* | \mathbf{x}^*, \mathbf{w}) p(\mathbf{w} | D) d\mathbf{w} \tag{4.28}$$

To calculate the integral in (4.28), we firstly compute $p(t^* | \mathbf{x}^*, \mathbf{w})$ and $p(\mathbf{w} | D)$. Here, it is known that $p(t^* | \mathbf{x}^*, \mathbf{w})$ is the likelihood function that describes the difference between the observed target t^* and the output $y(\mathbf{x}^*, \mathbf{w})$ of NN with **w**.

$$p(t^* | \mathbf{x}^*, \mathbf{w}) = \left(\frac{\beta}{2\pi}\right)^{1/2} \exp\left(t^* - \frac{\beta}{2}\{y(\mathbf{x}^*, \mathbf{w})\}^2\right) \tag{4.29}$$

where β denotes the precision of the noise, and we have $1/\beta \equiv \sigma_t^2$. $p(\mathbf{w} | D)$ describes the posterior distribution of the weights and can be estimated by using the Bayes' rule. It can be formulated by

$$\begin{aligned} p(\mathbf{w} | D) &= \frac{1}{p(D)} p(D | \mathbf{w}) p(\mathbf{w}) \\ &= \frac{1}{Zs} \exp\left(-\frac{\beta}{2} E_{\mathrm{D}} - \frac{\alpha}{2} E_{\mathrm{W}}\right) \end{aligned} \tag{4.30}$$

where the term E_{D} is the contribution from the likelihood $p(D | \mathbf{w})$, which is evaluated by the training data and assumed that the data is independent. The second term E_{W} is the contribution from the prior over the weights.

The above statement is the abbreviated description of the principle of Bayesian regularization technique for parameters optimization. The detailed implementation procedure will be introduced in Chap. 7. Based on the Bayesian regularization, the parameters of a NN can be estimated and then its output is calculated.

4.3.2 T-S Fuzzy Model-Based Prediction

Takagi-Sugeno fuzzy rule-based model requires only a small number of rules to describe complex and nonlinear models [9, 10]. The number of rules could be significantly smaller than those when Mamdani fuzzy model type is applied [11]. A shortcoming of TS fuzzy model lies in its less understandable presentation when compared to Mamdani fuzzy modeling version. The basic idea of TS model is the fact that an arbitrarily complex system is a combination of mutually inter-linked subsystems. Let K regions, corresponding to individual subsystems, be determined in the state-space under consideration. The behavior of the system in these regions can then be described with simpler functional dependencies. If the dependence is linear and one rule is assigned to each subsystem, a TS fuzzy model can be represented by K rules in the following form.

$$\begin{aligned} R_i : &\text{If } x_1 \text{ is } A_{i,1} \text{ and } x_2 \text{ is } A_{i,2} \text{ and } \cdots \text{ and } x_n \text{ is } A_{i,n} \\ &\text{Then } y_i = \mathbf{a}_i \cdot \mathbf{x} + b_i, i = 1, 2, \ldots, K \end{aligned} \tag{4.31}$$

where R_i is the ith rule and $x_1, x_2, \ldots, x_n$ are the input variables, $A_{i,1}, A_{i,2}, \ldots, A_{i,n}$ are the fuzzy sets assigned to corresponding input variables, y_i represents the value of the ith rule output, and $\mathbf{a}_i$, b_i are parameters of the consequent function. The final output of the TS fuzzy model for an arbitrary input $\mathbf{x}_k$ can be calculated by

$$\widehat{y}_k = \frac{\sum_{i=1}^{K} [\beta_i(x_k) y_i(x_k)]}{\sum_{i=1}^{K} \beta_i(x_k)}, \quad k = 1, 2, \ldots, N \tag{4.32}$$

where $\beta_i(\mathbf{x}_k)$ represents the firing strength of the ith rule, and $y_i(\mathbf{x}_k)$ is output of ith rule for $\mathbf{x}_k$ input sample.

The training method of TS fuzzy model is usually of global nature since the parameters of model are determined on the basis of all the samples in one algorithmic step. Assuming that the available samples cover the complete modeled problem, any of the already cited algorithms can yield a model of high or satisfactory accuracy. However, a global model generated in this way cannot guarantee the representation of the actual system that is satisfactory under all circumstances. Regions may appear in the state-space, which are not represented with required quality of the generated model. Such a situation may arise due to some a-posterior changes in real system, due to incompleteness of the training set or the condition of too small number of rules. It is therefore desirable to correct the model parameters using new samples, obtained from the current operating regime of the system. In other words, it means that the correction of the parameters of the consequent part of the rule, that accounts for the rule firing strengths and the results of global learning, could influence the quality of the model output. Eq. (4.32) can be written in a form that shows more explicitly the concept of the system description using local models [12], i.e.,

$$\widehat{y}_k = \sum_{i=1}^{K} [w_i(\mathbf{x}_k)(\mathbf{x}_k \cdot \mathbf{a}_i + b_i)], \quad k = 1, 2, \ldots, N \tag{4.33}$$

where $w_i(\mathbf{x}_k)$ represents the normalized firing strength of ith rule for kth sample. Such strength given with (4.34) shows that (4.32) and (4.33) are identical for the final output

$$w_i(\mathbf{x}_k) = \beta_i(\mathbf{x}_k) / \sum_{j=1}^{K} \beta_i(\mathbf{x}_k), \quad k = 1, 2, \ldots, N \tag{4.34}$$

For predicting an industrial indicator, some improvement should be considered in the TS fuzzy model. Here, we will take the BFG system of steel plant as an example to illustrate the application of the TS fuzzy model on industrial prediction. According to the operational characteristics of the BFG system, i.e., at least one gasholder must be connected to the gas pipeline networks and two gasholders cannot be connected to the gas pipeline simultaneously, we consider that the level of gasholder at time $k + 1$ is determined by the gasholder level at time k and the gas flow of other users connected to the gas pipeline at k. Thus, to avoid the influence of switch of these two gasholders, the BFG system can be described by

$$y(k+1) = y(k) + \eta f(x_1(k), x_2(k), \ldots, x_m(k), \mathbf{a}(k)) \tag{4.35}$$

where f is the nonlinear function and η is a coefficient of proportionality. According to the detailed requirement of TS fuzzy model, such a formula can be rewritten as

$$\Delta y(k) = y(k+1) - y(k) = \eta f(x_1(k), x_2(k), \ldots, x_m(k), \mathbf{a}(k)) \tag{4.36}$$

Then, we can use a TS fuzzy model to describe the nonlinear system in (4.36). The detailed procedure can be implemented following from (4.31) to (4.34). The TS fuzzy rule-based model, as a set of local models, enables application of a linear least square (LS) method [13, 14]. Combining the training data and the prior distribution of parameters, the model can be realized by using the Bayesian linear regression technique, referenced in Chap. 7.

4.3.3 Multi-Kernels Least Square Support Vector Machine

In this subsection, we introduce a multi-kernel learning-based LSSVM regression algorithm, deducing the dual form of the parameters optimization, and discuss how to solve the multi-kernel learning problem by using the reduced gradient method. The multi-kernels LSSVM is built based on the single-kernel learning-based LSSVM, which is introduced and interpreted in Chap. 3.

Given a training dataset $\{\mathbf{x}_i, y_i\}_{i=1}^{N}$ and a set of kernel functions $\{k_k\}_{k=1}^{m}$, we formulate a multi-kernel learning-based LSSVM model for regression

$$y_i = \sum_{k=1}^{m} f_k(\mathbf{x}_i) + b \tag{4.37}$$

where $\mathbf{x}_i \in R^p$ is the ith input and $y_i \in R$ is its corresponding output.

According to the way of dealing with the single-kernel learning-based LSSVM, the regression task based on the multi-kernel LSSVM can be transformed into a convex optimization problem.

$$\min J\{f_k, b, e, d\} = \frac{1}{2}\sum_{k=1}^{m} \frac{1}{d_k}\|f_k\|_{H_k}^2 + C\sum_{i=1}^{N} e_i^2 \tag{4.38}$$

$$\begin{aligned} \text{s.t.} \quad y_i &= \sum_{k=1}^{m} f_k(x_i) + b + e_i \quad \forall i \\ \sum_{k=1}^{m} d_k &= 1, d_k \geq 0 \quad \forall k \end{aligned} \tag{4.39}$$

where $f(x) = \sum_{k=1}^{m} f_k + b$ is the regression function to be learned, and f_k is estimated from different Hilbert kernel space H_k. H_k has a relationship to the kernel k_k that could be the RBF kernel or the polynomial kernel. d_k denotes the weight of the kernel k_k, and satisfies the sparse equality constraints of the l_1 norm, which controls the two norm of the function f_k of the objective. e_i is the fitting error of the ith training sample and $\sum_{i=1}^{N} e_i^2$ can be used to describe the empiric risk. C is the penalty factor employed here to balance the empiric risk and complexity. Compared to the single-kernel-based LSSVM, the combination of multiple kernels is capable of representing the feature space of the samples and extracts the important information from samples. Therefore, multi-kernel-based regression function can perform a better explanatory ability for the training samples and can fit the regression problem more effectively.

To solve such optimization problem formulated in (4.38) and (4.39), we introduce the Lagrangian multiplier α^i, β, and η_k to transform the constraint optimization to be an unconstraint Lagrangian optimization.

$$\begin{aligned} L(\{f_k\}, b, e, d, \alpha, \lambda, \eta) &= \frac{1}{2}\sum_{k=1}^{m} \frac{1}{d}\|f_k\|_{H_k}^2 + C\sum_{i=1}^{N} e_i^2 \\ -\sum_{i=1}^{N} \alpha_i \left(\sum_{k=1}^{m} f_k(x_i) + b + e_i - y_i\right) &- \beta\left(1 - \sum_{k=1}^{m} d_k\right) - \sum_{k=1}^{m} \eta_k d_k \end{aligned} \tag{4.40}$$

The optimal solution is obtained based on the following conditions:

$$\begin{cases} \dfrac{\partial L}{\partial f_k} = 0 \rightarrow \dfrac{1}{d_k} f_k(\cdot) = \sum_{i=1}^{N} \alpha_i k_k(\cdot, \mathbf{x}_i), \forall k \\ \dfrac{\partial L}{\partial b} = 0 \rightarrow \sum_{i=1}^{N} \alpha_i = 0 \\ \dfrac{\partial L}{\partial e^i} = 0 \rightarrow \alpha_i = 2Ce^i \\ \dfrac{\partial L}{\partial d_k} = 0 \rightarrow -\dfrac{1}{2} \dfrac{\|f_k\|_{H_k}^2}{d_k^2} + \beta - \eta_k = 0 \end{cases} \tag{4.41}$$

Substituting the conditions described in (4.41) into the Lagrangian function, the dual representation of the multi-kernel learning LSSVM regression problem can be written as

$$\begin{aligned} & \max J(\alpha, \beta) = \sum_{i=1}^{N} \alpha_i y_i - \beta - \frac{1}{4C} \sum_{i=1}^{N} \alpha_i^2 \\ & \text{s.t.} \quad \sum_{i=1}^{N} \alpha_i = 0, \beta \geq \frac{1}{2} \sum_{i,j=1}^{N} \alpha_i k_k(\cdot, \mathbf{x}_i) \alpha_j^{\mathrm{T}} \end{aligned} \tag{4.42}$$

If we assume $\beta + \frac{1}{4C} \sum_{i=1}^{n} \alpha_i^2 = \lambda$, then the above dual representation can be rewritten as

$$\begin{aligned} & \max J(\alpha, \lambda) = \sum_{i=1}^{N} \alpha_i y_i - \lambda \\ & \text{s.t.} \quad \sum_{i=1}^{N} \alpha_i = 0, \lambda \geq \frac{1}{2} \alpha^{\mathrm{T}} \left(k_k + \frac{1}{2C} I \right) \alpha \quad \forall k \end{aligned} \tag{4.43}$$

Based on the minimum optimization of the original problem and its differentiability, we consider the following constraint optimization problem.

$$\begin{aligned} & \min J(d) \\ & \text{s.t.} \quad \sum_{k=1}^{m} d_k = 1, d_k \geq 0 \end{aligned} \tag{4.44}$$

where

$$J(d) = \begin{cases} \min J\{f_k, b, e, d\} = \dfrac{1}{2} \sum_{k=1}^{m} \dfrac{1}{d_k} \|f_k\|_{H_k}^2 + C \sum_{i=1}^{N} e_i^2 \\ \text{s.t.} \quad y_i = \sum_{k=1}^{m} f_k(\mathbf{x}_i) + b + e_i \quad \forall i \end{cases} \tag{4.45}$$

From (4.44) and (4.45), one can see that the optimization problem of the multi-kernel learning-based LSSVM aims to solve a nonlinear function optimization problem with simple constraints. Thus, we use the general methods for the constraint optimization problem to solve the optimization described in (4.44), such as the reduced gradient method. Noticeable, the problem described in (4.46) is the first three terms in the Lagrangian function in (4.40), the corresponding dual problem of (4.44) and (4.45) based on the conditions in (4.41) can be written as

$$\begin{aligned} \max \ & J(\alpha) = -\frac{1}{2}\alpha\left(\sum_{k=1}^{m} d_k k_k + \frac{1}{2C}I\right)\alpha^{\mathrm{T}} + \sum_{i=1}^{N}\alpha_i y_i \\ \text{s.t.} \ & \sum_{i=1}^{N}\alpha_i = 0 \end{aligned} \tag{4.46}$$

It can be found that the objective function $J(d)$ defined by (4.44) is in fact the objective function defined by (4.45). Based on the strong duality, the objective function $J(d)$ is equal to the objective function defined by (4.46). As for any kernel weights satisfying the constraints, the dual problem of (4.46) is strictly concave, which determines the uniqueness of solution α^*. Combining the differentiability of the objective function and the uniqueness of the solution α^*, $J(d)$ should be differential. Thus, the gradient of $J(d)$ with respect to the weights of kernel can easily be calculated.

$$\frac{\partial J}{\partial d_k} = -\frac{1}{2}\alpha^* k_k \alpha^{*\mathrm{T}} \quad \forall k \tag{4.47}$$

According to the convex feature and the differentiability of the objective function under the Lipschitz gradient, the reduced gradients of the objective function are listed as given below:

$$\begin{cases} r_k = \dfrac{\partial J}{\partial d_k} - \dfrac{\partial J}{\partial d_u} & \forall k \neq u \\ r_u = \sum\limits_{k \neq u}\left(\dfrac{\partial J}{\partial d_k} - \dfrac{\partial J}{\partial d_u}\right) & u = \underset{\{k|d_k \neq 0\}}{\operatorname{argmax}}\,(d_k) \end{cases} \tag{4.48}$$

To satisfy the constraints of the weights d and guarantee the variables can be convergent to the KKT point, the feasible directions of the update of the weights are [15]

$$D_k = \begin{cases} -d_k r_k & \forall k \neq u, \ \text{if} \ r_k > 0 \\ -r_k & \forall k \neq u, \ \text{if} \ r_k \leq 0 \\ r_u & \text{for} \ k = u \end{cases} \tag{4.49}$$

Given a step s, the update equations of weights d can be written as

$$d \leftarrow d + sD \tag{4.50}$$

Based on the characteristics of the convex optimization, the convergence of the multi-kernel learning-based LSSVM regression can be evaluated by the duality gap. Considering the original problem of (4.43) and the dual problem of (4.46), the duality gap can be described as

$$\text{DualGap} = J(d^*) - \sum_{i=1}^{n} \alpha_i^* y_i + \frac{1}{2} \max_k \left(\alpha^* \left(K_k + \frac{1}{2C} I \right) \alpha^{*\mathrm{T}} \right) \tag{4.51}$$

where d^* and α^* are the optimal original variable and the optimal dual variable, respectively. $J(d^*)$ can be obtained by the single-kernel learning LSSVM regression. If $J(d^*)$ is replaced by the objective function of the dual problem of (4.46), the above duality gap can be rewritten as

$$\text{DualGap} = -\frac{1}{2} \alpha^* \left(\sum_{k=1}^{m} d_k^* k_k + \frac{1}{2C} I \right) \alpha^{*\mathrm{T}} + \frac{1}{2} \max_k \left(\alpha^* \left(k_k + \frac{1}{2C} I \right) \alpha^{*\mathrm{T}} \right) \tag{4.52}$$

Given the above reduced gradient and the update of weights and duality gap, the detailed implementation steps of the reduced gradient-based multi-kernel learning LSSVM are listed as follows. The entire process contains two loops. In the outside loop, the reduced gradient is used to solve (4.44) and update the weights to obtain the optimal combination of kernels. In the inside loop, based on the optimal combination of the kernels, the single-kernel learning LSSVM solves (4.45) to obtain the optimal solution α^*. The objective function defined by (4.44) is also updated.

Algorithm 4.1: The Reduced Gradient-Based Multi-Kernel Learning LSSVM

Step 1: Set the number of the kernels m and initialize the weight of each kernel $d_k = 1/m$;

Step 2: Use the single-kernel learning algorithm to solve the multi-kernel $K = \sum_{k=1}^{m} d_k K_k$ *learning based one.*

Step 3: Compute the objective function defined by (4.46) *based on the optimal solution* α^*.

Step 4: Compute the gradient of J(d) with respect to the weights. Compute the reduced gradient of J(d) at d_k *by using* (4.48). *Compute the feasible directs* D_k *of the update of the weights by using* (4.49).

Step 5: Set a max step $s_{\max} = -\frac{d_v}{D_v} \left\{ v = \operatorname*{argmin}_{\{k|D_k<0\}} -\frac{d_k}{D_k} \right\}$.

Step 6: Update the weights of different kernels d_k *by using* (4.50). *Based on the updated weights, update the combination of K.*

Step 7: Repeat the Step 2 and 3 to obtain the new objective function.

Step 8: Set $d = d_{\text{new}}$, $D_u = D_u - D_v$, $D_v = 0$, *repeat the steps from 6 to 8 until the value of objective function will not decrease.*

Step 9: Use Armijo to conduct the linear search $s^* = \underset{\{s \in [0, s_{\max}]\}}{\text{argmin}} J(d + sD)$ *to obtain the optimal step s*, and update the weights d based on s*.*

Step 10: If the duality gap between the original problem and the dual one is inferior to the given threshold, or the iterative steps reach to the predefined value, the learning process stops; otherwise, the learning process returns to Step 2.

From the above solving process, it is unnecessary to compute the gradient of objective function with respect to weights after the update of weights. In the inside loop, the objective function is computed by using the single-kernel learning LSSVM, which increases the velocity of linear search. In total, the computational cost of the above learning process is also greatly reduced. Besides the advantage of the computational cost, compared to the SVM regression, in the inside loop the single-kernel learning LSSVM regression is adopted by using a set of linear equations, which simplifies the solving process of the complex quadratic programming (QP) problem. The computational process that can obtain a better objective is much simpler and accurate. The feasible direction based on the better objective and the reduced gradient make the weights d convergent to the KKT point of the problem more quickly, in which only a few kernels are with large weights and the others are with minor ones and even zero. The combined kernel enriches the feature space of samples. Based on the reduced gradient-based multi-kernel learning LSSVM, more effective information can be extracted to build a regression model, which can also be used to estimate or predict the nonlinear problems with non-flat variation accurately.

4.3.4 Case Study

The prediction of the gasholder levels, related to the BFG system in steel industry that are presented in Chap. 3 (see Fig. 3.6), is very significant for the workers to schedule the production. This case study verifies the effectiveness of the multi-kernel LSSVM prediction method for gasholder level. In BFG system, only one gasholder is to store the gas in most cases; therefore, we only consider the prediction of a single gasholder level. Table 4.3 lists the related modeling parameters, where the time delay indicates the delay reflecting the influence of the inputs on holder level, and there are 29 inputs for predicting the gasholder level, containing four gas production flows, 12 gas consumption flows, 12 gas adjustment flows as well as the gasholder levels of previous time. The sampling frequency is set as 1 min. Since the variation

Table 4.3 Modeling parameters for BFG gasholder level prediction model

Gas	Input dimension	Input time delay (min)	Kernel function
BFG	29	1	$a =$ RBF $+ b =$ Polynomial ($a = b = 0.5$)

of the BFG data is with high frequency, the number of samples of BFG is up to 3500 covering various conditions as much as possible. The kernel function of the prediction model is determined by a large amount of experiments. The multiple kernel functions composed of RBF and polynomial are used for BFG system. This case study uses the genetic algorithm (GA) to optimize the parameters C and σ, of which their ranges is [1,1000] and [0.1, 1000], respectively; the degree of polynomial kernel function is set as 2; the population size of the GA is 30; and the termination condition is chosen as the MAE less than 1.5 km^3.

To demonstrate the advantages of the multi-kernel LSSVM method, a comparison result using the practical data is illustrated. The experiments consist of three classes of tested changing tendency of gasholder level, ascending, stationary, and descending with the prediction time of 30 min. In this subsection, four approaches are used including manual reasoning (MR), BP network [16], standard LSSVM with tenfold-cross-validation (10fc-LSSVM), and the proposed multi-kernel learning method. Note that the MR (see [17]) was widely used in literature. The BP network was commonly adopted to solve the practical regression before LSSVM was reported. The 10fc-LSSVM is often adopted for modeling many practical problems. The average results based on 50 groups of testing samples randomly selected are

Table 4.4 Comparison of BFG gasholder level prediction results with four models

Model	Gasholder	Tendency	MAE	UC	CT(s)
Manual reasoning	#1 BFG	Ascending	10.9928	2.0011	10.7472
		Stationary	9.5447	1.5368	
		Descending	12.6753	3.5775	
	#2 BFG	Ascending	11.2511	1.9031	
		Stationary	8.1723	2.1516	
		Descending	17.0247	1.3235	
BP network	#1 BFG	Ascending	8.6060	1.2142	116.1070
		Stationary	5.0002	1.3540	
		Descending	6.1905	1.2201	
	#2 BFG	Ascending	6.2162	1.1334	
		Stationary	6.3137	1.2134	
		Descending	9.5645	1.1010	
10kc-LSSVM	#1 BFG	Ascending	4.5571	1.0142	327.425
		Stationary	3.1226	1.0971	
		Descending	3.7085	1.0434	
	#2 BFG	Ascending	4.3785	1.0318	
		Stationary	2.7310	1.0415	
		Descending	4.1506	1.0160	
The multi-kernel learning	#1 BFG	Ascending	1.4123	1.0016	60.0219
		Stationary	1.3831	1.0184	
		Descending	0.8423	1.0030	
	#2 BFG	Ascending	0.9518	1.0005	
		Stationary	0.6286	1.0036	
		Descending	1.0353	1.0004	

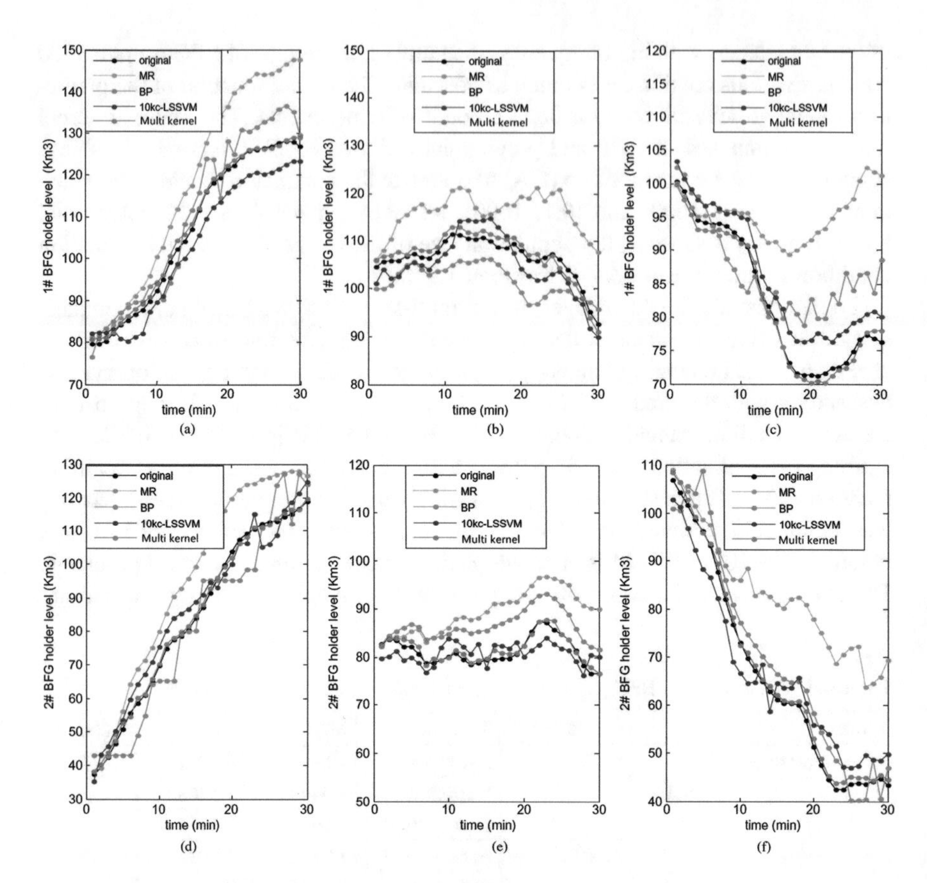

Fig. 4.3 #1 & #2 BFG gasholder level prediction result. (**a**) Ascending tendency of #1 BFG holder. (**b**) Stable tendency of #1 BFG holder. (**c**) Descending tendency of #1 BFG holder. (**d**) Ascending tendency of #2 BFG holder. (**e**) Stable tendency of #2 BFG holder. (**f**) Descending tendency of #2 BFG holder

listed in Table 4.4, and the graphical illustration comparison of a certain group of samples is shown in Fig. 4.3.

In Table 4.4 and Fig. 4.3, it is obvious that the multi-kernel model predicts the three-tendency of BFG gasholder level with best performance compared to MR, BP network, and 10fc-LSSVM. It is difficult for MR to establish a reasonable model when many practical influence factors are ignored. With respect to BP network, a model effectively reflecting the practical gasholder level with high complexity and nonlinearity cannot be established because the weight coefficient computed by gradient descent could usually fall into local optimum. And based on the empirical minimization principle, the generalization performance of BP-based network is very poor. However, considering the influence of gas production and consumption on gasholder level, the latter two methods based on structure risk minimization principle can transform the nonlinearity to linearity in high dimension by a mapping function, and make a compromise between the model complexity and the generalization to avoid the over-fit and under-fit of gasholder level with parameter optimization. Therefore, the latter two ones exhibit better performance than the former ones. Nevertheless, 10fc-LSSVM usually spends a lot of computation time since the kernel matrix is composed of the various training samples in each validation, and its inverse need to be computed repeatedly.

In view of the performance statistic (see Table 4.4) for COG and BFG prediction, the multi-kernel model has less MAE, uncorrelated coefficient (UC), and computing time (CT) compared to the 10fc-LSSVM, and such results can be obtained within 1 min. The differences between the predicted and the practical level are all less than 1.5 km^3 (the required MAE). Although 10fc-LSSVM seems to have a similar performance sometimes, it is difficult to meet the requirement of real time and high accuracy. After all, it spends a lot of time for the model training (about 5 min). The MAE by BP network is less than manual reasoning, but larger than 10fc-LSSVM and the multi-kernel model. And the fluctuant prediction results with larger UC cannot track the gasholder level's changing tendency, especially the ascending and descending situation, shown in Fig. 4.3. The computational time is up to 2 min.

4.4 Factor-Based Multi-Output Model

4.4.1 Multi-Output Least Square Support Vector Machine

In industrial application, due to the situations that the two gasholders are always interacted by each other, we added a combined fitting error to the prediction model in order to evaluate the effect between these multiple outputs. Thus, a multi-output LSSVM model is constructed in this section. Given a training set denoted as $S = \{\mathbf{x}_i, \mathbf{y}_i\}_{i=1}^{n}$, where $\mathbf{x}_i \in \mathbb{R}^p$ is the input vector that affects the holder levels, $\mathbf{y}_i \in \mathbb{R}^{n_y}$ is the corresponding holder level vector with dimensionality n_y. As for the value of p, it can be set as 3 here, considering the gas flow difference between the generation and consumption amount at current point, and the two holder levels at previous time point, respectively. As such, the multi-output LSSVM can be formulated by

$$\mathbf{F}(\mathbf{X}) = \mathbf{W}^{\mathrm{T}}\mathbf{\Phi}(\mathbf{X}) + \mathbf{B} \tag{4.53}$$

where the nonlinear mapping $\mathbf{\Phi}(\mathbf{X})$ converts the input data into a high-dimensional space. Compared to the single-output model, the data of the variables in this model are in the form of matrix. And, it is very crucial to determine their dimensionalities.

Enlightened by the computation of single-output LSSVM, we similarly discuss an optimization problem with constraints of the multi-output one, where the errors come from not only each single output, but also the combined fitting error. The optimization problem can be described as

$$\begin{aligned} \min\ & J\left(w^j, b^j, e^j, E_i\right) = \frac{1}{2}\sum_{j=1}^{n_y}\left(w^j\right)^{\mathrm{T}}\left(w^j\right) + \frac{1}{2}\sum_{j=1}^{n_y}\sum_{i=1}^{n}\gamma^j\left(e^j\right)^2 + \gamma^0\sum_{i=1}^{n}E_i \\ \text{s.t.}\ & y^j = \left(w^j\right)^{\mathrm{T}}\varphi(\mathbf{X}) + b^j + e^j, \quad j = 1, 2, \ldots, n_y \end{aligned} \tag{4.54}$$

$$\mathbf{E}_i = \left\|\mathbf{Y}_i - \mathbf{W}_i\boldsymbol{\Phi}(\mathbf{X}_i) - \mathbf{B}\right\|^2, \quad i = 1, 2, \ldots, n \tag{4.55}$$

where $\mathbf{W}_i$ is a diagonal matrix denoted as $\operatorname{diag}\left(w_i^1, w_i^2, \ldots, w_i^{n_y}\right)$, $\mathbf{Y}_i = \left[y_i^1, y_i^2, \ldots, y_i^{n_y}\right]^{\mathrm{T}}$, $\mathbf{B} = \left[b^1, b^2, \ldots, b^{n_y}\right]^{\mathrm{T}}$, and $\boldsymbol{\Phi}(X_i) = [\varphi(X_i), \varphi(X_i), \ldots, \varphi(X_i)]^{\mathrm{T}}$. w^j and b^j are the weight and bias, e^j is the corresponding single fitting error, and γ^0, larger than zero, is the penalty coefficient for the combined fitting errors.

The structure of this model is of vital importance. It can be noticed that [18] and [19] considered the single fitting error presented as (4.55). As for the two holders of the LDG system, the combined fitting error for the multi-outputs should be further considered to make this model more practical. With the introduction of Lagrangian multipliers, the established optimization model can be turned into a constraints-free problem.

$$\begin{aligned} L\left(w^j, b^j, e^j, E_i, \lambda_i, \mu_i^j\right) = & \frac{1}{2}\sum_{j=1}^{n_y}\left(w^j\right)^{\mathrm{T}}\left(w^j\right) + \gamma^0\sum_{i=1}^{n}\mathbf{E}_i \\ & + \frac{1}{2}\sum_{j=1}^{n_y}\sum_{i=1}^{n}\gamma^j\left(e_i^j\right)^2 - \sum_{i=1}^{n}\lambda_i\left(\mathbf{E}_i - \mathbf{Y}_i - \mathbf{W}_i\boldsymbol{\Phi}(\mathbf{X}_i) - \mathbf{B}^2\right) \\ & - \sum_{j=1}^{n_y}\sum_{i=1}^{n}\mu_i^j\left(\left(w_i^j\right)^{\mathrm{T}}\varphi(\mathrm{X}_i) + b^j + e^j + y_i^j\right) \end{aligned} \tag{4.56}$$

A similar situation had been reported in [20], where calculating a pseudo matrix inverse might result in the unsatisfactory solution. Here, the KKT conditions are first introduced [21], and the partial derivatives of the variables are calculated as follows:

$$\begin{cases} \dfrac{\partial L}{\partial w^j} = 0 \rightarrow w^j - 2\boldsymbol{\varphi}^{\mathrm{T}}\mathbf{D}_\lambda\left[y^j - w^j\varphi - \vec{1}^{\mathrm{T}}b^j\right] - \boldsymbol{\varphi}^{\mathrm{T}}\boldsymbol{\mu}^j = 0 \\ \dfrac{\partial L}{\partial b^j} = 0 \rightarrow -2\Lambda^{\mathrm{T}}\left[y^j - w^j\varphi - \vec{1}^{\mathrm{T}}b^j\right] - \vec{1}\,\boldsymbol{\mu}^j = 0 \\ \dfrac{\partial L}{\partial E_i} = 0 \rightarrow \gamma^0 - \lambda_i = 0 \\ \dfrac{\partial L}{\partial e^j} = 0 \rightarrow \gamma^j\vec{1}\,e^j - \vec{1}\,\boldsymbol{\mu}^j = 0 \\ \dfrac{\partial L}{\partial \lambda_i} = 0 \rightarrow \left\|\mathbf{Y}_i - \mathbf{W}_i\boldsymbol{\Phi}(\mathbf{X}_i) - \mathbf{B}\right\|^2 = \mathbf{E}_i \\ \dfrac{\partial L}{\partial \mu^j} = 0 \rightarrow \vec{\mathbf{1}}\,\mathbf{y}^j = \mathbf{w}^j\boldsymbol{\varphi} + n_y b^j + \vec{\mathbf{1}}\,\mathbf{e}^j \end{cases} \tag{4.57}$$

where the row vector $\vec{\mathbf{1}} = [11\cdots1]_{1\times n}$, the column vector $\boldsymbol{\Lambda} = [\lambda_1, \lambda_2, \ldots, \lambda_n]^{\mathrm{T}}$, and the diagonal matrix $D_\lambda = \operatorname{diag}(\lambda_1, \lambda_2, \ldots, \lambda_n)$. Considering the undetermined expression of the nonlinear mapping, this study transforms w into a mapping combination on high-dimension space by using the Representer Theorem [22], i.e.,

$$w^j = \sum_{i=1}^{n} \alpha_i^j \varphi(X_i') = \boldsymbol{\varphi}^{\mathrm{T}} \boldsymbol{\alpha}^j \tag{4.58}$$

And introducing the kernel function with dimensionality $n \times n$, $K^j(x_i, x_k) = \boldsymbol{\varphi}(x_i)\boldsymbol{\varphi}(x_k)^{\mathrm{T}}$.

Thus, combining (4.57) and (4.58), a full-ranked equation set is obtained to solve α, b, and μ

$$\begin{bmatrix} \mathbf{I} + 2\mathbf{D}_\lambda & 2\mathbf{D}_\lambda \vec{\mathbf{1}} & \mathbf{I} \\ -2\lambda^{\mathrm{T}}\mathbf{K} & -2\vec{\mathbf{1}}\lambda & \vec{\mathbf{1}}(\gamma^j)^{-1}\vec{\mathbf{1}} \\ \mathbf{K} & n_y \vec{\mathbf{1}}^T & \end{bmatrix} \begin{bmatrix} \boldsymbol{\alpha}^j \\ \mathbf{b}^j \\ \boldsymbol{\mu}^j \end{bmatrix} = \begin{bmatrix} 2\mathbf{D}_\lambda \mathbf{y}^j \\ -2\lambda^{\mathrm{T}}\mathbf{y}^j \\ \mathbf{y}^j \end{bmatrix} \tag{4.59}$$

Since the above formula can be viewed as the form of $\mathbf{AX} = \mathbf{B}$ and $\mathbf{A}$ is a invertible matrix, one can calculate (4.59) as $\mathbf{X} = \mathbf{A}^{-1}\mathbf{B}$. The regression function of the holder levels prediction based on the multi-output LSSVM can be formulated as

$$y^j = \sum_{i=1}^{n} \alpha_i^j \mathbf{K}(x, x_i) + b^j, \quad j = 1, 2, \ldots, n_y \tag{4.60}$$

where j is the number of outputs. As such, the generation-consumption gas flow difference on current time point and the holder levels on previous time are taken as the inputs, and the holder levels on current time point as the outputs. Repeating the above process, one can iteratively obtain the predicted holders levels in the future.

The parameters selection plays a crucial role on the whole model solving [23–25]. The parameters of the above model are primarily divided into two classes, the widths of Gaussian kernel on the corresponding outputs, and the penalty coefficients of the errors. In literature, a number of existing approaches were used, such as cross-validation with grid search [26] and Bayesian networks optimization [27]. However, those approaches exhibited a series of drawbacks. For example, cross-validation expensed a long period of time on calculating inverse matrices, which usually brought about a high computing cost when facing with a high real-time demand. And, grid search is usually applicable to optimize the parameters with one or two dimensionalities, but unavailable to the one more than three parameters. With respect to Bayesian networks optimization, it usually experiences a complicated solving process. When a gradient descent algorithm was applied to such parameter selection [28], a rather complex calculation for the partial derivatives would occur, which make the solving more difficult because of the consideration of combined fitting errors.

Based on the analysis above and the parameters amount regarding the established model, we would like to try the meta-heuristic approach such as a class of intelligence algorithms [29–31]. Particle swarm optimization, an evolutionary computation method of bionics [32], successfully overcame the drawbacks of generic algorithm (GA) when dealing with the optimization with continuous search scope [33, 34]. In this section, we described the PSO briefly, and we will detail it in Chap. 7. PSO has a quick convergence rate and good accuracy [35], and its iteration formula can be expressed as

$$\begin{aligned} v_{i,d}(t+1) = \omega v_{i,d}(t) + c_1 \cdot \mathrm{rand}(1)(P_{i,d} - x_{i,d}(t)) \\ + c_2 \cdot \mathrm{rand}(2)\left(P_{g,d} - x_{i,d}(t)\right) \end{aligned} \tag{4.61}$$

$$x_{i,d}(t+1) = x_{i,d}(t) + v_{i,d}(t+1) \tag{4.62}$$

where $x_{i,\ d}$, $v_{i,\ d}$, $P_{i,\ d}$, and $P_{g,\ d}$ refer to the current location, the current speed, the best location in history of the particle in the searching space, and the best location of the particles, respectively. The inertia weight ω controls the influence from the previous speed to the current one, c_1 and c_2 are the learning factors, and rand(1) and rand (2) denote the random values in the range of [0, 1].

4.4.2 Case Study

To verify the effectiveness of the multi-output LSSVM and its solving approach, the real data from the LDG system presented in Fig. 3.18 are randomly selected. From Fig. 3.18, one can see that two subsystems are involved in the whole LDG system, and each of them consists of two gasholders.

The time interval of sampling is 1 min, which can be viewed as the sampling frequency in the SCADA system. While the time interval of triggering the predictor is determined as 5 min, which is the prediction frequency in the application system. Furthermore, we select the prediction horizon as 30 min since the current scheduling workers usually pay their attentions to the gas flow circumstance of future 30 min. Regarding the amount of data points used as the model inputs, the system adopts the previous 30 data points (embedded dimension) for the real-time prediction.

With a long-term on-site simulation, it is discovered that one of the two holders in a subsystem is always with a status of off-line, which means it temporarily stops recycling the gas when #1 or #2 valve in Fig. 3.18 is turned off. Such situations do have a great impact on the prediction of the two holder levels because of the interaction from each other, and this is also the reason why the combined fitting error in the model is the main concern. To indicate the adaptability of the multi-output prediction method, several scenarios that cover all the situations in practice are summarized and further studied, respectively. Scenario 1, one holder normally runs while another is transforming from off-line to online; Scenario 2, one holder normally runs while another is transforming from online to off-line; Scenario 3, one holder normally runs while another is with the status of off-line; and Scenario 4 both holders normally run.

This subsection carries out a series of comparative experiments, including single-out SVM (S-SVM), the single-output LSSVM (S-LSSVM), the multi-kernel learning LSSVM (M-LSSVM [20]) which is introduced in Sect. 4.3 and the multi-output one. And, all the selected parameters of these approaches are optimized by the PSO. For accelerating the searching process, we experimentally narrow the boundaries of the parameters, where the penalty coefficients are initially set as an integer in [1,500] since integer is sensible enough to the predicted results. The widths of kernels are initially set as a decimal ranged in [0.10–1.50]. In the experiments, the MAPE and the RMSE are adopted as the evaluation indexes of accuracy.

Scenario 1: If #1 holder keeps online while #2 holder transforms from off-line into online, i.e., #1 valve keeps on while #2 valve turns from off to on. This scenario occurs frequently in practice, which gives a large impact on the online holder because of the connection between the two holders. The comparative prediction results, using the mentioned methods, are illustrated in Fig. 4.4. Similar results are illustrated in Fig. 4.5. These results indicate that the multi-output LSSVM obviously performs better than the other methods for such scenario. As for the prediction accuracy, although the S-SVM predicts a rough trend of the holder levels, the

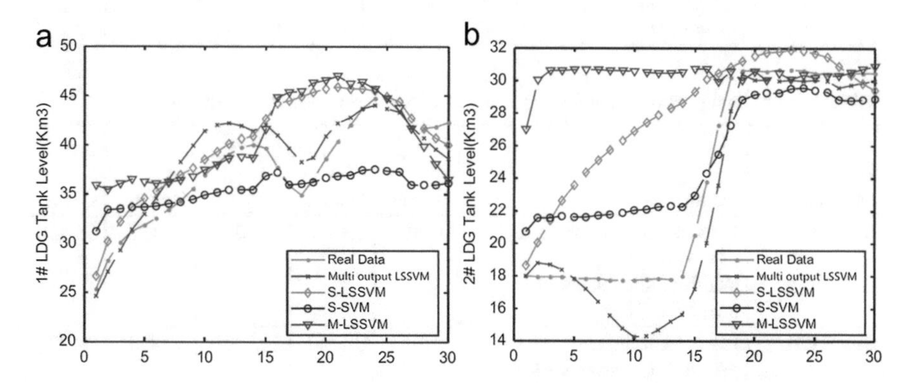

Fig. 4.4 Prediction results of Scenario 1 for #1 LDG subsystem. (**a**) #1 LDG Holder level and (**b**) #2 LDG Holder level

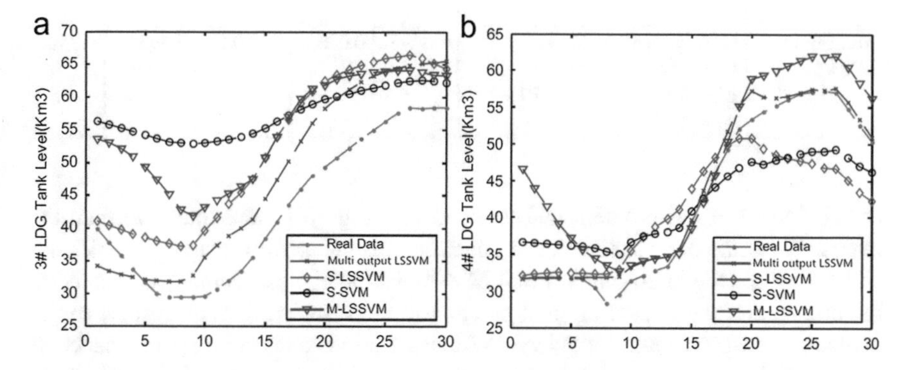

Fig. 4.5 Prediction results of Scenario 1 for #2 LDG subsystem. (**a**) #3 LDG Holder level and (**b**) #4 LDG Holder level.

Table 4.5 Optimal parameters and error statistics of Scenario 1 for LDG subsystem 1

Predict methods	Parameters	MAPE #1 Holder	MAPE #2 Holder	RMSE #1 Holder	RMSE #2 Holder	CT(s)
S-SVM	Loss function: ε-insensitive, insensitivity: 0.1 Width of Gaussian kernel: $\sigma = 50$	0.0882	0.0986	4.7688	3.2672	29.063
S-LSSVM	Penalty coefficient: $\gamma = 98$ Width of Gaussian kernel: $\sigma = 90$	0.0655	0.1377	3.9384	5.6121	0.367
M-LSSVM	Penalty coefficients: $\gamma_0 = 198$ $\gamma_1 = 289$ $\gamma_2 = 99$ Kernel widths: $\sigma_1 = 1.04$ $\sigma_2 = 0.97$	0.0891	0.2149	5.1933	8.8400	0.259
Multi-output LSSVM	Penalty coefficients: $\gamma_0 = 201$ $\gamma_1 = 298$ $\gamma_2 = 105$ Kernel widths: $\sigma_1 = 1.05$ $\sigma_2 = 0.97$	0.0336	0.0308	2.0558	1.4150	0.223

Table 4.6 Optimal parameters and error statistics of Scenario 1 for LDG subsystem 2

Predict methods	Parameters	MAPE #3 Holder	MAPE #4 Holder	RMSE #3 Holder	RMSE #4 Holder	CT(s)
S-SVM	Loss function: ε-insensitive, insensitivity: 0.1 Width of Gaussian kernel: $\sigma = 47$	0.2395	0.1020	16.2318	5.6187	35.258
S-LSSVM	Penalty coefficient: $\gamma = 102$ Width of Gaussian kernel: $\sigma = 85$	0.2647	0.1142	10.8635	6.1853	0.422
M-LSSVM	Penalty coefficients: $\gamma_0 = 119$ $\gamma_1 = 388$ $\gamma_2 = 106$ Kernel widths: $\sigma_1 = 1.01$ $\sigma_2 = 1.01$	0.2057	0.0788	12.5338	5.5646	0.369
Multi-output LSSVM	Penalty coefficients: $\gamma_0 = 115$ $\gamma_1 = 392$ $\gamma_2 = 105$ Kernel widths: $\sigma_1 = 1.01$ $\sigma_2 = 1.01$	0.0406	0.0232	4.3108	2.3935	0.352

calculation for a QP problem makes the computing cost more intensive than the others. For some clear statistics, the quantified results are also listed as Tables 4.5 and 4.6, where the multi-output LSSVM exhibits the highest accuracy compared to the others, and the related optimal parameters are also listed. With respect to the computing cost, the multi-output LSSVM also exhibits outstanding performance. It is apparently shown in the tables that outstanding performance. It is apparently shown in the tables that the generic SVM consumed the longest CT. Although it is

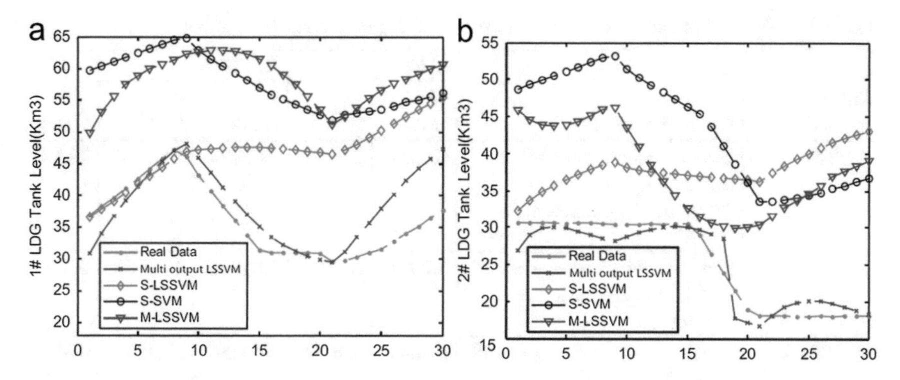

Fig. 4.6 Prediction results of Scenario 2 for #1 LDG subsystem. (**a**) #1 LDG Holder level and (**b**) #2 LDG Holder level

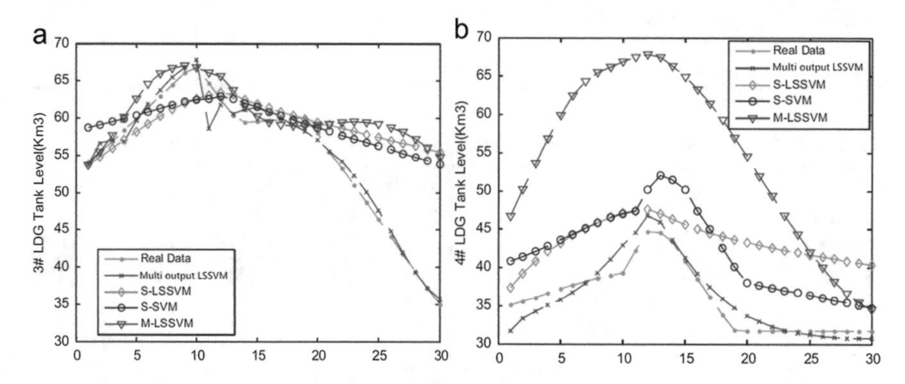

Fig. 4.7 Prediction results of Scenario 2 for #2 LDG subsystem. (**a**) #3 LDG Holder level and (**b**) #4 LDG Holder level

somewhat hard to clarify the remarkable difference on CT for the other three methods, the multi-output one also perform best on these comprehensive evaluation indices.

Scenario 2: This situation is somewhat analogous with Scenario 1, and it concerns the transformation from online to off-line by one of the two holders in a subsystem. As Figs. 4.6 and 4.7 shown below, we randomly chose holders #2 and #4 to be the one which includes process of transition. Figs. 4.6 and 4.7 and Table 4.7 and 4.8 clearly demonstrate the outstanding performance of the multi-output method. The above two scenarios can cover the practical circumstances that the holders status transit.

Scenario 3: For saving the energy and production cost, a class of situation that one of the two holders keeping off-line with relatively low level is often arranged in practice. Without loss of generality, one can assume #1 holder keeps online while #2 holder is off-line, viz., #1 valve keeps on while #2 valve keeps off. The comparative results are illustrated as Fig. 4.8 by using these approaches. Similarly, such situation

Table 4.7 Optimal parameters and error statistics of Scenario 2 for LDG subsystem #1

Predict methods	Parameters	MAPE #1 Holder	MAPE #2 Holder	RMSE #1 Holder	RMSE #2 Holder	CT(s)
S-SVM	Loss function:ε-insensitive, insensitivity: 0.1 Width of Gaussian kernel: $\sigma = 57$	0.4552	0.5957	21.5774	18.3494	28.105
S-LSSVM	Penalty coefficient: $\gamma = 95$ Width of Gaussian kernel: $\sigma = 93$	0.2331	0.4103	13.2846	14.5968	0.449
M-LSSVM	Penalty coefficients: $\gamma_0 = 188$ $\gamma_1 = 291$ $\gamma_2 = 105$ Kernel widths: $\sigma_1 = 1.04$ $\sigma_2 = 0.91$	0.4641	0.4074	22.4456	13.4832	0.374
Multi-output LSSVM	Penalty coefficients: $\gamma_0 = 70$ $\gamma_1 = 150$ $\gamma_2 = 100$ Kernel widths: $\sigma_1 = 1.01$ $\sigma_2 = 0.95$	0.0635	0.0756	4.5077	6.7854	0.356

Table 4.8 Optimal parameters and error statistics of Scenario 2 for LDG subsystem #2

Predict methods	Parameters	MAPE #3 Holder	MAPE #4 Holder	RMSE #3 Holder	RMSE #4 Holder	CT(s)
S-SVM	Loss function: ε-insensitive, insensitivity: 0.1 Width of Gaussian kernel: $\sigma = 55$	0.0931	0.1407	9.1552	8.5049	28.466
S-LSSVM	Penalty coefficient: $\gamma = 89$ Width of Gaussian kernel: $\sigma = 93$	0.0885	0.1770	9.9176	9.9896	0.372
M-LSSVM	Penalty coefficients: $\gamma_0 = 123$ $\gamma_1 = 338$ $\gamma_2 = 126$ Kernel widths: $\sigma_1 = 1.05$ $\sigma_2 = 1.01$	0.0981	0.4147	9.0955	20.0342	0.289
Multi-output LSSVM	Penalty coefficients: $\gamma_0 = 72$ $\gamma_1 = 298$ $\gamma_2 = 251$ Kernel widths: $\sigma_1 = 1.02$ $\sigma_2 = 0.98$	0.0450	0.0275	4.0019	1.7515	0.247

occurs in #2 subsystem where #3 holder keeps online (#1 valve keeps on) and holder #4 is off-line (#2 valve keeps off), illustrated as Fig. 4.9. Moreover, the quantified statistics are listed in Tables 4.9 and 4.10. In the comparisons, it can be easily figured out that as for such scenario the multi-output model also exhibits excellent prediction results. In terms of computing time, the multi-output method requires relatively less time cost and assures better accuracy on RMSE and MAPE than the other methods.

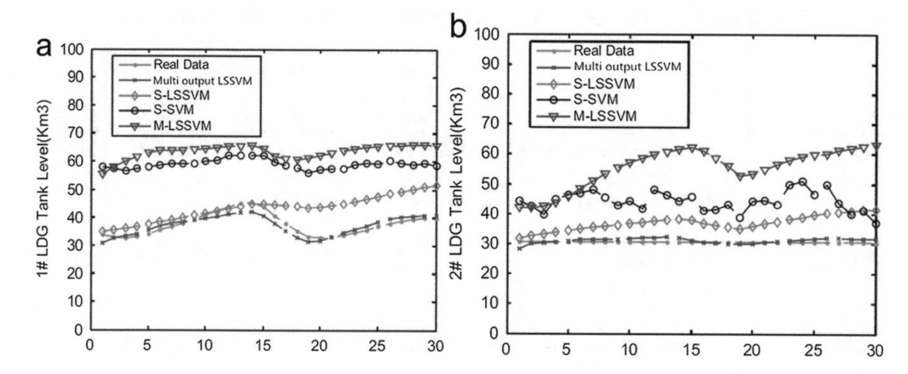

Fig. 4.8 Prediction results of Scenario 3 for #1 LDG subsystem. (**a**) #1 LDG Holder level and (**b**) #2 LDG Holder level

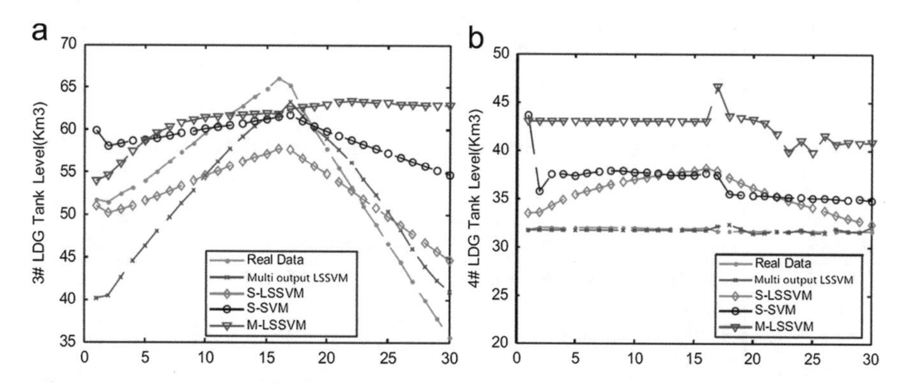

Fig. 4.9 Prediction results of Scenario 3 for #2 LDG subsystem. (**a**) #3 LDG Holder level and (**b**) #4 LDG Holder level

Scenario 4: At last, one can consider a class of situation that does not frequently appear, where both holders in a subsystem are in the status of online, i.e., both connected valves keep on. The results are comparatively illustrated as Figs. 4.10 and 4.11, and the quantified statistics are listed in Tables 4.11 and 4.12. It is clear that the proposed method performs best both in the prediction accuracy and the computing time. Based on the observations from the long-term experiments, it was found that the above four scenarios can completely cover all the possible situations in the operation of the LDG system, and the comparative analysis indicates that the proposed method exhibits the excellent generalization ability compared to the others.

With respect to the parameters optimization, some comparative experiments are also taken to show the effectiveness of the meta-heuristic PSO-based method presented in Sect. 4.4.1. The fitness function is designated by MAPE. The statistic results of the prediction and the corresponding computing time are listed in Table 4.13, where the accuracy by PSO presents an obvious advantage compared

Table 4.9 Optimal parameters and error statistics of Scenario 3 for LDG subsystem #1

Predict methods	Parameters	MAPE #1 Holder	MAPE #2 Holder	RMSE #1 Holder	RMSE #2 Holder	CT(s)
S-SVM	Loss function: ε-insensitive, insensitivity: 0.1 Width of Gaussian kernel: $\sigma = 51$	0.4729	0.4462	21.6280	21.6280	32.754
S-LSSVM	Penalty coefficient: $\gamma = 127$ Width of Gaussian kernel: $\sigma = 73$	0.1302	0.2077	7.3919	7.3939	0.399
M-LSSVM	Penalty coefficients: $\gamma_0 = 76$ $\gamma_1 = 278$ $\gamma_2 = 299$ Kernel widths: $\sigma_1 = 0.98$ $\sigma_2 = 0.99$	0.5682	0.8121	26.0076	26.0076	0.454
Multi-output LSSVM	Penalty coefficients: $\gamma_0 = 115$ $\gamma_1 = 392$ $\gamma_2 = 105$ Kernel widths: $\sigma_1 = 1.01$ $\sigma_2 = 1.01$	0.0256	0.0195	1.8627	1.0443	0.476

Table 4.10 Optimal parameters and error statistics of Scenario 3 for LDG subsystem #2

Predict methods	Parameters	MAPE #3 Holder	MAPE #4 Holder	RMSE #3 Holder	RMSE #4 Holder	CT(s)
S-SVM	Loss function: ε-insensitive, insensitivity: 0.1 Width of Gaussian kernel: $\sigma = 52$	0.0903	0.1522	7.8565	5.1530	29.367
S-LSSVM	Penalty coefficient: $\gamma = 82$ Width of Gaussian kernel: $\sigma = 55$	0.0663	0.·1175	5.0351	4.1120	0.459
M-LSSVM	Penalty coefficients: $\gamma_0 = 76$ $\gamma_1 = 282$ $\gamma_2 = 299$ Kernel widths: $\sigma_1 = 1.01$ $\sigma_2 = 0.06$	0.1174	0.3350	11.0679	10.7917	0.394
Multi-output LSSVM	Penalty coefficients: $\gamma_0 = 76$ $\gamma_1 = 272$ $\gamma_2 = 302$ Kernel widths: $\sigma_1 = 1.02$ $\sigma_2 = 0.05$	0.0221	0.0074	2.8832	0.2803	0.403

to that by GA. Furthermore, it is noticeable that the computing cost by PSO is apparently lower than that by GA.

To demonstrate that the multi-outputs prediction method is applicable and performs better than other ones, we further present a 3-month running statistics on the average prediction accuracy after implementing the developed software system in a steel plant in China, see Table 4.14. Both the values of MAPE and RMSE are obviously superior to the other methods, where the MAPE can be maintained to be

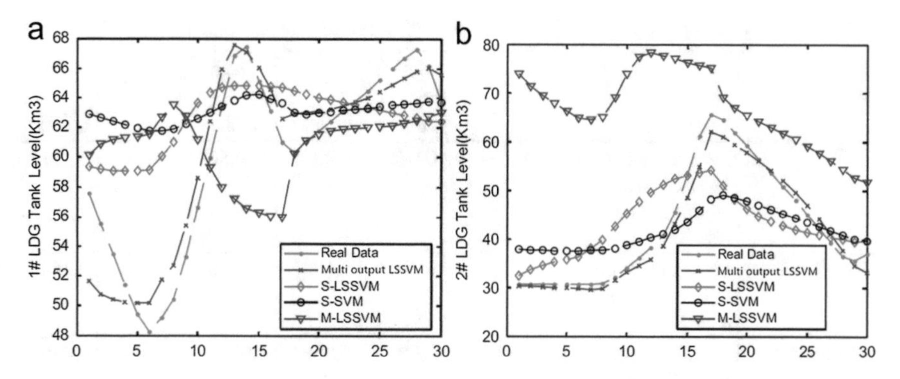

Fig. 4.10 Prediction results of Scenario 4 for #1 LDG subsystem. (**a**) #1 LDG Holder Level and (**b**) #2 LDG Holder Level

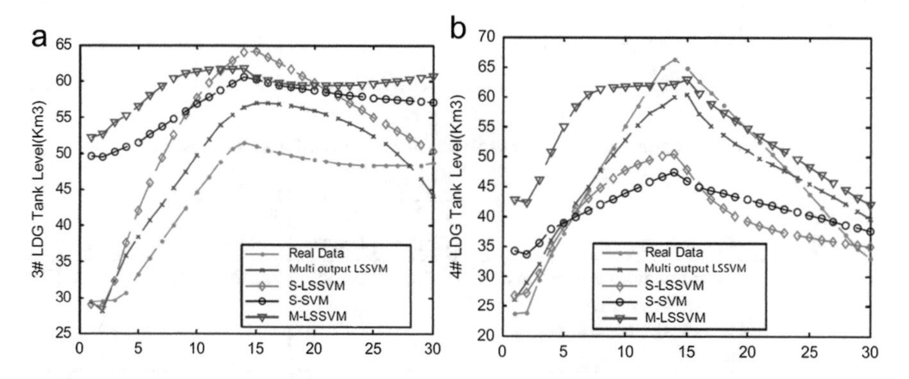

Fig. 4.11 Prediction results of Scenario 4 for #2 LDG subsystem. (**a**) #3 LDG Holder Level and (**b**) #4 LDG Holder Level

less than 0.1 and the RMSE can be less than 5. As such, the practical requirements on accuracy can be completely satisfied.

4.5 Discussion

In this chapter, we discussed the factor-based prediction methods for industrial process. First, we introduced some techniques about how to find the effective factors for the prediction model. Some practical cases are employed here to deeply explain these mentioned techniques. In Sect. 4.3, we introduced three methods for factor-based prediction with only one single output, including neural networks-based prediction models, TS fuzzy prediction models, and multi-kernels learning-based LSSVM. In Sect. 4.4, we studied the factor-based multi-output prediction model.

Table 4.11 Optimal parameters and error statistics of Scenario 4 for LDG subsystem #1

Predict methods	Parameters	MAPE #1 Holder	MAPE #2 Holder	RMSE #1 Holder	RMSE #2 Holder	CT(s)
S-SVM	Loss function: ε-insensitive, insensitivity: 0.1 Width of Gaussian kernel: $\sigma = 39$	0.0755	0.0021	6.0856	7.9605	33.773
S-LSSVM	Punish coefficient: $\gamma = 118$ Width of Gaussian kernel: $\sigma = 77$	0.0614	0.1087	5.2650	8.2002	0.454
M-LSSVM	Penalty coefficients: $\gamma_0 = 99$ $\gamma_1 = 300$ $\gamma_2 = 102$ Kernel widths: $\sigma_1 = 1.02$ $\sigma_2 = 0.98$	0.0819	0.3684	6.9792	27.5490	0.476
Multi-output LSSVM	Penalty coefficients: $\gamma_0 = 101$ $\gamma_1 = 298$ $\gamma_2 = 108$ Kernel widths: $\sigma_1 = 1.02$ $\sigma_2 = 0.98$	0.0254	0.0261	2.1404	2.1997	0.425

Table 4.12 Optimal parameters and error statistics of Scenario 4 for LDG subsystem 2

Predict methods	Parameters	MAPE #3 Holder	MAPE #4 Holder	RMSE #3 Holder	RMSE #4 Holder	CT(s)
S-SVM	Loss function: ε-insensitive, insensitivity: 0.1 Width of Gaussian kernel: $\sigma = 50$	0.2336	0.1350	12.7209	10.6890	31.259
S-LSSVM	Penalty coefficient: $\gamma = 62$ Width of Gaussian kernel: $\sigma = 45$	0.1407	0.1045	9.7073	10.3288	0.323
M-LSSVM	Penalty coefficients: $\gamma_0 = 76$ $\gamma_1 = 275$ $\gamma_2 = 305$ Kernel widths: $\sigma_1 = 0.94$ $\sigma_2 = 0.98$	0.2851	0.1114	15.6361	9.7828	0.276
Multi-output LSSVM	Penalty coefficients: $\gamma_0 = 76$ $\gamma_1 = 288$ $\gamma_2 = 301$ Kernel widths: $\sigma_1 = 0.93$ $\sigma_2 = 0.98$	0.0508	0.0352	5.5381	5.3438	0.253

These factor-based methods aforementioned in this chapter are all assumed that the input and output factor are continuous variables, and when encountering a discrete variable in a specific industrial case these methods cannot handle it directly. Therefore, for future work, the discrete variable-based prediction method can be developed.

Table 4.13 Results of comparison of the optimization methods

LDG holder	Optimization method	MAPE (%)	RMSE	CT (s)
#1	GA	0.0303	2.9345	94
	PSO	0.0282	2.0196	36
#2	GA	0.0262	2.1637	107
	PSO	0.0255	1.5530	38
#3	GA	0.0390	4.4536	89
	PSO	0.0378	4.2440	35
#4	GA	0.0223	3.1425	115
	PSO	0.0219	2.6725	40

Table 4.14 Statistical accuracy of the prediction methods

Prediction methods	LDG holder no.	MAPE (%)	RMSE
S-SVM	#1	0.2730	15.9914
	#2	0.3802	12.3976
	#3	0.3542	9.8754
	#4	0.2853	7.7895
S-LSSVM	#1	0.1429	8.2056
	#2	0.2519	11.3721
	#3	0.2236	12.3754
	#4	0.1716	10.0110
M-LSSVM	#1	0.3738	17.8822
	#2	0.4781	15.4626
	#3	0.3728	16.5423
	#4	0.3284	12.2513
Multi-output LSSVM	#1	0.0509	3.8087
	#2	0.0520	2.3231
	#3	0.0459	4.9869
	#4	0.0294	2.8562

References

1. Deng, J. L.. (1989). Introduction to Grey system theory. *The Journal of Grey System*, (1):1–24.
2. Hsu, L. C., & Wang, C. H. (2009). Forecasting integrated circuit output using multivariate grey model and grey relational analysis. *Expert Systems with Applications, 36*(2), 1403–1409.
3. Mackay, D. J. C. (1996). *Bayesian methods for backpropagation networks. Models of neural networks III* (pp. 211–254). New York: Springer.
4. Neal, R. M. (1996). *Bayesian learning for neural networks*. New York: Springer.
5. Li, Y., Campbell, C., & Tipping, M. (2002). *Bayesian automatic relevance determination algorithms for classifying gene expression data*. Bioinformatics, *18*(10):1332–1339.
6. Tipping, M. E. (2000). The relevance vector machine. *Advances in Neural Information Processing Systems, 12*, 652–658.
7. Bishop, C. M. (1995). *Neural networks for pattern recognition* (Chap. 10). Oxford: Oxford University Press.

8. Tipping, M. E. (2001). Sparse Bayesian learning and the relevance vector machine. *Journal of Machine Learning Research, 1*, 211–244.
9. Takagi, T., & Sugeno, M. (1985). Fuzzy identification of systems and its application to modeling and control. *IEEE Transactions on Systems, Man, and Cybernetics, 15*, 116–132.
10. Sugeno, M., & Tanaka, K. (1991). Successive identification of a fuzzy model and its applications to prediction of a complex system. *Fuzzy Sets and Systems, 42*(3), 315–334.
11. Lee, C. C. (1990). Fuzzy logic in control systems: Fuzzy logic controller. I. *IEEE Transactions on Systems, Man, and Cybernetics, 20*(2), 404–418.
12. Yen, J., Wang, L., & Gillespie, C. W. (1998). Improving the interpretability of TSK fuzzy models by combining global learning and local learning. *IEEE Transactions on Fuzzy Systems, 6*(4), 530–537.
13. Jang, J. R., Sun, C., & Mizutani, E. (1997). *Neuro-fuzzy and soft computing*. Englewood Cliffs, NJ: Prentice-Hall.
14. Passino, K. M., & Yurkovich, S. (1998). *Fuzzy control*. Menlo Park, CA: Addison-Wesley.
15. Mcormick, G. P. (1983). *Nonlinear programming*. New York: Wiley.
16. Haykin, S. (1999). *Neural networks: A comprehensive foundation* (2nd ed.). Upper Saddle River, NJ: Prentice-Hall.
17. Han, C., Chu, Y. H., Kim, J. H., Moon, S. J., Kang, I. S., & Qin, S. J.. (2004). Control of gasholder level by trend prediction based on time-series analysis and process heuristics. In *The seventh international symposium on advanced control of chemical processes*, Hong Kong, China.
18. Pérez-Cruz, F., Camps-Valls, G., Soria-Olivas, E., et al. (2002). *Multi-dimensional function approximation and regression estimation. Artificial Neural Networks—ICANN 2002* (pp. 757–762). Berlin: Springer.
19. Tuia, D., Verrelst, J., Alonso, L., et al. (2011). Multioutput support vector regression for remote sensing biophysical parameter estimation. *IEEE Geoscience and Remote Sensing Letters, 8*(4), 804–808.
20. Zhang, X., Zhao, J., Wang, W., et al. (2011). An optimal method for prediction and adjustment on byproduct gas holder in steel industry. *Expert Systems with Applications, 38*(4), 4588–4599.
21. Fletcher, R. (1987). *Practical methods of optimization* (2nd ed.). Chichester: Wiley.
22. Smola, A. J., & Scholkopf, B. (2001). *Learning with kernels*. Cambridge: MIT Press.
23. Friedrichs, F., & Igel, C. (2005). Evolutionary tuning of multiple SVM parameters. *Neurocomputing, 64*, 107–117.
24. Pai, P. F., & Hong, W. C. (2005). Support vector machines with simulated annealing algorithms in electricity load forecasting. *Energy Conversation Management, 46*(17), 2669–2688.
25. Guo, X. C., Yang, J. H., Wu, C. G., et al. (2008). A novel LS-SVM's hyper-parameter selection based on particle swarm optimization. *Neurocomputing, 71*(16), 3211–3215.
26. An, S., Liu, W., & Venkatesh, S. (2007). Fast cross-validation algorithms for least square support vector machine and kernel ridge regression. *Pattern Recognition, 40*(8), 2154–2162.
27. Van Gestel, T., Suykens, J. A. K., et al. (2001). Financial time series prediction using least squares support vector machines within thee vidence framework. *IEEE Transactions on Neural Networks, 12*(4), 809–821.
28. Perkins, S. (2003). Grafting: Fast, incremental feature selection by gradient descent in function space. *Journal of Machine Learning Research, 3*(3), 1333–1356.
29. Fleming, P. J., & Purshouse, R. C. (2002). Evolutionary algorithms in control systems engineering: A survey. *Control Engineering Practice, 10*(11), 1223–1241.
30. Ho, S., Shu, L., & Chen, J. (2004). Intelligente volutionary algorithms for large parameter optimization problems. *IEEE Transactions on Evolutionary Computation, 8*(6), 522–541.
31. Aslantas, V., Ozer, S., & Ozturk, S. (2009). Improving the performance of DCT-based fragile water marking using intelligent optimization algorithms. *Optics Communications, 282*(14), 2806–2817.
32. Kennedy, J., & Eberhart, R. (1995). Particle swarm optimization (pp. 2–8). In *Proceedings IEEE International Conference on Neural Networks*, 1995.

33. McGookin, E. W., & Murray-Smith, D. J. (2006). Submarine manoeuvring controllers' optimisation using simulated annealing and genetic algorithms. *Control Engineering Practice, 14* (1), 1–15.
34. McGookin, E. W., Murray-Smith, D. J., Li, Y., & Fossen, T. I. (2000). Ship steering control system optimisation using genetic algorithms. *Control Engineering Practice, 8*(4), 429–443.
35. Marinaki, M., Marinakis, Y., & Stavroulakis, G. E. (2010). Fuzzy control optimized by PSO for vibration suppression of beams. *Control Engineering Practice, 18*(6), 618–629.

Chapter 5
Industrial Prediction Intervals with Data Uncertainty

Abstract Prediction intervals (PIs) construction is a comprehensive prediction technique that provides not only the point estimates of the industrial variables, but also the reliability of the prediction results indicated by an interval. Reviewing the conventional PIs construction methods (e.g., delta method, mean and variance-based estimation method, Bayesian method, and bootstrap technique), we provide some recently developed approaches in this chapter. Here, a bootstrapping-based ESN ensemble (BESNE) model is specially proposed to produce reliable PIs for industrial time series, in which a simultaneous training method based on Bayesian linear regression is developed. Besides, to cope with the error accumulation caused by the traditional iterative mode of time series prediction, a non-iterative granular ESN is also reported for PIs construction, where the network connections are represented by the interval-valued information granules. In addition, we present a mixed Gaussian kernel-based regression model to construct PIs, in which a gradient descent algorithm is derived to optimize the hyper-parameters of the mixed Gaussian kernel. In order to tackle the incomplete testing input problem, a kernel-based high order dynamic Bayesian network (DBN) model for industrial time series is then proposed, which directly deals with the missing points involved in the inputs. Finally, we provide some case studies to verify the effectiveness of these approaches.

5.1 Introduction

It is noticeable that point-oriented prediction can only provide pointwise results without any indication of their reliability. The reliability and the accuracy of such predictions can hardly be guaranteed when the size of the training dataset is relatively small, or the values of the targets are affected by probabilistic events. In industrial production, to make a better decision and operational plan, the workers intend to be aware of the uncertainty degree of a forecasting besides the predicted values. Prediction intervals (PIs) construction is a more comprehensive methodology compared to the point-oriented version, since PIs do not only provide the prediction results, but their uncertainties indicated by an interval as well. When talking about

J. Zhao et al., *Data-Driven Prediction for Industrial Processes and Their Applications*, Information Fusion and Data Science,
https://doi.org/10.1007/978-3-319-94051-9_5

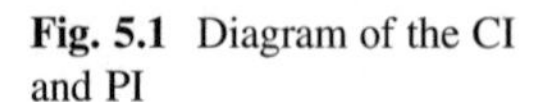

Fig. 5.1 Diagram of the CI and PI

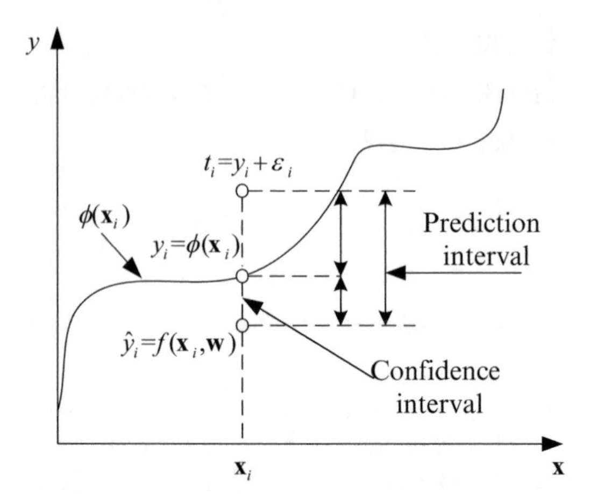

prediction intervals, one has to also notice another concept, confidence intervals (CIs), which pay attention to the uncertainties caused by prediction model. In literature, PIs is viewed as a much broader concept that considers not only the model uncertainties but also the data uncertainties [1]. The difference between a CI and a PI can be illustrated in Fig. 5.1.

In this book, we define the detailed concept of PIs. Supposing that there is a set of observations $D = (\mathbf{x}_i, t_i)$, $1 \leq i \leq n$, there is a basic nonlinear mapping

$$t_i = f(\mathbf{x}_i; \mathbf{w}) + \varepsilon_i \tag{5.1}$$

where t_i is the target, $f(\mathbf{x}_i; \mathbf{w})$ is the nonlinear model, and ε_i is a Gaussian noise with mean zero and constant variance σ_ε^2. In this framework, a confidence interval is concerned with the accuracy of the estimation of the true but unknown function $\phi(\mathbf{x}_i)$, i.e., it is concerned with the distribution of $\phi(\mathbf{x}_i) - f(\mathbf{x}_i; \mathbf{w})$. On the other hand, a PI is concerned with the accuracy of the estimation of the predicted output of nonlinear model, i.e., it is concerned with the distribution of $t_i - f(\mathbf{x}_i; \mathbf{w})$.

If $\widehat{y}_i$ is defined by $\widehat{y}_i = f(\mathbf{x}_i; \mathbf{w})$, then a relationship between the CI and the PI can be formulated by

$$(y_i - \widehat{y}_i) = (\phi(\mathbf{x}_i) - \widehat{y}_i) + \varepsilon_i \tag{5.2}$$

As seen in (5.2), the CI is enclosed in the PI.

In this chapter, we will introduce different types of PIs in industrial process or systems. First, we explain some commonly used techniques for the PIs construction in Sect. 5.2, including the delta method, the mean-variance estimation technique (MVE), the Bayesian method, and the Bootstrap technique. In Sect. 5.2, we will also analyze and compare the advantages and disadvantages of these methods and illustrate the features of each mentioned method. In Sect. 5.3, we introduce the NNs-based PIs construction for industrial time series. And in Sect. 5.4, we discuss another non-iterative NNs for PIs construction for industrial time series. In Sect. 5.5, we further introduce the mixed Gaussian kernels-based PIs construction technique for factors-based prediction. In Sect. 5.6, we consider one special case of PIs construction in industrial process, which is PIs construction with noisy inputs. As for this special case, four kinds of uncertainty are involved, that is the uncertainty coming from the output data noise, the uncertainty coming from the reliability of the prediction model, the uncertainty coming from the feedback, the uncertainty coming from the noise contained in the input. In Sect. 5.7, we will consider another case of PIs construction in industrial process, i.e., PIs construction with missing input. Finally, we summary the above section and actively discuss the possible development of PIs construction in industrial processes and systems.

5.2 Commonly Used Techniques for Prediction Intervals

For constructing the PIs, although there are many existing methods in literature, the basic and commonly used methods can be boiled down to four kinds, including the delta method, the mean and variance-based method, the Bayesian method, and the bootstrap method. These methods can be suitable for many kinds of linear and nonlinear prediction models, such as NNs, Gaussian kernels-based model, support vector regression, and so on. In what follows, we will introduce the basic principle of these methods and analyze their characteristics, merits, and demerits.

5.2.1 Delta Method

The delta method is a relative early-used PIs construction approach based on asymptotic theories [2]. We consider that $\mathbf{w}^*$ is the parameters of nonlinear prediction model that approximates the true regression function, i.e., $y_i = f(\mathbf{x}_i, \mathbf{w}^*)$. Based on the first-order Taylor series expansion, a nonlinear model can be linearized by

$$\widehat{y}_0 = f(\mathbf{x}_0, \mathbf{w}^*) + \mathbf{g}_0^{\mathrm{T}}(\widehat{\mathbf{w}} - \mathbf{w}^*) \tag{5.3}$$

where $\mathbf{g}_0^{\mathrm{T}}$ is the gradient of nonlinear function f with respect to the parameters $\mathbf{w}^*$, i.e.,

$$\mathbf{g}_0^{\mathrm{T}} = \left[\frac{\partial f(\mathbf{x}_0, \mathbf{w}^*)}{\partial \mathbf{w}_1^*} \frac{\partial f(\mathbf{x}_0, \mathbf{w}^*)}{\partial \mathbf{w}_2^*} \cdots \frac{\partial f(\mathbf{x}_0, \mathbf{w}^*)}{\partial \mathbf{w}_p^*} \right] \tag{5.4}$$

and p denotes the number of unknown parameters. In practice, the parameters $\widehat{\mathbf{w}}$ are adjusted through minimizing the sum of squared error (SSE) in order to obtain the expected $\mathbf{w}^*$. Under some regularity conditions, we have

$$\begin{aligned} t_0 - \hat{y}_0 &\approx [y_0 + \varepsilon_0] - \left[f(\mathbf{x}_0, \mathbf{w}^*) + \mathbf{g}_0^{\mathrm{T}}(\hat{\mathbf{w}} - \mathbf{w}^*)\right] \\ &= \varepsilon_0 + \mathbf{g}_0^{\mathrm{T}}(\hat{\mathbf{w}} - \mathbf{w}^*) \end{aligned} \tag{5.5}$$

Because of the statistical independence between the parameters $\hat{\mathbf{w}}$ and the noise ε_0, the predictive covariance of the error can be calculated by the sum of the covariance of the noise and the uncertainties caused by the prediction model.

$$\mathrm{var}(t_0 - \hat{y}_0) = \mathrm{var}(\varepsilon_0) + \mathrm{var}\left(\mathbf{g}_0^{\mathrm{T}}(\hat{\mathbf{w}} - \mathbf{w}^*)\right) \tag{5.6}$$

Assuming that the error ε_0 is assigned to a normal distribution with mean zero and variance σ_ε^2 ($\varepsilon \approx \mathcal{N}(0, \sigma_\varepsilon^2)$), the distribution over $\hat{\mathbf{w}} - \mathbf{w}^*$ can be approximated to have the statistical characteristics $\mathcal{N}\left(0, \sigma_\varepsilon^2 (\mathbf{F}^{\mathrm{T}}\mathbf{F})^{-1}\right)$, where $\mathbf{F}$ is the Jacobian matrix of the nonlinear model with respect to its parameters

$$\mathbf{F} = \begin{bmatrix} \frac{\partial f(x_1, \hat{\mathbf{w}})}{\partial \hat{w}_1} & \frac{\partial f(x_1, \hat{\mathbf{w}})}{\partial \hat{w}_2} & \cdots & \frac{\partial f(x_1, \hat{\mathbf{w}})}{\partial \hat{w}_p} \\ \frac{\partial f(x_2, \hat{\mathbf{w}})}{\partial \hat{w}_1} & \frac{\partial f(x_2, \hat{\mathbf{w}})}{\partial \hat{w}_2} & \cdots & \frac{\partial f(x_2, \hat{\mathbf{w}})}{\partial \hat{w}_p} \\ \vdots & \vdots & \ddots & \vdots \\ \frac{\partial f(x_n, \hat{\mathbf{w}})}{\partial \hat{w}_1} & \frac{\partial f(x_n, \hat{\mathbf{w}})}{\partial \hat{w}_2} & \cdots & \frac{\partial f(x_n, \hat{\mathbf{w}})}{\partial \hat{w}_p} \end{bmatrix} \tag{5.7}$$

Then, the second term $\mathrm{var}\left(\mathbf{g}_0^{\mathrm{T}}(\hat{\mathbf{w}} - \mathbf{w}^*)\right)$ in the right hand side of (5.6) can be computed by

$$\sigma_{\hat{y}_0}^2 = \sigma_\varepsilon^2 \mathbf{g}_0^{\mathrm{T}} (\mathbf{F}^{\mathrm{T}}\mathbf{F})^{-1} g_0 \tag{5.8}$$

By replacing the variance σ_ε^2 and $\sigma_{\hat{y}_0}^2$ in (5.6), the total variance of the predictive error can be expressed as

$$\sigma_0^2 = \sigma_\varepsilon^2 \left(1 + \mathbf{g}_0^{\mathrm{T}} (\mathbf{F}^{\mathrm{T}}\mathbf{F})^{-1} \mathbf{g}_0\right) \tag{5.9}$$

According to the above inference, the $(1-\alpha)\%$ PI for $\hat{y}_i$ is computed as [3]

$$\hat{y}_0 \pm t_{n-p}^{1-(\alpha/2)} \sigma_\varepsilon^2 \sqrt{1 + \mathbf{g}_0^{\mathrm{T}} (\mathbf{F}^{\mathrm{T}}\mathbf{F})^{-1} \mathbf{g}_0} \tag{5.10}$$

where $t_{n-p}^{1-\alpha/2}$ is the $(\alpha/2)$ quantile of a cumulative t-distribution function with $n-p$ degrees of freedom, σ_ε^2 can be represented by an unbiased estimation

$$\sigma_\varepsilon^2 \approx s_\varepsilon^2 = \frac{1}{n-p} \sum_{i=1}^{n} (t_i - \hat{y}_i)^2 \tag{5.11}$$

The advantage of the delta method lies in that it does not require the computation of the second derivative of the nonlinear prediction model. However, the delta method assumes that s_ε^2 is constant for all samples, which is not always suitable for the real-world data. Moreover, there are cases in practice in which the level of noise is systematically correlated by the target magnitude, and for these cases, it is not unexpected that the delta method will generate low-quality PIs.

5.2.2 Mean and Variance-Based Estimation

MVE method, also known as the maximum likelihood estimation method, was designed based on the assumption that the predictive errors are normally distributed around the true value of targets $y(\mathbf{x})$. Therefore, PIs can easily be constructed if the parameters of this distribution are estimated [4]. Particularly, if the distribution is a Gaussian one, only the mean and covariance need to be estimated for PIs construction. Different from the delta technique, two nonlinear prediction models are required for the PIs construction. Taking the NNs-based prediction model as an example, the structural diagram is shown in Fig. 5.2, where one NN is employed to

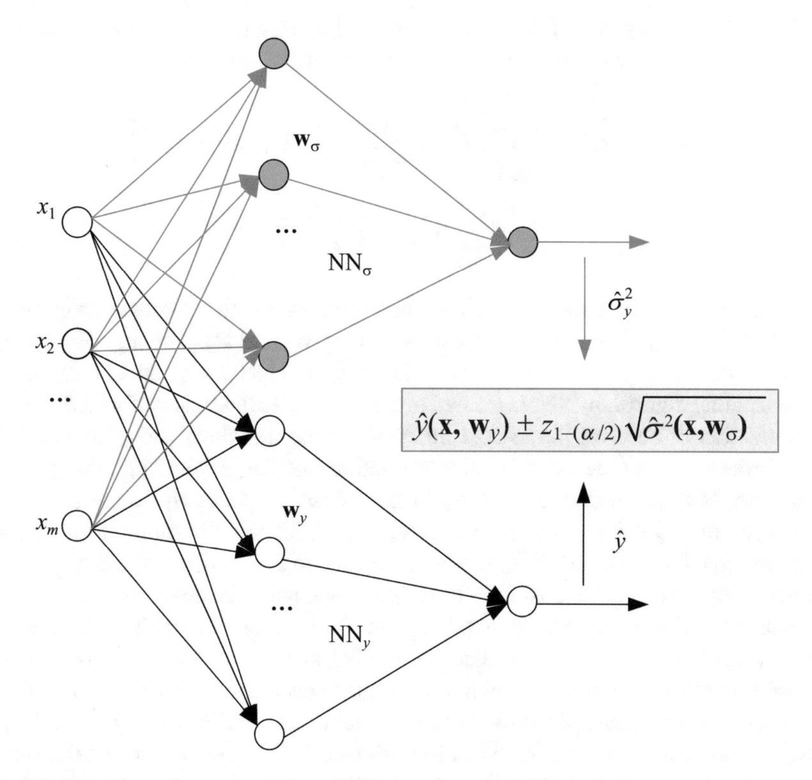

Fig. 5.2 The schematic representation of NNs-based on the MVE method

estimate the mean of targets and another is used to estimate the predictive variance. To guarantee strictly positive estimation of the variance $\widehat{\sigma}^2$, there is a specific constraint of the activation function of the neural network NN_σ. In general, the exponential function is chosen for the neurons of the NN_σ. In addition, there is no limitation on the size and structure of these two networks. If we assume that NN_y can accurately estimate the expected target $y(\mathbf{x})$, the approximate PIs with a confidence level $(1 - \alpha)\%$ can be constructed as follows:

$$\widehat{y}(\mathbf{x}, \mathbf{w}_y) \pm z_{1-\frac{\alpha}{2}} \sqrt{\widehat{\sigma}^2(\mathbf{x}, \mathbf{w}_\sigma)} \tag{5.12}$$

where $\mathbf{w}_y$ and $\mathbf{w}_\sigma$ are parameters of these two NNs for estimation of $\widehat{y}$ and $\widehat{\sigma}^2$, respectively. The priori of the target variance value σ_i are not known, which excludes the application of the error-based minimization techniques for training NN_σ. Instead, a maximum likelihood estimation approach can be applied for training these NNs. Based on the assumption of normally distributed errors around y_i, the delta conditional distribution will be

$$P(t_i | \mathbf{x}_i, \mathrm{NN}_y, \mathrm{NN}_\sigma) = \frac{1}{\sqrt{2\pi\widehat{\sigma}_i^2}} \exp\left(-\frac{\left(t_i - \widehat{y}_i\right)^2}{2\widehat{\sigma}_i^2}\right) \tag{5.13}$$

Taking the natural log of the distribution and ignoring the constant terms result in the following cost function, it will be minimized for all samples,

$$\begin{aligned} C(\mathbf{w}_y, \mathbf{w}_\sigma) = & \frac{1}{2}\sum_{i=1}^{n}\left\{\frac{\left[t_i - y(\mathbf{x}_i; \mathbf{w}_y)\right]^2}{\sigma^2(\mathbf{x}_i; \mathbf{w}_\sigma)} + \ln \sigma^2(\mathbf{x}_i; \mathbf{w}_\sigma)\right\} \\ & + \frac{\alpha_{\mathbf{w}_y}}{2}\sum_{i=1}^{W_y}\mathbf{w}_{y,i}^2 + \frac{\alpha_{\mathbf{w}_\sigma}}{2}\sum_{i=1}^{W_\sigma}\mathbf{w}_{\sigma,i}^2 \end{aligned} \tag{5.14}$$

Using this cost function, an indirect three-phase training technique proposed in [4] can be employed for simultaneously adjusting $\mathbf{w}_y$ and $\mathbf{w}_\sigma$. The algorithm needs two datasets, namely D_1 and D_2, for training NN_y and NN_σ, respectively. In phase I of the training algorithm, NN_y is trained to estimate y_i. Training is performed through minimization of an error-based cost function for the first dataset D_1. To avoid the over-fitting, D_2 can be used as the validation set for terminating the training algorithm. Nothing is done with NN_σ in this phase. In phase II, $\mathbf{w}_y$ are fixed, and D_2 is used for adjusting parameters of NN_σ. $\mathbf{w}_\sigma$ is achieved through minimizing the cost function defined in [2]. NN_y and NN_σ are used to approximate y_i and σ^2 for each sample, respectively. The cost function is then evaluated for the current set of NN_σ weights $\mathbf{w}_\sigma$. These weights then are updated by using the traditional gradient-descent-based methods. D_1 can also be applied as the validation set to limit the over-fitting effects. In phase III, two new training sets are resampled and applied for simultaneous adjustment of both network parameters. The retraining of NN_y and NN_σ is again carried out through minimization of [5]. As before, one of the sets is used as the validation set.

The main advantages of this method are its simplicity and that there is no need to calculate complex derivatives and the inversion of the Hessian matrix. Nonstationary variances can be approximated by employing more complex structures for NN_σ or through proper selection of the set of inputs. The main drawback of the MVE method is that it assumes NN_y precisely estimates the true mean of the targets y_i. Such an assumption can be violated in practice because of the existence of a bias in fitting the data due to a possible under-fitting of the NN model or due to omission of important attributes affecting the target behavior. In these cases, the NN generalization ability is weak, resulting in accumulation of uncertainty in the estimation of y_i. Therefore, the constructed PIs using [6] will underestimate the actual $(1 - \alpha)\%$ PIs, leading to a low coverage probability. Assuming $\widehat{y}_i \simeq y_i$ implies that the MVE method only considers one portion of the total uncertainty for construction of PIs. The considered variance is only due to errors, not the misspecification of model parameters (either $\mathbf{w}_y$ or $\mathbf{w}_\sigma$). This can result in misleadingly narrow PIs with a low coverage probability. This critical drawback has been theoretically identified and practically demonstrated in [7].

5.2.3 Bayesian Method

In Bayesian framework, the nonlinear prediction model is trained based on a regularized cost function in the general form

$$E(\mathbf{w}) = \alpha E_W + \beta E_D \tag{5.15}$$

where α and β are the hyper-parameters of the cost function determining the training purpose. E_W is the sum of squares of the network weight ($\mathbf{w}^{\mathrm{T}}\mathbf{w}$), and E_D is the sum of the square error. The method assumes that the set of parameters $\mathbf{w}$ is a random set of variables with an assumed priori distributions. A set of training data is used for adjusting the prior information of the parameters $\mathbf{w}$. Generally, if the training data is sufficient, the function will play a leading role in the cost function. In contrast, if the training data is insufficient, the priori information of the parameters will be more effective. Upon availability of a training dataset and the prediction model, the density function of the weights can be updated using the Bayes' rule

$$P(\mathbf{w}|\alpha,\beta,D,M) = \frac{P(D|\mathbf{w},\beta,M)P(\mathbf{w}|\alpha,M)}{P(D|\alpha,\beta,M)} \tag{5.16}$$

where M and D are the nonlinear model and the training dataset, respectively. $P(D|\mathbf{w},\beta,M)$ and $P(\mathbf{w}|\alpha,M)$ are the likelihood function of data occurrence and the prior density of parameters, respectively. Representing our knowledge, $P(D|\alpha,\beta,M)$ is a normalization factor enforcing that local probability is equal to one.

Assuming that ε_i are normally distributed and meanwhile $P(D|\mathbf{w},\beta,M)$ and $P(\mathbf{w}|\alpha,M)$ have normal distributions, we can write

$$P(D|\mathbf{w}, \beta, M) = \frac{1}{Z_D(\beta)} \exp(-\beta E_D) \quad (5.17)$$

and

$$P(\mathbf{w}|\alpha, M) = \frac{1}{Z_{\mathbf{w}}(\alpha)} \exp(-\alpha E_{\mathbf{w}}) \quad (5.18)$$

with $Z_D(\beta) = (\pi/\beta)^{n/2}$ and $Z_{\mathbf{w}}(\alpha) = (\pi/\alpha)^{(p/2)}$. n and p are the number of training samples and unknown parameters, respectively. By substituting (5.17) and (5.18) into (5.16), we have

$$P(\mathbf{w}|D, \alpha, \beta, M) = \frac{1}{Z_F(\beta, \alpha)} \exp(-(\alpha E_{\mathbf{w}} + \beta E_D)) \quad (5.19)$$

The purpose of Bayesian training technique is to maximize the posterior probability $P(\mathbf{w}|D, \alpha, \beta, M)$, which corresponds to the minimization of (5.15) that makes the connection between Bayesian methodology and regularized cost functions. By taking derivatives with respect to the logarithm of (5.19) and setting it equal to zero, the optimal values for alpha and beta are obtained [8, 9].

$$\alpha^{\mathrm{MP}} = \frac{n - \gamma}{E_{\mathbf{w}}(\mathbf{w}^{\mathrm{MP}})} \quad (5.20)$$

$$\beta^{\mathrm{MP}} = \frac{\gamma}{E_D(\mathbf{w}^{\mathrm{MP}})} \quad (5.21)$$

where $\gamma = p - 2\alpha^{\mathrm{MP}}\mathrm{tr}(\mathbf{H}^{\mathrm{MP}})^{-1}$ is the so-called effective number of unknown parameters, and p is the total number of the model parameters. $\mathbf{w}^{\mathrm{MP}}$ is the most probable estimation of the unknown parameters. $\mathbf{H}^{\mathrm{MP}}$ is the Hessian matrix of $E(\mathbf{w})$ defined as

$$\mathbf{H}^{\mathrm{MP}} = \alpha \nabla^2 E_{\mathbf{w}} + \beta \nabla^2 E_D \quad (5.22)$$

In usual, the Levenberg-Marquardt optimization algorithm approximates the Hessian matrix [10]. Application of this technique for training results in NNs can provide the predictive variance of the form

$$\sigma_i^2 = \sigma_D^2 + \sigma_{\mathbf{w}^{\mathrm{MP}}}^2 = \frac{1}{\beta} + \nabla_{\mathbf{w}^{\mathrm{MP}}}^T \widehat{y}_i (\mathbf{H}^{\mathrm{MP}})^{-1} \nabla_{\mathbf{w}^{\mathrm{MP}}} \widehat{y}_i \quad (5.23)$$

where the first term in the right hand side of (5.23) quantifies the amount of uncertainty in the training data (the intrinsic noise), the second term corresponds to the misspecification of the model parameters and their contribution to the variance of predictions.

As the total variance of ith future sample is known, a $(1 - \alpha)\%$ PI can be constructed

$$\widehat{y}_i \pm z^{1-\frac{\alpha}{2}}\left(\frac{1}{\beta}+\nabla_{\mathbf{w}^{\mathrm{MP}}}^{T}\widehat{y}_i\left(\mathbf{H}^{\mathrm{MP}}\right)^{-1}\nabla_{\mathbf{w}^{\mathrm{MP}}}\widehat{y}_i\right)^{\frac{1}{2}} \tag{5.24}$$

where $z^{1-\alpha/2}$ is the $1-\alpha/2$ quantile of a normal distribution function with mean zero and unit variance. Also, $\nabla_{\mathbf{w}^{\mathrm{MP}}}\widehat{y}_i$ is the gradient of the output function of the prediction model with respect to its parameters $\mathbf{w}^{\mathrm{MP}}$.

The Bayesian method for PIs construction has a strong mathematical foundation. Nonlinear models trained by using the Bayesian learning technique typically have a better generalization performance than other networks. This minimizes the effects of $\sigma^2_{\mathbf{w}^{\mathrm{MP}}}$ on the width of PIs. Furthermore, it eliminates the hassle for optimal determination of the regularizing parameters. The Bayesian method is computationally demanding in the development stage, similar to the delta technique. It requires calculation of the Hessian matrix in (5.22), which is time-consuming and cumbersome for large-scale models and datasets. However, the computational load decreases in the PIs construction stage as we only need to calculate the gradient of the output function.

5.2.4 *Bootstrap Technique*

Bootstrap technique is by far the most commonly used one documented in literature for the construction of CIs and PIs [11]. It cannot be applied for PIs construction without an ensemble model and the ensemble is generally constructed by the NNs. Such a method assumes that an ensemble model will produce a less biased estimation of the true regression of the targets [12]. Theoretically, as generalization errors of the NN models are made on different subsets of the parameter space, the collective decision produced by the ensemble is less likely to be in error than the decision made by any of the individual models. Generally, B training datasets $\{D\}_{b=1}^{B}$ resampled from the original dataset with replacement are required for building B individual models of the ensemble. According to this assumption, the real regression is estimated by averaging the point forecasts of B models.

$$\widehat{y}_i = \frac{1}{B}\sum_{b=1}^{B}\widehat{y}_i^{b} \tag{5.25}$$

where $\widehat{y}_i^{b}$ is the prediction of ith sample generated by bth bootstrap model.

As for the model misspecification variance, i.e., the uncertainties caused by the prediction model, it can be estimated by using the variance of B model outcomes based on the assumption that NN models are unbiased.

$$\sigma_{\widehat{y}_i}^2 = \frac{1}{B-1} \sum_{b=1}^{B} \left(\widehat{y}_i^b - \widehat{y}_i\right)^2 \tag{5.26}$$

This variance is mainly due to the random initialization of parameters and using different datasets for training NNs. CIs can be constructed using the approximation of $\sigma_{\widehat{y}_i}^2$ in (5.26). To construct PIs, we need to estimate the variance $\sigma_{\widehat{\varepsilon}_i}^2$ of errors. From (5.6), $\sigma\widehat{\varepsilon}^2$ can be calculated by

$$\sigma\widehat{\varepsilon}^2 \simeq E\left\{\left(t - \widehat{y}\right)^2\right\} - \sigma\widehat{y}^2 \tag{5.27}$$

To estimate the variance $\sigma_{\widehat{\varepsilon}_i}^2$, we should develop a set of variances squared residuals as the samples. The commonly-used method is to find the maximum value of $\left(t_i - \widehat{y}_i\right)^2 - \sigma_{\widehat{y}_i}^2$ and 0 as the observed targets of the noise variance.

$$r_i^2 = \max\left(\left(t_i - \widehat{y}_i\right)^2 - \sigma_{\widehat{y}_i}^2, 0\right) \tag{5.28}$$

where $\widehat{y}_i$ and $\sigma_{\widehat{y}_i}^2$ are obtained from (5.25) and (5.26). These residuals are linked by the set of corresponding inputs to form a new dataset

$$D_{r^2} = \left\{\left(\mathbf{x}_i, r_i^2\right)\right\}_{i=1}^{n} \tag{5.29}$$

A new nonlinear model can be indirectly trained to estimate the unknown values of $\sigma_{\widehat{\varepsilon}_i}^2$ so as to maximize the probability of observing the samples in D_{r^2}. The training cost function is defined by

$$C_{\text{BS}} = \frac{1}{2} \sum_{i=1}^{n} \left[\ln\left(\sigma_{\widehat{\varepsilon}_i}^2\right) + \frac{r_i^2}{\sigma_{\widehat{\varepsilon}_i}^2}\right] + \frac{\alpha}{2} \sum_{i=1}^{W} \mathbf{w}_i^2 \tag{5.30}$$

where α is a hyper-parameter and $\mathbf{w}_i$ is the unknown parameters of the nonlinear model. As noted before, the output activation function is selected to be exponential, enforcing a positive value for $\sigma_{\widehat{\varepsilon}_i}^2$. The minimization of C_{BS} can be done using a variety of methods, including traditional gradient descent methods, which can be found in Chap. 7.

For the construction of PIs by using the bootstrap method, generally $B + 1$ individual models are required in total, where B individuals are used for the estimation of $\sigma_{\widehat{y}_i}^2$ and another one is used for the estimation of $\sigma_{\widehat{\varepsilon}_i}^2$. Therefore, this method is computationally more demanding than other methods in its training stage. However, once the models are trained off-line, the online computational load for PI construction is only limited to point forecasts of the $B + 1$ individuals. This is in contrast with the claim in the literature than bootstrap PIs are computationally more intensive than other methods. Simplicity is another advantage of using the bootstrap method for PIs construction. There is no need to calculate complex matrices and derivatives, as required by the delta and Bayesian techniques.

The main disadvantage of the bootstrap technique lies in its dependence on B individual models. Frequently, if some of these models are biased, an inaccurate estimation of $\sigma^2_{\hat{y}_i}$ will be lead. Therefore, the total variance will be underestimated resulting in narrow PIs with a low coverage probability.

5.3 Neural Networks-Based PIs Construction for Time Series

If we do not consider the uncertainties coming from the input data, the PIs are only related to the uncertainties of the output data and the prediction model. Thus, as for the time series or factors-based prediction tasks, there is no essential difference. NNs are the most popularly used model for prediction, which can be found in many reported literatures. In this section, we will introduce one kind of NNs-based PIs construction methods.

5.3.1 ESNs Ensemble-Based Prediction Model

In this subsection, a bootstrapping ESNs ensemble (BESNE) model [13] is established by combining ESNs and bootstrap resampling and its structure can be designed as shown in Fig. 5.3, where B training datasets $\{D_1, D_2, \ldots, D_b, \ldots, D_B\}$ are resampled by using the original dataset $D_I = \{\mathbf{u}_i, t_i\}_{i=1}^n$, and $D_b = \left\{\mathbf{u}_b^i, t_b^i\right\}_{i=1}^n$ serves to train the bth ESN. For $\forall \mathbf{u}^i$, $\widehat{y}^i$ is the output of the ensemble, and $\widehat{y}_b^i$ denotes the output of the bth ESN driven by $\mathbf{u}_b^i$.

Based on the structure of the ensemble, the ESNs ensemble can be formulated as

$$\mathbf{x}_b^i = f\left[\mathbf{W}_b^{\text{in}} \mathbf{u}_b^i + \mathbf{W}_b \mathbf{x}_b^{i-1}\right] \tag{5.31}$$

$$y_b^i = \mathbf{W}_b^{\text{out}} \cdot \left[\mathbf{u}_b^i; \mathbf{x}_b^i\right] \tag{5.32}$$

$$t^i = y^i + \varepsilon^i \approx (1/B) \sum_{i=1}^{B} y_b^i + \varepsilon^i \tag{5.33}$$

$$\sigma_{y^i}^2 \simeq E\left\{\left(y^i - \widehat{y}^i\right)^2\right\} \approx [1/(B-1)] \sum_{i=1}^{B} \left(y^i - y_b^i\right)^2 \tag{5.34}$$

where ε^i is a Gaussian white noise with mean zero. $\sigma_{y^i}^2$ denotes the error calculated by the squared residual between y^i and y_b^i. Note that $\mathbf{W_b}^{\text{out}}$ here is a vector since we only consider the single output. Considering the fact that the parameters of individual ESNs in the ensemble are required to be trained simultaneously, especially the individual interactions, the ensemble is not just the simple sum or the average of these independent ESNs. As for the structure shown in Fig. 5.3, which integrates the

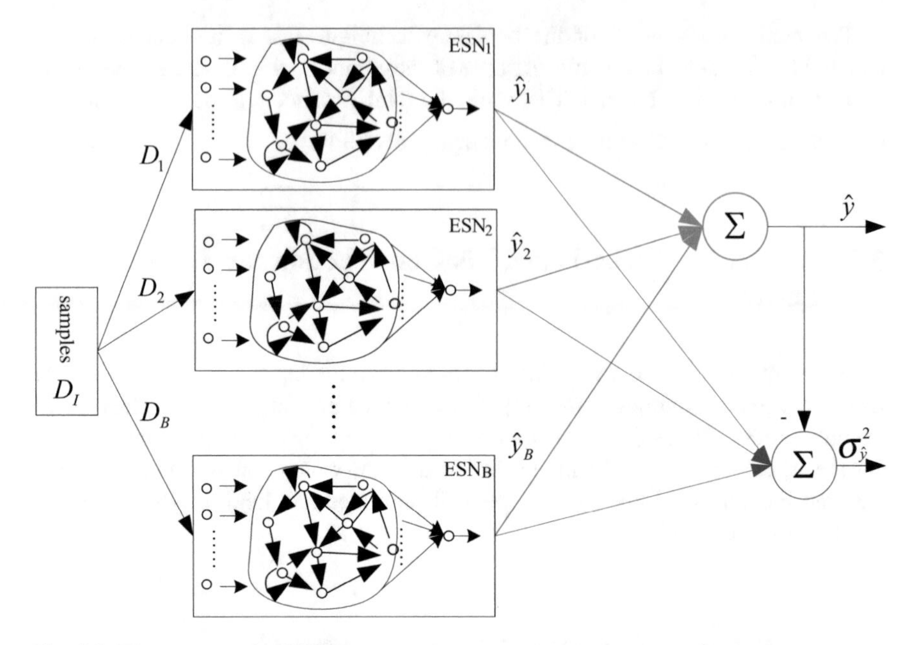

Fig. 5.3 The structure of a BESNE

multiple ESNs as an ensemble instead of using a single ESN, can reduce the randomness impact on the accuracy; the prediction variability can also be improved when dealing with the noisy nonlinear time series.

In general, besides the B ESNs, another one usually is required for estimating the uncertainties caused by the data noise, where a set of variance squared residuals that are the approximation of $\sigma_{\varepsilon *}^2$ should be constructed as the samples [14]. However, the estimate is commonly time-consuming because the network model should be indirectly trained based on these samples [15]. In such a model, the uncertainties $\sigma_{\varepsilon *}^2$ coming from data noise is simultaneously estimated as a hyper-parameter β in the parameters learning. Generally, we have

$$\sigma_{\varepsilon *}^2 = 1/\beta \tag{5.35}$$

If the value of hyper-parameter β is available, the PI with a $(1 - \alpha)100\%$ confidence level can be constructed by $\left(\widehat{y}^* \pm t_{\alpha/2}(B) \cdot \left(\sigma_{\widehat{y*}}^2 + \sigma_{\varepsilon *}^2\right)\right)$, where $t_{\alpha/2}(B)$ is the $(\alpha/2)$ quantile of a cumulative t-distribution function with B degrees of freedom.

5.3.2 *Bayesian Estimation of the Uncertainties*

As for the aforementioned BESNE model, the most important task is to estimate its parameters. To simultaneously estimate the hyper-parameters, the Bayesian estimation technique which will be introduced in details in Chap. 7, can be employed here. Given the resampled training datasets $\{D_1, D_2, \ldots, D_B\}$, where $D_b = \{\mathbf{u}_b^i, t_b^i\}_{i=1}^n$, the target output t_b^i reads as

$$t_b^i = \mathbf{W}_b^{\text{out}} \cdot [\mathbf{u}_b^i, \mathbf{x}_b^i] + \varepsilon_b^i \tag{5.36}$$

where ε_b^i is the Gaussian random variable with distribution $\mathcal{N}(0, \sigma^2)$. The probability density of the output can be expressed by

$$p(t_b^i | \mathbf{u}_b^i, \mathbf{W}_b^{\text{out}}) \propto \exp\left[-\frac{\beta}{2}(\mathbf{W}_b^{\text{out}} \cdot [\mathbf{u}_b^i, \mathbf{x}_b^i] - t_b^i)^2\right] \tag{5.37}$$

where $\beta = 1/\sigma^2$ is a hyper-parameter related to the output distribution. If we consider the samples in the ensemble as independent, the likelihood function can be expressed as the joint probability of a set of $B \times n$ target outputs

$$\begin{aligned} p(D|\boldsymbol{\theta}, \beta) &= \prod_{b=1}^{B}\prod_{i=1}^{n} p(t_b^i | \mathbf{u}_b^i, \mathbf{W}_b^{\text{out}}) \\ &= \frac{1}{Z_D(\beta)} \exp\left(-\frac{\beta}{2}\sum_{b=1}^{B}\sum_{i=1}^{n}\{\mathbf{W}_b^{\text{out}} \cdot [\mathbf{u}_b^i, \mathbf{x}_b^i] - t_b^i\}^2\right) \end{aligned} \tag{5.38}$$

where D represents the original training dataset and $\boldsymbol{\theta}$ is a collection matrix of the output weights in the ensemble $\boldsymbol{\theta} = [\mathbf{W}_1^{\text{out}}, \mathbf{W}_2^{\text{out}}, \ldots, \mathbf{W}_B^{\text{out}}]$. Because the noise is Gaussian, the normalized factor can be analytically expressed as

$$Z_D(\beta) = (2\pi/\beta)^{(B \times n)/2} \tag{5.39}$$

According to the Bayesian theorem, when the prior probability density $p(\boldsymbol{\theta})$ of the output weights and the evidence term $p(D)$ are known, the posterior probability density is given by

$$p(\boldsymbol{\theta}|\boldsymbol{\alpha}, \beta, D) = \frac{p(D|\boldsymbol{\theta}, \beta)p(\boldsymbol{\theta}|\boldsymbol{\alpha})}{p(D|\boldsymbol{\alpha}, \beta)} \tag{5.40}$$

If one has little understanding of the prior output weights, then the prior output weights can be expressed by a Gaussian distribution with a large variance. We model the distribution of each ESN output weights as a Gaussian distribution here. Assuming that the output weights of each ESN are independent, the distribution of the ensemble parameters can be formulated by a joint distribution,

$$\begin{aligned} p(\boldsymbol{\theta}|\boldsymbol{\alpha}) &= p\left(\mathbf{W}_1^{\text{out}}, \mathbf{W}_2^{\text{out}}, \ldots, \mathbf{W}_B^{\text{out}}|\boldsymbol{\alpha}\right) \\ &= \prod_{b=1}^{B} p\left(\mathbf{W}_b^{\text{out}}|\alpha_b\right) = \frac{1}{Z_W(\boldsymbol{\alpha})}\exp\left(\sum_{b=1}^{B}\left[-\frac{\alpha_b}{2}\left\|\mathbf{W}_b^{\text{out}}\right\|^2\right]\right) \end{aligned} \tag{5.41}$$

where the normalized factor $Z_W(\boldsymbol{\alpha})$ has the value

$$Z_W(\boldsymbol{\alpha}) = \prod_{b=1}^{B}\left[(2\pi/\alpha_b)^{W/2}\right] \tag{5.42}$$

and $W \times B$ is the dimension of $\boldsymbol{\theta}$. With the Gaussian approximation of the posterior output weights, the evidence term can be expressed as

$$p(D|\boldsymbol{\alpha}, \beta) = \int p(D|\boldsymbol{\theta}, \beta) p(\boldsymbol{\theta}|\boldsymbol{\alpha}) d\boldsymbol{\theta} \tag{5.43}$$

The posterior probability density becomes

$$p(\boldsymbol{\theta}|D) \propto \exp\left(-\frac{\beta}{2}\sum_{b=1}^{B}\sum_{i=1}^{n}\left(\mathbf{W}_b^{\text{out}} \cdot \left[\mathbf{u}_b^i, \mathbf{x}_b^i\right] - t_b^i\right)^2 - \sum_{b=1}^{B}\left\{\frac{\alpha_b}{2}\left\|\mathbf{W}_b^{\text{out}}\right\|^2\right\}\right) \tag{5.44}$$

Without loss of generality, its optimal value can be computed through maximizing the posterior distribution in the logarithm form, rewritten as

$$\ln p(\boldsymbol{\theta}|D) = -\frac{\beta}{2}\sum_{b=1}^{B}\sum_{i=1}^{n}\left(\mathbf{W}_b^{\text{out}} \cdot \left[\mathbf{u}_b^i, \mathbf{x}_b^i\right] - t_b^i\right)^2 - \sum_{b=1}^{B}\frac{\alpha_b}{2}\left\|\mathbf{W}_b^{\text{out}}\right\|^2 + \text{const} \tag{5.45}$$

As addressed above, the hyper-parameters $\boldsymbol{\alpha}$, β and the parameters $\boldsymbol{\theta}$ are all unknown. To obtain the optimal output weights, the hyper-parameters are usually computed by first maximizing the posterior probability of the hyper-parameters. Replacing the posterior maximization by the likelihood maximization, the logarithm form of the likelihood becomes

$$\begin{aligned} \ln p(D|\boldsymbol{\alpha}, \beta) &= -\sum_{b=1}^{B}\alpha_b E_{W,b}^{\text{MP}} - \beta E_D^{\text{MP}} - \frac{1}{2}\ln(\det\mathbf{A}) \\ &\quad + \frac{W}{2}\sum_{b=1}^{B}\ln\alpha_b + \frac{B \times n}{2}\ln\beta - \frac{B \times n}{2}\ln(2\pi) \end{aligned} \tag{5.46}$$

where

$$E_{W,b}^{\text{MP}} = \frac{1}{2}\left\|\mathbf{W}_{b,\text{MP}}^{\text{out}}\right\|^2 \tag{5.47}$$

$$E_D^{\mathrm{MP}} = \frac{1}{2}\sum_{b=1}^{B}\sum_{i=1}^{n}\left(\mathbf{W}_b^{\mathrm{out}} \cdot \left[\mathbf{u}_b^i, \mathbf{x}_b^i\right] - t_b^i\right)^2 \tag{5.48}$$

and $\mathbf{A}$ is the Hessian of $\sum_{b=1}^{B} \alpha_b E_{W,b}^{\mathrm{MP}} + \beta E_D^{\mathrm{MP}}$ with respect to the optimal parameters $\boldsymbol{\theta}_{\mathrm{MP}}$. Then,

$$\mathbf{A} = \beta \nabla\nabla E_D^{\mathrm{MP}} + \sum_{b=1}^{B} \alpha_b \left(\nabla\nabla E_{W,b}^{\mathrm{MP}}\right) = \beta \sum_{i=1}^{n} \mathbf{c}_i \cdot \mathbf{c}_i^{\mathrm{T}} + \sum_{b=1}^{B} \alpha_b \mathbf{I} \tag{5.49}$$

where $\mathbf{c}_i = \left[\mathbf{u}_1^i, \mathbf{x}_1^i; \mathbf{u}_2^i, \mathbf{x}_2^i; \ldots; \mathbf{u}_B^i, \mathbf{x}_B^i\right]$. So far, we can use the optimization methods introduced in Chap. 7 to evaluate the optimal weights and the hyper-parameters.

As for the estimation of the uncertainties, it can be conducted based on the assumption that the structural parameters of the bootstrap ESNs ensemble are known. However, if we don't design a method to identify the structure, the number of ESNs, and the reservoir dimensionality are all unknown. In the following subsection, we will introduce one method based on 0.632 bootstrap cross-validation for the model selection and structure identification.

5.3.3 *Model Selection and Structural Optimization*

The number of ESNs and the reservoir dimension are considered as the structural factors in the ESNs ensemble. Out-of-sample methods, such as the cross-validation and the bootstrap, are commonly used for model selection and can work reasonably well [16]. However, the cross-validation usually shows large variability when estimating the network structures [17]. Thus, the bootstrap is introduced into cross-validation for model selection [18].

Notably, independent data are required for cross-validation [19]. While the bootstrap is a resampling technique, it cannot guarantee whether the training data are independent from the testing data. Since the error introduced by non-independent data will be small, the 0.632 bootstrap cross-validation that considers the independence of data is adopted. The error of the 0.632 bootstrap cross-validation is the weighted sum of the ones based on independent and non-independent data [20].

For the ith input sample of the bth training set, the output of the bth ESN can be denoted as

$$\widehat{y}_b^i = \widehat{\mathbf{W}}_b^{\mathrm{out}}\left(\mathbf{u}_b^i; \mathbf{x}_b^i\right) \tag{5.50}$$

where $\widehat{\mathbf{W}}_b^{\mathrm{out}}$ can be estimated from bootstrap samples. $\mathbf{x}_b^i$ denotes the states of the bth reservoir driven by $\mathbf{u}_b^i$. Thus, the bootstrap error of the network ensemble is given by

$$\begin{aligned}\operatorname{err}\left(D_b, \widehat{\mathbf{W}}^{\text{out}}\right) &= \frac{1}{B}\sum_{b=1}^{B}\operatorname{err}\left(D_b, \widehat{\mathbf{W}}_b^{\text{out}}\right) \\ &= \frac{1}{n}\sum_{i=1}^{n}\sum_{b=1}^{B}\left(\widehat{y}_b^i - t_b^i\right)^2 / B\end{aligned} \tag{5.51}$$

where $\widehat{y}_b^i$ is the predicted value at $\mathbf{u}_b^i$ from the model estimated on bth bootstrap samples. Let the original set D_{I} be the input of the ensemble, the apparent error can be approximated by

$$\begin{aligned}\operatorname{err}\left(D_{\text{I}}, \widehat{\mathbf{W}}^{\text{out}}\right) &= \frac{1}{B}\sum_{b=1}^{B}\operatorname{err}\left(D_{\text{I}}, \widehat{\mathbf{W}}_b^{\text{out}}\right) \\ &= \frac{1}{n}\sum_{i=1}^{n}\sum_{b=1}^{B}\left(\widehat{y}^i\left(\widehat{\mathbf{W}}_b^{\text{out}}, \mathbf{u}^i\right) - t^i\right)^2 / B\end{aligned} \tag{5.52}$$

Since $\widehat{\mathbf{W}}^{\text{out}}$ is an approximate, the apparent error $\operatorname{err}\left(D_{\text{I}}, \widehat{\mathbf{W}}^{\text{out}}\right)$ should be biased against the prediction error $\operatorname{err}(D_{\text{I}}, \mathbf{W}^{\text{out}})$, which means $\operatorname{err}\left(D_{\text{I}}, \widehat{\mathbf{W}}^{\text{out}}\right)$ should be corrected by a difference $\omega(\mathbf{W}^{\text{out}})$ that constantly is equal to the expectation of the difference between the prediction error and the apparent error. Because $\mathbf{W}^{\text{out}}$ cannot be estimated, the bootstrap estimation of $\omega(\mathbf{W}^{\text{out}})$ is obtained.

$$\widehat{\omega}\left(\widehat{\mathbf{W}}^{\text{out}}\right) = \operatorname{err}\left(D_{\text{I}}, \widehat{\mathbf{W}}^{\text{out}}\right) - \operatorname{err}\left(D_b, \widehat{\mathbf{W}}_b^{\text{out}}\right) \tag{5.53}$$

As for the 0.632 bootstrap cross-validation, a more accurate compensation for the above difference can be redefined by $\omega^{0.632} = 0.632\left[e_0 - \operatorname{err}\left(D_{\text{I}}, \widehat{\mathbf{W}}^{\text{out}}\right)\right]$, where e_0 is the average error obtained from the bootstrap datasets excluding the predicted samples [18]. Given B bootstrap sample sets, e_0 can be estimated by

$$\widehat{e}_0 = \frac{1}{n}\sum_{i=1}^{n}\sum_{b \in C_i}\left[\widehat{\mathbf{W}}_b^{\text{out}}(\mathbf{u}_{b,i}, \mathbf{x}_{b,i}) - y_{b,i}\right]^2 / B_i \tag{5.54}$$

where C_i denotes the index set of the bootstrap sample sets excluding the ith data in the initial set D_{I}, and B_i is the number of such sample sets. The predicted error of the 0.632 bootstrapping estimation is given by

$$\text{Perr}^{0.632} = \operatorname{err}\left(D_{\text{I}}, \widehat{\mathbf{W}}^{\text{out}}\right) + \widehat{\omega}^{0.632} \tag{5.55}$$

In this optimization problem, the independent variables are the number of individual ESNs and reservoir dimensionality. The objective function is defined as the predicted error, given in (5.55). The value "0.632" comes from a theoretical argument indicating that the bootstrap samples used in computing $\widehat{e}_0$ are further away on the average than a typical test sample, by roughly a factor of 1/0.632. The adjustment in $\omega^{0.632}$ makes correction for this bias and makes $\text{Perr}^{0.632}$ roughly unbiased for the true error rate. We will not give the theoretical argument here, but note that the value 0.632 arises because it is approximately the probability that a given observation appears in bootstrap sample of size n.

5.3.4 Theoretical Analysis of the Prediction Performance

The conception of the model shown in Fig. 5.3 lies in describing a complex nonlinearity by the inter-connected network individuals. Only the output weights need to be estimated in an ESN, thus we could regard the training process as a linear regression problem. Therefore, different from the previous network ensemble, the model exhibits some advantages. First, it is easier to design a simultaneous training to learn the weight parameters of the ensemble since the parameters learning becomes in nature to solve a large-scale linear equations based on (5.36). Second, according to (5.49), the Hessian matrix for the Bayesian regularization is easy to be calculated. Thus, the ensemble requires less computational load than others, which is favorable for some practical problems. Third, more importantly, the ensemble can obtain a higher accuracy than the network with only one ESN. Here, the accuracy of the BESNE will be mathematically analyzed as follows.

The parameters of the ensemble model can be obtained by the simultaneous training method. Then, using Bayesian rule, the conditional distribution of t^* over the input $\mathbf{u}^*$ can be

$$p(t^*|\mathbf{u}^*, D) = \int p(t^*|\mathbf{u}^*, \boldsymbol{\theta}) p(\boldsymbol{\theta}|D) d\boldsymbol{\theta} \tag{5.56}$$

where $p(t^*|\mathbf{u}^*, \boldsymbol{\theta})$ is the posterior over the model. With the known the parameter $\boldsymbol{\theta}$, see (5.48). Then,

$$p(t^*|\mathbf{u}^*, D) \propto \int \exp\left(-\frac{\beta}{2}(y^* - t^*)^2\right) \cdot \exp\left(-\frac{1}{2}\Delta\boldsymbol{\theta}^{\mathrm{T}}\mathbf{A}\Delta\boldsymbol{\theta}\right) d\boldsymbol{\theta} \tag{5.57}$$

where

$$y^* = \frac{1}{B}\sum_{b=1}^{B} y_b^* = \frac{1}{B}\sum_{b=1}^{B} \mathbf{W}_b^{\mathrm{out}} \cdot [\mathbf{u}^*; \mathbf{x}_b^*] \tag{5.58}$$

According to the property of Gaussian distribution, the posterior of the output can still be expressed as a Gaussian form, i.e.,

$$p(t^*|\mathbf{u}^*, D) = \frac{1}{(2\pi\sigma_t)^{1/2}} \exp\left(-\frac{(t^* - y_{\mathrm{MP}}^*)^2}{2\sigma_t^2}\right) \tag{5.59}$$

where y^*_{MP} is the value of y^* with $\boldsymbol{\theta} = \boldsymbol{\theta}_{\mathrm{MP}}$. The function y^* can be linearly approximated by Taylor expanding around the optimal parameters $\boldsymbol{\theta}_{\mathrm{MP}}$ with neglecting the high-order terms.

$$\begin{aligned} y^* &= y^*_{\mathrm{MP}} + \mathbf{c}^{\mathrm{T}}\Delta\boldsymbol{\theta} \\ &= \frac{1}{B}\sum_{b=1}^{B}\mathbf{W}^{\mathrm{out}}_{b,\mathrm{MP}} \cdot \left[\mathbf{u}^*; \mathbf{x}^*_b\right] + \mathbf{c}^{\mathrm{T}}\Delta\boldsymbol{\theta} \end{aligned} \tag{5.60}$$

where $\mathbf{c} = \nabla_{\boldsymbol{\theta}} y^*|_{\boldsymbol{\theta}_{\mathrm{MP}}} = \left[\mathbf{u}^*; \mathbf{x}^*_1; \mathbf{u}^*; \mathbf{x}^*_2; \ldots; \mathbf{u}^*; \mathbf{x}^*_B\right]^{\mathrm{T}}$, $\Delta\boldsymbol{\theta} = \boldsymbol{\theta} - \boldsymbol{\theta}_{\mathrm{MP}}$. Based on (5.57)–(5.60), the variance σ_t^2 can be calculated by $\sigma_t^2 = 1/\beta + \mathbf{c}^{\mathrm{T}}\mathbf{A}^{-1}\mathbf{c}$. According to (5.49), $\mathbf{A}^{-1}$ can be formulated by a diagonal matrix with its diagonal elements $(\lambda_i + \alpha)^{-1}$, where $(\lambda_i + \alpha)^{-1}$ is the eigenvalue of $\mathbf{A}^{-1}$. Then,

$$\mathbf{c}^{\mathrm{T}}\mathbf{A}^{-1}\mathbf{c} = \begin{bmatrix} \mathbf{c}_1 \\ \mathbf{c}_2 \\ \vdots \\ \mathbf{c}_W \end{bmatrix}^{\mathrm{T}} \begin{bmatrix} \frac{1}{\lambda_{\mathrm{MP}} + \alpha} & 0 & \cdots & 0 \\ 0 & 0 & \cdots & 0 \\ \vdots & \vdots & \ddots & \vdots \\ 0 & 0 & \cdots & 0 \end{bmatrix} \begin{bmatrix} \mathbf{c}_1 \\ \mathbf{c}_2 \\ \vdots \\ \mathbf{c}_W \end{bmatrix} \tag{5.61}$$

Since $\mathbf{c}_i$ is a combination of $\left[\mathbf{u}^i_b; \mathbf{x}^i_b\right]$, $b = 1, 2, \ldots, B$, i.e., $\mathbf{c}_i = \left[\mathbf{u}^i_1; \mathbf{x}^i_1; \mathbf{u}^i_2; \mathbf{x}^i_2; \ldots; \mathbf{u}^i_B; \mathbf{x}^i_B\right]$, then

$$\sum_{i=1}^{n}\mathbf{c}_i \cdot \mathbf{c}_i^{\mathrm{T}} \gg \sum_{i=1}^{n}\left[\mathbf{u}^i; \mathbf{x}^i\right] \cdot \left[\mathbf{u}^i; \mathbf{x}^i\right]^{\mathrm{T}} \tag{5.62}$$

Then, we have

$$\mathbf{c}^{\mathrm{T}}\mathbf{A}^{-1}\mathbf{c} < [\mathbf{u}^*; \mathbf{x}^*]^{\mathrm{T}}\mathbf{A}_{\mathrm{s}}^{-1}[\mathbf{u}^*; \mathbf{x}^*] \tag{5.63}$$

where $\mathbf{A}_{\mathrm{s}} = \beta\sum_{i=1}^{n}\left[\mathbf{u}^i; \mathbf{x}^i\right] \cdot \left[\mathbf{u}^i; \mathbf{x}^i\right]^{\mathrm{T}} + \alpha\mathbf{I}$ is the Hessian matrix for a single ESN.

Finally, one can draw a conclusion that $\sigma_t^2 < \sigma_r^2$, where σ_r^2 is the posterior variance of the output for the single ESN. Since σ_t^2 can be used to measure how far the prediction y^*_{MP} lies from the target t^*. Then, the bootstrap ESNs ensemble can obtain a higher accuracy than the network with only one ESN and will exhibit a better performance for PIs construction.

5.3.5 *Case Study*

To verify the effectiveness of the BESNE model shown in Fig. 5.3, two classes of prediction tasks are considered in this subsection. The first one is to construct the prediction intervals for the noisy multiple superimposed oscillator (MSO) problem that mentioned in [21]. The second one is a real-world problem that involves real-time flow prediction for by-product blast furnace gas (BFG) in steel industry. Based on the two problems, the BESNE method is verified to obtain high accuracy, low computational load, and high stability.

Multiple Superimposed Oscillator

To demonstrate the capability of the BESNE method in solving the MSO problem, we conduct the following experiments, in which the desired signal consists of two sine waves, $\sin(0.2n) + \sin(0.311n)$, $n = 1, 2, \ldots$. First, the BESNE method is employed to construct PIs for noisy MSO, in which a Gaussian white noise with the variance 0.01, regarded as the uncertainty, is added to the original MSO. Besides, the other three methods, including the Bayesian-based single ESN (Bayesian ESN), the Bayesian-based single multi-layer perceptron (MLP) (Bayesian MLP) and the bootstrapping MLP ensemble (BMLPE), are employed. Here, the iterative prediction is adopted, which is one-step ahead of prediction with a rolling forecast origin. Before the experiments for PIs construction, some experimental analysis are performed to determine the structure and the parameters of these prediction methods. The optimal parameters experimentally determined off-line are listed as Table 5.1.

Figure 5.4a shows the results by using the BESNE with the confidence level 95%, in which the target values can be completely covered by the constructed PIs. Considering that the Bayesian-based neural network could be an effective alternative reported in literature, Fig. 5.4b indicates the results by using the Bayesian ESN. It is apparent that there are a few of points observed outside the interval range, which indicates that the Bayesian ESN is inferior to the BESNE in terms of the coverage probability. In order to report the structural advantage of the ensemble, the multi-

Table 5.1 Parameters setting of the comparative experiments for noisy MSO problem

Model	Items	Values
BESNE	Number of ESNs in an ensemble	33
	Size of individual reservoirs	10
	Sparseness of internal weights	0.02
	Spectral radius of internal weights	0.8
	Number of training samples	600
Bayesian ESN	Size of individual reservoirs	10
	Sparseness of internal weights	0.02
	Spectral radius of internal weights	0.8
	Number of training samples	600
BMLPE	Number of MLPs in an ensemble	8
	Number of hidden units	10
	Number of input units	30
Bayesian MLP	Number of hidden units	10
	Number of input units	30

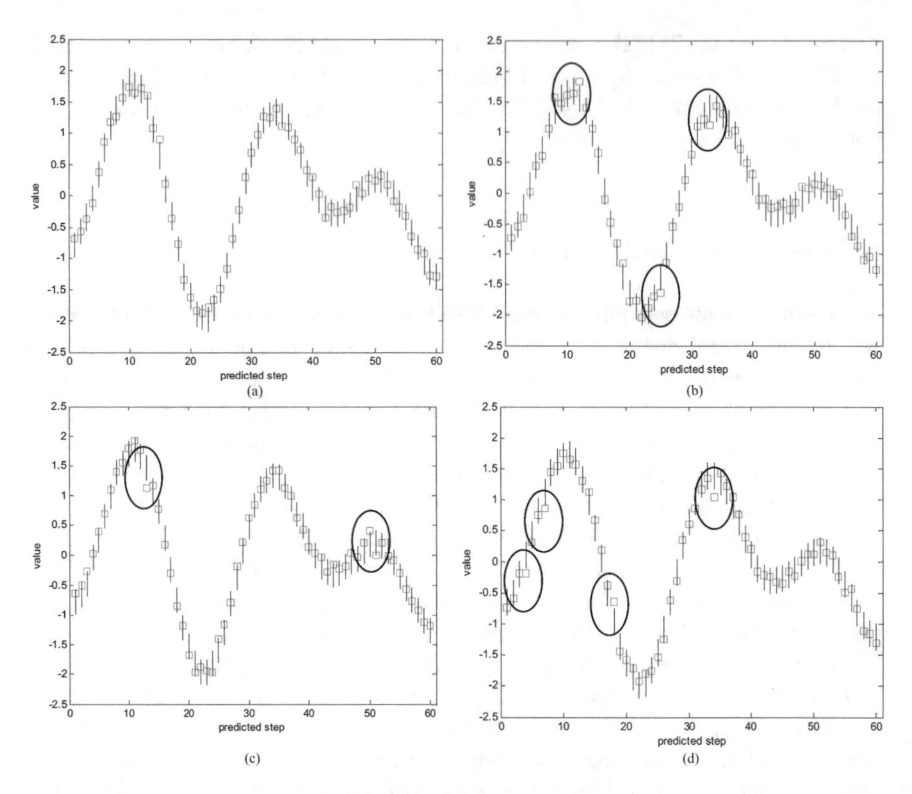

Fig. 5.4 (**a**) PIs constructed by the BESNE. (**b**) PIs constructed by the Bayesian ESN. (**c**) PIs constructed by the BMLPE. (**d**) PIs constructed by the Bayesian MLP

layer perceptron-based networks, the Bayesian MLP and the BMLPE, are also employed to perform the comparisons, and the results shown in Fig. 5.4c, d also illustrate that there are some data points beyond the interval range by these methods. A part of comparative PIs results are depicted in Fig. 5.5, where Fig. 5.5a gives the results by the BESNE and Fig. 5.5b–d comparatively provides the results by the above-mentioned methods. One can easily draw the conclusion that the BESNE exhibits the better performance on the prediction accuracy than the others. Thus, from Figs .5.4 and 5.5, it is obvious that the BESNE is more remarkable for the noisy time series prediction regarding both the coverage probability and the accuracy.

For a further comparison on the multiple evaluation indices, a series of statistical experimental results are reported in Table 5.2. Here, we use the RMSE defined by (2.59) in Chap. 2 to measure the prediction quality, and RMSE is frequently used to measure the difference between the predicted values and the targets. Observing from Table 5.2, the RMSE indicates that the BESNE model can effectively predict the noisy MSO problem with satisfactory accuracies. In addition, three evaluation criterions of PIs are also adopted here, including the PI coverage probability (PICP), the Mean PIs width (MPIW), and the combined index (CWC).

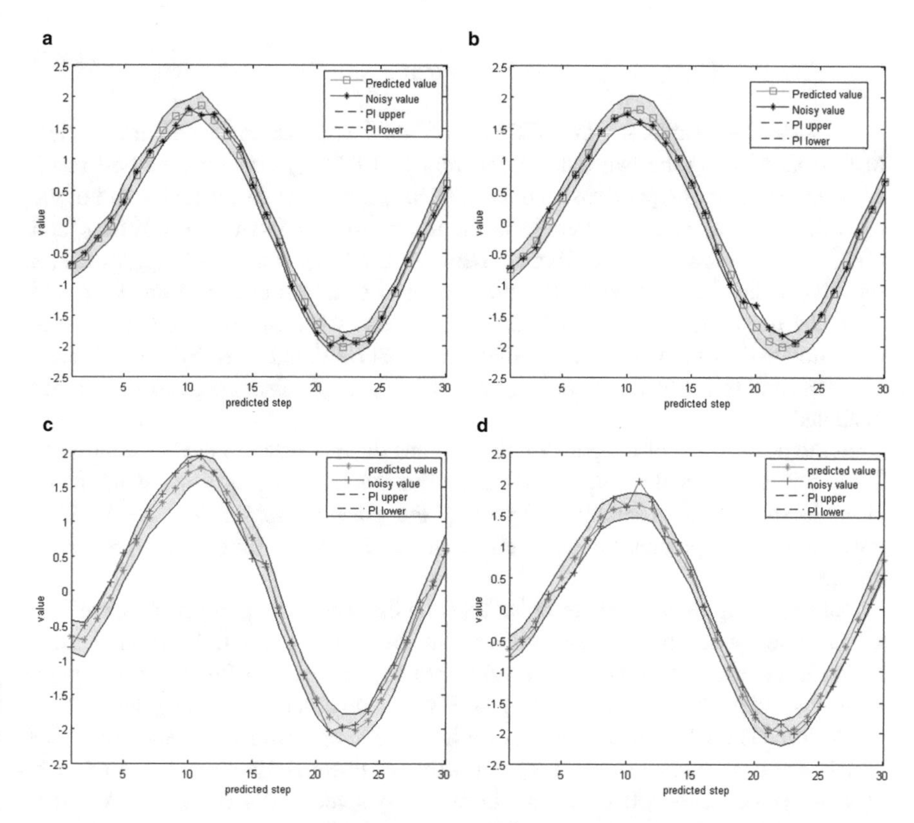

Fig. 5.5 (**a**) Prediction results based on BESNE. (**b**) Prediction results based on the Bayesian ESN. (**c**) Prediction results based on BMLPE. (**d**) Prediction results based on the Bayesian MLP

Table 5.2 Results of PIs: a comparative analysis

Method	CWC_{Best}	$\text{CWC}_{\text{Median}}$	PICP	MPIW	RMSE	Time (s)
Single ESN	0.3891	0.9054	0.9117	0.4180	0.1176	0.7009
Single MLP	0.3845	0.9569	0.9117	0.4056	0.1169	83.023
BMLPE	0.4307	0.7259	0.9209	0.4241	0.1183	440.9199
BESNE	0.3875	0.6157	0.9700	0.4025	0.1012	13.6078

$$\text{PICP} = \frac{1}{n_{\text{test}}} \sum_{i=1}^{n_{\text{test}}} c_i \tag{5.64}$$

$$\text{MPIW} = \frac{1}{n_{\text{test}}} \sum_{i=1}^{n_{\text{test}}} (U_i - L_i) \tag{5.65}$$

$$\text{CWC} = \text{MPIW}(1 + \gamma(\text{PICP})\exp(-\eta(\text{PICP} - \mu))) \tag{5.66}$$

where c_i is equal to 1 when the target is placed in the interval range; elsewise, c_i is equal to 0; η and μ are two hyper-parameters that control the location and the amount of CWC jump; γ(PICP) is given by

$$\gamma = \begin{cases} 0 & \text{PICP} \geq \mu \\ 1 & \text{PICP} < \mu \end{cases} \tag{5.67}$$

Besides, from Table 5.2, the BESNE exhibits the highest PICP and the smallest MPIW. Combining the two indices, the values of CWC_{Best} denotes the best result among the ten times repetitions, and the $\text{CWC}_{\text{Median}}$ denotes the average result of the repetitions. As such, one can evaluate the results from the values of CWC_{Best} and $\text{CWC}_{\text{Median}}$; moreover, the difference between CWC_{Best} and $\text{CWC}_{\text{Median}}$ is also a sensitive indication that represents the stability of the PIs construction. From the statistical results reported in Table 5.2, the CWCs by the BESNE are the best one compared to the others. And, the stability of the BESNE, indicated by the difference between the two CWCs (CWC_{Best} and $\text{CWC}_{\text{Median}}$), also outperformed other methods.

In the perspective of the computational loads that are listed in the last column of Table 5.2, although the single ESN costs the least computing time, it might be unstable sometimes. Thereby, considering the practical application, the BESNE presents the comprehensive good effectiveness for noisy time series prediction problems.

For analyzing the impacts of noise level on the prediction performance, we also give a series of comparative statistical results by using the multiple additive noise levels. In general, when the noise level increases, the width of PIs will enlarge. As such, it could be limitative to evaluate the PIs construction by only using the PICP or the MPIW. The CWC that considers both the coverage probability and the width variation is a comprehensive index to measure the uncertainty of the data. In such a way, the experiments with different additive noisy level are discussed here. We also conduct each method by ten repeated experiments, and the values of CWCs are reported in Table 5.3. From this table, the BESNE shows the least CWC_{Best}, and the median are also much closer to the CWC_{Best} than the others.

Table 5.3 Comparative experiments for the noisy MSO problem with different noisy level

Method	Index	$\sigma_t^2 = 0.005$	$\sigma_t^2 = 0.01$	$\sigma_t^2 = 0.015$
Single ESN	CWC_{Best}	0.5179	0.3891	0.6643
	$\text{CWC}_{\text{Median}}$	0.5219	0.9054	1.6714
	RMSE	0.0781	0.1176	0.1452
Single MLP	CWC_{Best}	0.2784	0.3845	0.5183
	$\text{CWC}_{\text{Median}}$	0.7423	0.9569	1.5709
	RMSE	0.0875	0.1169	0.1477
BMLPE	CWC_{Best}	0.3137	0.4307	0.6103
	$\text{CWC}_{\text{Median}}$	0.4872	0.7259	0.8758
	RMSE	0.0756	0.1183	0.1460
BESNE	CWC_{Best}	0.2759	0.3875	0.4653
	$\text{CWC}_{\text{Median}}$	0.4894	0.6157	0.7798
	RMSE	0.0737	0.1012	0.1318

Table 5.4 Parameters setting of the comparative experiments for generation flow of BFG

Model	Items	Values
BESNE	Number of ESNs in an ensemble	18
	Size of individual reservoirs	10
	Sparseness of internal weights	0.02
	Spectral radius of internal weights	0.8
	Number of training samples	500
Single ESN	Size of individual reservoirs	10
	Sparseness of internal weights	0.02
	Spectral radius of internal weights	0.8
	Number of training samples	500
BMLPE	Number of MLPs in an ensemble	8
	Number of hidden units	10
	Number of input units	30
Single MLP	Number of hidden units	10
	Number of input units	30

Application on Prediction for Generation Flow of BFG

In this subsection, we present a real-world prediction problem in industry by using the BESNE. We also take the BFG prediction as the example, which is the one of the useful energy resources in steel industry. And, it is very meaningful to predict the BFG flow for energy scheduling in a steel plant. However, the historical data on-site are always accompanied with noise, and the existing point-oriented predictions without indication of the accuracy are less reliable for the industrial applications.

Similarly, we consider the issue of coverage probability for the BFG generation flow, and the optimal parameters of each method experimentally determined off-line are listed as Table 5.4. The iterative prediction is adopted here for PIs construction, which is one-step ahead of prediction with a rolling forecast origin. Figure 5.6a shows the results based on the BESNE with the confidence level 95%, in which the target values can be completely covered by the constructed PIs. Figure 5.6b shows the results by the Bayesian ESN, where a few of points are observed outside the interval range, and Fig. 5.6c, d are those by the Bayesian MLP and the BMLPE, respectively. There are also some data points beyond the interval range, and it is apparent that the BESNE has the highest coverage probability among these four methods. For a clear presentation, a part of PIs is depicted in Fig. 5.7 by these methods. One can easily draw the conclusion that the BESNE exhibits the better performance on the accuracy than the others for the BFG generation flow prediction.

To further analyze the performance of the BESNE method for the generation flow prediction, a set of statistical experiments is conducted, and the results are shown in Table 5.5. In order to guarantee the indication of the statistical experiments, the PIs construction is repeated ten times. Viewing the RMSE of different methods listed in this table, the BESNE has the highest accuracy. The PICP of the BESNE is 0.9501

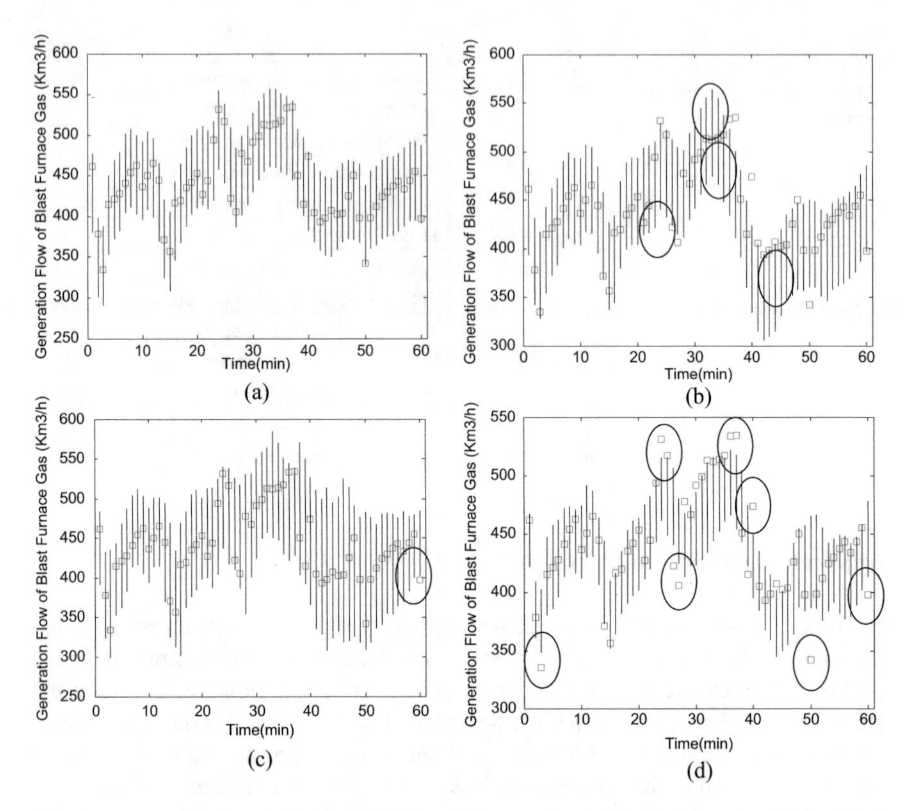

Fig. 5.6 (**a**) PIs constructed by the BESNE. (**b**) PIs constructed by the Bayesian ESN. (**c**) PIs constructed by the BMLPE. (**d**) PIs constructed by the Bayesian MLP

and the MPIW is 114.8689. Although these two values are not the best, the BESNE still outperforms other models when considering a middle course of the PICP and the MPIW denoted by the index CWC. As such, one can evaluate the results from the values of CWC_{Best} and CWC_{Median}; moreover, the difference between CWC_{Best} and CWC_{Median} is also a sensitive indication that represents the stability of the PIs construction. From the statistical results reported in Table 5.5, the CWCs by the BESNE are the best one compared to the others. And, the stability of the BESNE, the difference between the two CWCs, is also presented the superior performance.

In the perspective of the computational loads listed in the last column of Table 5.5, although the single ESN and the single MLP cost less computing time than the BESNE, they might be usually unstable and with low accuracy.

To sum up, from Figs. 5.6 and 5.7 and Table 5.5, considering the accuracy, the stability and the computational load, the BESNE exhibits the comprehensive good effectiveness for solving this industrial prediction, which can guide the decision-makers or the schedulers to make a reasonable operation with low risk.

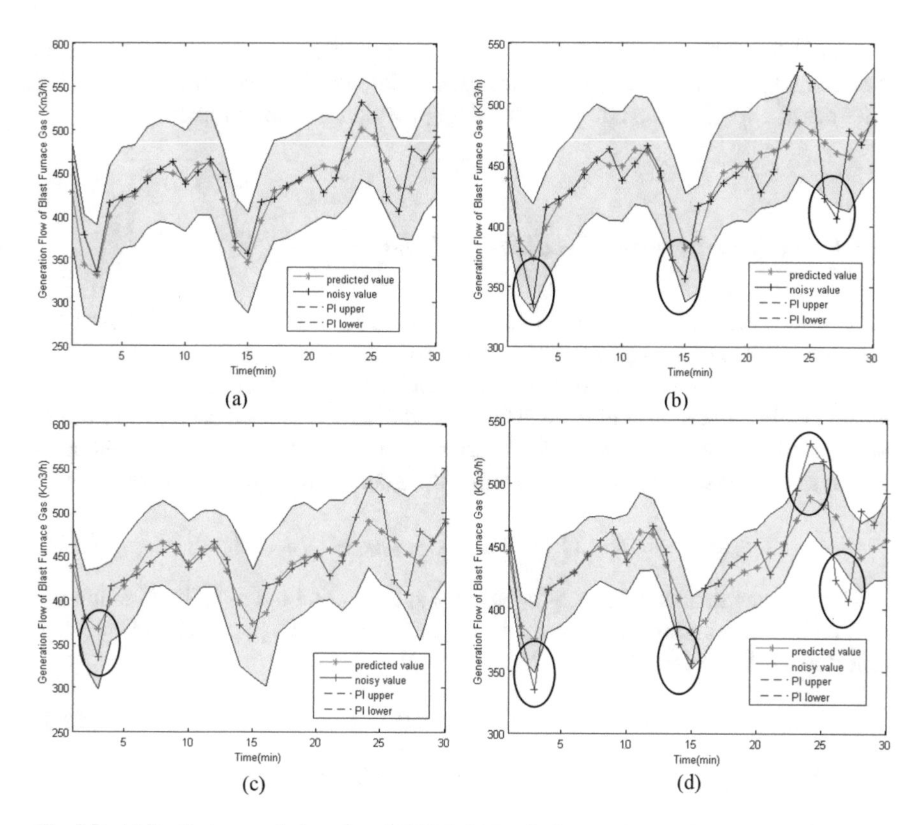

Fig. 5.7 (**a**) Prediction results based on BESNE. (**b**) Prediction results based on the Bayesian ESN. (**c**) Prediction results based on BMLPE. (**d**) Prediction results based on the Bayesian MLP

Table 5.5 Results of PIs: a comparative analysis

Method	CWC_{Best}	CWC_{Median}	PICP	MPIW	RMSE (km^3/h)	Time (s)
Single ESN	156.4658	269.7409	0.8445	90.6730	29.7990	0.7615
Single MLP	269.8002	675.9000	0.8056	88.6931	36.5910	46.7607
BMLPE	145.8961	202.3750	0.9800	185.5712	30.2600	470.9155
BESNE	115.5666	186.2997	0.9501	114.8689	27.1948	50.4813

5.4 Non-iterative NNs for PIs Construction

When using interval-weighted NNs for PIs construction, the iterative prediction mode is always accompanied with error accumulation that is rather negative to the reliability of the PIs. In this subsection, an interval-weighted ESN (IWESN) is developed for PIs construction, in which the network connections are represented by the interval-valued information granules. To cope with the error accumulation caused by the iterative mode, a non-iterative prediction mode for IWESN

(NI-IWESN) [22] is presented here. The training process of the IWESN can be viewed as the optimization of the allocation of information granularity, in which a PSO-based approach is employed for solving the optimization problem.

5.4.1 A Non-iterative Prediction Mode

Here, a non-iterative prediction mode is proposed for the granular ESN-based PIs construction. Considering a sequence of numeric data $\mathbf{x} = \{x(l)\}$, $l = 1, 2, \ldots, L$, it can be split into a collection of temporal segments with finite length of w, each of which can be viewed as a granule. The prediction model is built based on information granularity for the granular time series. In such a way, a number of temporal segments $\mathbf{s}_k(x)$ is constructed for modeling the granular time series, where

$$\mathbf{s}_k(x) = [x((k-1)w+1), x((k-1)w+2), \ldots, x(kw)]^{\mathrm{T}} \tag{5.68}$$

For a given granular sequence $\{\mathbf{s}_k(x)\}$, $k = 1, 2, \ldots, L/w$ and its split of granularity w, a nonlinear relationship between $\mathbf{s}_{k-n}(x), \ldots, \mathbf{s}_{k-2}(x), \mathbf{s}_{k-1}(x)$ and $\mathbf{s}_k(x)$ can reflect its dynamic characteristic [23], i.e.,

$$\mathbf{s}_k(x) = f(\mathbf{s}_{k-m}(x), \ldots, \mathbf{s}_{k-2}(x), \mathbf{s}_{k-1}(x)) + \mathbf{q}_k \tag{5.69}$$

Given the formula (5.68), one can redescribe this model by

$$x((k-1)w+i) = \mathbf{s}_k(x) \cdot \mathbf{e}_i, \quad 1 \le i \le w \tag{5.70}$$

If defining $y_i(k) = x(kw + i)$, $i = 1, 2, \ldots, w$, a new state-space representation can be formulated for a positive integer h. Then,

$$\begin{pmatrix} \mathbf{s}_{k+1}(x) \\ \mathbf{s}_{k+2}(x) \\ \vdots \\ \mathbf{s}_{k+h}(x) \end{pmatrix} = f\begin{pmatrix} \mathbf{s}_{k-m}(x), \ \ldots, \mathbf{s}_{k-1}(x), \mathbf{s}_k(x) \\ \mathbf{s}_{k+1-m}(x), \ \ldots, \mathbf{s}_k(x), \mathbf{s}_{k+1}(x) \\ \vdots \\ \mathbf{s}_{k+h-m}(x), \ \ldots, \mathbf{s}_{k+h-2}(x), \mathbf{s}_{k+h-1}(x) \end{pmatrix} + \begin{pmatrix} \mathbf{q}_{k+1} \\ \mathbf{q}_{k+2} \\ \vdots \\ \mathbf{q}_{k+h} \end{pmatrix} \tag{5.71}$$

$$\begin{pmatrix} y_i(k) \\ y_i(k+1) \\ \vdots \\ y_i(k+h-1) \end{pmatrix} = \begin{pmatrix} \mathbf{s}_{k+1}(x) \\ \mathbf{s}_{k+2}(x) \\ \vdots \\ \mathbf{s}_{k+h}(x) \end{pmatrix} \cdot \mathbf{e}_i, \quad 1 \le i \le w \tag{5.72}$$

where $\mathbf{e}_i \in \mathbb{R}^w$ denotes the ith elementary column vector and $\mathbf{q}_k$ are uncorrelated process Gaussian white noise with zero means. As for the above state-space, an ESN can be employed for the modeling task. Since each segment $\mathbf{s}_k(x)$ is a sequence with length w, a non-iterative prediction can be completed when the parameter h is equal to 1. And then, a sequence with length w can be predicted.

$$y_i(k) = s_{k+1}(x) \cdot \mathbf{e}_i, \quad i = 1, 2, \ldots, w \tag{5.73}$$

where the parameter w can be regarded as the prediction length.

As for the above non-iterative prediction mode, the selection of w is very significant. Generally, the best value of w is found simply through trial and error. In this section, a Gamma test-based selection method is proposed for optimizing the value of w.

5.4.2 Interval-Weighted ESN and Its Iterative Prediction

Interval-weighted NNs have been reported in [24, 25], whose connections are not numeric values but interval-valued granularity. As mentioned in Chap. 3, ESN is a kind of recurrent neural network (RNN), whose input and internal weights are randomly generated and fixed in the training process [26, 27]. Only the output weights need to be trained for ESN, thus the number of the unknown parameters is generally less than other NNs. Since ESN has a lower computational complexity, an interval-weighted ESN is developed in this section for PIs construction, whose structure is shown in Fig. 5.8. The connections of the interval-weighted ESN are represented by the interval-based information granules, i.e., the output weights $\mathbf{W}^{\text{out}}$ are no longer numeric values but an interval $[\mathbf{W}^{\text{out},-}, \mathbf{W}^{\text{out},+}]$. When a numeric input $\mathbf{u}(k)$ is given, the network output can also be an interval, denoted by $y_i(k) = \left[y_i^-(k), y_i^+(k)\right]$. In such a way, the recursive formula of an interval-weighted ESN reads as

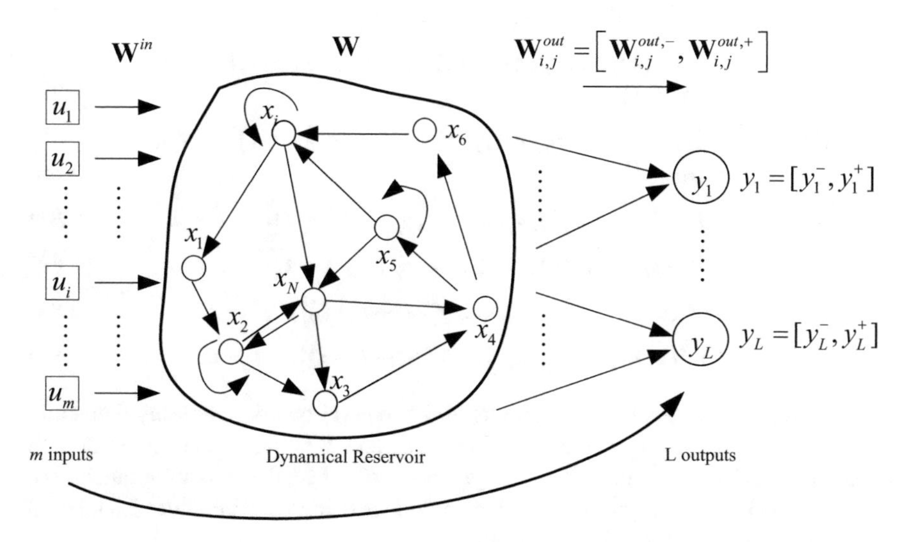

Fig. 5.8 The structure of the interval-weighted ESN

$$\mathbf{x}(k) = f\left(\mathbf{W}^{\text{in}}\mathbf{u}(k) + \mathbf{W}\mathbf{x}(k-1)\right) \tag{5.74}$$
$$y_i^-(k) = f^{\text{out}}\left(\mathbf{W}_i^{\text{out},-}(\mathbf{u}(k), \mathbf{x}(k))\right) \tag{5.75}$$
$$y_i^+(k) = f^{\text{out}}\left(\mathbf{W}_i^{\text{out},+}(\mathbf{u}(k), \mathbf{x}(k))\right) \tag{5.76}$$

where $\mathbf{u}(k)$ is the exogenous input; $\mathbf{x}(k)$ is the internal states; and $\mathbf{y}(k)$ is the output. $\mathbf{W}^{\text{in}} = \left(w_{i,j}^{\text{in}}\right)$ denotes the input weights with dimensionality $N \times m$; $\mathbf{W} = (w_{i,\,j})$ denotes the internal weights of the neurons in the reservoir with dimensionality $N \times N$. To provide proper memorization capabilities, $\mathbf{W}$ should be sparse whose connectivity level ranges from 1 to 5% and its spectral radius should be less than 1. $[\mathbf{W}^{\text{out},-}, \mathbf{W}^{\text{out},+}] = \left[w_{i,j}^{\text{out},-}, w_{i,j}^{\text{out},+}\right]$ denotes the interval-based output weights; f and f^{out} are the activation functions of the internal neurons and the output neurons, respectively. $y_i^-(k)$ and $y_i^+(k)$ are the lower and upper bounds of the PIs constructed by this non-iterative ESN.

Iterative prediction is always considered for ESN-based time series prediction, in which along with the iterative predicting, the predicted outputs are continually regarded as the input of the network for the next-time prediction [28–30]. In this case, the error accumulation with the increase of the iterative steps cannot be avoided unless the predicted targets can be fully estimated without any bias. However, this is almost impossible. As for the point-oriented prediction, the accumulated error might be accepted with respect to some real-world problems. However, as for the PIs construction, it is the upper and lower bounds of PIs that are used as the new input of the network for the next-time prediction (see (5.77)–(5.84)), which might lead to a serious error accumulation.

$$\mathbf{u}^-(k+l) = \left[u_{l+1}(k), \ldots, u_m(k), y_i^-(k)\right], \quad l = 1 \tag{5.77}$$
$$\mathbf{u}^+(k+l) = \left[u_{l+1}(k), \ldots, u_m(k), y_i^+(k)\right], \quad l = 1 \tag{5.78}$$
$$\mathbf{u}^-(k+l) = \left[u_{l+1}(k), \ldots, u_m(k), y_i^-(k), \ldots, y_i^-(k+l-1)\right], \quad 1 < l < (m-1) \tag{5.79}$$
$$\mathbf{u}^+(k+l) = \left[u_{l+1}(k), \ldots, u_m(k), y_i^+(k), \ldots, y_i^+(k+l-1)\right], \quad 1 < l < (m-1) \tag{5.80}$$
$$\mathbf{u}^-(k+l) = \left[u_m(k), y_i^-(k), \ldots, y_i^-(k+l-1)\right], \quad l = m \tag{5.81}$$
$$\mathbf{u}^+(k+l) = \left[u_m(k), y_i^+(k), \ldots, y_i^+(k+l-1)\right], \quad l = m \tag{5.82}$$
$$\mathbf{u}^-(k+l) = \left[y_i^-(k), \ldots, y_i^-(k+l-1)\right], \quad l > m \tag{5.83}$$
$$\mathbf{u}^+(k+l) = \left[y_i^+(k), \ldots, y_i^+(k+l-1)\right], \quad l > m \tag{5.84}$$

Considering the algebraic operations of the interval-based granularity defined in [31], one can assume two intervals $\mathbf{X} = [a, b]$ and $\mathbf{Y} = [c, d]$, and their addition operation of $\mathbf{X}$ and $\mathbf{Y}$ follows $\mathbf{X} + \mathbf{Y} = [a + c, b + d]$. Then, the activation function of the neurons in the reservoir is monotone non-decreasing function, thus the internal states at the $k + l$ step can be computed by

$$\mathbf{x}^{-}(k+l)=f\left(\mathbf{W}^{\text{in}}\mathbf{u}^{-}(k+l)+\mathbf{W}\mathbf{x}^{-}(k+l-1)\right) \tag{5.85}$$

$$\mathbf{x}^{+}(k+l)=f\left(\mathbf{W}^{\text{in}}\mathbf{u}^{+}(k+l)+\mathbf{W}\mathbf{x}^{+}(k+l-1)\right) \tag{5.86}$$

If the activation function of the output neurons is generalized identity function, the outputs of the granular network at $k + l$ step can be described as follows, according to the multiplication operation $\mathbf{X} \times \mathbf{Y} = [\min(ac, ad, bc, bd), \max(ac, ad, bc, bd)]$.

$$y_i^{--}(k+l) = \mathbf{W}_i^{\text{out},-}(\mathbf{u}^{-}(k+l), \mathbf{x}^{-}(k+l)) \tag{5.87}$$

$$y_i^{-+}(k+l) = \mathbf{W}_i^{\text{out},-}(\mathbf{u}^{+}(k+l), \mathbf{x}^{+}(k+l)) \tag{5.88}$$

$$y_i^{+-}(k+l) = \mathbf{W}_i^{\text{out},+}(\mathbf{u}^{-}(k+l), \mathbf{x}^{-}(k+l)) \tag{5.89}$$

$$y_i^{++}(k+l) = \mathbf{W}_i^{\text{out},+}(\mathbf{u}^{+}(k+l), \mathbf{x}^{+}(k+l)) \tag{5.90}$$

The bounds of the output interval read as

$$y_i^{-}(k+l) = \min\left(y_i^{--}(k+l), y_i^{-+}(k+l), y_i^{+-}(k+l), y_i^{++}(k+l)\right) \tag{5.91}$$

$$y_i^{+}(k+l) = \max\left(y_i^{--}(k+l), y_i^{-+}(k+l), y_i^{+-}(k+l), y_i^{++}(k+l)\right) \tag{5.92}$$

It is apparent that based on the above inference, the iterative prediction is almost a curse on the interval-based granular computing.

5.4.3 Gamma Test-Based Model Selection

The Gamma test technique tends to determine the embedding dimension of a time series by estimating its effective noise variance [32]. The idea of the Gamma test can be briefly described here. As for a time series $\mathbf{x} = \{x(l)\}$, $l = 1, 2, \ldots, L$, a set of input-output sample pairs can be formed by $\{\mathbf{x}_i, y_i\}$, $1 \le i \le N$, where $\mathbf{x}_i$ is the input vector with dimensionality m. First, one can calculate the average distance between the kth-nearest-neighbor of $\mathbf{x}_i$ in the input set $\{\mathbf{x}_1, \ldots, \mathbf{x}_N\}$ and $\mathbf{x}_i$,

$$\delta_N(k) = \frac{1}{N}\sum_{i=1}^{N}\left\|\mathbf{x}_{\text{NN}[i,k]} - \mathbf{x}_i\right\|^2 \tag{5.93}$$

where $\mathbf{x}_{\text{NN}[i,k]}$ denotes the kth-nearest-neighbor of $\mathbf{x}_i$, NN$[i, k]$ denotes the possible multiple nearest neighbors of $\mathbf{x}_i$; and $1 \le k \le p$, where p stands for some integer with the values ranging from 10 to 50 [33].

Second, the average distance between the corresponding outputs can be calculated by

$$\gamma_N(k) = \frac{1}{2N}\sum_{i=1}^{N}\left|y_{\text{NN}[i,k]} - y_i\right|^2 \tag{5.94}$$

where $y_{\mathrm{NN}[i,k]}$ is the output value associated to $\mathbf{x}_{\mathrm{NN}[i,k]}$, but not necessarily the kth-nearest-neighbor y_i of in output space.

Third, the variance of noise should be calculated, then p pairs of input-output points denoted as $(\delta_{N(k)}, \gamma_{N(k)})$ can be obtained. The Gamma test employs a linear regression technique to estimate the intercept value of these p points when $\delta_{N(k)} \to 0$ with the assumption that the relationship between the points $(\delta_{N(k)}, \gamma_{N(k)})$ is approximately linear if N is sufficiently large. The intercept value is viewed as the effective noise variance Γ of the time series, one can see details in [34]. When Γ obtains the minimum value, the optimal parameters are achieved. In this case, the representation of the phase space is the most effective to reflect the dynamic characteristics of the time series.

In this section, a novel model selection method based on the Gamma test is reported to estimate the optimal value of w, and its solution process can be addressed as follows. For a set of granules $\mathbf{s}_k(x)$ with the representation of (5.68), one can choose a suitable range for the parameters w, where $w \in [w_0, w_n]$. According to the state-space represented in (5.71), a new phase space can be rewritten as

$$\begin{pmatrix} \mathbf{s}_{k-m+1}(x) & \cdots & \mathbf{s}_{k-1}(x) & \mathbf{s}_k(x) & \to & \mathbf{s}_{k+1}(x) \\ \mathbf{s}_{k-m}(x) & \cdots & \mathbf{s}_k(x) & \mathbf{s}_{k+1}(x) & \to & \mathbf{s}_{k+2}(x) \\ \mathbf{s}_{k-m+1}(x) & \cdots & \mathbf{s}_{k+1}(x) & \mathbf{s}_{k+2}(x) & \to & \mathbf{s}_{k+3}(x) \\ \vdots & & \vdots & \vdots & \vdots & \vdots \\ \mathbf{s}_{k-m+h-2}(x) & \cdots & \mathbf{s}_{k+h-2}(x) & \mathbf{s}_{k+h-1}(x) & \to & \mathbf{s}_{k+h}(x) \end{pmatrix} \tag{5.95}$$

According to (5.68), $\mathbf{s}_k(x)$ is a temporal segment with the length w. Substituting (5.68) into (5.95), an input $\mathbf{s}_{i-\eta}(x), \ldots, \mathbf{s}_{i-2}(x), \mathbf{s}_{i-1}(x)$ can be formed as a new input $\mathbf{x}_i$ with dimensionality of $\eta \times w$, and the output $\mathbf{s}_i(x)$ can be represented as $\mathbf{y}_i$ with the dimensionality of w.

In this section, the Gamma test is employed to determine the best values of m and w with an improvement of (5.95).

$$\gamma_N(k) = \frac{1}{2N} \sum_{i=1}^{N} \left\| \mathbf{y}_{\mathrm{NN}[i,k]} - \mathbf{y}_i \right\|^2 \tag{5.96}$$

Given that the parameter p in the Gamma test has a strong effect on the linear mapping of the input-output pairs $(\delta_{N(k)}, \gamma_{N(k)})$, we estimate the effective noise variance by employing the cross-validation-based approach to determine the appropriate value of p in its setting range.

To achieve a good performance of the PIs, interval-valued output weights $\mathbf{W}^{\mathrm{out}} = [\mathbf{W}^{\mathrm{out},-}, \mathbf{W}^{\mathrm{out},+}]$ are very important, which depends on the optimal allocation of information granularity controlled by the parameters $\varepsilon_{i,j}$. It is assumed that the information granularity is allocated based on a set of original parameters $w_{i,j}^{\mathrm{out}}$ that can be estimated by the SVD. Given a set of $\varepsilon_{i,j}$, the corresponding bounds of the interval-valued output weights can be described as

$$w_{i,j}^{\text{out},-} = w_{i,j}^{\text{out}} - (-\varepsilon_{i,j}^{-}|w_{i,j}^{\text{out}}|) = w_{i,j}^{\text{out}} + \varepsilon_{i,j}^{-}\left|w_{i,j}^{\text{out}}\right| \tag{5.97}$$

$$w_{i,j}^{\text{out},+} = w_{i,j}^{\text{out}} + \varepsilon_{i,j}^{+}\left|w_{i,j}^{\text{out}}\right| \tag{5.98}$$

where $w_{i,j}^{\text{out}}$ denotes the weight of the *j*th interval unit to the *i*th output unit, $1 \le i \le L$ and $1 \le j \le N$. $\varepsilon_{i,j}^{-}$ and $\varepsilon_{i,j}^{+}$ control the lower and upper bounds of the output weights, respectively.

Optimizing the allocation of information granularity is a very challenging work. Since the information granularity of the output weights is designed for PIs construction, the optimal granular allocation can be achieved by optimizing the performance evaluation indexes of the PIs including the *PICP* and the *MPIW*, which are introduced in Sect. 5.3.5.

In general, a higher PIs coverage probability and a smaller PIs width are required for a set of PIs. However, *PICP* and *MPIW* are two conflicted evaluation indexes, i.e., the higher the *PICP* is, and the larger the *MPIW* is. The evaluation index *CWC* that a function of the *PICP* and *MPIW* can address this confliction [35]. Although *CWC* is an effective criterion for PIs evaluation, a small *CWC* might be obtained once the values of the *NMPIW* and the PICP are simultaneously very low or very high. In this case, the abnormal solution for the allocation of information granularity could happen. Thus, we add the constraints to the optimization objectives, i.e., the values of *PICP* and *NMPIW* must be constrained in the process of optimization. If they are not satisfied with the constraints, a solution with a much smaller objective during one evolution cannot be chosen as the local or global optimal solution. In this case, the optimization process of the allocation of information granularity is more reasonable.

In this subsection, a PSO-based optimization method is designed here. A particle i can be represented by a position vector $\mathbf{x}_i$ and a velocity vector $\mathbf{v}_i$, where $\mathbf{x}_i = [x_{i,\,1}, x_{i,\,2}, \ldots, x_{i,\,D}]$, $\mathbf{v}_i = [v_{i,\,1}, v_{i,\,2}, \ldots, v_{i,\,D}]$ and $i = 1, 2, \ldots, N$. N is the number of particles in a population and D is the number of decision variables. Each particle in the population should update its own velocity and position according to CWC during evolution [36].

$$\begin{cases} \mathbf{v}_i(t+1) = \omega\mathbf{v}_i(t) + c_1 r_1(\mathbf{pBest}_i - \mathbf{x}_i(t)) + c_2 r_2(\mathbf{gBest} - \mathbf{x}_i(t)) \\ \mathbf{x}_i(t+1) = \mathbf{x}_i(t) + \mathbf{v}_i(t+1) \end{cases} \tag{5.99}$$

where t is the iteration time, $\omega \ge 0$ is the inertia weight coefficient, c_1, $c_2 \ge 0$ is the acceleration coefficient, r_1 and r_2 are uniformly distributed random numbers in the closed interval [0,1], $\mathbf{pBest}_i$ is the local optimal solution of the *i*th particle and $\mathbf{gBest}$ is the global optimal solution.

5.4.4 Case Study

To verify the performance of the NI-IWESN method, two benchmark prediction problems, the noisy MSO and the noisy Mackey Glass time series are employed in this section. The desired MSO signal has been introduced in Sect. 5.3.5. And the Mackey Glass time series is generated by a time delay differential system reported in [37]. To simulate the noisy environment of the dataset, the Gaussian white noises with variance 0.001 are added to these two benchmark problems. In addition, as for the real-world practical data, we also employ two industrial tasks for the validation experiments.

Two Benchmark Prediction Problems

First, we perform the experiments for the optimal parameters w by using the Gamma test technique, where the number of the validation data samples is set to 1200 and the parameter p is set to 50. To reduce the computational complexity, the embedded dimensionality of the network is set to several times the value of the parameter w. The results are shown in Fig. 5.9, where Fig. 5.9a is for the noisy MSO problem and Fig. 5.9b is for the noisy Mackey Glass problem. From Fig. 5.9a when $m = 3 \times w$ and $w = 21$, a very little validation error can be achieved. From Fig. 5.9b, when $m = 4 \times w$ and $w = 61$, the least validation error can be achieved for the noisy Mackey Glass problem.

Second, the PSO technique is used to training the NI-IWESN, where the size of dynamic reservoir is set as 200, the sparseness of interval weights is set as 0.02, and the spectral radius of interval weights is set as 0.8. The number of the training samples is set to 900 and 1100 for the noisy MSO and Mackey Glass problems, respectively. As for the parameters of the PSO method, the size of the particle swarm i s set as 200, since the number of the unknown parameters is around 200. And the count of the loops of the PSO for parameters is set to 300 according to the trial and

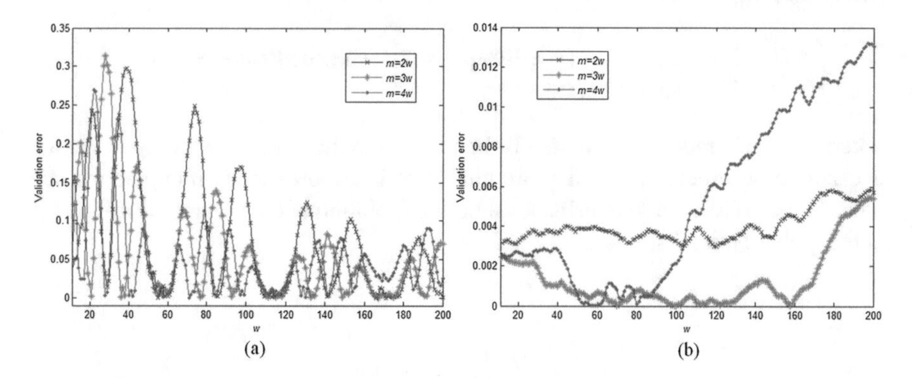

Fig. 5.9 Chart of Gamma test-based estimation for the size of the granule. (**a**) The noisy MSO problem. (**b**) The noisy Mackey Glass problem

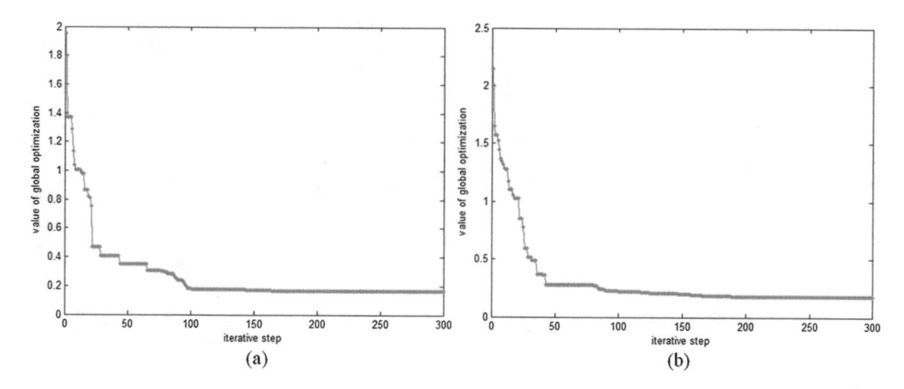

Fig. 5.10 The training error curve of the PSO for the NI-IWESN. (**a**) The noisy MSO problem. (**b**) The noisy Mackey Glass problem

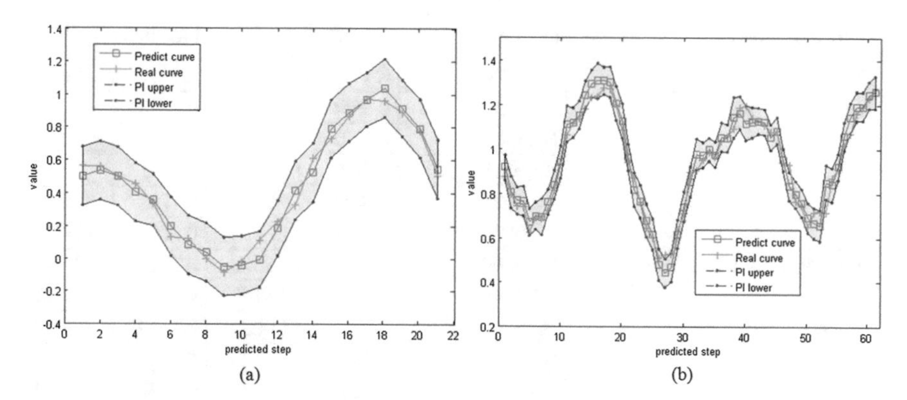

Fig. 5.11 PIs constructed by the NI-IWESN for the noisy MSO and noisy Mackey Glass problem. (**a**) The noisy MSO problem. (**b**) The noisy Mackey Glass problem

error procedure. The training error curves by using the PSO technique for those noisy problems are shown in Fig. 5.10, where the training processes of those two problems are both convergent within less than 300 steps.

Other 500 samples different from the training samples are adopted for validating the prediction model, see Fig. 5.11. The length of the PIs is set as the determined parameter w, which can be seen in Fig. 5.9. From Fig. 5.11, we see that the PIs constructed by the NI-IWESN have 100% coverage probability, and the width of the PIs is also very reasonable. Since the NI-IWESN is a kind of non-iterative prediction mode without any error accumulation, we try to construct much longer term PIs for these two prediction problems. Here, we set the number of the output units is equal to twice w, and the results are shown in Fig. 5.12, where the long-term PIs constructed by the NI-IWESN exhibit nearly 100% coverage probability and a reasonable width of the PIs. Especially for the noisy Mackey Glass problem, although the prediction length is equal to 122, the PIs are not diverging from the real value.

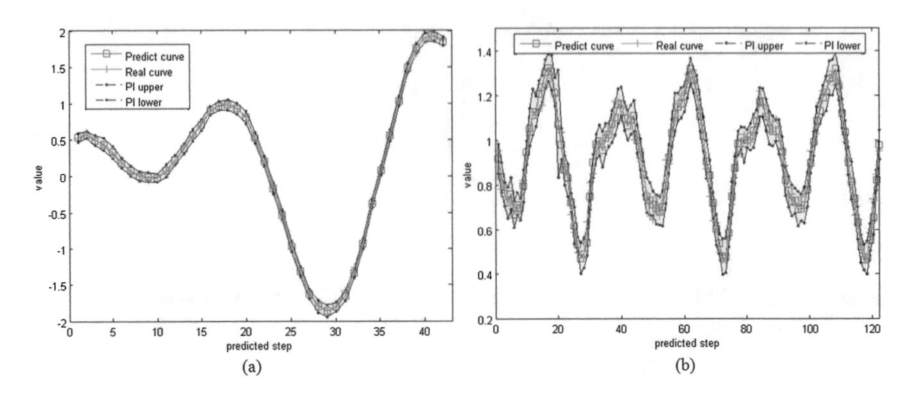

Fig. 5.12 A much longer term PIs constructed by the NI-IWESN. (**a**) The noisy MSO problem. (**b**) The noisy Mackey Glass problem

Table 5.6 Comparison of the statistical results of PIs constructed by different methods

Problems	Methods	CWC	PICP	MPIW	Time (s)
Noisy MSO problem	Delta MLP	0.0851	0.9618	0.2518	21.76
	MVE MLP	0.0825	0.9500	0.2306	20.56
	Bayesian MLP	0. 0643	1.0000	0.1945	50.48
	Bootstrap MLP	0.0418	0.9876	0.1791	466.71
	NI-IWESN	0.0416	1.0000	0.1726	814.24
Noisy Mackey Glass time series	Delta MLP	0.2311	0.9600	0.2141	27.71
	MVE MLP	0.2401	0.9753	0.2314	21.25
	Bayesian MLP	0.1789	1.0000	0.1689	59.43
	Bootstrap MLP	0.1842	0.9828	0.1746	487.61
	NI-IWESN	0.1695	0.9859	0.1672	1251.19

To further evaluate the performance, a series of statistical experimental results are reported in Table 5.6. To guarantee the full indication of the statistical experiments, the *PICP*, *MPIW*, *CWC,* and computing time are employed here for statistic. To facilitate the comparison, the prediction lengths for all those comparative methods are set to 60. From Table 5.6, it can be observed that the NI-IWESN model presents the best performance for those two noisy problems with the overall consideration of computing time, *PICP*, *MPIW,* and *CWC*. In Table 5.6, the performance of the non-iterative granular ESN is superior in most cases.

Interval Prediction on the By-Product Gas System in Steel Industry

Here, we also address two real-world industrial problems by using the NI-IWESN. Steel industry is usually accompanied with high energy consumption and environmental pollution, in which the by-product gas is one of the useful energy resources.

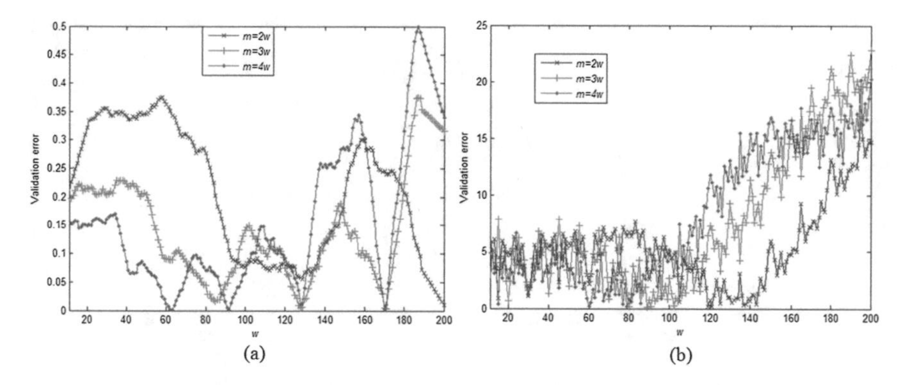

Fig. 5.13 Chart of Gamma test-based estimation for the size of the granule. (**a**) Generation gas flow of #3 blast furnace. (**b**) The BFG flow consumed by the #1,2 coke oven

There is a need to predict the gas generation and consumption flow for energy scheduling in a steel plant. The experimental data, with a sample interval of 1 min, come from a steel plant in China, and cover the generation gas flow of the blast furnace and the consumption flow of BFG by the ovens in 2015. The original dataset consists of 4000 data samples. First, we conduct the experiments to determine w, and the results are shown in Fig. 5.13. As for the generation gas flow of the #3 blast furnace, when $m = 4 \times w$ and $w = 63$, a very little validation error can be achieved from Fig. 5.13a. From Fig. 5.13b, when $m = 3 \times w$and $w = 89$, the least validation error can be achieved for the BFG flow consumed by the #1,2 coke oven. The other parameters of the non-iterative model for the gas flow are set as follows. The size of dynamic reservoir of ESN is set as 100, the sparseness of interval weights is set as 0.02, and the spectral radius of interval weights is set as 0.8. As for those two industrial problems, 1100 training samples are used to train the model. As for the parameters of the PSO, the size of the particle swarm is set as 400, since the number of the unknown parameters of those two problems is inferior to 400. And the count of the loops of the PSO for parameters is equal to 500 according to trial and error. The training error curves of the PSO for those two industrial problems are shown in Fig. 5.14, where it can be seen that both the training processes are convergent.

Other 500 samples are randomly selected from the original set of the samples for testing the NI-IWESN, and the PIs constructed for the by-product gas flow are shown in Fig. 5.15. The length of the PIs is set as the determined parameter w, which can be seen in Fig. 5.13. From Fig. 5.15, we see that the PIs constructed by the NI-IWESN have a very high coverage probability, and the width of the PIs is also reasonable. Since the NI-IWESN is a kind of non-iterative prediction mode without any error accumulation, we try to construct much longer term PIs for these two tasks. Here, we set the number of the output units is equal to twice w, and the results are shown in Fig. 5.16, where the long-term PIs constructed by the NI-IWESN still exhibit a high coverage probability and a reasonable width of the PIs.

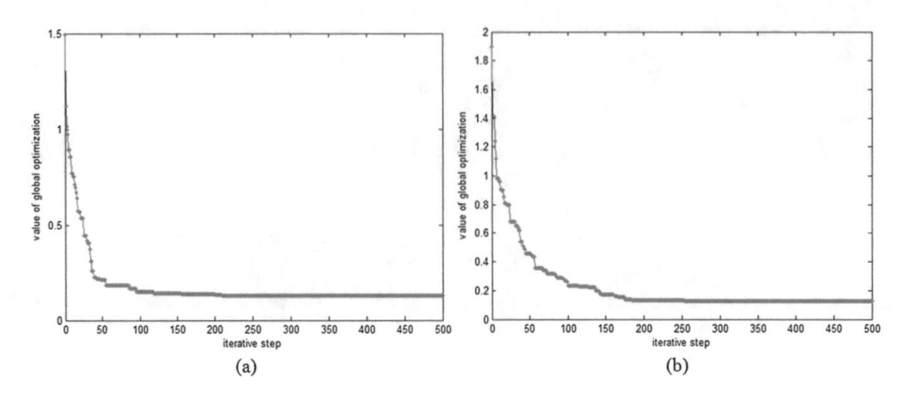

Fig. 5.14. The training error curve of the PSO for the NI-IWESN. (**a**) Generation gas flow of #3 blast furnace. (**b**) The BFG flow consumed by the #1,2 coke oven

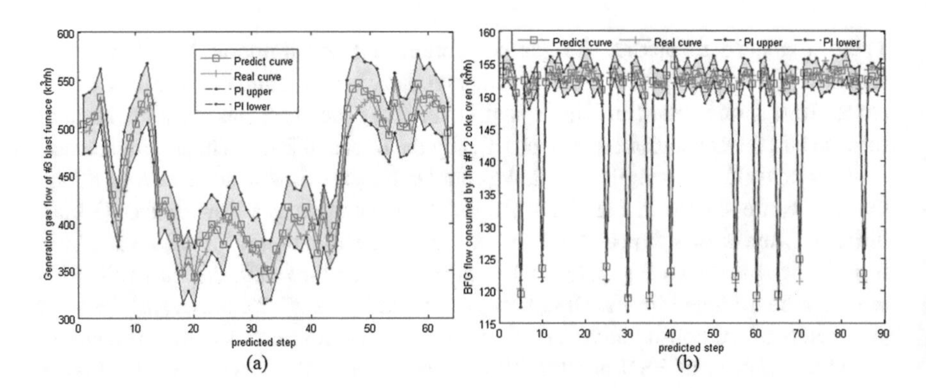

Fig. 5.15 PIs constructed by the NI-IWESN for two industrial time series. (**a**) Generation gas flow of #3 blast furnace. (**b**) The BFG flow consumed by the #1,2 coke oven

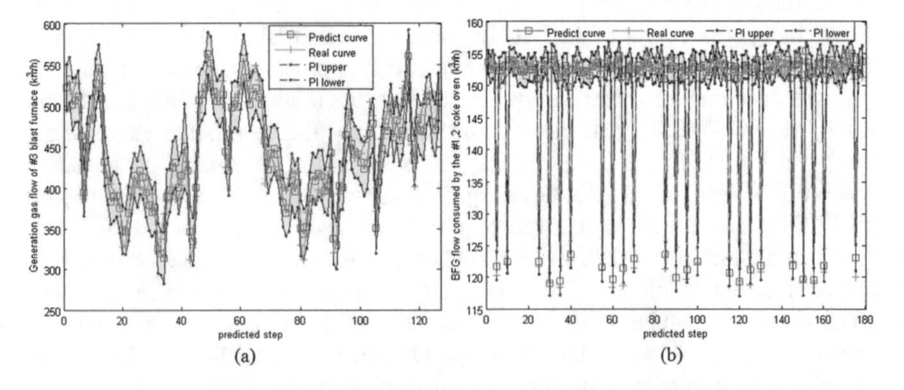

Fig. 5.16. A much longer term PIs constructed for two industrial time series. (**a**) Generation gas flow of #3 blast furnace. (**b**) The BFG flow consumed by the #1,2 coke oven

Table 5.7 Comparison of the statistical results of PIs constructed by different methods

	Method	CWC	PICP	MPIW	Time (s)
Generation gas flow of the #3 blast furnace	Delta MLP	1748.52	0.8167	126.8344	34.12
	MVE MLP	7703.33	0.7667	126.0083	21.02
	Bayesian MLP	247.31	0.8095	95.4028	64.76
	Bootstrap MLP	9077.06	0.8000	124.8035	470.92
	NI-IWESN	1724.87	0.8167	125.1189	1168.39
BFG flow consumed by the #1,2 coke oven	Delta MLP	0.1985	0.9333	3.6064	26.71
	MVE MLP	0.2014	0.9333	3.6593	20.15
	Bayesian MLP	0.2093	0.9167	3.5076	75.49
	Bootstrap MLP	0.1908	0.9326	3.5785	486.76
	NI-IWESN	0.1069	0.9500	4.0287	1860.09

A series of statistical experimental results are also reported in Table 5.7 for comparing the performance of these methods. To guarantee the full indication of the experiments, the *PICP*, *MPIW*, *CWC,* and the computing time are employed here. To facilitate the comparison, the prediction lengths for all those comparative methods are set to 60. From Table 5.7, it is observed that the NI-IWESN shows the best performance for the BFG flow consumed by the #1,2 coke oven with the overall consideration of these evaluation criterions. Based on the MapReduce framework, the computational precision and efficiency of the non-iterative granular ESN can be greatly improved. In this chapter, four problems are employed to verify the performance of the NI-IWESN, in which the NI-IWESN shows the best performance for three problems. As for the generation gas flow of the #3 blast furnace, the performance of the NI-IWESN is not the best but superior to the MLP-based Delta, MVE, and Bootstrap methods. All the experimental results demonstrate that the NI-IWESN can construct effective PIs for noisy time series without iteration.

5.5 Gaussian Kernel-Based Causality PIs Construction

As introduced in Chap. 4, besides the NNs-based models, the commonly used factors-based prediction models involve support vector regression (SVR) model. However, since the parameters of the SVM include the bias parameter *b*, it will raise the difficulty when using SVM for PIs construction. In [38], it reported a method based on SVM for PIs construction, from which we can see that the method costs a very high computational load. In this section, we consider another class of kernel-based model for PIs construction, a mixed Gaussian kernel-based model.

5.5.1 Mixed Gaussian Kernel for Regression

Let us first recall a generalized kernel-based learning. Given the noise widely existing in practical systems, the observed value of a system is expressed as

$$t_n = f(\mathbf{x}_n, \mathbf{w}) + \gamma_n \tag{5.100}$$

where t_n is the observed output, $f(\mathbf{x}_n, \mathbf{w})$ is the model output, $\mathbf{x}_n$ is the input, $\mathbf{w}$ is the parameter of the regression model, and γ_n denotes a Gaussian white noise with mean zero. Assuming that $f(\mathbf{x}_n, \mathbf{w}) = \mathbf{w}^{\mathrm{T}}\phi(\mathbf{x}_n)$, where $\phi(\cdot)$ is a mapping function to a high dimensional space, the parameters can be calculated by minimizing a regularization expression [39], i.e.,

$$J(w) = \frac{1}{2}\sum_{n=1}^{N}\left[\mathbf{w}^{\mathrm{T}}\phi(\mathbf{x}_n) - t_n\right]^2 + \frac{\lambda}{2}\mathbf{w}^{\mathrm{T}}\mathbf{w} \tag{5.101}$$

where $\lambda \geq 0$ represents a penalty coefficient and N is the number of the sample data. We compute the partial derivative of (5.32) with respect to $\mathbf{w}$ and setting it equal to zero. Then, the dual expression of the original objective function reads as

$$J(\mathbf{a}) = \frac{1}{2}\mathbf{a}^{\mathrm{T}}\mathbf{K}\mathbf{K}\mathbf{a} - \mathbf{a}^{\mathrm{T}}\mathbf{K}\mathbf{t} + \frac{1}{2}\mathbf{t}^{\mathrm{T}}\mathbf{t} + \frac{\lambda}{2}\mathbf{a}^{\mathrm{T}}\mathbf{K}\mathbf{a} \tag{5.102}$$

where $\mathbf{t} = (t_1, t_2, \ldots, t_N)^{\mathrm{T}}$, $\mathbf{K}$ is a matrix, its each element reads as $k(\mathbf{x}_n, \mathbf{x}_m) = \phi(\mathbf{x}_n)^{\mathrm{T}}$ $\phi(\mathbf{x}_m)$, and $\mathbf{a} = (\mathbf{K} + \lambda\mathbf{I}_N)^{-1}\mathbf{t}$. The regression model is formulated as

$$f(\mathbf{x}, \mathbf{w}) = \mathbf{w}^{\mathrm{T}}\phi(\mathbf{x}) = \sum_{n=1}^{N} a_n\phi(\mathbf{x}_n)\phi(\mathbf{x}) = \mathbf{k}(\mathbf{x})^{\mathrm{T}}\mathbf{a} \tag{5.103}$$

where $\mathbf{k}(\mathbf{x})$ is a N-dimensional vector (N kernel functions), each of which can be expressed as $k_n(\mathbf{x}) = k(\mathbf{x}_n, \mathbf{x})$. And, the commonly used kernel function is the Gaussian one in various applications [28, 29]. However, the assumed regression $f(\mathbf{x}_n, \mathbf{w}) = \mathbf{w}^{\mathrm{T}}\phi(\mathbf{x}_n)$ lacks the bias term when compared to the generic SVM [40], i.e., the accuracy could be largely impacted.

One can incorporate two different Gaussian kernels to form a new mixed Gaussian kernel for constructing the regression model, each of which can be regarded as a coordinator of another one. That is,

$$k(\mathbf{x}_i, \mathbf{x}_j) = \beta k_1(\mathbf{x}_i, \mathbf{x}_j) + (1-\beta)k_2(\mathbf{x}_i, \mathbf{x}_j) \tag{5.104}$$

where $0 < \beta < 1$ used to exhibit the different kernel widths of the sub-kernels $k_1(\mathbf{x}_i, \mathbf{x}_j)$ and $k_2(\mathbf{x}_i, \mathbf{x}_j)$, i.e.,

$$k_l(\mathbf{x}_i, \mathbf{x}_j) = \exp\left(-\frac{\|\mathbf{x}_i - \mathbf{x}_j\|^2}{2\sigma_l^2}\right), \quad l = 1, 2 \tag{5.105}$$

Why we choose two Gaussian kernels to construct one mixed kernel is rooted in two thoughts. First, when one kernel is dominated, another one will be a coordinator. The coordinated one can be used the bias variable set in the SVM model. Second, the mixed kernel might have a much stronger learning ability than that with only one Gaussian kernel.

5.5.2 Mixed Gaussian Kernel for PIs Construction

Given that the prediction reliability is significantly concerned by the energy schedulers or operators, a PIs construction method is reported here based on the mixed Gaussian kernel model. Here, one can convert (5.103) into a new function $y(\cdot)$ with respect to $\mathbf{a}$. Assuming that the estimated parameters are denoted by $\widehat{\mathbf{a}}$, the real ones of the system are expressed as $\mathbf{a}^*$, and a Gaussian white noise, ε_n, indicating the difference between the real system output and the model one, then we have a new model of the system, i.e.,

$$y(\mathbf{x}_n, \mathbf{a}^*) = y(\mathbf{x}_n, \widehat{\mathbf{a}}) + \varepsilon_n \tag{5.106}$$

One can still use noise γ_n to specify the difference between the observed value and the real output, see (5.100). Then, the observed system's output reads as

$$t_n = y(\mathbf{x}_n, \widehat{\mathbf{a}}) + \varepsilon_n + \gamma_n \tag{5.107}$$

If $y(\mathbf{x}_0, \widehat{\mathbf{a}})$ (with respect to an input $\mathbf{x}_0$) comes as a result of the first-order Taylor series expansion around $\mathbf{a}^*$, then

$$y(\mathbf{x}_0, \widehat{\mathbf{a}}) \approx y(\mathbf{x}_0, \mathbf{a}^*) + \mathbf{g}_0^{\mathrm{T}}(\widehat{\mathbf{a}} - \mathbf{a}^*) \tag{5.108}$$

where

$$\mathbf{g}_0^{\mathrm{T}} = \left[\frac{\partial y(\mathbf{x}_0, \mathbf{a}^*)}{\partial a_1^*}, \frac{\partial y(\mathbf{x}_0, \mathbf{a}^*)}{\partial a_2^*}, \ldots, \frac{\partial y(\mathbf{x}_0, \mathbf{a}^*)}{\partial a_N^*}\right] \tag{5.109}$$

and $\mathbf{a}^* = (a_1^*, a_2^*, \ldots, a_N^*)^{\mathrm{T}}$. It is apparent that if the system is modeled by a kernel-based method, see (5.103), then $\partial y(\mathbf{x}_0, \mathbf{a}^*)/\partial a_1^* = k(\mathbf{x}_0, \mathbf{x}_1)$. Also, we assume $\mathbf{k}(\mathbf{x}_0) = [k(\mathbf{x}_0, \mathbf{x}_1), k(\mathbf{x}_0, \mathbf{x}_2), \ldots, k(\mathbf{x}_0, \mathbf{x}_N)]$, and can simplify the expression $\mathbf{g}_0^{\mathrm{T}}$ as $\mathbf{g}_0^{\mathrm{T}} = \mathbf{k}(\mathbf{x}_0)$. As such, integrating the (5.100), (5.104), and (5.105), we obtain

$$t_0 - \widehat{y}_0 \approx \gamma_0 - \mathbf{g}_0^{\mathrm{T}}(\widehat{\mathbf{a}} - \mathbf{a}^*) \tag{5.110}$$

Furthermore, assuming that the value of $\widehat{\mathbf{a}}$ can be considered as the unbiased estimation of $\mathbf{a}^*$, the expectation of the difference comes in the form

$$E[t_0 - \widehat{y}_0] \approx E[\gamma_0] - \mathbf{g}_0^{\mathrm{T}} E[(\widehat{\mathbf{a}} - \mathbf{a}^*)] \approx 0 \tag{5.111}$$

Based on the principle of statistical independence [39], its variance

$$\mathrm{var}[t_0 - \widehat{y}_0] \approx \mathrm{var}[\gamma_0] + \mathrm{var}[\mathbf{g}_0^{\mathrm{T}}(\widehat{\mathbf{a}} - \mathbf{a}^*)] \tag{5.112}$$

Considering that $\gamma_0 \sim \mathcal{N}(0, \sigma_\gamma^2)$, the distribution of $\widehat{\mathbf{a}} - \mathbf{a}^*$ is governed by $\mathcal{N}\left(0, \sigma_\gamma^2 \left[\mathbf{F}(\widehat{\mathbf{a}})^{\mathrm{T}} \mathbf{F}(\widehat{\mathbf{a}})\right]^{-1}\right)$, where $\mathbf{F}(\widehat{\mathbf{a}})$ is the Jacobian matrix of $y(\mathbf{x}_0, \widehat{\mathbf{a}})$ taken with respect to $\widehat{\mathbf{a}}$, i.e.,

$$\mathbf{F}(\widehat{\mathbf{a}}) = \frac{\partial \mathbf{y}(\mathbf{x}, \widehat{\mathbf{a}})}{\partial \widehat{\mathbf{a}}} = \begin{bmatrix} \left(\frac{\partial y_1(\mathbf{x}_1, \widehat{\mathbf{a}})}{\partial \widehat{a}_1}\right) & \left(\frac{\partial y_1(\mathbf{x}_1, \widehat{\mathbf{a}})}{\partial \widehat{a}_2}\right) & \cdots & \left(\frac{\partial y_1(\mathbf{x}_1, \widehat{\mathbf{a}})}{\partial \widehat{a}_N}\right) \\ \left(\frac{\partial y_2(\mathbf{x}_2, \widehat{\mathbf{a}})}{\partial \widehat{a}_1}\right) & \left(\frac{\partial y_2(\mathbf{x}_2, \widehat{\mathbf{a}})}{\partial \widehat{a}_2}\right) & \cdots & \left(\frac{\partial y_2(\mathbf{x}_2, \widehat{\mathbf{a}})}{\partial \widehat{a}_N}\right) \\ \vdots & \vdots & \vdots & \vdots \\ \left(\frac{\partial y_N(\mathbf{x}_N, \widehat{\mathbf{a}})}{\partial \widehat{a}_1}\right) & \left(\frac{\partial y_N(\mathbf{x}_N, \widehat{\mathbf{a}})}{\partial \widehat{a}_2}\right) & \cdots & \left(\frac{\partial y_N(\mathbf{x}_N, \widehat{\mathbf{a}})}{\partial \widehat{a}_N}\right) \end{bmatrix} \tag{5.113}$$

Besides, given $\partial y(\mathbf{x}_i, \mathbf{a}^*)/\partial a_j^* = k(\mathbf{x}_i, \mathbf{x}_j)$, (5.81) can be formulated as a kernel matrix, i.e., $\mathbf{F}(\widehat{\mathbf{a}}) = \mathbf{K}$. Thus, the advantage of the proposed kernel-based method becomes apparent, which greatly reduces the computational complexity and satisfies the real-time demand of applications. The variance of (5.112) can be further reformulated as

$$\mathrm{var}[t_0 - \widehat{y}_0] \approx \sigma_\gamma^2 + \sigma_\gamma^2 \mathbf{g}_0^{\mathrm{T}} (\mathbf{K}^{\mathrm{T}}\mathbf{K})^{-1} \mathbf{g}_0 \tag{5.114}$$

where the first term comes from the inherent noise of the sample data and the second one denotes the variance of model estimation. Thus, one can form the PI with confidence degree $(1 - \alpha)\%$ of the predicted value $\widehat{y}_0$ as

$$\widehat{y}_0 \pm t_{n-p}^{\alpha/2} \left(\sigma_\gamma^2 + \sigma_\gamma^2 \mathbf{g}_0^{\mathrm{T}} (\mathbf{K}^{\mathrm{T}}\mathbf{K})^{-1} \mathbf{g}_0\right)^{1/2} \tag{5.115}$$

where α is the quantile of a t-distribution and $n - p$ is the freedom degree of the t-distribution.

5.5.3 Estimation of Effective Noise-Based Hyper-Parameters

It is required to determine the hyper-parameters in advance before modeling, i.e., the penalty factor λ, the weight β, and the widths of the two kernels σ_1 and σ_2. We consider that the variance of the predicted error should be closer to that of the noise. A hyper-parameters optimization based on "effective noise" estimation is reported here. In Chap. 7, we give an illustration of the parameter optimization for the LSSVM based on noise estimation.

Given that the prediction can be well modeled if only the variance of the predicted error becomes closer to that of noise, an objective is formulated as

$$\min_{\lambda, \beta, \sigma_l}: e = |\mathrm{var}(\mathrm{pred}) - \mathrm{var}(\mathrm{noise})| \tag{5.116}$$

$$\begin{aligned} \mathrm{var}(\mathrm{pred}) &= \mathrm{var}\left(t - y\left(\mathbf{x}, \widehat{\mathbf{a}}\right)\right) \\ &= \mathrm{var}\left(y\left(\mathbf{x}, \widehat{\mathbf{a}}\right) + \varepsilon + \gamma - y\left(\mathbf{x}, \widehat{\mathbf{a}}\right)\right) \\ &= \mathrm{var}(\varepsilon) + \mathrm{var}(\gamma) \end{aligned} \tag{5.117}$$

where var(pred) denotes the variance of the predicted error that can be derived with the use of (5.114), and the noise variance, var(noise) can be denoted as var(γ). As such, the objective function can be formulated as follows:

$$e = \mathrm{var}(\varepsilon) = \frac{1}{2} \sum_{n=1}^{N} \left[t_n - \mathbf{k}(\mathbf{x}_n)^{\mathrm{T}} \widehat{\mathbf{a}}\right]^2 - \sigma_\gamma^2 \tag{5.118}$$

where ε is an implicit variable, the training set is $\{\mathbf{x}_n, t_n\}_{n=1,2,\ldots,N}$, and the noise variance σ_γ^2 can be obtained by the above Gamma test, see details in [32, 34].

Here, a conjugate gradient algorithm elaborated in Chap. 7 is employed to determine the hyper-parameters of the model, i.e., the penalty factor λ, the kernel weight β, and the widths of the Gaussian kernels σ_l $(l = 1, 2)$. One can calculate the partial derivatives of the objective function with respect to these parameters, i.e.,

$$\begin{cases} \dfrac{\partial e}{\partial \sigma_l} = \sum_{n=1}^{N} \left[t_n - \mathbf{k}(\mathbf{x}_n)^{\mathrm{T}} \widehat{\mathbf{a}}\right] \left[-\dfrac{\partial \mathbf{k}(\mathbf{x}_n)^{\mathrm{T}}}{\partial \sigma_l} \widehat{\mathbf{a}} - \mathbf{k}(\mathbf{x}_n)^{\mathrm{T}} \dfrac{\partial \widehat{\mathbf{a}}}{\partial \sigma_l}\right] \\ \dfrac{\partial e}{\partial \beta} = \sum_{n=1}^{N} \left[t_n - \mathbf{k}(\mathbf{x}_n)^{\mathrm{T}} \widehat{\mathbf{a}}\right] \left[\dfrac{\partial \mathbf{k}(\mathbf{x}_n)^{\mathrm{T}}}{\partial \beta} \widehat{\mathbf{a}} + \mathbf{k}(\mathbf{x}_n)^{\mathrm{T}} \dfrac{\partial \widehat{\mathbf{a}}}{\partial \beta}\right] \\ \dfrac{\partial e}{\partial \lambda} = \sum_{n=1}^{N} \left[t_n - \mathbf{k}(\mathbf{x}_n)^{\mathrm{T}} \widehat{\mathbf{a}}\right] \mathbf{k}(\mathbf{x}_n)^{\mathrm{T}} \dfrac{\partial \widehat{\mathbf{a}}}{\partial \lambda} \end{cases} \tag{5.119}$$

Then, to derive the expression of (5.119), we have

$$\begin{cases} \dfrac{\partial \mathbf{k}(\mathbf{x}_n)^{\mathrm{T}}}{\partial \sigma_l} = \left[\dfrac{\partial k(\mathbf{x}_n, \mathbf{x}_1)}{\partial \sigma_l}, \dfrac{\partial k(\mathbf{x}_n, \mathbf{x}_2)}{\partial \sigma_l}, \ldots, \dfrac{\partial k(\mathbf{x}_n, \mathbf{x}_N)}{\partial \sigma_l}\right] \\ \dfrac{\partial \widehat{\mathbf{a}}}{\partial \sigma_l} = \dfrac{\partial (\mathbf{K} + \lambda \mathbf{I}_N)^{-1}}{\partial \sigma_l}\mathbf{t} = -\left[(\mathbf{K} + \lambda \mathbf{I}_N)^{\mathrm{T}}\right]^{-1} \cdot \dfrac{\partial \mathbf{K}}{\partial \sigma_l} \cdot (\mathbf{K} + \lambda \mathbf{I}_N)^{-1}\mathbf{t} \\ \dfrac{\partial \mathbf{k}(\mathbf{x}_n)^{\mathrm{T}}}{\partial \beta} = \left[\dfrac{\partial k(\mathbf{x}_n, \mathbf{x}_1)}{\partial \beta}, \dfrac{\partial k(\mathbf{x}_n, \mathbf{x}_2)}{\partial \beta}, \ldots, \dfrac{\partial k(\mathbf{x}_n, \mathbf{x}_N)}{\partial \beta}\right] \\ \dfrac{\partial \widehat{\mathbf{a}}}{\partial \beta} = -\left[(\mathbf{K} + \lambda \mathbf{I}_N)^{\mathrm{T}}\right]^{-1} \cdot \dfrac{\partial \mathbf{K}}{\partial \beta} \cdot (\mathbf{K} + \lambda \mathbf{I}_N)^{-1}\mathbf{t} \\ \dfrac{\partial \widehat{\mathbf{a}}}{\partial \lambda} = -\left[(\mathbf{K} + \lambda \mathbf{I}_N)^{\mathrm{T}}\right]^{-1} (\mathbf{K} + \lambda \mathbf{I}_N)^{-1}\mathbf{t} \end{cases} \tag{5.120}$$

where $\partial \mathbf{K}/\partial \sigma_l$ is a matrix of dimensionality $N \times N$, and each of the elements can be denoted as $\partial k(\mathbf{x}_i, \mathbf{x}_j)/\partial \sigma_l$; similarly, $\partial \mathbf{K}/\partial \beta$ consists of the elements denoted as $\partial k(\mathbf{x}_i, \mathbf{x}_j)/\partial \beta$. Then,

$$\begin{cases} \dfrac{\partial k(\mathbf{x}_i, \mathbf{x}_j)}{\partial \sigma_1} = \beta k_1(\mathbf{x}_i, \mathbf{x}_j) \dfrac{\|\mathbf{x}_i - \mathbf{x}_j\|^2}{\sigma_1^3} \\ \dfrac{\partial k(\mathbf{x}_i, \mathbf{x}_j)}{\partial \sigma_2} = (1 - \beta) k_2(\mathbf{x}_i, \mathbf{x}_j) \dfrac{\|\mathbf{x}_i - \mathbf{x}_j\|^2}{\sigma_2^3} \\ \dfrac{\partial k(\mathbf{x}_i, \mathbf{x}_j)}{\partial \beta} = k_1(\mathbf{x}_i, \mathbf{x}_j) - k_2(\mathbf{x}_i, \mathbf{x}_j) \end{cases} \tag{5.121}$$

Although such type of the algorithm might tend to be trapped in some local minimum [41], it has been experimentally observed that it exhibits sound convergence properties as long as the initial values have been carefully determined. We specify the initial values off-line using the cross-validation.

5.5.4 *Case Study*

To verify the effectiveness of the Gaussian kernel-based causality PIs construction (GK-CPIs) method, we conduct the following experiments. All of the experimental data come from the energy data center of a steel plant, China, where the by-product gases consists of BFG, COG, and LDG, as presented in Fig. 3.6, Fig. 3.1, and Fig. 3.18, respectively. There are four blast furnaces in the BFG system, which can supply into the gas transportation network on average 1.8 million m^3 BFG per hour; six coke ovens and six converters are, respectively, comprised in the COG system and the LDG system. In particular, the LDG system is divided into two subsystems, each of which supply the LDG to the different users, named as #1 LDG subsystem and #2 LDG subsystem.

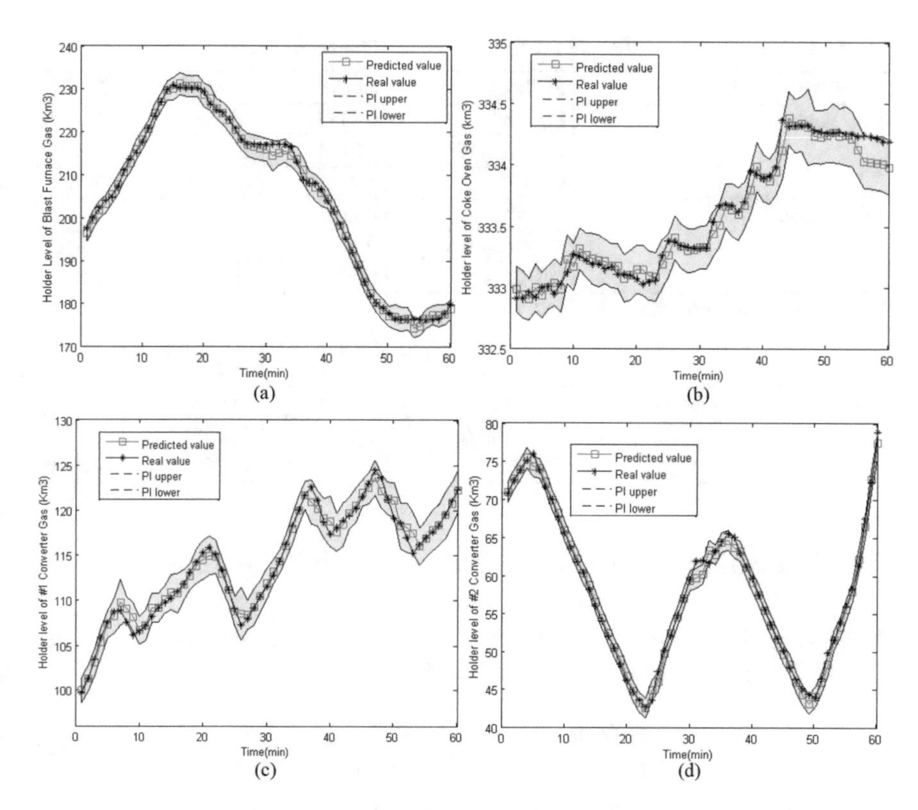

Fig. 5.17 PIs of gasholder levels (**a**) gasholder level of blast furnace, (**b**) gasholder level of coke oven, (**c**) gasholder level of #1 converter and (**d**) gasholder level of #2 converter

To indicate the performance of the GK-CPIs, Fig. 5.17 illustrates the PI results of the gasholders, in which the intervals can basically cover the real values of the holder levels. For a further analysis on its advantage, a series of quantitative comparisons with the existing PIs based on LSSVM [38] and the bootstrap NN [14] are reported in Table 5.8, where the RMSE describes the accuracy of the predicted values. To evaluate the quality of the constructed PIs, one can refer to the CWC that integrates the PICP and the MPIW as the evaluation criterion [14]. From Table 5.8, it is apparent that the GK-CPIs exhibit the remarkable accuracy compared to the others. In the perspective of computing time, although the solving efficiency of the GK-CPIs is lower than that by the LSSVM-based method because of transforming the Jacobian matrix into the kernel computing, it can completely meet the practical requirement for online rolling dynamic scheduling

Table 5.8 The comparison of the PI results by using the different methods

Gasholders	Methods	RMSE	Time (s)	CWC
BFG holder	LSSVM	0.8757	4.093	8.9145
	Bootstrap NN	0.8104	29.122	4.3120
	GK-CPIs	0.7646	16.201	3.9299
COG holder	LSSVM	0.1215	4.047	2.1922
	Bootstrap NN	0.1144	22.230	1.0783
	GK-CPIs	0.1058	15.237	1.0215
LDG holder of #1 subsystem	LSSVM	0.9752	4.063	11.1254
	Bootstrap NN	0.9073	37.492	7.2240
	GK-CPIs	0.8953	15.819	5.9341
LDG holder of #2 subsystem	LSSVM	0.8065	4.072	6.3521
	Bootstrap NN	0.7939	36.643	6.2893
	GK-CPIs	0.5778	14.568	2.8206

5.6 Prediction Intervals Construction with Noisy Inputs

A typical ESN contains the input layer, the reservoir, and the output layer. And, an ESN with output feedback can be formulated as

$$\begin{aligned} \mathbf{x}_k &= f\left(\mathbf{W}^{\text{in}}\mathbf{u}_k + \mathbf{W}\mathbf{x}_{k-1} + \mathbf{W}^{\text{back}}y_{k-1}\right) \\ t_k &= y_k + n_k = f^{\text{out}}(\mathbf{W}^{\text{out}}(\mathbf{u}_k, \mathbf{x}_{k-1}, y_{k-1})) + n_k \end{aligned} \tag{5.122}$$

where $\mathbf{u}_k$ is the exogenous input with dimensionality m, $\mathbf{x}_k$ is the internal states with dimensionality N, and y_k is the output. $\mathbf{W}^{\text{in}} = \left(W_{i,j}^{\text{in}}\right) \in \mathbb{R}^{N\times m}$ denotes the input weights, and $\mathbf{W} = (W_{i,j}) \in \mathbb{R}^{N\times N}$ denotes the internal weights of the neurons in the reservoir. To provide proper memorization capabilities, $\mathbf{W}$ should be sparse whose connectivity level ranges from 1 to 5% and its spectral radius should be less than 1. $\mathbf{W}^{\text{back}} = \left(W_{i,j}^{\text{back}}\right) \in \mathbb{R}^{N\times 1}$ denotes the output feedback weights, and $\mathbf{W}^{\text{out}} = \left(W_{i,j}^{\text{out}}\right) \in \mathbb{R}^{1\times(m+N+1)}$ denotes the output weights. f and f^{out} are the activation functions of internal neurons and output neurons, respectively. n_k is an independent white Gaussian noise sequence reflecting the output uncertainty.

Here, t_k is the noisy output and y_k is the output of the ESN. Given the output feedback is the prior output of the ESN, the feedback uncertainty should be considered in this model. As for the input uncertainties, one can assume a random vector $\mathbf{z}_k$ as the noisy input, i.e.,

$$\mathbf{z}_k = f(\mathbf{u}_k, \boldsymbol{\varepsilon}_k) \tag{5.123}$$

where $\mathbf{u}_k$ is the hidden input, and $\boldsymbol{\varepsilon}_k$ is a random noise vector, independent of $\mathbf{u}_k$. Because in industrial practice the levels of the noises for each input dimension may be different, the covariance of $\boldsymbol{\varepsilon}_k$ can be denoted as $\boldsymbol{\Sigma} = \text{diag}\left(\sigma_1^2, \sigma_2^2, \ldots, \sigma_m^2\right)$.

5.6.1 Bayesian Estimation of the Output Uncertainty

Based on the regression described in (5.122), the output $y(\mathbf{u}^*, \mathbf{W}^{\text{out}})$ of the ESN depends on the new input $\mathbf{u}^*$ and a set of model weights $\mathbf{W}^{\text{out}}$, then the conditional distribution can be written as the integral over these parameters

$$p(t^*|\mathbf{u}^*, D) = \int p(t^*|\mathbf{u}^*, \mathbf{W}^{\text{out}})p(\mathbf{W}^{\text{out}}|D)d\mathbf{W}^{\text{out}} \tag{5.124}$$

where $p(t^*|\mathbf{u}^*, \mathbf{W}^{\text{out}})$ is a likelihood function that denotes the difference between the real observed targets t^* and the output y^* of the network given the parameters $\mathbf{W}^{\text{out}}$

$$p(t^*|\mathbf{u}^*, \mathbf{W}^{\text{out}}) = \left(\frac{\beta}{2\pi}\right)^{1/2} \exp\left(\frac{\beta}{2}\{f^{\text{out}}(\mathbf{u}^*, \mathbf{W}^{\text{out}}) - t^*\}^2\right) \tag{5.125}$$

where β is a hyper-parameter with the value $1/\beta = \sigma_t^2$.

On the right hand side of (5.124), another posterior distribution $p(\mathbf{W}^{\text{out}}| D)$ is still unknown. Using Bayes' rule, $p(\mathbf{W}^{\text{out}}| D)$ can be written as

$$p(\mathbf{W}^{\text{out}}|D) = \frac{1}{p(D)} p(D|\mathbf{W}^{\text{out}})p(\mathbf{W}^{\text{out}}) \tag{5.126}$$

If we assume the priori of $p(\mathbf{W}^{\text{out}})$ is a Gaussian distribution with covariance of $1/\alpha$, the distribution $p(\mathbf{W}^{\text{out}}| D)$ is written as

$$\begin{aligned} p(\mathbf{W}^{\text{out}}|D) &= \frac{1}{Z_S}\exp\left(-\frac{\beta}{2}E_D - \frac{\alpha}{2}E_W\right) \\ &= \frac{1}{Z_S}\exp(-S(\mathbf{W}^{\text{out}})) \end{aligned} \tag{5.127}$$

where Z_S is a normalizing constant given by the integral $Z_S = \int \exp(-S(\mathbf{W}^{\text{out}}))d\mathbf{W}^{\text{out}}$.

The term E_D is the contribution from likelihood $p(D| \mathbf{W}^{\text{out}})$, which assumes that the data is independent can be written as the product

$$\begin{aligned} p(D|\mathbf{W}^{\text{out}}) &= \prod_{i=1}^{N} p\left(t_i^n|\mathbf{u}_i^n, \mathbf{W}^{\text{out}}\right) \\ &= \frac{1}{Z_D(\beta)}\exp\left(-\frac{\beta}{2}\sum_{i=1}^{N}\{f^{\text{out}}(\mathbf{u}_i^n, \mathbf{W}^{\text{out}}) - t_i^n\}^2\right) \\ &= \frac{1}{Z_D(\beta)}\exp\left(-\frac{\beta}{2}E_D\right) \end{aligned} \tag{5.128}$$

where Z_D is the normalizing constant which gives $Z_D(\beta) = (2\pi/\beta)^{N/2}$.

The second term E_W is the contribution from the prior over the weights

$$p(\mathbf{W}^{\text{out}}) = \frac{1}{Z_W(\alpha)} \exp\left(-\frac{\alpha}{2} E_W\right) \tag{5.129}$$

where again $Z_W(\alpha)$ is a normalizing constant which gives $Z_W(\alpha) = \int \exp(-(\alpha/2) E_W) d\mathbf{W}^{\text{out}}$.

The prior can be interpreted as giving the regularizing term to the integral. Although there are several forms of prior that can be used, in this model a simple quadratic regularization function is used $E_W = \|\mathbf{W}^{\text{out}}\|^2$.

To evaluate the most probable weights $\mathbf{W}^{\text{out}}_{\text{MP}}$, we maximize the posterior distribution $p(\mathbf{W}^{\text{out}}|D)$ that is equivalent to minimize the function $S(\mathbf{W}^{\text{out}})$. First, the function $S(\mathbf{W}^{\text{out}})$ can be linearized by Taylor series expansion and neglect the third-order terms, this leads to the approximation

$$S(\mathbf{W}^{\text{out}}) \approx S_{\text{MP}} + \Delta(\mathbf{W}^{\text{out}})^{\text{T}} \mathbf{A} \Delta(\mathbf{W}^{\text{out}}) \tag{5.130}$$

where $\mathbf{A}$ is the Hessian matrix $\mathbf{A} = \nabla\nabla S_{\text{MP}}$, $\Delta(\mathbf{W}^{\text{out}}) = \mathbf{W}^{\text{out}} - \mathbf{W}^{\text{out}}_{\text{MP}}$, and $S\left(\mathbf{W}^{\text{out}}_{\text{MP}}\right)$ has been written as $S(\mathbf{W}^{\text{out}})$.

Substituting (5.125) and (5.127) into (5.124), it leads to the relationship

$$\begin{aligned} p(t^*|\mathbf{u}^*, D) \propto \int \Bigg\{ & \exp\left(-\frac{\beta}{2}\sum_{i=1}^{N} \{f^{\text{out}}(\mathbf{u}_i^n, \mathbf{W}^{\text{out}}) - t_i^n\}^2\right) \\ & \cdot \exp\left(-\frac{1}{2}\Delta(\mathbf{W}^{\text{out}})^{\text{T}} \mathbf{A} \Delta(\mathbf{W}^{\text{out}})\right)\Bigg\} d\mathbf{W}^{\text{out}} \end{aligned} \tag{5.131}$$

The function $y\left(\mathbf{u}_i^n, \mathbf{W}^{\text{out}}\right)$ may now be linearly approximated by Taylor expanding about $\mathbf{W}^{\text{out}}_{\text{MP}}$, i.e.,

$$y\left(\mathbf{u}_i^*, \mathbf{W}^{\text{out}}\right) \approx y\left(\mathbf{u}_i^*, \mathbf{W}^{\text{out}}_{\text{MP}}\right) + \mathbf{g}^{\text{T}} \Delta \mathbf{W}^{\text{out}} \tag{5.132}$$

with $\mathbf{g} = \nabla_W y\left(\mathbf{u}_i^*, \mathbf{W}^{\text{out}}\right)\big|_{\mathbf{W}^{\text{out}} = \mathbf{W}^{\text{out}}_{\text{MP}}}$.

Substituting (5.132) into (5.131) and evaluating the integral over $\mathbf{W}^{\text{out}}$, we have

$$p(t^*|\mathbf{u}^*, D) = \frac{1}{\left(2\pi\sigma_{t^*}^2\right)^{1/2}} \exp\left(\frac{\{t^* - f^{\text{out}}(\mathbf{u}^*; \mathbf{W}^{\text{out}})\}^2}{2\sigma_{t^*}^2}\right) \tag{5.133}$$

where $\sigma_{t^*}^2 = 1/\beta + \mathbf{g}^{\text{T}}\mathbf{A}\mathbf{g}$, β is a hyper-parameter related to the distribution of the output. And $\mathbf{A}$ is a Hessian matrix of the activation function of the output neurons. $\mathbf{g}$ is the partial derivative of function $f^{\text{out}}(\mathbf{u}^*, \mathbf{W}^{\text{out}})$ with respect to $\mathbf{W}^{\text{out}}$, when $\mathbf{W}^{\text{out}} = \mathbf{W}^{\text{out}}_{\text{MP}}$.

5.6.2 Estimation of the External Input Uncertainties

If the influence of the input noise is considered for prediction, the predictive distribution $p(t^* | \mathbf{z}^*, D^{'})$ is represented by

$$p(t^* | \mathbf{z}^*, D') = \int p(t^* | \mathbf{u}^*, D') p(\mathbf{u}^* | \mathbf{z}^*) d\mathbf{u}^* \tag{5.134}$$

where $\mathbf{u}^*$ is the noiseless input and $\mathbf{z}^*$ is the real input with noise. $p(t^* | \mathbf{u}^*, D^{'})$ is the predictive distribution when the noiseless input $\mathbf{u}^*$ is available that can be defined as

$$p(t^* | \mathbf{u}^*, D') = \int p(t^* | \mathbf{u}^*, \mathbf{W}^{\text{out}}) p(\mathbf{W}^{\text{out}} | D') d\mathbf{W}^{\text{out}} \tag{5.135}$$

If the variance of the noise is assumed to be very small, the function $f^{\text{out}}(\mathbf{u}^*; \mathbf{W}^{\text{out}})$ may be linearly approximated by Taylor expanding about $\mathbf{z}^*$ and neglect the third-order terms.

$$f^{\text{out}}(\mathbf{u}^*; \mathbf{W}^{\text{out}}) = f^{\text{out}}(\mathbf{z}^*; \mathbf{W}^{\text{out}}) + \mathbf{h}_{\mathbf{z}}^{\text{T}} \delta \mathbf{z}^* \tag{5.136}$$

where $\delta \mathbf{z}^* = \mathbf{u}^* - \mathbf{z}^*$, and $\mathbf{h}_{\mathbf{z}}^{\text{T}}$ is the partial derivatives of $f^{\text{out}}(\mathbf{u}^*; \mathbf{W}^{\text{out}})$ with respect to $\mathbf{z}^*$

$$\mathbf{h}_{\mathbf{z}}^{\text{T}} = \nabla_{\mathbf{u}^*} f^{\text{out}}(\mathbf{u}^*, \mathbf{x}^*; \mathbf{W}^{\text{out}}) \big|_{\mathbf{u}^* = \mathbf{z}^*} \tag{5.137}$$

Thus, according to the Bayes' rule, the predictive distribution can be rewritten as

$$\begin{aligned} p(t^* | \mathbf{u}^*, \mathbf{W}^{\text{out}}) &= \int p(t^* | \mathbf{z}^*, \mathbf{W}^{\text{out}}) p(\mathbf{u}^* | \mathbf{z}^*) d\mathbf{z}^* \\ &= \int \exp\left(\frac{\beta}{2} \{ f^{\text{out}}(\mathbf{z}^*; \mathbf{W}^{\text{out}}) + \mathbf{h}_{\mathbf{z}}^{\text{T}} \partial \mathbf{z} - t^* \}^2 - \frac{1}{2} (\partial \mathbf{z}^*)^{\text{T}} \sum (\partial \mathbf{z}^*) \right) d\mathbf{z}^* \end{aligned} \tag{5.138}$$

where $\beta = 1/\sigma_t^2$. According to the characteristics of the Gaussian distribution inference in [39], the distribution in (5.138) can be approximated by a Gaussian distribution.

$$p(t^* | \mathbf{u}^*, \mathbf{W}^{\text{out}}) = \frac{1}{Z_{\mathbf{z}}} \exp\left(-\frac{\beta'}{2} \{ t^* - f^{\text{out}}(\mathbf{z}^*; \mathbf{W}^{\text{out}}) \}^2 \right) \tag{5.139}$$

where $Z_{\mathbf{z}}$ is a normalizing constant and $1/\beta' = 1/\beta + \mathbf{h}_{\mathbf{z}}^{\text{T}} \sum \mathbf{h}_{\mathbf{z}}$.

As for the second term $p(\mathbf{W}^{\text{out}} | D^{'})$ on the right hand side of (5.135), we can expand the dataset $D^{'} = \{t_i, \mathbf{z}_i\}|_{i=1,2,...,n}$ and marginalize over the noiseless input $\mathbf{u}_i$, then

$$p(\mathbf{W}^{\text{out}}|D') = \int p(\mathbf{W}^{\text{out}}, \mathbf{u}_i|\mathbf{z}_i, t_i)d\mathbf{u}_i \tag{5.140}$$

Based on the Bayes' rule and the fact that the observed target t_i and the real inputs $\mathbf{z}_i$ are independent given the noiseless $\mathbf{u}_i$, the (5.140) can be written as

$$p(\mathbf{W}^{\text{out}}|D') = \frac{p(\mathbf{W}^{\text{out}})}{p(t_i|\mathbf{z}_i)} \int p(t_i|\mathbf{u}_i|, \mathbf{W}^{\text{out}})p(\mathbf{u}_i|\mathbf{z}_i)d\mathbf{u}_i \tag{5.141}$$

where $p(\mathbf{W}^{\text{out}}) \propto \exp(-\alpha/2\|\mathbf{W}^{\text{out}}\|^2)$ is the prior distribution of weights and α is a hyper-parameter related to the distribution. The posterior distribution in (5.141) can be formulated by

$$\begin{aligned} p(\mathbf{W}^{\text{out}}|D') &\propto \prod_{i=1}^{N} \exp\left(\frac{\beta'}{2}\{t_i - f^{\text{out}}(\mathbf{z}_i; \mathbf{W}^{\text{out}})\}^2 - \frac{\alpha}{2}\|\mathbf{W}^{\text{out}}\|^2\right) \\ &= \exp\left(-\frac{\beta'}{2}\sum_{i=1}^{n}\{t_i - f^{\text{out}}(\mathbf{z}_i; \mathbf{W}^{\text{out}})\}^2 - \frac{\alpha}{2}\|\mathbf{W}^{\text{out}}\|^2\right) \end{aligned} \tag{5.142}$$

Substituting (5.142) and (5.135) into (5.134), the predictive distribution is written as

$$\begin{aligned} p(t^*|\mathbf{z}^*, D) = \frac{1}{Z_{\mathbf{u}}} \int & \exp\left(-\frac{\beta'}{2}\{t^* - f^{\text{out}}(\mathbf{z}^*; \mathbf{W}^{\text{out}})\}^2\right) \\ & \cdot \prod_{i=1}^{n} \exp\left(-\frac{\beta'}{2}\{t_i - f^{\text{out}}(\mathbf{z}_i; \mathbf{W}^{\text{out}})\}^2 - \frac{\alpha}{2}\|\mathbf{W}^{\text{out}}\|^2\right) d\mathbf{W}^{\text{out}} \end{aligned} \tag{5.143}$$

where $Z_{\mathbf{u}}$ is a normalizing constant. To transform (5.143) to be the form of Gaussian, we define the form of $M(\mathbf{W}^{\text{out}})$ as follows:

$$M(\mathbf{W}^{\text{out}}) = \frac{\beta'}{2} \sum_{i=1}^{n} \{t_i - f^{\text{out}}(\mathbf{z}_i; \mathbf{W}^{\text{out}})\}^2 + \frac{\alpha}{2}\|\mathbf{W}^{\text{out}}\|^2 \tag{5.144}$$

To make the predictive distribution, (5.144) can be linearly approximated by the Taylor series expansion with respect to $\mathbf{W}_{\text{MP}}^{\text{out}}$. Then,

$$M(\mathbf{W}^{\text{out}}) = M\left(\mathbf{W}_{\text{MP}}^{\text{out}}\right) + \frac{1}{2}\left(\mathbf{W}^{\text{out}} - \mathbf{W}_{\text{MP}}^{\text{out}}\right)^{\text{T}} \mathbf{A}\left(\mathbf{W}^{\text{out}} - \mathbf{W}_{\text{MP}}^{\text{out}}\right) \tag{5.145}$$

where $\mathbf{A}$ is the Hessian matrix of $M(\mathbf{W}^{\text{out}})$ with respect to $\mathbf{W}^{\text{out}}$.

Substituting (5.145) into (5.143), the predictive distribution can be approximated by the following Gaussian distribution.

$$p(t^*|\mathbf{z}^*,D) = \frac{1}{\left(2\pi\sigma_{t^*}^2\right)^{1/2}} \exp\left(\frac{\{t^* - f^{\text{out}}(\mathbf{z}^*;\mathbf{W}^{\text{out}})\}^2}{2\sigma_{t^*}^2}\right) \tag{5.146}$$

where the covariance of the distribution is

$$\sigma_{t^*}^2 = \frac{1}{\beta'} + \mathbf{g}^{\text{T}}\mathbf{A}^{-1}\mathbf{g} = \frac{1}{\beta} + \mathbf{h}_{\mathbf{z}}^{\text{T}}\Sigma\mathbf{h}_{\mathbf{z}} + \mathbf{g}^{\text{T}}\mathbf{A}^{-1}\mathbf{g} \tag{5.147}$$

and the diagonal element of Σ is the covariance of the noise of the input data. Compared to the variance in (5.133), the variance in (5.147) has another term $\mathbf{h}_{\mathbf{z}}^{\text{T}}\Sigma\mathbf{h}_{\mathbf{z}}$, which quantified the influence of the input data noise to the prediction.

5.6.3 *Estimation of the Output Feedback Uncertainties*

Considering the modeling without output feedback, the predicted distribution $p(t^*|\mathbf{u}^*,D)$ can be similarly approximated by a Gaussian distribution in terms of the marginal distribution

$$p(t^*|\mathbf{u}^*,D) = \int p(t^*|\mathbf{u}^*,t^-,\mathbf{W}^{\text{out}})p(\mathbf{W}^{\text{out}}|D)d\mathbf{W}^{\text{out}} \tag{5.148}$$

where t^- denotes the noisy output feedback.

If y^- is the output feedback with noise, it is possible to linearize $f^{\text{out}}(\mathbf{u}^*,\mathbf{x}^*,y^-;\mathbf{W}^{\text{out}})$ around t^-, then one can neglect the second-order terms, i.e.,

$$f^{\text{out}}(\mathbf{u}^*,\mathbf{x}^*,y^-;\mathbf{W}^{\text{out}}) = f^{\text{out}}(\mathbf{u}^*,\mathbf{x}^*,t^-;\mathbf{W}^{\text{out}}) + \mathbf{h}_{t^-}^{\text{T}}\delta t^- \tag{5.149}$$

where $\delta t^- = y^- - t^-$, $\mathbf{h}_{t^-}^{\text{T}}$ is the partial derivative of $f^{\text{out}}(\mathbf{u}^*,\mathbf{x}^*,y^-;\mathbf{W}^{\text{out}})$ with respect to y^-, i.e.,

$$\begin{aligned}\mathbf{h}_{t^-}^{\text{T}} &= \nabla_{y^-} f^{\text{out}}(\mathbf{u}^*,\mathbf{x}^*,y^-;\mathbf{W}^{\text{out}})\big|_{y^-=t^-} \\ &= \mathbf{W}_{y^-}^{\text{out}} + \mathbf{W}_{\mathbf{x}}^{\text{out}} \cdot \mathbf{W}^{\text{back}}\left([\cosh(\mathbf{x}^*)]^{\text{T}}[\cosh(\mathbf{x}^*)]\right)^{-1}\end{aligned} \tag{5.150}$$

where $\mathbf{W}_{y^-}^{\text{out}}$ is a block matrix of the output weights corresponding to y^-, and $\mathbf{W}_{\mathbf{x}}^{\text{out}}$ is a block matrix of the output weights corresponding to the internal states. As such, using the Bayesian rule, the output distribution is

$$\begin{aligned}p(t^*|\mathbf{u}^*,t^-,\mathbf{W}^{\text{out}}) &= \int p(t^*|\mathbf{u}^*,y^-,\mathbf{W}^{\text{out}})p(y^-|t^-)dy^- \\ &= \int \exp\left(\frac{\beta}{2}\{\mathbf{W}^{\text{out}}(\mathbf{u}^*,\mathbf{x}^*,y^-) - t^*\}^2 - \frac{1}{2\sigma_t^2}(\delta t^-)^{\text{T}}(\delta t^-)\right)dy^-\end{aligned} \tag{5.151}$$

where $\beta = 1/\sigma_t^2$. Then, the output distribution can be further approximated by a Gaussian distribution

$$p(t^*|\mathbf{u}^*, t^-, \mathbf{W}^{\text{out}}) = \frac{1}{Z_{t^-}} \exp\left(-\frac{\beta'}{2}\{\Delta t^*\}^2\right) \tag{5.152}$$

where $\Delta t^* = t^* - f^{\text{out}}(\mathbf{u}^*, \mathbf{x}^*, t^-; \mathbf{W}^{\text{out}})$, Z_{t^-} is the normalizing constant and

$$1/\beta' = 1/\beta + \sigma_t^2 = \left(1 + \mathbf{h}_{t^-}^{\text{T}}\mathbf{h}_{t^-}\right)/\beta \tag{5.153}$$

As for the $p(\mathbf{W}^{\text{out}}| D)$ term in (5.148), one can expand $D = \{t_i, \mathbf{u}_i\}|_{i\,=\,1,\,2,\,\ldots,\,n}$ and marginalize over the output feedback y_i^-, then

$$p(\mathbf{W}^{\text{out}}|D) = \int p\left(\mathbf{W}^{\text{out}}, y_i^-|\mathbf{u}_i, t_i, t_i^-\right)dy_i^- \tag{5.154}$$

Using the Bayesian rule and the conditional independence of t_i on t_i^- given y_i^-, (5.154) can be rewritten as

$$p(\mathbf{W}^{\text{out}}|D) = \frac{p(\mathbf{W}^{\text{out}})}{p(t_i|t_i^-)} \cdot \int p\left(t_i|\mathbf{u}_i, y_i^-, \mathbf{W}^{\text{out}}\right)p\left(y_i^-|t_i^-\right)dy_i^- \tag{5.155}$$

where $p(\mathbf{W}^{\text{out}}) \propto \exp\left[(-\alpha/2)\|\mathbf{W}^{\text{out}}\|^2\right]$, α is a hyper-parameter. Based on (5.155), the posterior distribution of the output weights reads as

$$\begin{aligned} p(\mathbf{W}^{\text{out}}|D) &\propto \prod_{i=1}^{n}\exp\left(-\frac{\beta'}{2}\{\Delta t_i\}^2 - \frac{\alpha}{2}\|\mathbf{W}^{\text{out}}\|^2\right) \\ &= \exp\left(-\frac{\beta'}{2}\sum_{i=1}^{n}\left[\{\Delta t_i\}^2 - \frac{\alpha}{2}\|\mathbf{W}^{\text{out}}\|^2\right]\right) \end{aligned} \tag{5.156}$$

where $\Delta t_i = t_i - f^{\text{out}}\left(\mathbf{u}_i, \mathbf{x}_i, t_i^-; \mathbf{W}^{\text{out}}\right)$. Substituting (5.151) and (5.156) into (5.148), we have

$$\begin{aligned} p(t^*|\mathbf{u}^*, D) = \frac{1}{Z_{t\mathbf{u}}}\int\Bigg\{&\exp\left(-\frac{\beta'}{2}\{\Delta t_i\}^2\right) \\ &\cdot\prod_{i=1}^{n}\exp\left(-\frac{\beta'}{2}\{\Delta t_i\}^2 - \frac{\alpha}{2}\|\mathbf{W}^{\text{out}}\|^2\right)\Bigg\}d\mathbf{W}^{\text{out}} \end{aligned} \tag{5.157}$$

where $Z_{t\mathbf{u}}$ is the normalizing constant, and

$$M(\mathbf{W}^{\text{out}}) = \frac{\beta'}{2}\{\Delta t_i\}^2 + \frac{\alpha}{2}\|\mathbf{W}^{\text{out}}\|^2 \tag{5.158}$$

For the prediction distribution, (5.158) can be linearly approximated by the Taylor series expansion with respect to $\mathbf{W}_{\text{MP}}^{\text{out}}$. That is,

$$M(\mathbf{W}^{\mathrm{out}}) = M(\mathbf{W}^{\mathrm{out}}_{\mathrm{MP}}) + \frac{1}{2}(\Delta\mathbf{W}^{\mathrm{out}})\mathbf{A}(\Delta\mathbf{W}^{\mathrm{out}}) \tag{5.159}$$

where $\Delta\mathbf{W}^{\mathrm{out}} = \mathbf{W}^{\mathrm{out}} - \mathbf{W}^{\mathrm{out}}_{\mathrm{MP}}$ and $\mathbf{A}$ is the Hessian matrix of $M(\mathbf{W}^{\mathrm{out}})$. Taking (5.159) into (5.157), a Gaussian approximation of the predictive distribution can be obtained

$$p(t^*|\mathbf{u}^*, D) = \frac{1}{\left(2\pi\sigma_{t^*}^2\right)^{1/2}} \exp\left(\frac{\{t^* - f^{\mathrm{out}}(\mathbf{u}^*, \mathbf{x}^*, t^-; \mathbf{W}^{\mathrm{out}})\}^2}{2\sigma_{t^*}^2}\right) \tag{5.160}$$

where $\sigma_{t^*}^2 = 1/\beta' + \mathbf{g}^{\mathrm{T}}\mathbf{A}^{-1}\mathbf{g} = (1/\beta)\left(1 + \mathbf{h}_{t^-}^{\mathrm{T}}\mathbf{h}_{t^-}\right) + \mathbf{g}^{\mathrm{T}}\mathbf{A}^{-1}\mathbf{g}$.

As such, the variance of the predictive distribution consists of three components. 1) The variance in the distribution over the observed targets, 2) an estimation of the output feedback uncertainty, and 3) the uncertainty induced by the weights.

5.6.4 Estimation of the Total Uncertainties and PIs Construction

Given that the input of the ESN is noisy, the predictive distribution can be formulated as

$$p(t^*|\mathbf{z}^*, D') = \int p(t^*|\mathbf{z}^*, t^-, \mathbf{W}^{\mathrm{out}}) p(\mathbf{W}^{\mathrm{out}}|D') d\mathbf{W}^{\mathrm{out}} \tag{5.161}$$

Given the noisy input and the feedback, the output distribution can be written as

$$\begin{aligned} p(t^*|\mathbf{z}^*, t^-, \mathbf{W}^{\mathrm{out}}) &= \int p(t^*|\mathbf{z}^*, t^-, \mathbf{W}^{\mathrm{out}}) p(\mathbf{z}^*|\mathbf{u}^*) d\mathbf{u}^* \\ &= \int \exp\left\{\frac{\beta'}{2}(\Delta t^*)^2 - \frac{1}{2}(\Delta\mathbf{z}^*)\mathbf{\Sigma}(\Delta\mathbf{z}^*)\right\} d\mathbf{z}^* \end{aligned} \tag{5.162}$$

where $\mathbf{u}^*$ is the hidden input without noise, $\Delta\mathbf{z}^* = \mathbf{u}^* - \mathbf{z}^*$. To simplify (5.162), $f^{\mathrm{out}}(\mathbf{u}^*; \mathbf{W}^{\mathrm{out}})$ can be linearly approximated by Taylor series expansion about $\mathbf{z}^*$, i.e.,

$$f^{\mathrm{out}}(\mathbf{u}^*; \mathbf{W}^{\mathrm{out}}) = f^{\mathrm{out}}(\mathbf{z}^*; \mathbf{W}^{\mathrm{out}}) + \mathbf{h}_{\mathbf{z}}^{\mathrm{T}}\delta\mathbf{z} \tag{5.163}$$

where $\delta\mathbf{z} = \mathbf{u}^* - \mathbf{z}^*$ and $\mathbf{h}_{\mathbf{z}}^{\mathrm{T}} = \nabla_{\mathbf{u}^*} f^{\mathrm{out}}(\mathbf{u}^*; \mathbf{W}^{\mathrm{out}})\big|_{\mathbf{u}^*=\mathbf{z}^*}$. Furthermore, (5.162) can be approximated by a Gaussian distribution, i.e.,

$$p(t^*|\mathbf{z}^*, t^-, \mathbf{W}^{\mathrm{out}}) = \frac{1}{Z_S} \exp\left(-\frac{\beta''}{2}\{\Delta t^*\}^2\right) \tag{5.164}$$

where $1/\beta'' = (1/\beta)\left(1 + \mathbf{h}_{t^-}^{\mathrm{T}}\mathbf{h}_{t^-}\right) + \mathbf{h}_{\mathbf{z}}^{\mathrm{T}}\mathbf{\Sigma}\mathbf{h}_{\mathbf{z}}$ and $\Delta t^* = t^* - \mathbf{W}^{\mathrm{out}} \cdot [\mathbf{z}^*, \mathbf{x}^*, t^-]$.

One can consider the $p(\mathbf{W}^{out}|D')$ term in (5.161). Expanding $D' = \{t_i, \mathbf{z}_i\}_{i=1}^{n}$ and marginalizing over $\mathbf{u}_i$, we have

$$p(\mathbf{W}^{out}|D') = \int p(\mathbf{W}^{out}, \mathbf{u}_i|t_i, \mathbf{z}_i)d\mathbf{u}_i \tag{5.165}$$

Using the Bayesian rule and the conditional independence of t_i on $\mathbf{z}_i$ conditioning on $\mathbf{u}_i$, (5.165) can be rewritten as

$$p(\mathbf{W}^{out}|D') = \frac{p(\mathbf{W}^{out})}{p(t_i|\mathbf{z}_i)} \int p(t_i|\mathbf{u}_i, \mathbf{W}^{out})p(\mathbf{u}_i|\mathbf{z}_i)d\mathbf{u}_i \tag{5.166}$$

Substituting (5.164) and (5.166) into (5.161), a Gaussian approximation of the predictive distribution can be obtained

$$p(t^*|\mathbf{z}^*, D) = \frac{1}{\left(2\pi\sigma_r^2\right)^{1/2}} \exp\left(\frac{\{t^* - f^{out}(\mathbf{z}^*, \mathbf{x}^*; \mathbf{W}^{out})\}}{2\sigma_r^2}\right) \tag{5.167}$$

where $\sigma_r^2 = 1/\beta'' + \mathbf{g}^T\mathbf{A}^{-1}\mathbf{g}$.

Then, according to [42], the variance σ_r^2 of the predictive distribution reads as

$$\sigma_r^2 = \frac{1}{\beta} + \frac{1}{\beta}\mathbf{h}_{t^-}^T\mathbf{h}_{t^-} + \mathbf{h}_{\mathbf{z}}^T\Sigma\mathbf{h}_{\mathbf{z}} + \mathbf{g}^T\mathbf{A}^{-1}\mathbf{g} \tag{5.168}$$

where $\mathbf{A}$ is the Hessian matrix of $S(\mathbf{W}^{out})$, which is the summation of the likelihood and the priori over the weights

$$S(\mathbf{W}^{out}) = \frac{\beta''}{2}\sum_{i=1}^{n}\{t_i - f^{out}(\mathbf{z}_i, \mathbf{x}_i; \mathbf{W}^{out})\}^2 + \frac{\alpha}{2}\|\mathbf{W}^{out}\|^2 \tag{5.169}$$

It is noticeable from the above formula that compared to the previous variance $\sigma_{t^*}^2 = (1/\beta)\left(1 + \mathbf{h}_{t^-}^T\mathbf{h}_{t^-}\right) + \mathbf{g}^T\mathbf{A}^{-1}\mathbf{g}$, the estimation of uncertainty contains an additional term $\mathbf{h}_{\mathbf{z}}^T\Sigma\mathbf{h}_{\mathbf{z}}$ that reflects the contribution of the prediction distribution from the variance of the input noise process. Given $\mathbf{z}^*$, the total variance of the predictive distribution is known, a $(1 - \alpha)\%$ PI can be constructed

$$y^* \pm \mathbf{z}^{1-\frac{\alpha}{2}}\left(\frac{1}{\beta} + \frac{1}{\beta}\mathbf{h}_{t^-}^T\mathbf{h}_{t^-} + \mathbf{h}_{\mathbf{z}}^T\Sigma\mathbf{h}_{\mathbf{z}} + \mathbf{g}^T\mathbf{A}^{-1}\mathbf{g}\right)^{1/2} \tag{5.170}$$

where $\mathbf{z}^{1-\alpha/2}$ is the $1 - \alpha/2$ quantile of a normal distribution function with mean zero and unit variance.

5.6.5 Case Study

To verify the effectiveness of the ESN model with output feedback (ESN-OF), the pipeline pressure data coming from the energy data center of a certain plant of China are employed for the experiments. The pipeline structure of the BFG system in steel industry is presented in Fig. 3.6. We randomly select the outlet pressure of #1, 2 gasholders as the validation point. The related parameters of the ESN-OF are listed in Table 5.9.

According to Fig. 3.6, the outlet pressure of #1 gasholder is relevant to the outlet flows of #2 holder, the generation of #2,3 blast furnace, the flow of diffusion tower, the consumptions of #1,2,3 power plants, the consumption of #4 power generator, and the consumption of CCPP. Similarly, the outlet pressure of #2 gasholder is relevant to the outlet flow of #1 gasholder, the generation of #2,3 blast furnace, the flow of diffusion tower, the consumption of #1,2,3 power plant, the consumption of #4 power generator, and the consumption of CCPP.

The dimensionality of the reservoir and other parameters of the ESN are empirically set. The initial value of hyper-parameters α and β'' is set as 5 and 50, respectively. For outlet pressure prediction of #1 gasholder, the final values of hyper-parameters $\alpha$and β'' are optimized as 14.4587 and 103.8986, respectively. And, for those of #2 gasholder, the values of α and β'' are 19.5674 and 112.6395, respectively.

Figure 5.18 shows the results of the ESN-OF, the predicted pressures of #1 and #2 gasholders, respectively. The ESN-OF exhibits a good performance. And, the constructed PIs can provide some indicated information about the reliability of the prediction accuracy and present the possible interval that the targets located. Figure 5.19 illustrates a clearer illustration of PIs based on the ESN-OF with the confidence level 95%, in which the target values can be basically covered by the constructed PIs.

Furthermore, to quantitatively evaluate the performance, a series of statistical results are also reported in Table 5.10. To guarantee the indication of the statistics, the PIs construction is independently repeated by 50 times. Here, the RMSE is used to measure the prediction quality, which is defined by (2.59). Although the RMSE of the pressure prediction surpasses 200 (Pa), the accuracy can completely meet the demands of the energy scheduling with consideration of the pressure order of 7000–8000 (Pa). Besides, the PICP and the MPIW are further adopted here. CWC that gives concerns to both the PICP and MPIW is a more comprehensive index.

Table 5.9 The parameters of the ESN-OF

Parameters	Value
Reservoir dimensionality	200
Sparse degree of the internal weights	0.02
Spectral radius of the internal weights	0.8
The number of training samples	1000
Initial value of α	5.0
Initial value of β''	50.0

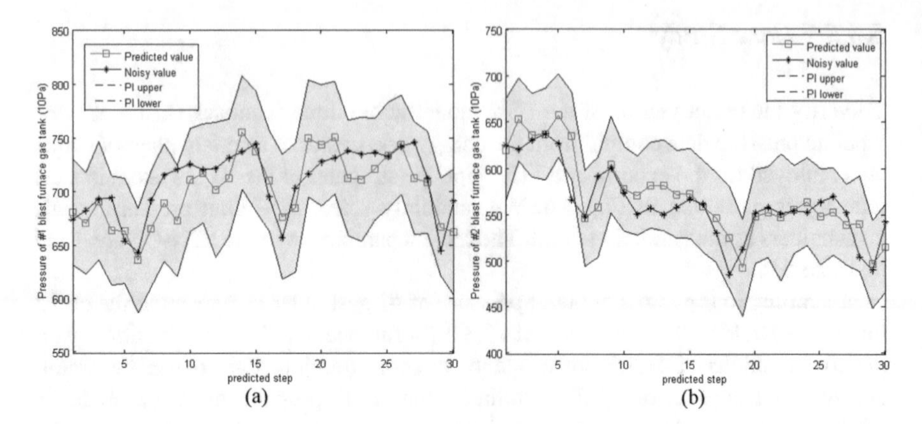

Fig. 5.18 The predicted pressure for #1 and #2 gasholder

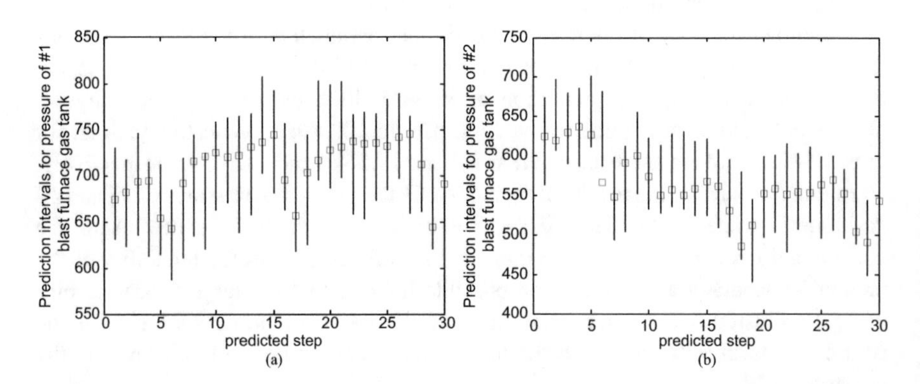

Fig. 5.19 PIs for the pressure of #1 and #2 gasholder

Table 5.10 Statistical analysis of the PIs for pipeline pressure

Index	Pipeline pressure of #1 gasholder	Pipeline pressure of #2 gasholder
RMSE (10 Pa)	22.5253	20.7134
PICP	1.0	0.9667
MPIW	104.4642	98.8186
CWC	104.4642	98.8186
Time (s)	3.65	2.47

From Table 5.10, it is apparent that the PI coverage probability of the pressures is superior to 95%. In this case, when the variable μ in (5.67) is set to 95%, the value of CWC is equal to that of MPIW. The PI coverage probability based on an acceptable interval width can meet the practical requirements. The last row in Table 5.10 shows the computational time of the ESN-OF, which indicates that the predicted horizon is set to be 30 min, and the computing efficiency is acceptable. All in all, the ESN-OF obtains a satisfactory performance on PIs for the pipeline pressure.

5.7 Prediction Intervals with Missing Input

As for the practical PIs construction in industry, the complex industrial environment, which often produces a lot of missing data points, leads to the incompleteness of the input vector of the prediction model. Figure 5.20 illustrates a time series data with missing points in input vector, where the circles in the solid line box denote the input vector of the model, and the hollow ones denote the value-missed variables in the input vector. In general, the PIs cannot be directly built up with the incomplete input vectors if using the generic approaches. In contrast, when facing with incomplete input vectors, dynamic Bayesian networks (DBNs) is still capable of constructing the PIs by using the probability distribution of the nodes in the future time slices, which benefits from the dependence of all the variables in the model.

5.7.1 Kernel-Based DBN Prediction Model

The kernel-based DBN model [43] is capable of offering prediction intervals for time series with missing inputs. It can be illustrated by Fig. 5.21, where every time slice contains only one node that denotes the value of a time series at a time slice. The order of this DBN is p, meaning that a node at the tth time slice has p parent nodes in previous time slices. And the structure and parameters of the nodes after the tth time slice share that at tth time slice.

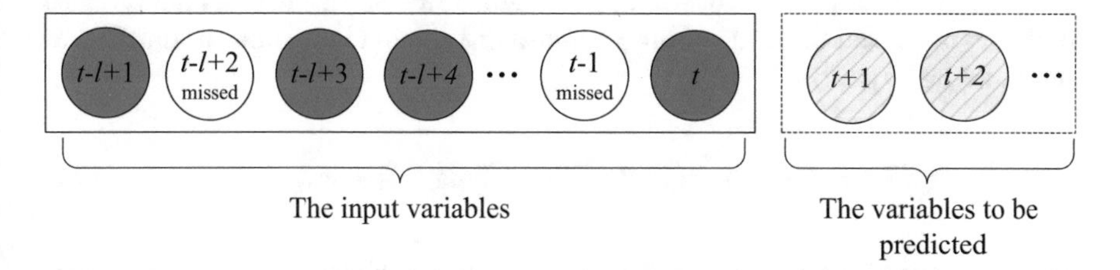

Fig. 5.20 A time series with missing points in the input vector

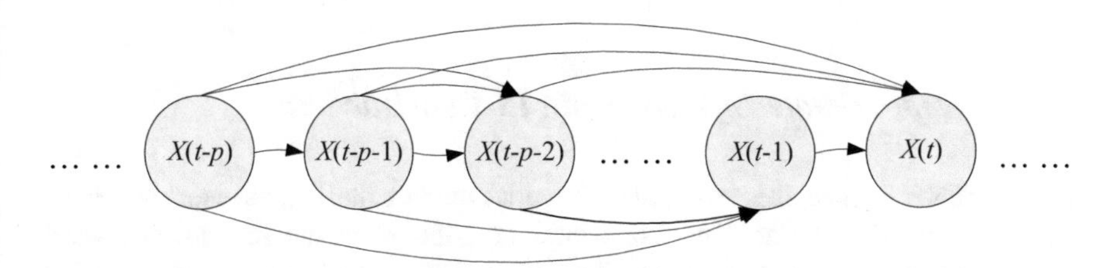

Fig. 5.21 The structure of a kernel-based DBN model for univariate time series

We assume a sample set $\{\mathrm{x}_i\}_{i=1}^N$, where $\mathbf{x}_i = [x_i(1), x_i(2), \ldots, x_i(l)]$ denotes a vector for the values of all nodes of ith sample with the length l. Besides, $\mathbf{x}(t)$ denotes the collection of the samples of the node at the tth time slice. The sample sequences read as a matrix with each row denoting a sample sequence

$$\begin{bmatrix} x_1(1) & x_1(2) & x_1(3) & \cdots & x_1(l) \\ x_2(1) & x_2(2) & x_2(3) & \cdots & x_2(l) \\ x_3(1) & x_3(2) & x_3(3) & \cdots & x_3(l) \\ \cdots & \cdots & \cdots & \cdots & \cdots \\ x_N(1) & x_N(2) & x_N(3) & \cdots & x_N(l) \end{bmatrix} \tag{5.171}$$

Supposing that the random variable $X(t)$ corresponds to the node at tth time slice, we consider its parents not only in the same time slice, but also in p former time slices. Thus, the values of its parents of the tth time slice node can be denoted by $\{\mathrm{pa}_i(t)\}_{i=1}^N$, where $\mathbf{pa}_i(t) = \left\{\mathrm{pa}_i^{(t)}(t), \ldots, \mathrm{pa}_i^{(t-k)}(t), \ldots, \mathrm{pa}_i^{(t-p)}(t)\right\}$. Note that $\mathbf{pa}_i^{(t-k)}(t)$ denotes the parents of $X(t)$ in $(t-k)$th time slice and $\mathbf{pa}(t)$ denotes the sample matrix of the parents of the tth time slice node.

Since the values of the nodes are all continuous, we give the probability density function of the tth time slice node in the Gaussian form, i.e.,

$$\begin{aligned} f^{<t>}(x_i(t)|\mathbf{pa}_i(t)) &= N\left(x_i(t)|\mu_{x_i(t)}, \sigma^2_{x_i(t)}\right) \\ &= \frac{1}{\sqrt{2\pi\sigma^2(t)}}\exp\left[-\frac{(x_i(t) - g^{<t>}(\mathbf{pa}_i(t)))}{2\sigma^2(t)}\right] \end{aligned} \tag{5.172}$$

where $x_i(t)$ denotes the value of the node variable at the tth time slice, $\mathbf{pa}_i(t)$ denotes the vector value of parent nodes of $x_i(t)$ with dimension p, $\sigma^2(t)$ denotes the variance of the tth node, and the mean of this Gaussian distribution is a nonlinear function of $\mathbf{pa}_i(t)$, i.e., $g^{<t>}(\mathbf{pa}_i(t))$.

$$g^{<t>}(\mathbf{pa}_i(t)) = \sum_{k=1}^N w_k(t) K^{<t>}(\mathbf{pa}_i(t), \mathbf{pa}_k(t)) + w_0(t) \tag{5.173}$$

where $K^{<t>}(\mathbf{pa}_i(t), \mathbf{pa}_k(t))$ denotes the kernel function at tth time slice, $\mathbf{w}(t) = [w_0(t), w_1(t), \ldots, w_N(t)]^{\mathrm{T}}$ is the weights vector of the kernel function, and $w_0(t)$ denotes the bias scalar.

5.7.2 Approximate Inference and PI Construction

As for DBN inference, there were two categories of methods, i.e., the exact inference and the approximate one. The exact inference, including the forward-backward algorithm and the junction tree algorithm, are partially suitable for discrete variables models and linear models. While the kernel-based DBN is a nonlinear model, its

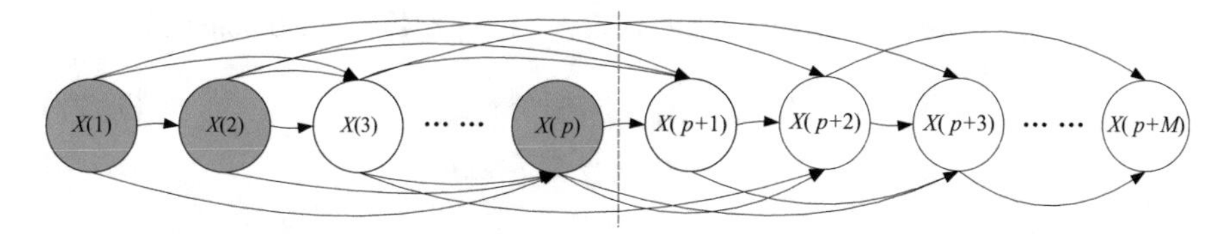

Fig. 5.22 Static version of a kernel-based DBN with incomplete evidence for time series prediction intervals

inference is therefore by the aid of approximate approach. The weighted likelihood algorithm is one of the efficient approximate approaches developed for static BN [44].

The DBN constructed for PIs is firstly converted to the corresponding static network version G_{static} described in Fig. 5.22, where the shaded nodes are observed variables, namely the evidence $\mathbf{E}$, and the rest are the unobserved ones, namely the query nodes $\mathbf{Q}$. There are a number of unobserved nodes in the p former time slices. The variables at the p former time slices serve as the incomplete inputs, and the probability distribution of the last M nodes are to be predicted by using the weighted likelihood algorithm. At first, we can find the topological sort of G_{static}, namely ρ, and assume that N_s samples are to be sampled.

In the ith sampling process, we initialize a sample $\mathbf{s}_i$ with weighting $\omega_i = 1$, where $i = 1, 2, 3, \ldots, N_s$. For each variable X in turn in the topological sort ρ, if X is in the evidence set $\mathbf{E}$, then it is just set to its instantiated value. If it is not the case, then it is sampled from the conditional density function $f(X| \mathbf{Par}(X))$, in which the conditioning variable X is set to their currently sampled values, where $\mathbf{Par}(X)$ represents the parents of X. And the weighting associated with the resulting sample $\mathbf{s}_i$ is then given by

$$\omega_i = \prod_{X \in \mathbf{E}} P(X|\mathbf{Par}(X)) \tag{5.174}$$

Considering that the values of the query node $Q(Q \in \mathbf{Q})$, $\mathbf{q}$ are extracted from the sample set $\mathbf{s}$, each element q_i in $\mathbf{q}$ is associated with the normalized weighting ω_i^{norm}, the ith element of the normalized weighting vector $\boldsymbol{\omega}^{\text{norm}}$. Then, the mean and the standard deviation of the query variable are given by

$$\mu_{\text{query}} = \sum_{i=1}^{N_s} q_i \omega_i^{\text{norm}} \tag{5.175}$$

$$\sigma_{\text{query}} = \sum_{i=1}^{N_s} \left(q_i - \mu_{\text{query}}\right) \omega_i^{\text{norm}} \left(q_i - \mu_{\text{query}}\right) \tag{5.176}$$

Finally, a PI with a nominal confidence level ($1-\alpha\%$) is constructed by

$$U_{\text{limit}} = \mu_{\text{query}} + z^{1-\frac{\alpha}{2}}\sigma_{\text{query}} \tag{5.177}$$

$$L_{\text{limit}} = \mu_{\text{query}} - z^{1-\frac{\alpha}{2}}\sigma_{\text{query}} \tag{5.178}$$

where U_{limit} and L_{limit} are the upper and the lower limitations of the constructed PI, respectively, and $z^{1-(\alpha/2)}$ is the $1-(\alpha/2)$ quantile of a Gaussian distribution function with zero mean and unit variance.

5.7.3 Learning a Kernel-Based DBN

The structure learning of this kernel-based DBN, which can be easily performed, reduces to selecting the order p of the DBN because there is only one node at each time slice. Therefore, for any tth time slice node $X(t)$, the parameters of its probability contain the variance $\sigma^2(t)$ and the weights vector $\mathbf{w}(t)$, requiring to be learned from the training data. These parameters can be learned by the sparse Bayesian learning method with hierarchical prior settings of $\sigma^2(t)$ and $\mathbf{w}(t)$. Besides, the parameters of the kernel function can be determined by cross-validation.

In order to employ the Bayesian-based approach, $\mathbf{w}(t)$ and $\sigma^2(t)(\beta(t)=\sigma^{-2}(t))$ are both viewed as the random variables. If each element in $\mathbf{w}(t)$ is a Gaussian distribution variable with mean zero, then

$$f_{\mathbf{w}}(\mathbf{w}(t)|\upalpha(t)) = \prod_{i=0}^{N} N\left(w_i(t)|0, \alpha_i(t)^{-1}\right) \tag{5.179}$$

where $\upalpha(t) = [\alpha_0(t), \alpha_1(t), \alpha_2(t), \ldots, \alpha_N(t)]^{\mathrm{T}}$. To specify the hierarchical prior of $\mathbf{w}(t)$, we define the hyper priors over $\upalpha(t)$, as well as $\beta(t)$. And these quantities are scale parameters, and the suitable priors over these parameters are Gamma distributions. Hence, we designate $\upalpha(t)$ and $\beta(t)$ to obey the gamma prior distributions, i.e.,

$$f_{\upalpha}(\upalpha(t)) = \prod_{i=0}^{N} \text{Gamma}(\alpha_i(t)|a, b) \tag{5.180}$$

$$f_{\beta}(\beta(t)) = \text{Gamma}\left(\beta_j(t)|c, d\right) \tag{5.181}$$

where $\text{Gamma}(\alpha|\, a, b) = \Gamma(\alpha)^{-1} b^a \alpha^{a-1} \mathrm{e}^{-b\alpha}$, and $\Gamma(a) = \int_0^{\infty} t^{a-1}\mathrm{e}^{-t}dt$.

Therefore, the log of the marginal likelihood of the tth time slice node is given by

$$\begin{aligned} \ln \mathrm{ML}_t &= \ln\left(f^{<t>}(\mathbf{x}(t)|\mathbf{pa}(t); \boldsymbol{\alpha}(t), \beta(t))\right) \\ &= \ln \boldsymbol{N}(\mathbf{x}(t)|0, \mathbf{C}(t)) \\ &= -\frac{1}{2}\left\{N \ln(2\pi) + \ln|\mathbf{C}(t)| + \mathbf{x}(t)^{\mathrm{T}}\mathbf{C}^{-1}(t)\mathbf{x}(t)\right\} \end{aligned} \tag{5.182}$$

where $\mathbf{C}(t) = \beta^{-1}(t)\mathbf{I} + \mathbf{\Phi}(t)\mathbf{A}^{-1}(t)\mathbf{\Phi}(t)^{\mathrm{T}}$, $\mathbf{A}(t) =$ diag $(\alpha(t))$, $\mathbf{I}$ is the identity matrix, and $\mathbf{\Phi}(t)$ is the kernel matrix and $\mathbf{\Phi}(t) = [\boldsymbol{\phi}_1(t), \boldsymbol{\phi}_2(t), \ldots, \boldsymbol{\phi}_i(t), \ldots, \boldsymbol{\phi}_N(t)]^{\mathrm{T}}$, and $\boldsymbol{\phi}_i(t) = [1, K^{<t>}(\mathbf{pa}_i(t), \mathbf{pa}_1(t)), K^{<t>}(\mathbf{pa}_i(t), \mathbf{pa}_2(t)), \ldots, K^{<t>}(\mathbf{pa}_i(t), \mathbf{pa}_N(t))]^{\mathrm{T}}$. And the log of marginal likelihood of the entire model can be computed by summing the log of the marginal likelihood of all time slice nodes.

$\ln\mathrm{ML}_t$ is approximately maximized by finding the maxima of $\alpha^*(t)$ and $\beta^*(t)$, see the details in [45]. Then, one can clearly see that most of the elements in $\alpha(t)$ tend to infinite values during the iterative learning process such that the corresponding elements in $\mathbf{w}(t)$ are converged to near zero, which results in a relatively sparse model. Therefore, we rewrite (5.173) as

$$g^{<t>}(\mathbf{pa}_i(t)) = \sum_{k=1}^{N_{\text{few}}} w_k(t)K^{<t>}(\mathbf{pa}_i(t), \mathbf{pa}_k(t)) + b^{<t>} \tag{5.183}$$

There are only small amount of training data associated with the kernel functions.

5.7.4 Case Study

In this subsection, practical industrial dataset is adopted to verify the performance of the kernel-based DBN for PIs construction (KDBN-PIs). The prediction of the COG flow can provide scientific guidance for energy scheduling workers so as to improve the utilization ratio of COG and reduce the environmental pollution. While because there exists lots of missing points in the data records, caused by the complex industrial environment, it makes the model input incomplete and leads to a difficulty in direct PI construction for such data. A period of COG flow data, coming from a steel plant in China in 2014, serve as the experimental data in this study. And, the sampling frequency of these data is one minute. The tenfold CV is employed to evaluate the model. Considering the fact that the flow variation within less than 1 h can fundamentally reflect the dynamics of COG flow, the order of the DBN is set to 50. The DBN constructed based on the COG data is shown in Fig. 5.23. In this experiment, 30 points of data in the future are predicted, one can consider the DBN containing 80 time slices, regarding the former 50 nodes as inputs and the later 30 nodes as variables to be predicted for PIs construction.

The PIs constructed for the COG data with complete and incomplete inputs are shown in Fig. 5.24, where 95% confidence levels are illustrated. It can be seen from this figure that as the increase of the missing proportion in inputs, the PIs results do decline on quality, but the real data values are still completely covered by the constructed PIs. And, it is fairly apparent that the KDBN-PIs method exhibits a high stability for the industrial data when facing with different levels of missing proportions.

To further illustrate the performance of the KDBN-PIs, a number of comparative experiments by using different methods are carried out, including the Bayesian methods based on a single ESN and the bootstrap method based on multiple

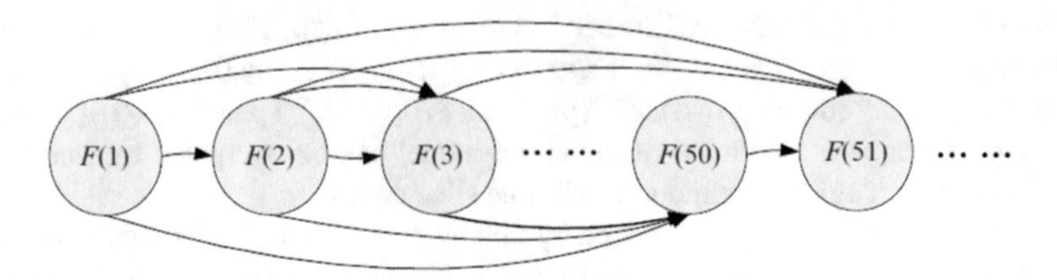

Fig. 5.23 DBN structure for COG prediction

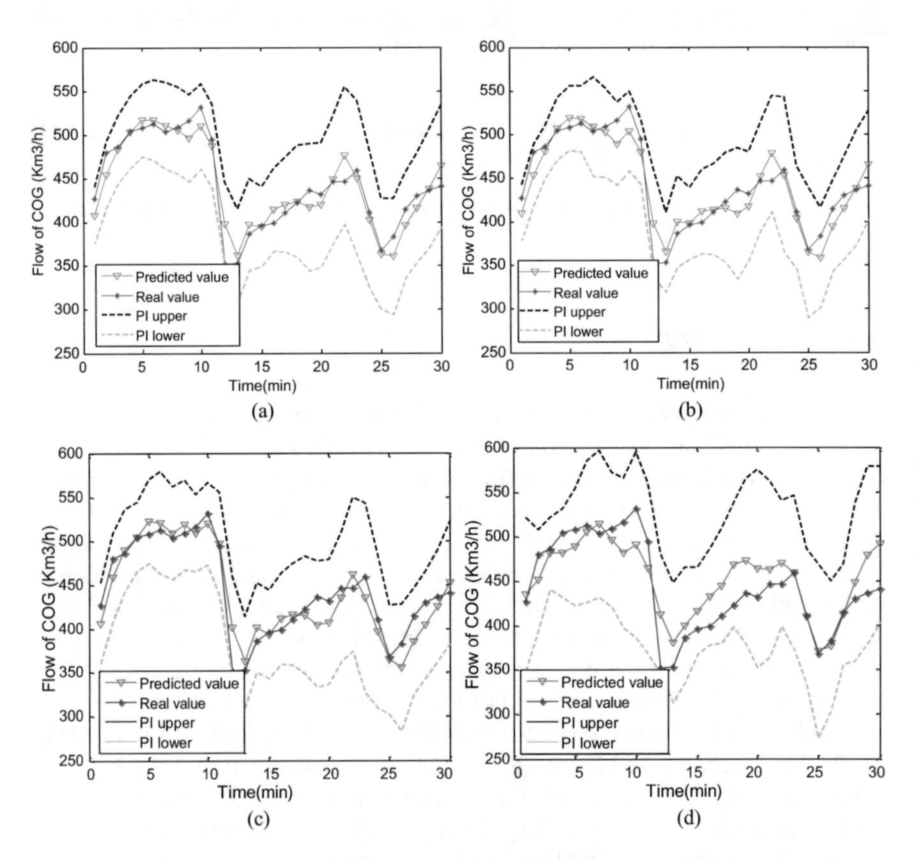

Fig. 5.24 PIs for the COG data in different levels of missing inputs. (**a**) Complete input, (**b**) 5% missing, (**c**) 10% missing, (**d**) 30% missing

ESNs, respectively. As for the incomplete inputs, these methods firstly use the well-known data imputations, such as the nearest neighbor imputation, the mean substitution, and the spline interpolation, to make the input data complete. For the predictive variances, Fig. 5.25. shows their trends over time for the competing models as well as the KDBN-PIs, where Fig. 5.25a, b illustrates the results when input is 5 and 30% missing, respectively. One can see that there is a remarkable increasing trend for low missing data ratio by the KDBN-PIs, while there are no such

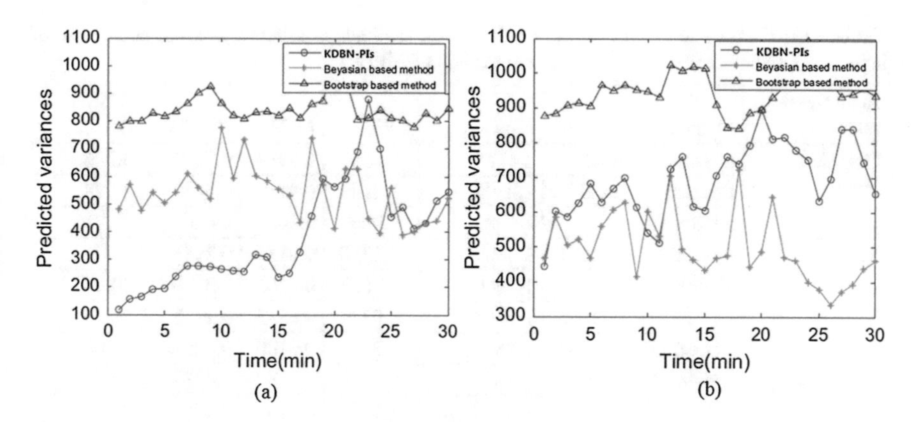

Fig. 5.25 Trends of predicted variances of COG data by all methods (**a**) 10% missing, (**b**) 30% missing

trends by the competing models. Besides, for higher missing data ratio, there are no increasing trends by all methods. The increasing trend is reasonable for low missing data ratio because uncertainty will accumulate in time; however, for higher missing data ratio, this linear trend is overshadowed with the additional uncertainty coming from missing data.

Furthermore, the comparative statistics on COG data (randomly independent experiments for 60 times) are listed in Table 5.11. When the input missing ratio is 10 and 30%, the values of CWC criterion of the KDBN-PIs are not the least among all methods. And the Bayesian-based ESN shows the least average PI width (see NMPIW listed in Table 5.11). While although the BESNE doesn't always present the best results compared to those of the others, the indicators of the KDBN-PIs, including CWC and PICP, demonstrate a high level stability with the increase of the input missing. In short, the performance of the PIs constructed by the KDBN-PIs is more stable when facing with different levels of incomplete inputs.

5.8 Discussion

In this chapter, we discussed the impact of data uncertainties on prediction problems of industry process. In Sect. 5.2, we firstly introduced some commonly used PIs construction methods, analyzed and compared the advantages and disadvantages of these methods, and illustrated the characteristics of each mentioned method. In Sects. 5.3 and 5.4, we discussed the PIs construction problems for industrial time series. And in Sect. 5.3, we introduced an iterative prediction mode for PIs construction while in Sect. 5.4 we introduced a non-iterative prediction model. In Sect. 5.5, a mixed Gaussian kernels-based PIs construction technique is further presented for factors-based prediction. In Sect. 5.6, we consider one special case of PIs construction in industrial process, which is PIs construction with noisy inputs.

Table 5.11 Comparison of prediction results for the methods in different missing percentage of inputs for COG (NMPIW is the normalized version of MPIW)

Input missing ratio	Methods		CWC_{Median}	CWC_{SD}	PICP	NMPIW	RMSE (km^3/h)
Complete input	Bayesian method		1.4148	0.9647	0.8988	0.3212	30.06
	Bootstrap method		1.2308	1.2130	0.9150	0.4440	37.28
	KDBN-PIs		1.0986	0.9066	0.9083	0.4204	26.71
5%	Bayesian method	NNI	1.4809	1.1338	0.8333	0.2945	28.86
		MS	1.7060	1.0653	0.8266	0.3330	30.49
		SI	1.3385	0.9344	0.8333	0.2963	28.24
	Bootstrap method	NNI	1.2641	1.2022	0.9100	0.4438	37.48
		MS	1.2149	0.8578	0.9083	0.4437	38.49
		SI	1.2676	1.2111	0.9100	0.4444	37.50
	KDBN-PIs		1.1302	0.8164	0.9050	0.4281	27.36
10%	Bayesian method	NNI	1.5081	0.7893	0.8200	0.2952	28.03
		MS	3.2187	5.5157	0.8177	0.3385	33.03
		SI	1.3898	0.9547	0.8377	0.2984	28.83
	Bootstrap method	NNI	1.2027	1.2253	0.9166	0.4426	37.70
		MS	1.3761	1.3166	0.8800	0.4427	40.36
		SI	1.1182	0.4722	0.9100	0.4425	38.00
	KDBN-PIs		1.3763	0.9760	0.8850	0.4437	30.54
30%	Bayesian method	NNI	2.0180	2.3936	0.7511	0.2993	32.26
		MS	2.5121	3.4642	0.7800	0.3942	42.04
		SI	3.0410	15.904	0.7177	0.3404	36.63
	Bootstrap method	NNI	1.4532	1.2940	0.8717	0.4391	38.28
		MS	1.2807	0.7339	0.8880	0.4434	44.66
		SI	1.4921	1.4116	0.8933	0.4514	39.66
	KDBN-PIs		1.5355	1.1504	0.8807	0.4434	31.87

As for this special case, four kinds of uncertainty are involved, that is the uncertainty coming from the output data noise, the uncertainty coming from the reliability of the prediction model, the uncertainty coming from the feedback, the uncertainty coming from the input data noise. In Sect. 5.7, we consider a case of PIs construction with missing input in industrial process.

Currently, we only consider the situations that the missing points in the testing input for PIs construction; however, the existing framework cannot handle the training data with missing points. For the future work, the PIs construction method can focus on training a model with incomplete training data.

References

1. Zapranis, A., & Livanis, E. (2005). Prediction intervals for neural network models. In *Proceedings of the 9th WSEAS International Conference on Computers*. World Scientific and Engineering Academy and Society (WSEAS)

2. De Veaux, R. D., Schumi, J., Schweinsberg, J., & Ungar, L. H. (1998). Prediction intervals for neural networks via nonlinear regression. *Technometrics, 40*(4), 273–282.
3. Hwang, J. T. G., & Ding, A. A. (1997). Prediction intervals for artificial neural networks. *Journal of the American Statistical Association, 92*(438), 748–757.
4. Nix, D. A., & Weigend, A. S. (1994). Estimating the mean and variance of the target probability distribution. In *Proceedings of the IEEE International Conference on Neural Networks*, Orlando, FL (Vol. 1, pp. 55–60).
5. Rivals, I., & Personnaz, L. (2000). Construction of confidence intervals for neural networks based on least squares estimation. *Neural Networks, 13*(4–5), 463.
6. Ding, A., & He, X. (2003). Backpropagation of pseudo-errors: Neural networks that are adaptive to heterogeneous noise. *IEEE Transactions on Neural Networks, 14*(2), 253–262.
7. Dybowski, R., & Roberts, S. (2000). Confidence intervals and prediction intervals for feed-forward neural networks. In R. Dybowski & V. Gant (Eds.), *Clinical applications of artificial neural networks*. Cambridge, U.K: Cambridge University Press.
8. Bishop, C. M. (1995). *Neural networks for pattern recognition*. London, UK: Oxford University Press.
9. MacKay, D. J. C. (1989). The evidence framework applied to classification networks. *Neural Computation, 4*(5), 720–736.
10. Hagan, M., & Menhaj, M. (2002). Training feedforward networks with the Marquardt algorithm. *IEEE Transactions on Neural Networks, 5*(6), 989–993.
11. Efron, B. (1979). Bootstrap methods: Another look at the jackknife. *Annals of Statistics, 7*(1), 1–26.
12. Heskes, T. (1997). Practical confidence and prediction intervals. In T. P. M. Mozer & M. Jordan (Eds.), *Neural information processing systems* (Vol. 9, pp. 176–182). Cambridge, MA: MIT Press.
13. Sheng, C., Zhao, J., Wang, W., et al. (2013). Prediction intervals for a noisy nonlinear time series based on a bootstrapping reservoir computing network ensemble. *IEEE Transactions on Neural Networks & Learning Systems, 24*(7), 1036–1048.
14. Tibshirani, R. (1996). A comparison of some error estimates for neural network models. *Neural Computation, 8*(1), 152–163.
15. Khosravi, A., Nahavandi, S., Creighton, D., et al. (2011). Comprehensive review of neural network-based prediction intervals and new advances. *IEEE Transactions on Neural Networks, 22*(9), 1341–1356.
16. Anguita, D., Ghio, A., Oneto, L., et al. (2012). In-sample and out-of-sample model selection and error estimation for support vector machines. *IEEE Transactions on Neural Networks & Learning Systems, 23*(9), 1390.
17. Efron, B., & Tibshirani, R. J. (1993). *An introduction to the bootstrap*. New York, USA: Chapman & Hall.
18. Efron, B., & Tibshirani, R (1995). Cross-validation and the bootstrap: Estimating the error rate of a prediction rule. Dept. Stat., Stanford Univ., Stanford, CA, USA, Tech. Rep. TR-477.
19. Arlot, S., & Celisse, A. (2010). A survey of cross-validation procedures for model selection. *Statistics Surveys, 4*, 40–79.
20. Efron, B., & Tibshirani, R. (1997). Improvements on cross-validation: The .632+ bootstrap method. *Journal of the American Statistical Association, 92*(438), 548–560.
21. Xue, Y., Yang, L., & Haykin, S. (2007). Decoupled echo state networks with lateral inhibition. *Neural Networks, 20*(3), 365–376.
22. Sheng, C., Zhao, J., & Wang, W. (2017). Map-reduce framework-based non-iterative granular echo state network for prediction intervals construction. *Neurocomputing, 222*, 116–126.
23. Dong, R., & Pedrycz, W. (2008). A granular time series approach to long-term forecasting and trend forecasting. *Physica A: Statistical Mechanics and its Applications, 387*(13), 3253–3270.
24. Song, M., & Pedrycz, W. (2013). Granular neural networks: Concepts and development schemes. *IEEE Transactions on Neural Networks & Learning Systems, 24*(4), 542–553.

25. Cimino, A., Lazzerini, B., Marcelloni, F., et al. (2011). Granular data regression with neural networks, Fuzzy logic and applications. *Lecture Notes in Computer Science, 6857*, 172–179.
26. Jaeger, H., & Haas, H. (2004). Harnessing nonlinearity: Predicting chaotic systems and saving energy in wireless communication. *Science, 304*, 78–80.
27. Jaeger, H. (2002). Tutorial on training recurrent neural networks, covering BPTT, RTRL, EKF and echo state network approach. German National Research Center for Information Technology, GMD Rep. 159.
28. Zhao, J., Wang, W., Liu, Y., et al. (2011). A two-stage online prediction method for a blast furnace gas system and its application. *IEEE Transactions on Control Systems Technology, 19* (3), 507–520.
29. Zhao, J., Liu, Q., Wang, W., et al. (2012). Hybrid neural prediction and optimized adjustment for coke oven gas system in steel industry. *IEEE Transactions on Neural Networks and Learning Systems, 23*(3), 439–450.
30. Liu, Y., Liu, Q., Wang, W., et al. (2012). Data-driven based model for flow prediction of steam system in steel industry. *Information Sciences, 193*, 104–114.
31. Pedrycz, W., & Homenda, W. (2013). Building the fundamentals of granular computing: A principle of justifiable granularity. *Applied Soft Computing, 13*(10), 4209–4218.
32. Jones, J. A., Evans, D., & Kemp, S. E. (2007). A note on the Gamma test analysis of noisy input/output data and noisy time series. *Physica D: Nonlinear Phenomena, 229*(1), 1–8.
33. Liitiainen, E., Verleysen, M., Corona, F., & Lendasse, A. (2009). Residual variance estimation in machine learning. *Neurocomputing, 72*(16–18), 3692–3703.
34. Evans, D., & Jones, A. J. (2002). A proof of the gamma test. *Society of London. Series A, 458*, 2759–2799.
35. Khosravi, A., Nahavandi, S., Creighton, D., et al. (2011). Lower upper bound estimation method for construction of neural network-based prediction intervals. *IEEE Transactions on Neural Networks, 22*(3), 337–346.
36. Kennedy, J., & Eberhart, R. (1995). Particle swarm optimization. In *Proceedings of the IEEE International Conference on Neural Networks* (pp. 1942–1948). Piscataway: IEEE Service Center.
37. Mackey, M. C., & Glass, L. (1977). Oscillation and chaos in physiological control systems. *Science, 197*(4300), 287–289.
38. De, B. K., De, B. J., Suykens, J. A., et al. (2011). Approximate confidence and prediction intervals for least squares support vector regression. *IEEE Transactions on Neural Networks, 22* (1), 110–120.
39. Bishop, C. M. (2006). *Pattern recognition and machine learning*. New York: Springer Press.
40. Vapnik, V. (1995). *The nature of statistical learning theory*. New York: Springer.
41. Boyd, V., & Faybusovich, L. (2006). Convex optimization. *IEEE Transactions on Automatic Control, 51*(11), 1859–1859.
42. Wright, W. A. (1999). Bayesian approach to neural-network modeling with input uncertainty. *IEEE Transactions on Neural Networks, 10*(6), 1261.
43. Chen, L., Liu, Y., Zhao, J., Wang, W., & Liu, Q. (2016). Prediction intervals for industrial data with incomplete input using kernel-based dynamic Bayesian networks. *Artificial Intelligence Review, 46*, 307–326.
44. Fung, R., & Chang, K. C. (1990). Weighting and integrating evidence for stochastic simulation in Bayesian networks. In P. P. Bonissone, M. Henrion, L. N. Kanal, & J. F. Lemmer (Eds.), *Uncertainty in artificial intelligence* (Vol. 5, pp. 208–219). North Holland: Elsevier.
45. Tipping, M. E. (2001). Sparse Bayesian learning and the relevance vector machine. *Journal of Machine Learning Research, 1*, 211–244.

Chapter 6
Granular Computing-Based Long-Term Prediction Intervals

Abstract In industrial practice, long-term prediction for process variables is fairly significant for the process industry, which is capable of providing the guidance for equipment control, operational scheduling, and decision-making. This chapter firstly introduces the basic principles of granularity partition, and a long-term prediction model for time series and factor-based prediction are developed in this chapter. In terms of time series prediction, the unequal-length temporal granules are constructed by exploiting dynamic time warping (DTW) technique, and a granular-computing (GrC)-based hybrid collaborative fuzzy clustering (HCFC) algorithm is proposed for the mentioned factor-based prediction problem. Besides, in this chapter, the long-term prediction approach is also combined with the PIs construction in order to provide the prediction reliability in the context of long-term time series task. Similarly, the PIs construction on multi-dimension problem is also introduced by employing the structure of the HCFC algorithm. To verify the effectiveness of these approaches, this chapter provides some experimental analysis on industrial data coming from an energy data center of a steel plant.

6.1 Introduction

In previous chapters, we focus on the short-term prediction (e.g., 30 min or 1 h) for the industrial process variables, including the time series prediction in Chap. 3 and the factor-based industrial process prediction in Chap. 4, which can satisfy the demand of decision-making in a short term. However, sometimes, the workers will also consider a scheduling operation in a longer term (e.g., 1 or 2 days). The long-term prediction for the process variables can provide guidance for making the production plan in a long term. For example, in the blast furnace iron-making process, the long-term prediction of the by-product gas can provide the flow variation tendency in advance, then a relevant energy scheduling scheme can be established ahead, which allows an efficient, low-cost, and sustainable manufacturing process of steel industry.

J. Zhao et al., *Data-Driven Prediction for Industrial Processes and Their Applications*, Information Fusion and Data Science,
https://doi.org/10.1007/978-3-319-94051-9_6

Data-driven approaches are considered as the first choice for industrial prediction, especially in steel industry. An online parameter optimization-based prediction was modeled for a converter gas system, and a parallel computing strategy based on GPU structure was implemented [1]. A multi-kernel support vector regressor (SVR) was designed to estimate the gasholder level of a steel plant [2]. Similarly, the complexity of a converter gas system was analyzed and the multiple holder levels were simultaneously predicted by a multi-output least square SVR [3]. Although these methods provided satisfactory results for predicting horizon limited within 1 h, their accuracies will largely degrade if the horizon has to be extended because of the usage of the iterative prediction. Discarding this iterative prediction, the long-term prediction employs a kind of granular computing-based method, which avoids the drawback of error accumulation in the iterative prediction.

Besides, the reliability of the long-term prediction results is also an important issue in industrial applications. The PIs construction can provide the quantified reliability of the results, and there are various methods for PIs construction in literature introduced in Chap. 5, such as the MVEs, the delta, and the Bayesian principle-based approach. Besides, hybrid Kalman-filter, particle swarm optimization (PSO) with a SVM, and the probabilistic forecasting are also utilized. However, these approaches only demonstrated medium- or short-term PIs. And not only the predicting horizon was relatively short, but also a large amount of model parameters was involved in these approaches, which came up with a low availability in dealing with real-time industrial applications. Therefore, it is necessary to develop a long-term PIs construction framework.

In a nutshell, in the perspective of long-term PIs construction with data-based methodologies, it is rather important to develop a long-term prediction scheme with reliability estimation for demanding applications on industrial environment. Therefore, a set of granular computing-based prediction models involving both pointwise and interval ones are reported so as to provide adequate prediction length and quantitatively illustrate the prediction reliability.

6.2 Techniques of Granularity Partition

The existing studies were mostly focused on partitioning the information granules via ascending or descending tendencies of data fluctuation. However, such a solving method is suitable for only a small number of artificial or traditional datasets, and usually fails to handle industrial datasets or application-oriented problems. Granular computing (GrC), inherently associated with computational intelligence, embraces fuzzy sets, rough sets, and interval analysis for data-driven modeling and optimization. It takes data granule, i.e., data segments with a series of points, as the analysis unit instead of single data point, which enables to extend the prediction horizon (long-term prediction).

6.2.1 Partition of Equal Length

Let us first consider time series composed of temporal segments of the same length (different lengths could be considered separately). Each segment comes with "n" temporal samples, and the number of segments is equal to "N", see Fig. 6.1. The corresponding time segments $\mathbf{s}_1, \mathbf{s}_2, \ldots, \mathbf{s}_N$, are vectors in $\mathbb{R}^n$, and the objective is to complete a long-term prediction of the next temporal segment $\widehat{\mathbf{s}}$, see also Fig. 6.1. The formation of the model comprises two main steps.

The first step is the granular representation of temporal segments of the time series. This is done by clustering $\mathbf{s}_1, \mathbf{s}_2, \ldots, \mathbf{s}_N$ into "C" clusters, which helps us reveal some general tendencies associated with the segments. The use of FCM (considered here as a sound representative fuzzy clustering) gives rise to "C" prototypes $\mathbf{v}_1, \mathbf{v}_2, \ldots, \mathbf{v}_C$. The membership degree of the temporal segment $\mathbf{s}_k$ to cluster "i" (described by prototype $\mathbf{v}_i$) is computed as follows:

$$z_{ki} = \frac{1}{\sum_{j=1}^{C} \left(\frac{\|\mathbf{s}_k - \mathbf{v}_i\|}{\|\mathbf{s}_k - \mathbf{v}_j\|} \right)^{2/(m-1)}} \quad m > 1; \quad k = 1, 2, \ldots, N; \quad i = 1, 2, \ldots, C \tag{6.1}$$

In the second step, the relationship between successive time segments $\mathbf{s}_1, \mathbf{s}_2, \ldots$, (whose knowledge is necessary to realize prediction) can be formed in different ways. As an example, we discuss second-order temporal dependencies between the current segments $\mathbf{s}_{k-2}, \mathbf{s}_{k-1}$, and the forecasted one $\mathbf{s}_k$. In other words, we envision relationships coming in the form, $\mathbf{s}_1, \mathbf{s}_2 \rightarrow \mathbf{s}_3$; $\mathbf{s}_2, \mathbf{s}_3 \rightarrow \mathbf{s}_4$; and …, etc. In terms of the representation provided by fuzzy clustering, we have the relationships expressed as $\mathbf{z}_1, \mathbf{z}_2 \rightarrow \mathbf{z}_3$; $\mathbf{z}_2, \mathbf{z}_3 \rightarrow \mathbf{z}_4$; and …, etc., where $\mathbf{z}_i = [z_{1i} \; z_{2i} \; \cdots \; z_{Ci}]^{\mathrm{T}}$.

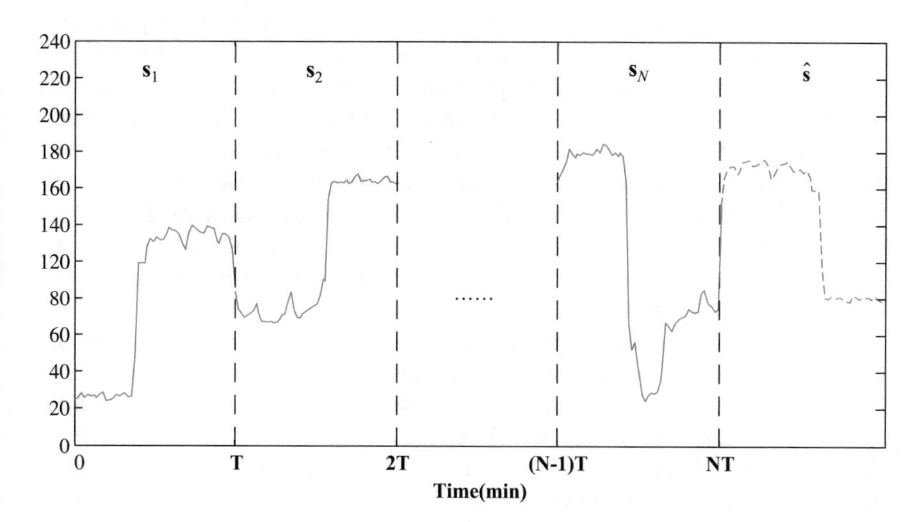

Fig. 6.1 Time series along with its information granules of equal size (length)

The formation of the overall prediction model can engage the concept of fuzzy associative memories. In essence, we form a union of the Cartesian products ($\times$) of the triples of the temporal segments: $\mathbf{z}_1 \times \mathbf{z}_2 \times \mathbf{z}_3, \mathbf{z}_2 \times \mathbf{z}_3 \times \mathbf{z}_4, \ldots$ and arrange them in the form of a single fuzzy relation being a union of the individual Cartesian products $R = (\mathbf{z}_1 \times \mathbf{z}_2 \times \mathbf{z}_3) \cup (\mathbf{z}_2 \times \mathbf{z}_3 \times \mathbf{z}_4)\ldots$ The Cartesian product is implemented by using algebraic product while the union uses the maximum operator. Note that in this way R is defined in the hypercube $[0, 1]^C \times [0, 1]^C \times [0, 1]^C$.

For prediction purposes, one infers $\widehat{\mathbf{z}}_k$ for a given $\mathbf{z}_{k-1}$ and $\mathbf{z}_{k-2}$. We have $\widehat{\mathbf{z}}_k = (\mathbf{z}_{k-2} \times \mathbf{z}_{k-1}) \circ R$, where R denotes a fuzzy relation capturing the relationships among information granules and "o" is a certain relational operator. In the sequel, the prediction results completed in the vector in $[0, 1]^C$ are transformed to the temporal segment $\widehat{\mathbf{s}}_k$ using the well-known relationship

$$\widehat{\mathbf{s}}_k = \sum_{i=1}^{C} \mathbf{v}_i \widehat{z}_{ki} \Big/ \sum_{i=1}^{C} \widehat{z}_{ki} \tag{6.2}$$

where $\widehat{\mathbf{z}}_k = [\widehat{z}_{k1} \quad \widehat{z}_{k2} \quad \cdots \quad \widehat{z}_{kC}]^{\mathrm{T}}$, and $\widehat{\mathbf{z}}_k \in [0, 1]^C$.

6.2.2 Partition of Unequal Length

We granulate time series, which results in a collection of information granules treated as sound processing units supporting all subsequent processing. In the existing research, the construction of generic granular semantics was discussed in [4] by looking at ascending or descending tendency of the data. However, with respect to real-world industrial data, the data variations, and capturing only ascending or descending tendencies, could not accurately reflect the related production situation or the equipment operating status. For instance, taking also the steel industry as example, the BFG consumption flow of hot blast stove generally consists of three categories of operational modes with the corresponding gas flows illustrated in Fig. 6.2. Although there are some similarities between these modes, the amplitude and the duration of each mode are substantially different. We can use three feature modes of the following semantically different information granules, see Modes #1–#3, to describe the operational characteristics of the equipment. Since there is a high level of noise involved in production processes, the acquired industrial data can hardly be fully matched with the modes described above.

In this section, a production process concerned granulation is presented, which considers the data variation of the operational processes. Without loss of generality, assuming an original industrial time series, $\mathbf{t}$, one can granulate it from the historic dataset by trial and error to generate N segments, i.e., $\mathbf{t} = \{\mathbf{t}_1, \mathbf{t}_2, \ldots, \mathbf{t}_N\}$, each of which might correspond to a certain operational stage. Taking the gas flow of hot blast stove as an example shown in Fig. 6.3, four possible granules ($N = 4$) with unequal-length time windows are being formed.

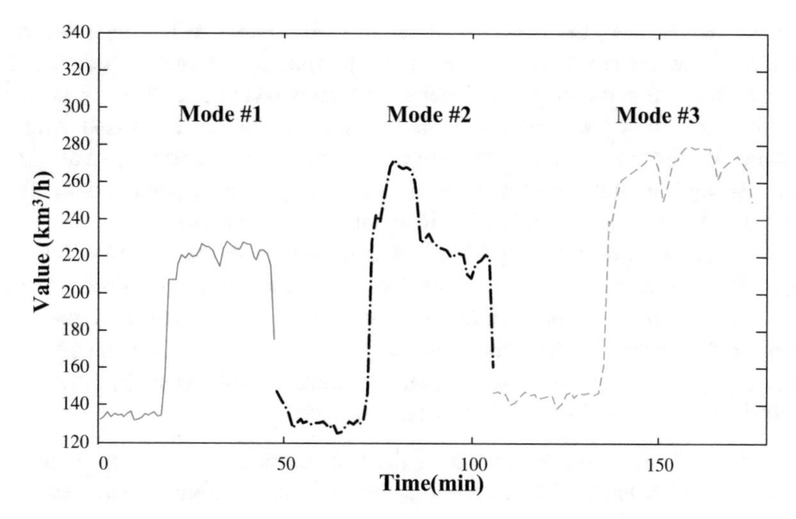

Fig. 6.2 Three operational modes of BFG consumption flows of hot blast stove

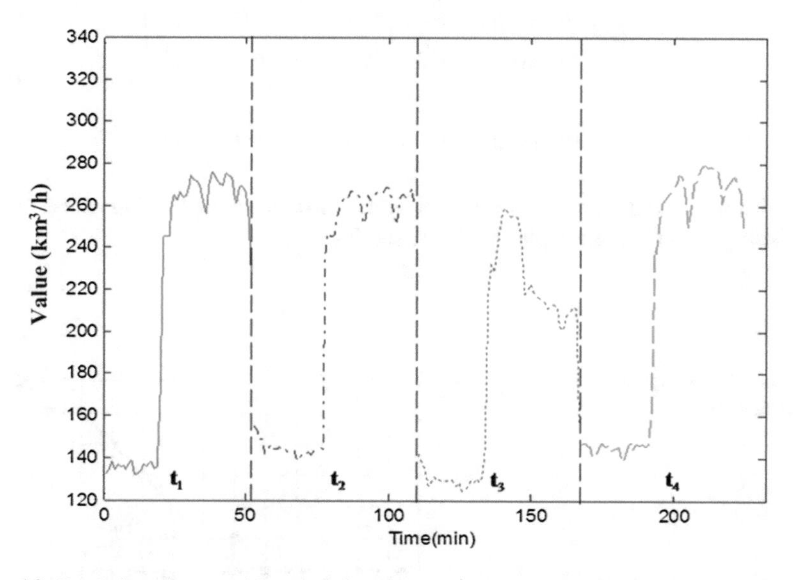

Fig. 6.3 Examples of unequal-length information granules formed for the burning-shutdown process of hot blast stove ($N = 4$)

It should be noted that the formed information granules have different lengths of time windows and different data amplitude, see the above example of BFG consumption of hot blast stove. Such unequal-length granules exactly reflect the practical operations of energy variation, and the lengths cannot be determined in advance or designate as a fixed value. If this is the case, the negative impact on the accuracy could occur. This effect will be visible in the series of experiments. In addition, these granules representing the essential semantic units require to be quantified by the distances between any two of them in the perspective of fuzzy modeling. Such distance calculation can be realized using the Minkowski or Euclidean distance, both of which are, however, applicable to temporal sequences expressed over time windows of the same length. Therefore, we construct a method to equalize the unequal-length granules on the basis of maintaining their respective dynamic features so as to make their similarities quantitatively comparable.

The dynamic time warping (DTW) was originally introduced to compare the similarity between two strings of characters. A time warping normalization (TWN) method is designed on a basis of DTW concept in order to equalize the scale of the established industrial information granules. Consider two data segments, $\mathbf{t}_1 = \{t_{11}, t_{12}, \ldots, t_{1n}\}$ and $\mathbf{t}_2 = \{t_{21}, t_{22}, \ldots, t_{2m}\}$, of dimensionality (length) of n and m, respectively. We introduce the following definition.

Definition 6.1 The distance matrix of the two segments is a $n \times m$ dimensional matrix, in which each of its element is the distance of arbitrary two data points present in these series, i.e.,

$$D = \begin{bmatrix} d(t_{11}, t_{21}) & d(t_{11}, t_{22}) & \cdots & d(t_{11}, t_{2m}) \\ d(t_{12}, t_{21}) & d(t_{12}, t_{22}) & \cdots & d(t_{12}, t_{2m}) \\ \vdots & \vdots & \ddots & \vdots \\ d(t_{1n}, t_{21}) & d(t_{1n}, t_{22}) & \cdots & d(t_{1n}, t_{2m}) \end{bmatrix} \tag{6.3}$$

Definition 6.2 A grid chart of size $n \times m$ describes a sequence of distances of the two strings based on a warping path, W, see Fig. 6.4.

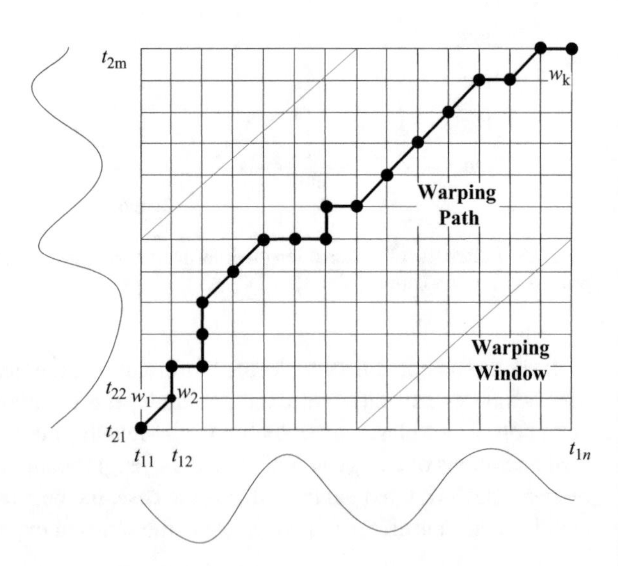

Fig. 6.4 An example of warping path

$$W = w_1(t_{11}, t_{21}) \rightarrow w_2(t_{12}, t_{22}), \rightarrow \cdots \rightarrow, w_k(t_{1n}, t_{2m}) \tag{6.4}$$

where w_i ($i = 1, 2, \ldots, k$) denotes the points positioned on the warping path starting from (t_{11}, t_{21}) and terminating at (t_{1n}, t_{2m}).

Definition 6.3 The DTW distance returns a minimum of the sum of distances optimized along the warping path.

$$\mathrm{DTW}(\mathbf{t}_1, \mathbf{t}_2) = \mathrm{Min}\left[\frac{1}{k}\sum_{i=1}^{k} d(w_i)\right] \tag{6.5}$$

where $d(w_i)$ is the corresponding distance of the two segments on the warping path, i.e., a certain element of the matrix D.

It is obvious from Definition 6.2 that the enumeration method produces the optimal warping path. However, such a method is computationally expensive if the lengths of the granules are large. A dynamic programming is therefore adopted here for determining the DTW distance. In such a way, one can adopt the DTW-based method to select a standard one in these N unequal-length formed granules [5], which can be regarded as the most general granule in the samples.

Standard Granule Selection

For the N information granules already formed and described by $\mathbf{t}_1$, $\mathbf{t}_2$, ..., $\mathbf{t}_N$, we select one of these granules as the standard (referential), denoted as $\mathbf{t}_s$, where $\mathbf{t}_s = \arg \underset{\mathbf{t}_i}{\mathrm{Min}} \sum_{j=1}^{N} \mathrm{DTW}(\mathbf{t}_i, \mathbf{t}_j)$ $(i = 1, 2, \ldots, N)$. In other words, we realize the summation of the DTW distances and determine the minimal one. Then, one defines $\mathbf{t}_i$ which leads to the minimal sum.

Time Warping Normalization

After obtaining the standard granule, the other granules become extended or suppressed to make their lengths identical to the standard one by employing a method of the time warping normalization (TWN). A detailed example can be presented as follows. Assuming that given are two similar granules **a** and **b** of different dimensionality, **a** is the standard one, and $\mathbf{b}'$ is the equalized sequence resulting from **b**. Then, the corresponding relationship based on warping paths is shown in Fig. 6.5. If the ith point of **a** corresponds to the jth point of **b**, then the relationship can be represented by $a_i \rightarrow b_j$. Corresponding to a series of points of **b**, the relationship can be represented by $a_i \rightarrow (b_j, b_{j+1}, \ldots, b_r)$ (note the vertical bold line of the warping paths involving a series of points in Fig. 6.4; here we regard $\mathbf{t}_1$ as

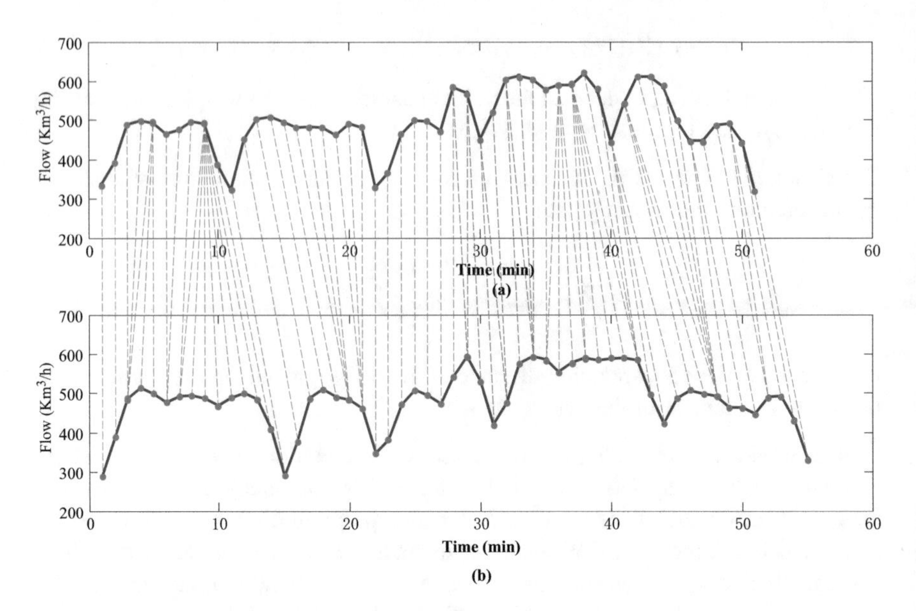

Fig. 6.5 Two similar sequences and their corresponding relationship determined through the use of the DTW. (**a**) The original sequence $\boldsymbol{a}$, (**b**) the original sequence $\boldsymbol{b}$

a, and $\mathbf{t}_2$ as **b**). We present a warping paths-based sequence equalization method in the form

- If $a_i \rightarrow (b_j, b_{j+1}, \ldots, b_r)$, then $b_i' = \text{Avg}\left(b_j, b_{j+1}, \ldots, b_r\right)$;
- If $(a_i, a_{i+1}, \ldots, a_r) \rightarrow b_j$, then $b_i' = b_j, b_{i+1}' = b_j, \ldots, b_r' = b_j$.

where the operator Avg stands for the average value of the multiple data points. It is clear that the stretched sequence $\mathbf{b}'$ is of equal length as **a**. Thus, by using the DTW-based method, the one-to-many or many-to-one relationship of two sequences can be transformed into the one-to-one relationship. Figure 6.6 illustrates an equalization of a period of real-world industrial time series, where the original sequence and the corresponding equalized one are both illustrated. It is observed that the equalized sequence retains the trend and amplitude of the original sequence without essential losses of the sequence similarity.

By engaging the above expansion or compression mechanism, one can consider the selected $\mathbf{t}_s$ as the standard granule and equalize the others to make all of the granules of the same length. It is noticeable that the computational complexity of TWN should be further considered when dealing with industrial data. Since this method has to calculate the DTW distance summation $N \times N$ times, the value of N should not be too large. In such a way, the calculation process can be carried out in real time.

Based on the TWN method, the unequal-length industrial granules can be equalized to exhibit the same length. Once completed, the granules are used to realize time series prediction. To highlight the role of the TWN, a schematic view to illustrate the

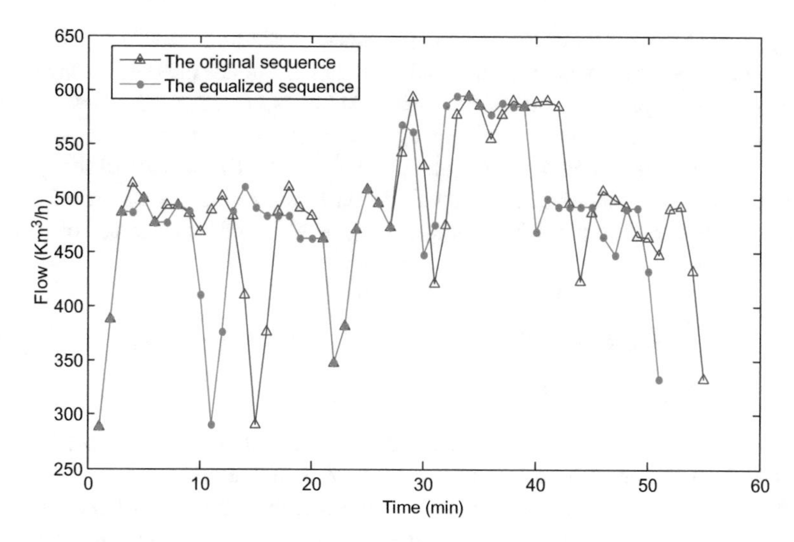

Fig. 6.6 The comparison of the original sequence and the equalized one for a period of real-world industrial time series

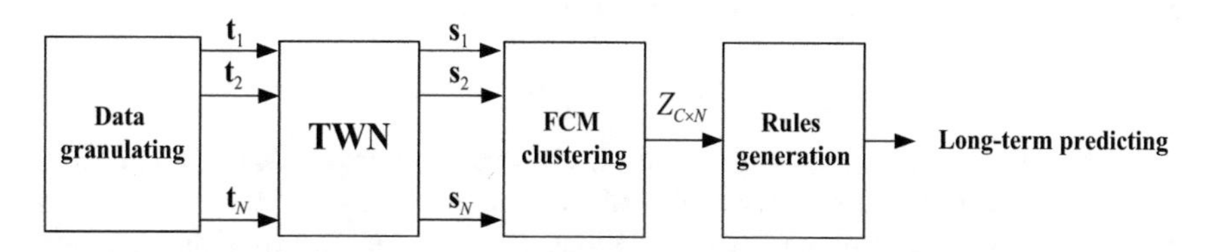

Fig. 6.7 An overall prediction procedure, shown are main functional modules

underlying prediction process is shown in Fig. 6.7. In what follows, the fuzzy rules generation will be further presented for the long-term prediction.

6.3 Long-Term Prediction Model

6.3.1 Granular Model for Time Series Prediction

To reflect the prototypes of the industrial operational modes, an effective method to form the semantic descriptors would be to cluster the data segments obtained for each time windows. The use of fuzzy clustering, and FCM, in particular, delivers here a viable alternative. In this section, the FCM method is adopted to determine the optimal partition matrix and the prototypes (centers) in order to build fuzzy rules for the long-term prediction. In other words, the centers produced in this way should be capable of reflecting the operational mode based on the underlying semantics, refer to Fig. 6.1. Instead of using the generic pointwise clustering, the equalized granules are regarded as the essential elements in FCM, which extends the prediction horizon.

Based on the preliminaries of granular time series presented so far, we can still assume that the equalized granules dataset $S = \{\mathbf{s}_1, \mathbf{s}_2, \ldots, \mathbf{s}_N\}$, where $\mathbf{s}_i = \{s_{i1}, s_{i2}, \ldots, s_{in}\}$ is a granule with the length n, and N denotes the number of the granules used for the fuzzy reasoning. Then, the FCM-based clustering centers are calculated, and the corresponding partition matrix, $Z_{C \times N}$, is obtained with the use of (6.1).

$$Z_{C\times N} = \begin{bmatrix} z_{11} & z_{12} & \cdots & z_{1N} \\ z_{21} & z_{22} & \cdots & z_{2N} \\ \cdots & \cdots & \cdots & \cdots \\ z_{C1} & z_{C2} & \cdots & z_{CN} \end{bmatrix} = \begin{bmatrix} \mathbf{z}_1 & \mathbf{z}_2 & \cdots & \mathbf{z}_N \end{bmatrix} \tag{6.6}$$

where $\mathbf{z}_i = \begin{bmatrix} z_{1i} & z_{2i} & \cdots & z_{Ci} \end{bmatrix}^{\mathrm{T}}$, and each of the elements of $\mathbf{z}_i$ is the membership degree (level of matching) of the information granule to a prototype.

We construct the fuzzy rules based on the membership level of the equalized granule that is the closest to the prototype. We consider the relationship of the n_Ith order meaning that the next state (information granule) depends upon a n_I-step history; refer to the illustration presented in Fig. 6.1. This form of the relationship can be schematically captured in the following way:

$$\mathbf{s}_{k-n_I}, \mathbf{s}_{k-n_I+1}, \ldots, \mathbf{s}_{k-1} \rightarrow \mathbf{s}_k \tag{6.7}$$

where $\mathbf{s}_{k-n_I}, \mathbf{s}_{k-n_I+1}, \ldots, \mathbf{s}_{k-1}$ are the current and past granules of the mapping, and $\mathbf{s}_k$ constitutes the next state (forecasting) of the mapping. In such a way, the fuzzy model is geared toward capturing relationships between information granules-fuzzy sets.

In what follows, we present the method using which a third-order granular time series is constructed, i.e., a (3 + 1)-tuple contains the current granules of the logical relationship and the dependent next one. The fuzzy rules can be easily derived from these tuples. They are organized in a form of a certain rule base conveying the current states of the logical relationships and assuming the form

$$R_r: \text{If } \mathbf{z}_{k-3} \text{ is } \mathbf{i}_{r3}, \mathbf{z}_{k-2} \text{ is } \mathbf{i}_{r2}, \text{and } \mathbf{z}_{k-1} \text{ is } \mathbf{i}_{r1}, \text{Then } \mathbf{z}_k \text{ is } V\mathbf{h}_r \quad k = 4, 5, \ldots, N \tag{6.8}$$

where $\mathbf{i}_{ri} \in \{\mathbf{i}_1, \mathbf{i}_2, \ldots, \mathbf{i}_C\}$, $i = 1, 2, 3$; and V is the matrix of centers of the clusters (prototypes), $V = \begin{bmatrix} \mathbf{v}_1 & \mathbf{v}_2 & \cdots & \mathbf{v}_C \end{bmatrix}$; and $\mathbf{h}_r = \begin{bmatrix} h_{r1} & h_{r2} & \cdots & h_{rC} \end{bmatrix}^{\mathrm{T}}$, where h_{ri} is the total number of times the three inputs can be tagged by the specific prototypes, and i denotes the position of the corresponding output. In this section, a t-norm (the product operation) is adopted to compute the matching degree of multiple inputs, i.e.,

$$z_r = z_{k-3}^{\max} \times z_{k-2}^{\max} \times z_{k-1}^{\max} \tag{6.9}$$

where $z_{k-3}^{\max}$ is the maximal value of the vector $\mathbf{z}_{k-3}$ that denotes the membership degree to which an information granule belongs to the closest prototype. Based on the reasoning relationship described in granular time series prediction, one can obtain the forecasted membership level

$$\widehat{\mathbf{z}}_k = z_r \mathbf{h}_r \tag{6.10}$$

Finally, the predicted granule can be obtained by using the defuzzification concept, see (6.2). For the overall prediction, it is clear that the essential units are the equalized granules, which make the computational efficiency largely improved by the extending data scale.

To make the architecture of the unequal-length GrC-based method more understandable, the prediction procedure can be summarized as follows:

Algorithm 6.1: Unequal-Length GrC-Based Long-Term Prediction

Step 1: *Granulate the original industrial time series data involving the energy process semantics, such as the example shown in* Fig. 6.3. *The result* $\mathbf{t} = \{\mathbf{t}_1, \mathbf{t}_2, \ldots, \mathbf{t}_N\}$ *is obtained here, where each* $\mathbf{t}_i$*refers to a granule exhibiting some practically sound semantics.*

Step 2: *Equalize these granules by using the standard granule standard based on the DTW concept and the TWN.*

Step 3: *Construct the fuzzy relations as shown in* (6.8) *based on GrC concept and calculate the membership degree* z_r*by* (6.9).

Step 4: *Obtain the final forecasted membership* $\widehat{\mathbf{z}}_k$ *by the defuzzification mechanism* (6.10).

Case Study

To demonstrate the quality of the unequal-length GrC-based prediction method, the comparative experiments by using the data-driven methods are reported involving traditional auto-regressive integrated moving average model (ARIMA), ESN, the kernel learning-based method (LSSVM), and the GrC-based method. In addition, the computing efficiencies of these methods are further compared. In this subsection, the experiments of these comparative methods are conducted on the BFG system, the COG system, and the LDG system in steel industry.

BFG System Experiments

As mentioned in Chap. 3, the BFG system data exhibit the large fluctuation and amplitude with no visible distinct regularity. We randomly select the BFG consumption of the hot blast stove as an example. The prediction results are illustrated as Fig. 6.8, in which the different prediction horizons, namely 480, 720, and 1440 min, are depicted. In order to present the comparative results more clearly, the average

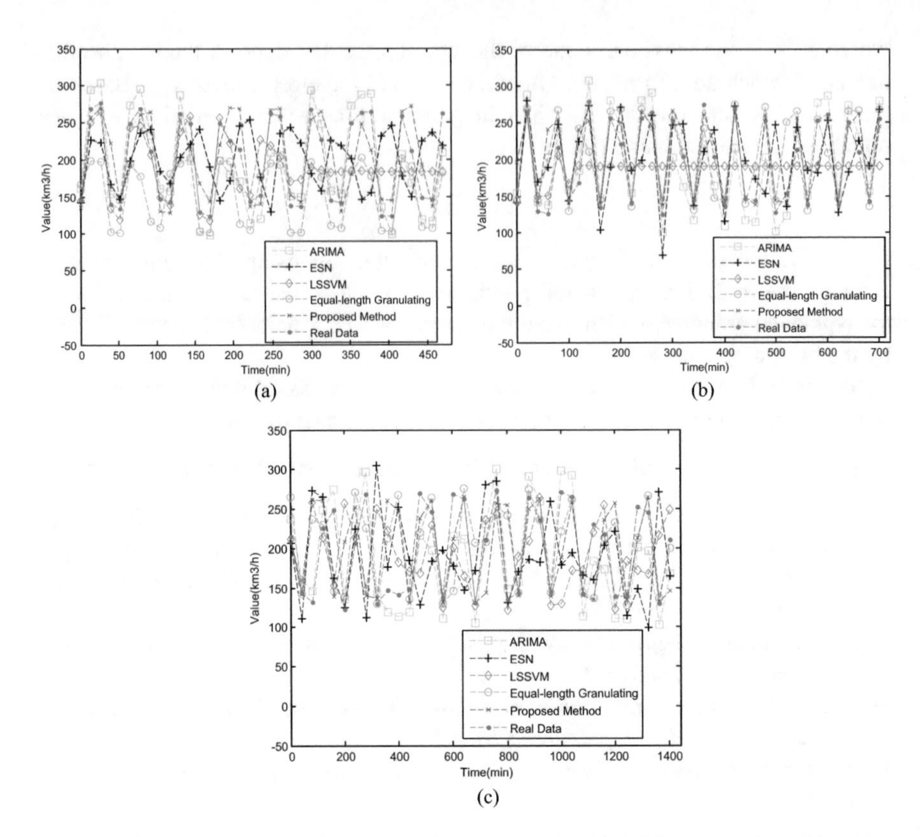

Fig. 6.8 The results of the different prediction horizons for the BFG consumption of hot blast stove: (**a**) 480, (**b**) 720, and (**c**) 1440

values of the estimated points by some methods are exhibited in this figure. It becomes apparent from this figure that although the accuracy level is basically acceptable within the former 60 data points when running the methods except the unequal-length GrC-based one, along with the prolongation of the time interval, the neural network and the kernel learning method obviously exhibit the rather poor performance when compared to the unequal-length GrC-based method.

To come up with a complete statistical analysis of the results, we illustrate the comparison of the real values and the predicted ones for the BFG consumption of hot blast stove by the unequal-length GrC-based method, see Fig. 6.9, in which various lengths of the prediction horizons are involved. It is clearly seen that these points are basically distributed around the line $y = x$, which indicates that the high accuracy has been achieved. To quantify the statistical comparison on the accuracy and the computing time (CT), the results for these horizons are listed in Table 6.1 by conducting 20-fold independent experiments, and the parameters of these methods, optimized as possible as we can by using the optimal algorithms or the trial and error approach, are listed in Table 6.2. With regard to the effectiveness, the unequal-length GrC-based method exhibits higher accuracy on the whole compared to the others,

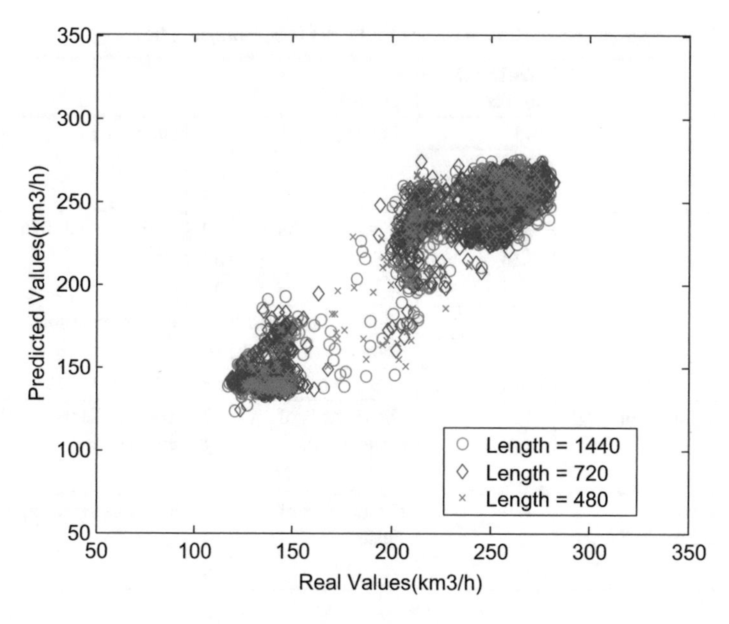

Fig. 6.9 Comparison of real and predicted values for BFG consumption flows

Table 6.1 The statistic prediction results for the BFG consumption flows

Methods	Prediction horizon	MAPE	RMSE	CT (s)
ARIMA	480	25.93	49.9569	0.098
	720	26.80	51.4922	0.258
	1440	27.66	53.5187	0.402
ESN	480	18.47	64.2827	57.551
	720	23.33	78.7632	60.002
	1440	27.85	90.0398	65.394
LSSVM	480	15.37	51.1434	7.998
	720	17.87	55.2321	14.234
	1440	20.22	60.0033	29.532
Equal-length granulating	480	29.88	92.1351	0.003
	720	35.28	102.602	0.005
	1440	16.37	61.2507	0.033
Unequal-length GrC-based method	480	6.14	28.2563	0.005
	720	8.08	37.6324	0.008
	1440	7.99	44.7231	0.021

and the errors by the unequal-length GrC-based method also does not worsen significantly along with the increase of the horizon. In contrast the three other methods show a clear deterioration of the quality of the results in terms of the both evaluation indices. In terms of computing time, the performance of the unequal-length GrC-based method is definitely superior to those associated with other

Table 6.2 The parameters of these methods for the BFG consumption flows

Methods	Prediction horizon	Parameters			
ARIMA	480	Order of AR	7	Order of MA	2
	720		11		3
	1440		14		3
ESN	480	Training samples amounts	30	Nodes amounts in DR	100
	720		30		110
	1440		35		100
LSSVM	480	Width of Gaussian Kernel	20	Penalty coefficient	50
	720		30		32
	1440		25		75
Equal-length granulating	480	The number of clusters	2	The number of input granules	3
	720		3		5
	1440		5		8
Unequal-length GrC-based method	480	The number of clusters	2	The number of input granules	4
	720		3		7
	1440		5		9

methods; meanwhile, the time cost is relatively insensitive to the changes of the prediction horizon, which can fully guarantee the satisfaction of real-time requirements of application as well as meet the prediction demand. It can be noted that because the unequal-length GrC-based method deals with the granulated data whose scale is usually less than 60 points and the size of the granules is about 100, it is obviously more efficient for the similar long-term prediction when compared to the pointwise (numeric) methods.

For further quantification of the performance, the results obtained when by the equal-length granulation are also reported in this table. It is evident that in this case the accuracy is far lower than the one produced by the unequal-length GrC-based method although its computing cost could be sustained in the approximately same range. This could be explained that the fixed equal-length granulating fails to flexibly adapt to the dynamic nature of the data (their variability) when the manufacturing process exhibits a slight variation.

COG System Experiments

The generation flow of the COG system is also the one of most concerns in the energy system because of its higher calorie compared to the other gases. If the COG generation flow of a long time period could be estimated in real time, then the by-product gases, even some other energy resource media such as the steam and the power, can be reasonably scheduled to reduce the energy cost for manufacturing. However, the real-world COG flow exhibits a high frequent variation due to the continuous burners switching, which makes the prediction accuracy hard to improve.

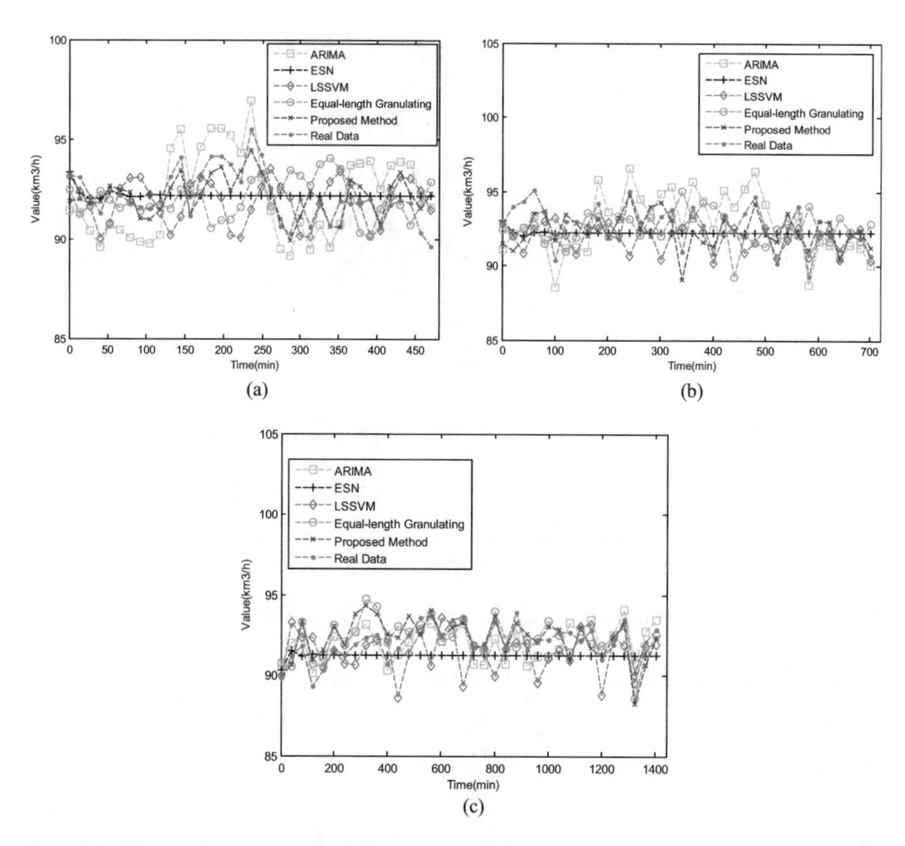

Fig. 6.10 The results obtained for different horizons for the COG generation flow: (**a**) 480, (**b**) 720, and (**c**) 1440

The real COG generation data in the plant are employed, and the illustrative comparison are presented in Fig. 6.10, where the performance gives the similar results on the accuracy that indicate the absolute superiority by the unequal-length GrC-based method to the others. By comparing the real values and the predicted ones, see Fig. 6.11, the accuracy can be further verified in a statistical fashion. The statistical results are listed in Table 6.3, where 20-fold independent experiments were conducted, while the relevant parameters are listed in Table 6.4. The effectiveness by the unequal-length GrC-based method also exhibits fairly superior to others in respect of accuracy. It is also noticeable that the equal-length granulating and the unequal-length GrC-based method exhibit basically the same computational efficiency, which is definitely better than that of some other methods.

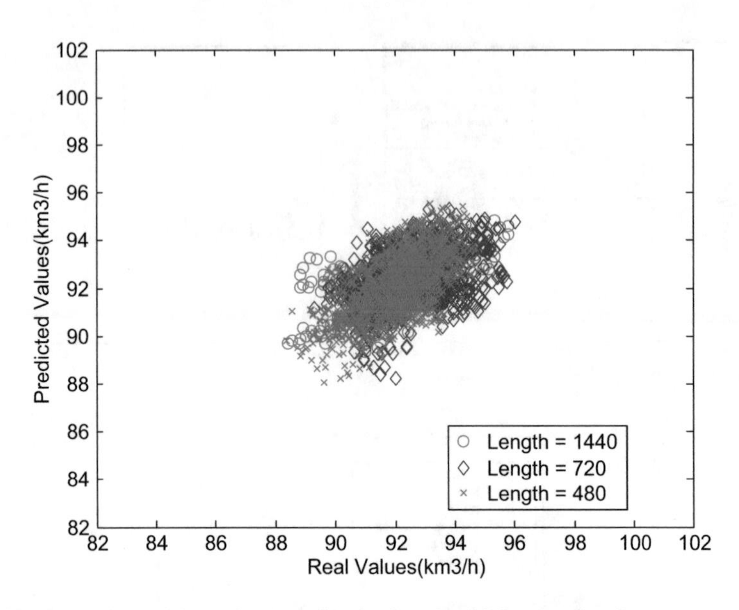

Fig. 6.11 Comparison of the real and predicted values for COG generation flows

Table 6.3 The statistic prediction results for the COG generation flows

Methods	Prediction horizon	MAPE	RMSE	CT (s)
ARIMA	480	0.83	1.9809	0.077
	720	0.95	2.1049	0.135
	1440	1.23	2.4279	0.212
ESN	480	1.97	2.7123	47.922
	720	3.14	3.1455	54.287
	1440	6.58	6. 6111	56.091
LSSVM	480	1.77	1.2324	5.789
	720	3.86	3.8768	10.033
	1440	10.98	8.3437	21.223
Equal-length granulating	480	2.73	3.9546	0.012
	720	3.95	4.5278	0.009
	1440	2.50	3.7805	0.014
Unequal-length GrC-based method	480	0.73	1.2333	0.021
	720	0.99	1.1155	0.009
	1440	0.84	1.0108	0.022

LDG System Experiments

As for the LDG flow in the energy system, we consider the LDG consumption of a blast furnace as well, which usually presents a sort of irregular pseudo-periodicity that is rather difficult to be estimated by equal-length time windows, especially for long-term prediction. Combining with the status switching of the equipment, one can

Table 6.4 The parameters of these methods for the COG generation flows

Methods	Prediction horizon	Parameters			
ARIMA	480	Order of AR	4	Order of MA	1
	720		5		2
	1440		8		2
ESN	480	Training samples amounts	30	Nodes amounts in DR	100
	720		40		100
	1440		25		100
LSSVM	480	Width of Gaussian Kernel	15	Penalty coefficient	80
	720		20		50
	1440		20		55
Equal-length granulating	480	The number of clusters	2	The number of input granules	4
	720		2		4
	1440		3		5
Unequal-length GrC-based method	480	The number of clusters	2	The number of input granules	3
	720		2		5
	1440		3		6

cluster a 60-min energy process and treat it as a data granule. By using the comparative methods, their performance is obviously worse than those of the unequal-length GrC-based method, refer to Fig. 6.12, where the different prediction horizons are also involved. Furthermore, the statistic experiments are similarly presented in Fig. 6.13 and Tables 6.5 and 6.6. These statistical results indicate that the long-term prediction by the unequal-length GrC-based method exhibit an apparent superiority in terms of accuracy and computational cost when compared with the generic data-driven methods, and the results also exhibit low sensitivity vis-a-vis the length of the prediction horizon. In addition, the results of the equal-length granulating are also looked at, see Table 6.5. The results are worse than those produced by the unequal-length GrC-based method. This is not surprising since the equal-length granulating ignores the semantics that reflects upon the actual manufacturing process.

Based on the above comparative experiments and analyses, the industrial semantics-based granulation can actually reflect the operational circumstance of the energy system; therefore, the unequal-length GrC-based prediction method can provide the energy scheduling workers with the scientific guidance of the long-term energy variation.

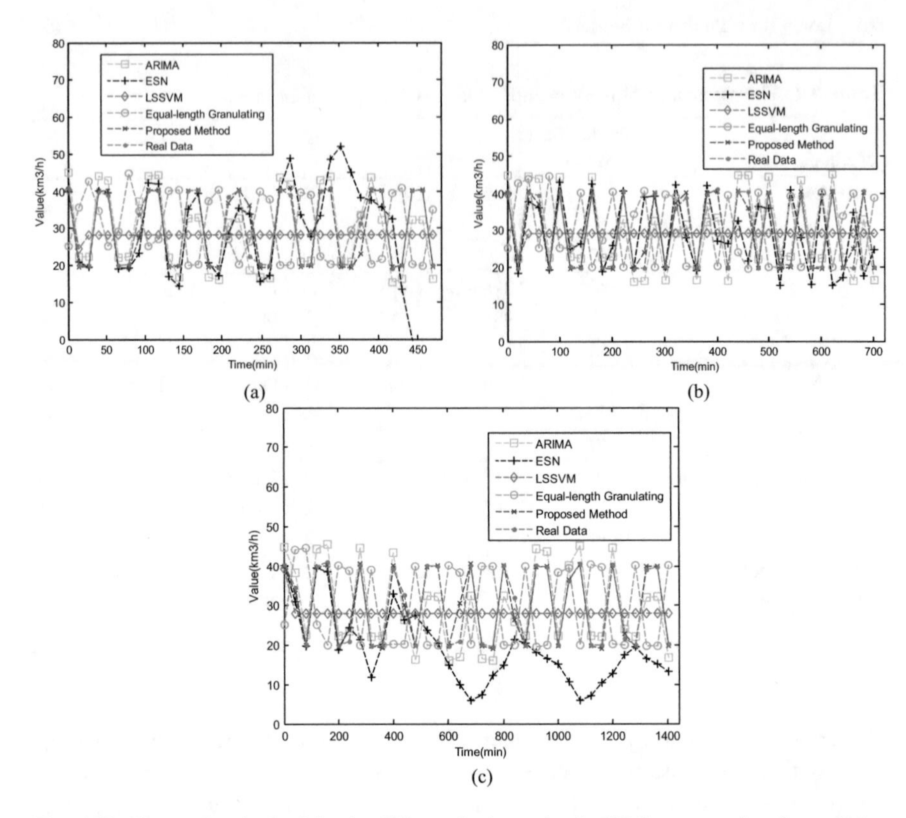

Fig. 6.12 The results obtained for the different horizons for the LDG consumption flow of blast furnace: (**a**) 480, (**b**) 720, and (**c**) 1440

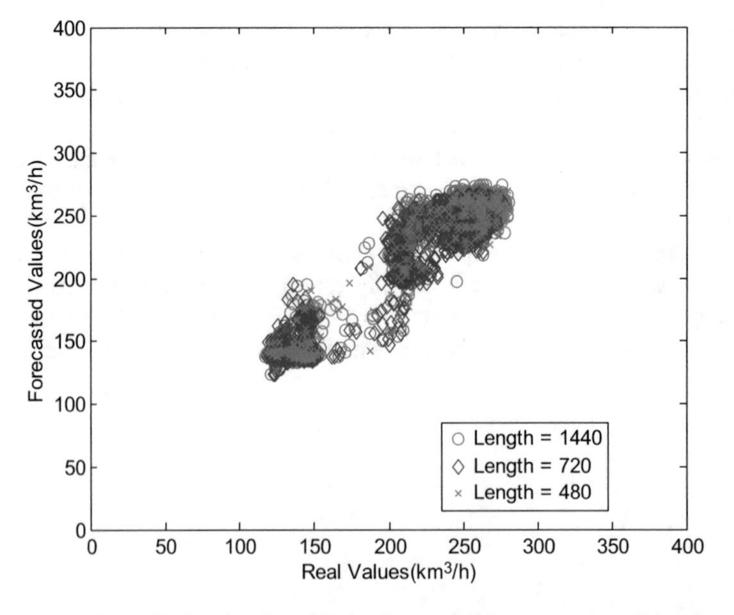

Fig. 6.13 Comparison of the real and predicted values for LDG consumption of blast furnace

Table 6.5 The statistic prediction results for the LDG consumptions of blast furnace

Methods	Prediction horizon	MAPE	RMSE	CT (s)
ARIMA	480	9.11	8.8933	0.084
	720	23.33	9.1261	0.098
	1440	35.42	9.3819	0.123
ESN	480	32.01	27.2312	45.863
	720	33.27	41.6122	46.896
	1440	34.26	49.1145	52.980
LSSVM	480	17.77	7.2334	8.111
	720	22.34	9.9924	12.215
	1440	30.98	12.1344	27.134
Equal-length granulating	480	39.66	17.4785	0.018
	720	33.62	16.2182	0.036
	1440	13.38	9.5450	0.025
Unequal-length GrC-based method	480	5.24	5.8433	0.021
	720	5.77	5.4936	0.019
	1440	8.11	8.4622	0.030

Table 6.6 The parameters of these methods for the LDG consumptions of blast furnace

Methods	Prediction horizon	Parameters			
ARIMA	480	Order of AR	8	Order of MA	3
	720		10		5
	1440		17		4
ESN	480	Training samples amounts	20	Nodes amounts in DR	100
	720		45		80
	1440		30		120
LSSVM	480	Width of Gaussian Kernel	35	Penalty coefficient	70
	720		20		90
	1440		30		85
Equal-length granulating	480	The number of clusters	4	The number of input granules	5
	720		4		6
	1440		3		8
Unequal-length GrC-based method	480	The number of clusters	4	The number of input granules	6
	720		4		8
	1440		3		8

6.3.2 Granular Model for Factor-Based Prediction

As for the factor-based prediction, there are two problems that should be resolved.

1. In respect of practical requirement, the long-term prediction for the multi-output model (two gasholder levels) is on high demand for the practical energy

application. It is largely helpful for the further energy optimization and scheduling so as to reduce the production cost for enterprises.
2. In terms of theoretical development, the mutual impacts in some multi-output system among the influence factors must be reasonably considered in the model establishment, such as the gas generation units or the users.

As we discussed before, the horizon of analysis by GrC is effectively extended for dealing with a large prediction scale. And among the algorithms that combined with GrC, FCM is always utilized for one dataset problem. Though capable of analyzing multiple datasets, FCM reveals no directly structure in the model for measuring the mutual impact between them. To overcome the shortcomings, the collaborative fuzzy clustering was then proposed in [6] to summarize them into three scenarios, i.e., horizontal, vertical, and also hybrid collaboration involving horizontal and vertical clustering.

Here, we take the LDG system in steel industry as an example to illustrate the hybrid collaborative fuzzy clustering based long-term prediction (HCFC-LTP) method. Considering the input datasets, i.e., the historical and present gasholder level, and the generation-consumption flow difference, it is apparent that the former two are the same objects in different time feature space which represent a horizontal structure. Whereas, the holder levels along with the generation flow, the consumption one and its difference values can be deemed as the different objects in the same time feature space, i.e., a vertical relationship. Bearing the above analysis in mind, a hybrid collaborative clustering approach is proposed to establish a novel multi-output prediction model. The following are the five datasets:

Set 1: 1# gasholder level (previous) $\boldsymbol{L}_0^{(1)} = \left\{\boldsymbol{l}_{0,1}^{(1)}, \boldsymbol{l}_{0,2}^{(1)}, \ldots, \boldsymbol{l}_{0,N}^{(1)}\right\}$

Set 2: 2# gasholder level (previous) $\boldsymbol{L}_0^{(2)} = \{\boldsymbol{l}_{0,1}^{2}, \boldsymbol{l}_{0,2}^{2}, \ldots, \boldsymbol{l}_{0,N}^{2}\}$

Set 3: 1# gasholder level (current) $\boldsymbol{L}^{(1)} = \left\{\boldsymbol{l}_1^{(1)}, \boldsymbol{l}_2^{(1)}, \ldots, \boldsymbol{l}_N^{(1)}\right\}$

Set 4: 2# gasholder level (current) $\boldsymbol{L}^{(2)} = \left\{\boldsymbol{l}_1^{(2)}, \boldsymbol{l}_2^{(2)}, \ldots, \boldsymbol{l}_N^{(2)}\right\}$

Set 5: the flow difference (current) $\boldsymbol{D} = \{\boldsymbol{d}_1, \boldsymbol{d}_2, \ldots, \boldsymbol{d}_N\}$

The elements of them, i.e., $\boldsymbol{l}_{0,i}^{(1)}, \boldsymbol{l}_{0,i}^{(2)}, \boldsymbol{l}_i^{(1)}, \boldsymbol{l}_i^{(2)}$, or d_i, are the data granules with length w. With practical concern, the structure between Set 1 and Set 3 (Set 2 and Set 4) is horizontal, while the relationships between Set 2 and Set 5 can be regarded as the vertical. As a result, the whole structure can be illustrated as Fig. 6.14, in which α_1, α_2 and β_1, β_2, β_3, β_4 are the corresponding collaboration coefficients, and the cubes are the results from clustering including the matrices of fuzzy membership or prototypes.

The algorithm begins with computing the initial clusters for each dataset with FCM. In such a way, the following optimization needs to be completed.

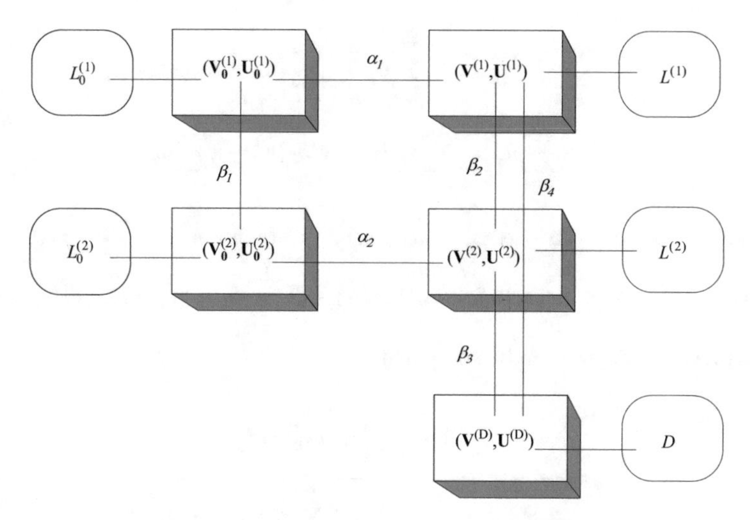

Fig. 6.14 Structure of hybrid collaborative clustering for LDG system

$$
\begin{aligned}
Q &= \sum_{k=1}^{N}\sum_{i=1}^{c} u_{ik}^2 d_{ik}^2 \\
\text{s.t.} \quad & \sum_{i=1}^{c} u_{ik} = 1
\end{aligned}
\tag{6.11}
$$

where N is the number of granules, and c refers to the number of clusters. u_{ik} represents the fuzzy membership of the kth granule to ith cluster, and the vector $\mathbf{v}_i$ is prototype of the ith cluster. To accommodate the collaboration mechanism in the optimization process, the objective function is extended as

$$
\begin{aligned}
Q = {} & \sum_{k=1}^{N}\sum_{i=1}^{c} u_{ik}^2 d_{ik}^2 + \sum_{p=1}^{D_1}\alpha_p \sum_{k=1}^{N}\sum_{i=1}^{c}\{u_{ik} - u_{ik}(S_1)\}^2 d_{ik}^2 \\
& + \sum_{q=1}^{D_2}\beta_q \sum_{k=1}^{N}\sum_{i=1}^{c} u_{ik}^2 \|v_i - v(S_2)\|^2 \text{s.t.} \quad \sum_{i=1}^{c} u_{ik} = 1
\end{aligned}
\tag{6.12}
$$

where α_p and β_q, respectively, represent the horizontal and vertical collaboration coefficients, and S_1 and S_2 refer to the corresponding datasets. Taking $\boldsymbol{L}^1$ as an example, $u_{ik}(S_1)$ is the element of $\mathbf{U}_0^{(1)}$, $\alpha_p = \{\alpha_1\}$ denotes the corresponding collaborative coefficient set, and $v(S_2)$ is the element of $\mathbf{V}^{(2)}$ and $\mathbf{V}^{(D)}$, $\beta_q = \{\beta_2, \beta_4\}$ denotes the corresponding collaborative coefficient set.

We give a thorough detailed derivation process, and then summarize the steps of applying this algorithm. Enlightened by the computation strategy of FCM, one can introduce a Lagrange Multiplier λ, then an unconstrained optimization problem can be formulated as

$$
\begin{aligned}
f = & \sum_{k=1}^{N}\sum_{i=1}^{c} u_{ik}^2 d_{ik}^2 + \sum_{p=1}^{S_1}\alpha_p \sum_{k=1}^{N}\sum_{i=1}^{c}\{u_{ik} - u_{ik}(S_1)\}^2 d_{ik}^2 \\
& + \sum_{q=1}^{S_2}\beta_q \sum_{k=1}^{N}\sum_{i=1}^{c} u_{ik}^2 \|v_i - v(S_2)\|^2 - \lambda \sum_{i=1}^{c} u_{ik}
\end{aligned}
\tag{6.13}
$$

With regard to the necessary condition $\frac{\partial f}{\partial u_{ik}} = 0$, i.e., $\frac{\partial f}{\partial u_{ik}} = 2u_{ik}d_{ik}^2 + \$\$2\sum_{p=1}^{S_1}\alpha_p\{u_{ik} - u_{ik}(S_1)\}d_{ik}^2 + 2\sum_{q=1}^{S_2}\beta_q u_{ik}\|v_i - v(S_2)\|^2 - \lambda = 0,$ the membership degree can be computed as

$$
u_{ik} = \frac{\lambda + 2\sum_{p=1}^{S_1}\alpha_p u_{ik}(S_1) d_{ik}^2}{2d_{ik}^2\left(1 + \sum_{p=1}^{S_1}\alpha_p\right) + 2\sum_{q=1}^{S_2}\beta_q\|v_i - v(S_2)\|^2}
\tag{6.14}
$$

Considering the constraint in (6.12), i.e., $\sum_{i=1}^{c} u_{ik} = 1$, we have $\sum_{i=1}^{c}\frac{\lambda + 2\sum_{p=1}^{S_1}\alpha_p u_{ik}(S_1) d_{ik}^2}{2d_{ik}^2\left(1 + \sum_{p=1}^{S_1}\alpha_p\right) + 2\sum_{q=1}^{S_2}\beta_q\|v_i - v(S_2)\|^2} = 1$. Introducing the notations $\varphi_{ik} = \sum_{p=1}^{S_1}\alpha_p u_{ik}(S_1)$, $\psi = \sum_{p=1}^{S_1}\alpha_p$ and $\eta_i = \sum_{q=1}^{S_2}\beta_q\|v_i - v(S_2)\|^2$, the computational formula for λ reads as

$$
\lambda = \left(1 - \sum_{i=1}^{c}\frac{\varphi_{ik}d_{ik}^2}{d_{ik}^2(1+\psi)+\eta_i}\right) \Bigg/ \sum_{i=1}^{c}\frac{1}{d_{ik}^2(1+\psi)+\eta_i}
\tag{6.15}
$$

Bringing (6.15) into (6.14), the final formula for computing the membership of ith granule to the kth cluster u_{ik} can be described as

$$
u_{ik} = \frac{\frac{1}{2}\left(1 - \sum_{j=1}^{c}\frac{\varphi_{jk}d_{jk}^2}{d_{jk}^2(1+\psi)+\eta_j}\right) + \sum_{j=1}^{c}\frac{\varphi_{ik}d_{ik}^2}{d_{jk}^2(1+\psi)+\eta_j}}{\sum_{j=1}^{c}\frac{d_{ik}^2(1+\psi)+\eta_i}{d_{jk}^2(1+\psi)+\eta_j}}
\tag{6.16}
$$

where $i = 1, 2, \ldots, c$, and $k = 1, 2, \ldots, N$.

In the calculation of the prototypes, the Euclidean distance between the patterns and the prototypes is employed, and then (6.13) can be rewritten in an explicit form,

$$f = \sum_{k=1}^{N}\sum_{i=1}^{c} u_{ik}^2 \sum_{j=1}^{n}\left(x_{kj} - v_{ij}\right)^2 + \sum_{p=1}^{S_1}\alpha_p \sum_{k=1}^{N}\sum_{i=1}^{c}\{u_{ik} - u_{ik}(S_1)\}^2 \sum_{j=1}^{n}\left(x_{kj} - v_{ij}\right)^2 + \sum_{q=1}^{S_2}\beta_q \sum_{k=1}^{N}\sum_{i=1}^{c} u_{ik}^2 \|v_i - v(S_2)\|^2 - \lambda \sum_{i=1}^{c} u_{ik} \tag{6.17}$$

where n denotes the quantity of the data points in a granule, x_{kj} is the patterns in $\boldsymbol{l}_i^{(1)}$ and $\boldsymbol{l}_i^{(2)}$. Similarly, let $\frac{\partial f}{\partial v_{ij}} = 0$,

$$\frac{\partial f}{\partial v_{ij}} = -2\sum_{k=1}^{N} u_{ik}^2 \left(x_{kj} - v_{ij}\right) - 2\sum_{p=1}^{S_1}\alpha_p \sum_{k=1}^{N}\{u_{ik} - u_{ik}(S_1)\}^2 \left(x_{kj} - v_{ij}\right) - 2\sum_{q=1}^{S_2}\beta_q \sum_{k=1}^{N} u_{ik}^2 \left(v_{ij} - v_{ij}(S_2)\right) = 0 \tag{6.18}$$

One can get the formula for computing the center of ith cluster v_{ij}, see

$$v_{ij} = \frac{\sum_{k=1}^{N} u_{ik}^2 x_{kj} + \sum_{p=1}^{S_1}\alpha_p \sum_{k=1}^{N}\{u_{ik} - u_{ik}(S_1)\}^2 x_{kj} - \sum_{q=1}^{S_2}\beta_q \sum_{k=1}^{N} u_{ik}^2 v_{ij}(S_2)}{\sum_{k=1}^{N} u_{ik}^2 + \sum_{p=1}^{S_1}\alpha_p \sum_{k=1}^{N}\{u_{ik} - u_{ik}(S_1)\}^2 - \sum_{q=1}^{S_2}\beta_q \sum_{k=1}^{N} u_{ik}^2} \tag{6.19}$$

where $i = 1, 2, \ldots, c$, and $j = 1, 2, \ldots, n$.

The HCFC-LTP model for hybrid fuzzy clustering should be summarized into two phases. Firstly, the FCM is utilized for each set to obtain the initial memberships and the prototypes. It is obvious that the number of the clusters has to be the same to all five datasets. Secondly, the HCFC approach is determined by computing (6.16) and (6.19) until a termination criterion is satisfied. These two computational phases can be listed as follows:

Algorithm 6.2: Hybrid Fuzzy Clustering by the HCFC-LTP

Phase 1: *Generating the clusters without collaboration*

For *each data*

repeat *compute prototypes and partition matrices for all subsets of patterns*

until *a termination criterion has been satisfied.*

Phase 2: *Collaborating between the clusters*

repeat *compute the prototypes and partition matrices using* (6.16) and (6.19) *with the given matrix of collaborative links* α_1, α_2*and* β_1, β_2, β_3, β_4,

until *a termination criterion has been satisfied.*

After the above modeling process, the memberships matrices $\mathbf{U}^{(1)}$, $\mathbf{U}^{(2)}$ and the prototypes $\mathbf{V}^{(1)}$, $\mathbf{V}^{(2)}$ are obtained. Here, one can integrate $\mathbf{V} = [\mathbf{V}^{(1)}; \mathbf{V}^{(2)}]$.

Based on the result of the HCFC, this section establishes a set of fuzzy rules with fuzzy inference. The common form of the fuzzy rule can be illustrated as

$$R: \text{If } x \text{ is } A_i, \text{then } y \text{ is } B_j$$

where x and y refer to the linguistic variables, and A_i, and B_j are the corresponding fuzzy sets within their own universe of discourse. With regard to the complex patterns of the data we discussed, it is hard to use 'up,' 'down,' or the alike for data descriptions. One can use "which cluster the granule to the most extent belongs to" to represent as the linguistic variable. As such, the rules can be obtained in the form of

$$\begin{aligned} R: \text{If } \boldsymbol{L}_0^{(1)}(t-1) \text{ is } c_{\boldsymbol{L}_0^{(1)}},\ \boldsymbol{L}_0^{(2)}(t-1) \text{ is } c_{\boldsymbol{L}_0^{(2)}}, \boldsymbol{D}(t) \text{ is } c_{\boldsymbol{D}}, \\ \text{then } \boldsymbol{L}^{(1)}(t) \text{ is } c_{\boldsymbol{L}^{(1)}}, \boldsymbol{L}^{(2)}(t-1) \text{ is } c_{\boldsymbol{L}^{(2)}} \end{aligned} \tag{6.20}$$

where $c_{\boldsymbol{L}_0^{(1)}}, c_{\boldsymbol{L}_0^{(2)}}, c_{\boldsymbol{D}}$ and $c_{\boldsymbol{L}^{(1)}}, c_{\boldsymbol{L}^{(2)}}$ are tags for identifying which cluster the input and output belongs to. For instance, $c_{\boldsymbol{L}_0^{(1)}}$ would be marked as 2 when the maximal membership of $\boldsymbol{L}_0^{(1)}(t-1)$ is toward the second cluster. In such a way, N fuzzy rules are attained.

As for the prediction process, one can search the rules that have the same tag as the input in the testing set. For summarizing the searching result, a set of variables considering the output of the rules and the corresponding membership grade $u_{ij}^{\boldsymbol{L}^{(1)}}$, $u_{ij}^{\boldsymbol{L}^{(2)}}$ are computed as

$$\mathbf{p}^{(1)} = \sum_{j=1}^{N} \mathbf{h}_j^{(1)} u_{ij}^{\boldsymbol{L}^{(1)}}, \quad \mathbf{p}^{(2)} = \sum_{j=1}^{N} \mathbf{h}_j^{(2)} u_{ij}^{\boldsymbol{L}^{(2)}} \tag{6.21}$$

where $\mathbf{h}_j^{(1)}$ and $\mathbf{h}_j^{(2)}$ are $c \times 1$ column vector of the tag result (e.g., $[0, 1, 0, \ldots, 0]$ which has the maximal membership toward second cluster), $i = 1, 2, \ldots, c$. Employing the centroid defuzzification given its both intuitive thinking and simple computation, and denoting $\mathbf{P} = [\mathbf{p}^{(1)}, \mathbf{p}^{(2)}]$, one can obtain the final long-term prediction result, see

$$\hat{\mathbf{P}} = \mathbf{P}^{\mathrm{T}}\mathbf{V} \tag{6.22}$$

As the computation demonstrated, the GrC-based HCFC long-term prediction model avoids to involve the iteration mechanism. Instead, the final results are solved by defuzzification as a granule, which eliminates the negative effect of accumulative error on the iterative prediction.

For clarifying the modeling process, a list of detailed procedures is shown as follows:

Algorithm 6.3: The GrC-Based HCFC Long-Term Prediction

Step 1: *Establish a hybrid collaborative fuzzy clustering structure (see* Fig. 6.14*), along with the related dataset preparation.*

Step 2: *Generate the clusters for each data set.*

Step 3: *For the given matrix of collaborative links* α_1, $\alpha_2$and β_1, β_2, β_3, β_4, *compute the prototypes and partition matrices by using* (6.16) and (6.20) *in order to collaborate the clusters, and then optimize them until a termination criterion has been satisfied.*

Step 4: *Construct the fuzzy rule base with technique of fuzzy inference by using* (6.20).

Step 5: *Obtain the long-term prediction result with defuzzification with* (6.21) and (6.22).

Case Study

Two groups of the experiments on the LDG system are conducted here, in which Group #1 aims to demonstrate the superior of the hybrid mechanism that takes the influence factors of the holder level into account. In this regard, the GrC-based single-output prediction model that only concerns the characteristics of time series is comparatively analyzed. Group #2 pays more attentions to indicate the advantage of GrC for the long-term prediction, which compares the HCFC-LTP with the iterative computing-based multi-output LSSVM, refer to [3]. All the parameters in the above approaches are off-line determined by trial and error.

Comparing with the Single-Output Model

The comparative results of the holder levels in #1 subsystem are presented in Fig. 6.15, where two holders are involved. Since the impacts coming from the

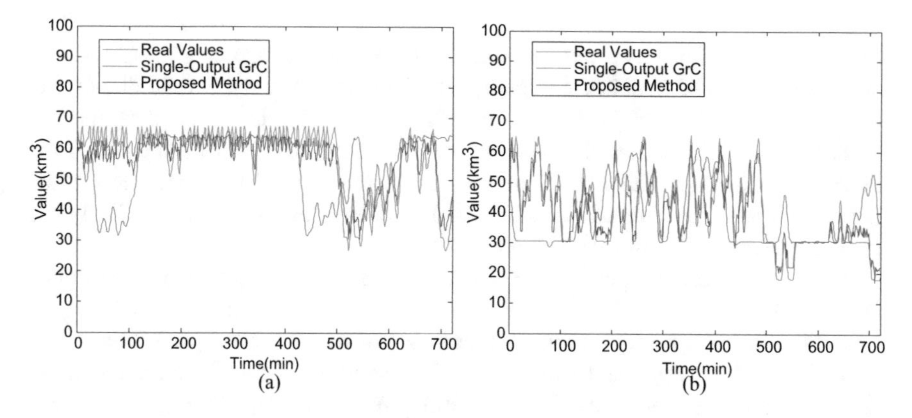

Fig. 6.15 Comparison of prediction results for #1 subsystem. (**a**) #1 Gasholder, (**b**) #2 Gasholder

Table 6.7 Comparative results of the prediction methods for #1 subsystem

Predict methods	#1 gasholder		#2 gasholder	
	MAPE	RMSE	MAPE	RMSE
Single-output GrC	0.1649	15.6574	0.1785	14.9295
HCFC-LTP	0.0393	2.9994	0.0735	5.5443

Table 6.8 Parameters of the prediction methods for #1 subsystem

Predict methods	#1 gasholder	#2 gasholder
Single-output GrC	Number of clusters: $c = 4$ Fuzzy coefficient: $m = 0.01$ Number of granules (as input): $n_{\text{in}} = 7(4)$	Number of clusters: $c = 3$ Fuzzy coefficient: $m = 0.01$ Number of granules (as input): $n_{\text{in}} = 7(5)$
HCFC-LTP	Collaboration coefficients: $\alpha_1 = 1.5$, $\alpha_2 = 1.0$ $\beta_1 = \beta_2 = 0.5$, $\beta_3 = 3.5$, $\beta_4 = 3.0$	
	Number of clusters: $c = 4$ Fuzzy coefficient: $m = 0.01$	Number of clusters: $c = 5$ Fuzzy coefficient: $m = 0.01$

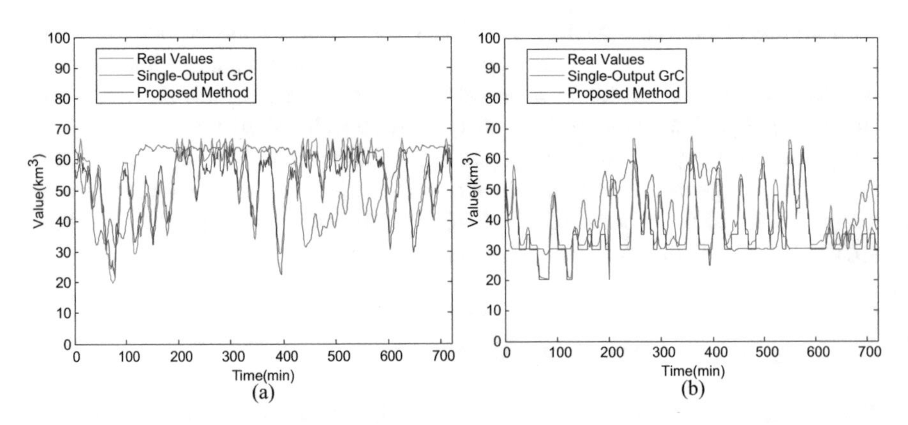

Fig. 6.16 Comparison of the prediction results for #2 subsystem. (**a**) #3 Gasholder, (**b**) #4 Gasholder

influence factors (the gas generation units and the users) are ignored, the GrC-based single-output model is out of the reality, and its outcomes are certainly dissatisfactory. While the HCFC-LTP presents a better result on the levels trend estimation. And, Table 6.7 illustrates the quantitative error statistics. It is apparent that the HCFC-LTP narrows the MAPE in 8%, and the lower RMSE also means that the results contain few outliers that possibly exhibit excessive errors. Table 6.8 lists the parameters of the algorithms.

Besides, a group of typical comparative result of #2 subsystem are also illustrated in Fig. 6.16. Comparing to those of #1 subsystem, the performance of the single-output model has no obvious improvement. Although some good prediction results on #4 holder level appear, it still fails for the overall tendency. Furthermore, these figures indicate that the HCFC-LTP can still give a much better result with acceptable errors (see details in Table 6.9), and the parameters are listed in Table 6.10.

Table 6.9 Comparative results of the prediction methods for #2 subsystem

Predict methods	#3 gasholder		#4 gasholder	
	MAPE	RMSE	MAPE	RMSE
Single-output GrC	0.1860	14.8761	0.1483	12.3655
HCFC-LTP	0.0459	3.3147	0.0542	3.8923

Table 6.10 Parameters of the prediction methods for #2 subsystem

Predict methods	#3 gasholder	#4 gasholder
Single-output GrC	Number of clusters: $c = 8$ Fuzzy coefficient: $m = 0.01$ Number of granules (as input): $n_{\text{in}} = 8(5)$	Number of clusters: $c = 4$ Fuzzy coefficient: $m = 0.01$ Number of granules (as input): $n_{\text{in}} = 6(4)$
HCFC-LTP	Collaboration coefficients: $\alpha_1 = \alpha_2 = 1.0$ $\beta_1 = \beta_{[2]2} = 1.0, \beta_3 = 0.5, \beta_4 = 2.0$	
	Number of clusters: $c = 5$ Fuzzy coefficient: $m = 0.01$	Number of clusters: $c = 5$ Fuzzy coefficient: $m = 0.01$

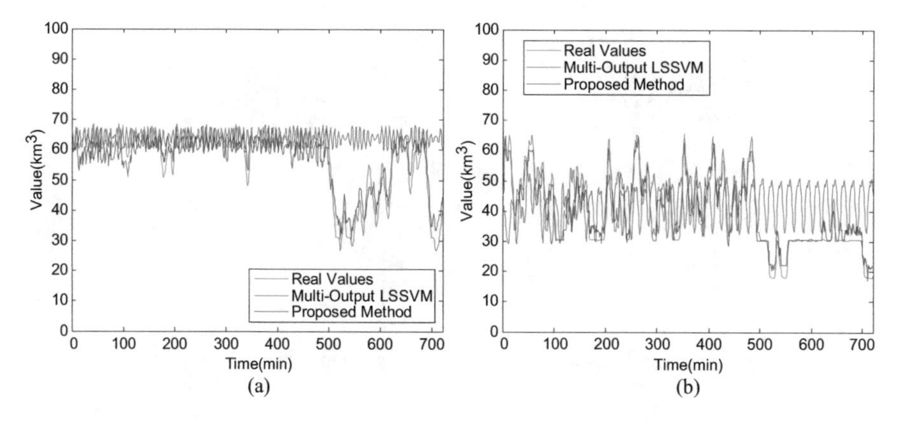

Fig. 6.17 Comparison of prediction results for #1 subsystem. (**a**) #1 Gasholder, (**b**) #2 Gasholder

Comparing with the Multi-output Model (Iteration Mechanism, [3])

The multi-output LSSVM model predicts only one point each time, and the long period prediction requires to iteratively run the model. Bearing this in mind, the computing time will be largely raised. Hence, we add an index of time cost for an explicitly comparison.

The results also include two sets for each LDG subsystem. Figure 6.17 presents the comparative results by using the different methods on #1 subsystem (Table 6.11). The result of multi-output LSSVM behaves fluctuation which cannot track the real values well. While the HCFC-LTP not only exhibits a better accuracy when facing with data fluctuations (e.g., about from 350th point to 600th point of #2 gasholder), but also performs the outstanding outcomes when forecasting the steady phase of the level (e.g., about from 150th point to 250 point of #2 gasholder). Besides, the HCFC-

Table 6.11 Comparative results of the prediction methods for #1 subsystem

	#1 gasholder			#2 gasholder		
Predict methods	MAPE	RMSE	CT (s)	MAPE	RMSE	CT (s)
Multi-output LSSVM	0.1185	12.8336	33.565	0.1796	14.1288	29.451
HCFC-LTP	0.0556	4.0255	4.012	0.0404	3.1294	4.365

Table 6.12 Parameters of the prediction methods for #1 subsystem

Predict methods	#1 gasholder	#2 gasholder
Multi-output LSSVM	Penalty coefficients: $\gamma_0 = 90$ $\gamma_1 = 255$ $\gamma_2 = 295$ Kernel widths: $\sigma_1 = 0.99$ $\sigma_2 = 0.95$	Penalty coefficients: $\gamma_0 = 85$ $\gamma_1 = 305$ $\gamma_2 = 280$ Kernel widths: $\sigma_1 = 0.94$ $\sigma_2 = 0.98$
HCFC-LTP	Collaboration coefficients: $\alpha_1 = \alpha_2 = 1.5$ $\beta_1 = 0.5, \beta_2 = 1.0, \beta_3 = 0.5, \beta_4 = 2.5$	
	Number of clusters: $c = 7$ Fuzzy coefficient: $m = 0.01$	Number of clusters: $c = 4$ Fuzzy coefficient: $m = 0.01$

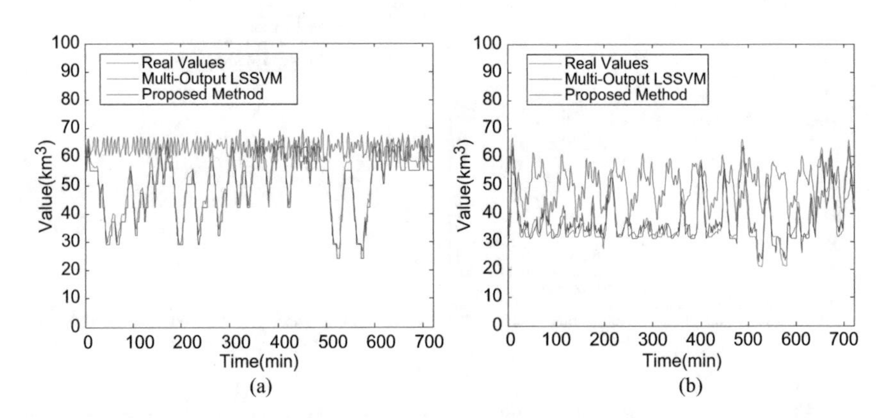

Fig. 6.18 Comparison of prediction results for #2 subsystem. (**a**) #3 Gasholder, (**b**) #4 Gasholder

Table 6.13 Comparative results of the prediction methods for #2 subsystem

	#3 gasholder			#4 gasholder		
Predict methods	MAPE	RMSE	CT (s)	MAPE	RMSE	CT (s)
Multi-output LSSVM	0.1549	14.4779	28.674	0.2561	18.8960	31.335
HCFC-LTP	0.0637	4.7049	3.938	0.0445	3.1025	3.886

LTP also consumes far less time than the LSSVM. The parameters are also listed in Table 6.12. Figure 6.18 is listed for #2 subsystem. With regard to the two gasholders, the multi-output LSSVM performs well at the very beginning because of the consideration of the influence factors. However, the accuracy gradually goes poor along with the iteration. In contrast, the HCFC-LTP shows excellent results, which can be concluded from both the figures and error statistics in Table 6.13.

Table 6.14 Parameters of the prediction methods for #2 subsystem

Predict methods	#3 gasholder	#4 gasholder
Multi-output LSSVM	Penalty coefficients: $\gamma_0 = 85$ $\gamma_1 = 285$ $\gamma_2 = 270$ Kernel widths: $\sigma_1 = 0.94$ $\sigma_2 = 0.96$	Penalty coefficients: $\gamma_0 = 75$ $\gamma_1 = 305$ $\gamma_2 = 295$ Kernel widths: $\sigma_1 = 0.95$ $\sigma_2 = 0.93$
HCFC-LTP	Collaboration coefficients: $\alpha_1 = 1.0, \alpha_2 = 1.5$ $\beta_1 = 0.5, \beta_2 = 1.0, \beta_3 = 2.5, \beta_4 = 2.0$	
	Number of clusters: $c = 5$ Fuzzy coefficient: $m = 0.01$	Number of clusters: $c = 4$ Fuzzy coefficient: $m = 0.01$

Table 6.15 Error statistics (average of the 6 month results)

Predict methods	#1 holder		#2 holder		#3 holder		#4 holder	
	MAPE	RMSE	MAPE	RMSE	MAPE	RMSE	MAPE	RMSE
Single-output GrC	0.1699	17.3452	0.1534	16.1386	0.1699	18.9841	0.1721	15.3332
Multi-output LSSVM	0.1954	15.5432	0.2123	21.4213	0.1954	17.5417	0.2008	20.7441
HCFC-LTP	0.0692	4.454	0.0499	4.983	0.0692	5.344	0.0601	4.287

In order to demonstrate the feasibility and the applicability of the reported method, we further provide here a 6-month running statistics on the average prediction accuracy after implementing it in practice compared to the other two approaches (Table 6.14). Table 6.15 gives the error statistics, which indicate that the accuracy of the HCFC-LTP is relatively stable, and capable of providing the remarkable outcomes compared to the others.

6.4 Granular-Based Prediction Intervals

PIs construction can be deemed as a vertical extension of numeric values estimation. The initial prototypes are extended as the interval-valued clustering centers. Aiming at maximizing the data amount covered by the granulation area along with the help of fuzzy inference and defuzzification, the PIs can be constructed.

6.4.1 Initial PIs Construction

Based on the numeric prototypes, the granular clustering centers are described by

$$\overline{v_{ij}} = \left[v_{ij}^-, v_{ij}^+\right] = \left[v_{ij} - \varepsilon_j, v_{ij} + \varepsilon_j\right] \tag{6.23}$$

where $i = 1, 2, \ldots, c, j = 1, 2, \ldots, n, v_{ij}$ refers to the initial prototypes, and ε_j denotes the level of information granularity. Then, the corresponding membership grades are computed by

$$\overline{u_{ik}} = \left[u_{ik}^{-}, u_{ik}^{+}\right] = \left[\left(\sum_{t=1}^{c} \frac{\left\|s_k - v_i^{-}\right\|}{\left\|s_k - v_t^{-}\right\|}\right)^{-\frac{2}{m-1}}, \left(\sum_{t=1}^{c} \frac{\left\|s_k - v_i^{+}\right\|}{\left\|s_k - v_t^{+}\right\|}\right)^{-\frac{2}{m-1}}\right] \tag{6.24}$$

where $k = 1, 2, \ldots, N$. The following fuzzy inference starts from the establishment of fuzzy rules on upper and lower boundaries of the intervals, respectively

$$\begin{aligned} R_t^{-} &: \text{If } s(t - n_I)^{-} \text{ is } c_{t-n_I}^{-}, \ldots, s(t-1)^{-} \text{ is } c_{t-1}^{-}, \text{then } s(t)^{-} \text{ is } c_t^{-} \\ R_t^{+} &: \text{If } s(t - n_I)^{+} \text{ is } c_{t-n_I}^{+}, \ldots, s(t-1)^{+} \text{ is } c_{t-1}^{+}, \text{then } s(t)^{+} \text{ is } c_t^{+} \end{aligned} \tag{6.25}$$

where $c_{t-n_I}^{-}, c_{t-1}^{-}, c_t^{-}$ and $c_{t-n_I}^{+}, c_{t-1}^{+}, c_t^{+}$ are the tags for identifying which cluster the granule belongs to, e.g., $c_{t-n_I}^{-}$ would be marked as 2 when the maximal membership of $s(t - n_I)^{-}$ is toward second cluster. On the basis of such rules, one can use two $c \times 1$ vectors $\mathbf{h}_j^{-}$ and $\mathbf{h}_j^{+}$ along with the related membership grades u_{ij}^{-} and u_{ij}^{+} to quantize the above statement and compute the following variables.

$$\mathbf{s}^{-} = \sum_{j=1}^{N} \mathbf{h}_j^{-} u_{ij}^{-}, \quad \mathbf{s}^{+} = \sum_{j=1}^{N} \mathbf{h}_j^{+} u_{ij}^{+} \tag{6.26}$$

where $i = 1, 2, \ldots, c$. With the help of center defuzzification technique, one can get a predicted value

$$\hat{\mathbf{X}} = \left[\widehat{\mathbf{x}^{-}}, \widehat{\mathbf{x}^{+}}\right] = \left[\mathbf{s}^{-T}\mathbf{V}^{-}, \mathbf{s}^{+T}\mathbf{V}^{+}\right] \tag{6.27}$$

It can also be deemed as a degranulation result of the granulated data.

6.4.2 *PIs Optimization*

As for the PIs optimization, the evaluation criteria should be reasonably considered as well-defined performance indices. Two indices of interest are involved here.

Coverage. The objective of this criterion is to maximize the data amount covered by the granulation area. Here, one can formulate this evaluation by

$$\min \ n\text{-card}\left\{s_{ij} \in \left[\mathbf{s}^{-T}\mathbf{V}^{-}, \mathbf{s}^{+T}\mathbf{V}^{+}\right]\right\} \tag{6.28}$$

where card{} refers to the cardinality, $s_{ij}(i = 1, 2, \ldots, N$, and $j = 1, 2, \ldots, n)$ is the data points in the testing dataset, n denotes the length of the testing dataset, i.e., the predicting horizon.

Specificity. In order to quantify the specificity of the information granules, it can be measured by an average of the intervals

$$\min \frac{1}{n}\left|\mathbf{s}^{-T}\mathbf{V}^{-} - \mathbf{s}^{+T}\mathbf{V}^{+}\right| \tag{6.29}$$

Combining the above two criteria, the objective function of the optimization problem can be constructed by

$$\min \ \left(n\text{-card}\left\{s_{ij} \in \left[\mathbf{s}^{-T}\mathbf{V}^{-}, \mathbf{s}^{+T}\mathbf{V}^{+}\right]\right\}\right) \times \left(\frac{1}{n}\left|\mathbf{s}^{-T}\mathbf{V}^{-} - \mathbf{s}^{+T}\mathbf{V}^{+}\right|\right) \tag{6.30}$$

Furthermore, one can consider the constraint condition as the boundaries of the vertical extension on the prototypes, i.e., the average value of the cumulative levels of the information granularity is restricted to a given ε_0

$$\sum_{j=1}^{n} \frac{\varepsilon_j}{n} = \varepsilon_0 \tag{6.31}$$

Also, every information granularity level should be constrained within the average value

$$\varepsilon_j > 0 \tag{6.32}$$

To solve this optimization problem, evolutionary algorithm is preferred as a trustworthy solution. Considering the convergence efficiency and the accuracy, a PSO algorithm is employed for the optimal allocation of information granularity. Introducing penalty coefficients γ, the constrained optimization problem is transformed into an unconstrained one

$$\left(n\text{-card}\left\{s_{ij} \in \left[\mathbf{s}^{-T}\mathbf{V}^{-}, \mathbf{s}^{+T}\mathbf{V}^{+}\right]\right\}\right) \times \left(\frac{1}{n}\left|\mathbf{s}^{-T}\mathbf{V}^{-} - \mathbf{s}^{+T}\mathbf{V}^{+}\right|\right) + \gamma\left(\sum_{j=1}^{n} \frac{\varepsilon_j}{n} - \varepsilon_0\right) \tag{6.33}$$

Deploying (6.33) as the fitness function, the prediction intervals can be finally obtained by iteratively calculating ε_j by PSO until a terminal criterion is satisfied.

6.4.3 Computing Procedure

The computing procedures of the granular computing-based prediction intervals (GrC-PIs) approach are as follows. And one can see a program flow illustrated as Fig. 6.19.

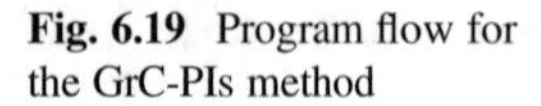

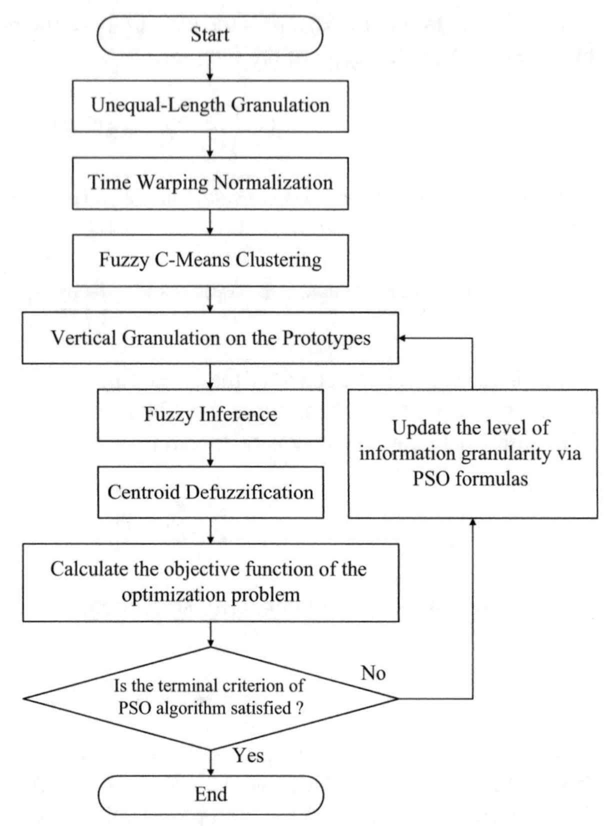

Fig. 6.19 Program flow for the GrC-PIs method

Algorithm 6.4: Granular Computing-Based Prediction Intervals (GrC-PIs)

Step 1: *Based on the industrial semantic characteristics, the original data is granulated to form a series of unequal-length granules.*

Step 2: *By using the time warping normalization, the formed granules are transformed into ones with identical length in order to be capable of applying fuzzy C Means clustering algorithm.*

Step 3: *Cluster the equalized granules to obtain the clustering center $v_{ij}$and the membership degree u_{ik}.*

Step 4: *Granulate the clustering center on the vertical direction as interval values. And, based on the obtained membership grades, establish fuzzy rules as* (6.25) *and construct the initial prediction intervals by using* (6.26) and (6.27).

Step 5: *Regard the formula* (6.33) *as the fitness function for updating $\varepsilon_j$and optimize the PIs until the terminal criterion is satisfied. Then, the long-term PIs for the energy flows are finally obtained.*

6.4.4 Case Study

Given a large amount of gas users in steel industry, this section selects two main gas users that have significant impacts on the entire gas system, the BFG consumption flow of hot blast stove and the BFG generation flow. The input and output of the models are the flow data in history and its future trend, respectively. The practical data range from August to December, 2015, and their sample interval is 1 min. Different from the existing predicting horizons (limited within 1–2 h) for the generic energy forecasting, the experiments designate the predicting horizons to be 480 and 720 min, respectively, owing to the shift switching (8 h) and a half of the production day (12 h). To indicate the applicability and the effectiveness of the GrC-PIs, a series of comparative experiments by using the delta, the MVE, and the GrC-PIs are conducted. The parameters of these algorithms are respectively optimized by trial and error. Considering that the evaluation of the PIs construction usually involves the interval coverage ratio and the interval width, the PICP and the MPIW, introduced in Chap. 5 are adopted here as the evaluation indices.

BFG Consumption Flow of Hot Blast Stove

As described in Chap. 3, the hot blast stove is one of the most important BFG users in steel industry, whose gas flow comes with two obvious phases, the normal running phase and the stove suspending one. In practice, each amplitudes and spans of these phases are rather different in various production procedures. The prediction results by using the comparative methods are illustrated as Fig. 6.20, where the predicting horizon is 8 h (480 predicted points). It is apparent that due to the iterative mechanism, the delta method and the MVE exhibit poor prediction accuracies when the predicting horizon increases. Especially, after 60 predicted points, both methods failed to approximately provide the flow tendencies, which become incapable of satisfying the practical requirements on the gas flow prediction.

With respect to the GrC-based approach, the accuracies are relatively average throughout the predicting horizon, and the real values of the gas flow can be accurately traced. In addition, it is clear from the figure that as for the intervals coverage index, these two methods provide poor performances compared to the GrC-PIs, where the real flows can be well covered by the GrC-based approach. To clarify the statistic results of these approaches, Table 6.16 lists their evaluation indices of the PIs construction and the corresponding computing cost. One can draw the conclusion that the GrC-PIs are superior to the others in terms of both prediction accuracy and computing cost. Thus, the real-time demands can be satisfied for the practical usage.

To further evaluate the effectiveness of the GrC-based approach, we extend the predicting horizon to 12 h (720 predicted points). Similar results are observed in Fig. 6.21 and the statistics indices are presented in Table 6.2. It is noticeable that

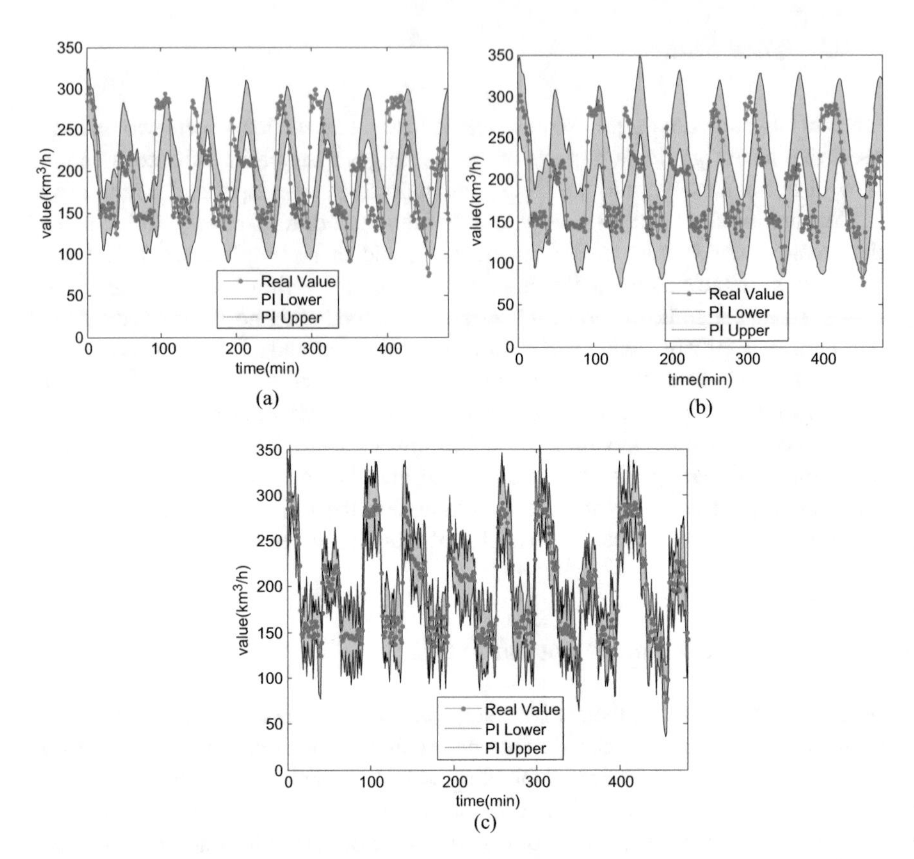

Fig. 6.20 PIs results of BFG consumption of a hot blast stove by using these methods (prediction horizon: 480 points). (**a**) Delta, (**b**) MVE, (**c**) GrC-PIs

Table 6.16 The statistic performance of these methods on BFG consumption of a hot blast stove (Prediction horizon: 480 points)

Methods	PICP	MPIW	CT (s)
Delta	0.275	42.503	49.88
MVE	0.373	60.557	19.23
GrC-PIs	0.903	41.041	4.42

although the covered data points by using the comparative methods are a little bit worse than that of the GrC-PIs, as shown in Table 6.17, the tendencies of data flows are poorly forecasted by the Delta and the MVE compared to the GrC-based one. In terms of computing time, the superiority of the GrC-PIs is obvious and the prediction task for a future 12 h only takes less than 10 s because of its one-time computation with no iteration.

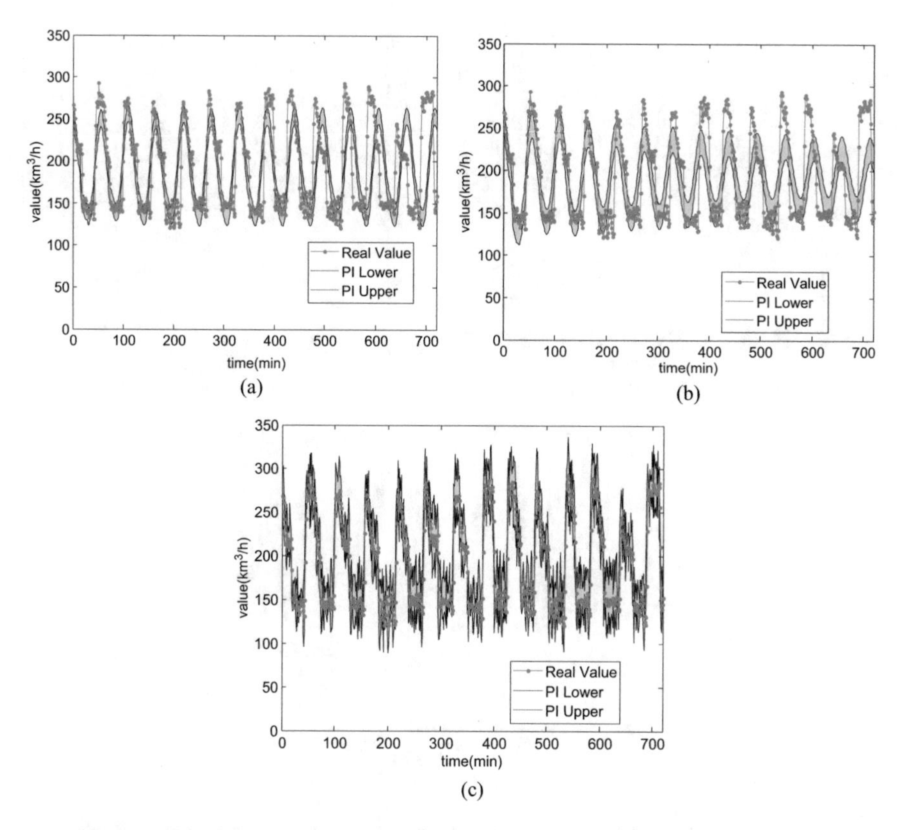

Fig. 6.21 PIs results on BFG consumption of a hot blast stove by using the methods (prediction horizon: 720 points). (**a**) Delta, (**b**) MVE, (**c**) GrC-PIs

Table 6.17 Error statistics of these methods on BFG consumption of a hot blast stove (Prediction horizon: 720 points)

Methods	PICP	MPIW	CT (s)
Delta	0.507	39.729	53.11
MVE	0.661	60.841	26.32
GrC-PIs	0.883	27.993	6.33

BFG Generation Flows

The BFG generation is the main gas flow in energy system concerned by energy scheduling operators, whose values vary largely and exhibit a class of quasi-periodicity that have an important impact on the entire gas balance. Given that the variation of BFG generation presents large amplitude and heavy fluctuation, it is rather hard to predict the gas flows accurately.

Two groups of experimental results are provided, whose data come from the practical database on-site. The predicting horizons involve 480 and 720 min, as

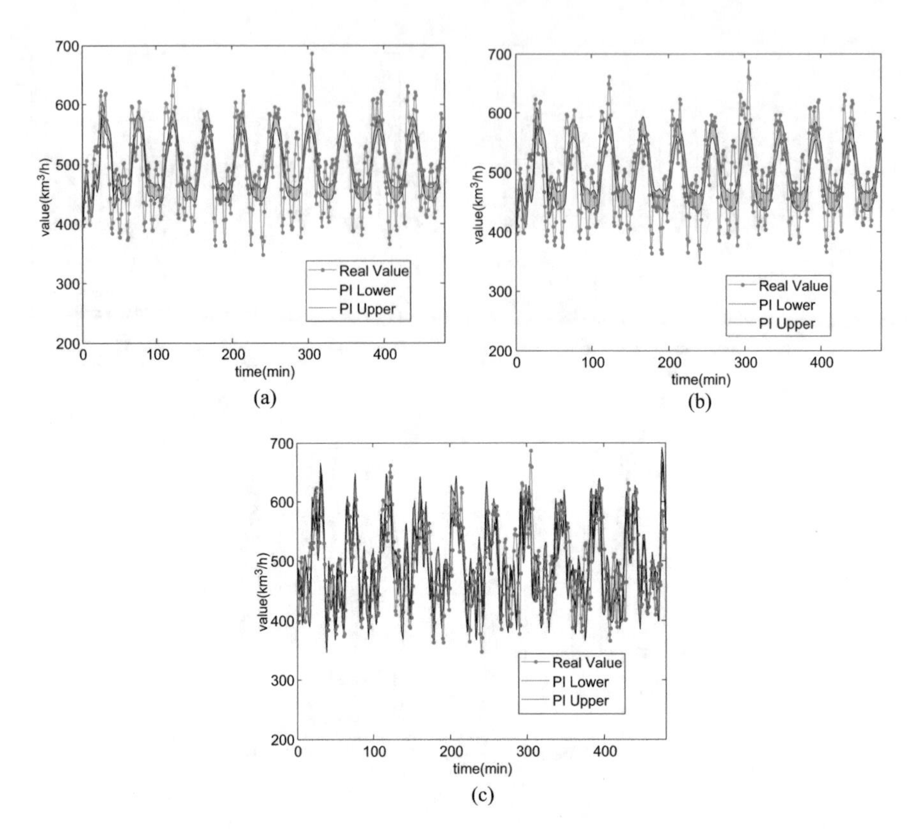

Fig. 6.22 PIs results on BFG generation flow with these methods (prediction horizon: 480 points). (**a**) Delta, (**b**) MVE, (**c**) GrC-PIs

illustrated in Figs. 6.22 and 6.23. The statistic results are listed in Tables 6.18 and 6.19, respectively. It can be seen from the tables that more than 80% real values can be covered by the GrC-PIs even though the predicting horizon extended to 12 h (720 points), and the flow tendency can also be successfully estimated with a relatively narrow interval. With respect to the approaches of the Delta and the MVE, the Delta demonstrates the lowest interval coverage index and the most expensive computing cost, while the MVE provides the poorest interval widths that means the difficulty of decision-making for the energy scheduling has to be largely increased.

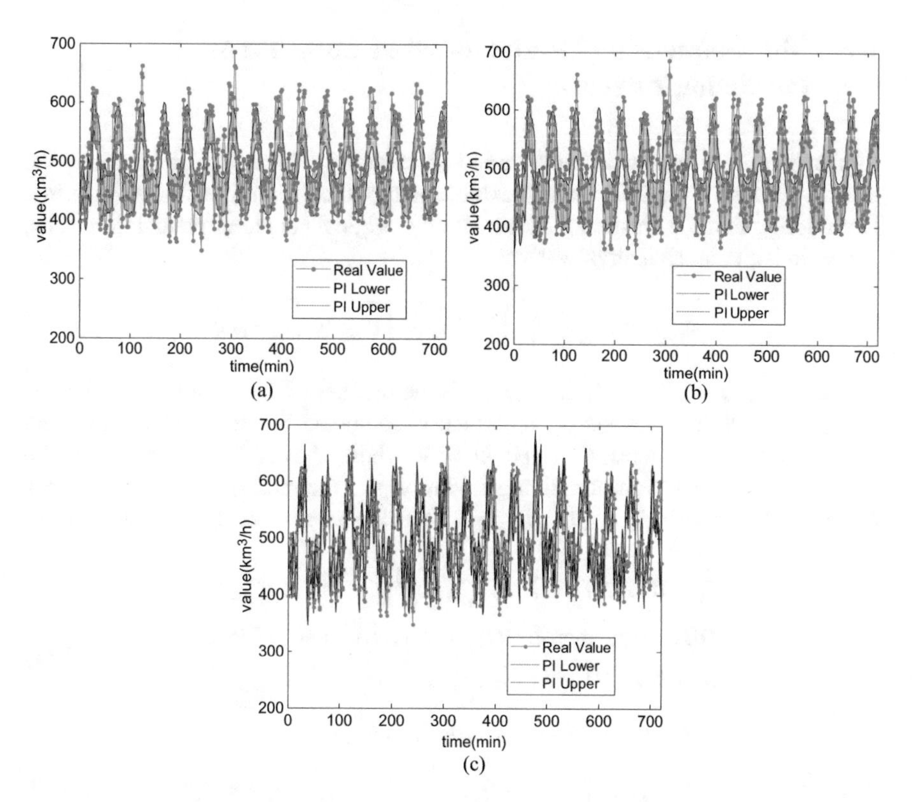

Fig. 6.23 PIs results on BFG generation flow with these methods (prediction horizon: 720 points). (**a**) Delta, (**b**) MVE, (**c**) GrC-PIs

Table 6.18 Error statistics of these methods on BFG generation flow (Prediction horizon: 480 points)

Methods	PICP	MPIW	CT (s)
Delta	0.254	44.556	44.58
MVE	0.369	67.588	13.65
GrC-PIs	0.906	49.055	3.11

Table 6.19 Error statistics of these methods on BFG generation flow (Prediction horizon: 720 points)

Methods	PICP	MPIW	CT (s)
Delta	0.332	42.187	61.94
MVE	0.468	60.023	18.51
GrC-PIs	0.846	36.874	9.44

6.5 Multi-dimension Granular-Based Long-Term Prediction Intervals

Similarly, the PIs construction on multi-dimension problem is also based on the structure shown in Fig. 6.14 which exhibits a hybrid collaborative fuzzy clustering. Considering that the clustering centers are the original source of prediction results, we extended them as interval values

$$\overline{v_{m,ij}} = \left[v_{m,ij}^{-}, v_{m,ij}^{+}\right] = \left[v_{m,ij} - \varepsilon_{m,j}, v_{m,ij} + \varepsilon_{m,j}\right] \tag{6.34}$$

where $i = 1, 2, \ldots, c, j = 1, 2, \ldots, n$. m is the number of the output which denotes $\{1, 2\}$. $v_{m,\ ij}$ is the elements of $\mathbf{V}_M$ which is computed by HCFC algorithm, here $m = 1$ when M denotes $L^{(1)}$, and $m = 2$ when M is $L^{(2)}$. $\varepsilon_{m,\ j}$ refers to the corresponding level of information granularity, and the related membership grades can be computed as $\overline{u_{m,ik}} = \left[u_{m,ik}^{-}, u_{m,ik}^{+}\right]$. Then, the fuzzy rules can be established as

$$\begin{aligned}
&R_t^- : \text{If } \boldsymbol{L}_0^{(1)}(t-1)^- \text{ is } c_{\boldsymbol{L}_0^{(1)}}^-, \boldsymbol{L}_0^{(2)}(t-1)^- \text{ is } c_{\boldsymbol{L}_0^{(2)}}^-, \\
&\boldsymbol{D}(t)^- \text{ is } c_{\boldsymbol{D}}^- \text{ then } \boldsymbol{L}^{(1)}(t)^- \text{ is } c_{\boldsymbol{L}^{(1)}}^-, \boldsymbol{L}^{(2)}(t-1)^- \text{ is } c_{\boldsymbol{L}^{(2)}}^- \\
&R_t^+ : \text{If } \boldsymbol{L}_0^{(1)}(t-1)^+ \text{ is } c_{\boldsymbol{L}_0^{(1)}}^+, \boldsymbol{L}_0^{(2)}(t-1)^+ \text{ is } c_{\boldsymbol{L}_0^{(2)}}^+, \\
&\boldsymbol{D}(t)^+ \text{ is } c_{\boldsymbol{D}}^+ \text{ then } \boldsymbol{L}^{(1)}(t)^+ \text{ is } c_{\boldsymbol{L}^{(1)}}^+, \boldsymbol{L}^{(2)}(t-1)^+ \text{ is } c_{\boldsymbol{L}^{(2)}}^+
\end{aligned} \tag{6.35}$$

where $c_{\boldsymbol{L}_0^{(1)}}^-$, $c_{\boldsymbol{L}_0^{(2)}}^-$, $c_{\boldsymbol{D}}^-$, $c_{\boldsymbol{L}^{(1)}}^-$, $c_{\boldsymbol{L}^{(2)}}^-$ and $c_{\boldsymbol{L}_0^{(1)}}^+$, $c_{\boldsymbol{L}_0^{(2)}}^+$, $c_{\boldsymbol{D}}^+$, $c_{\boldsymbol{L}^{(1)}}^+$, $c_{\boldsymbol{L}^{(2)}}^+$ are tags for identifying which cluster the input and output belong to. For instance, $c_{\boldsymbol{L}_0^{(1)}}^-$ would be marked as 5 when the maximal membership of $\boldsymbol{L}_0^{(1)}(t-1)^-$ is toward the fifth cluster. In such a way, N fuzzy rules are attained. Then, one can search the rules that have the same tag as the input in the testing set. In order to summarize the searching result, a set of variables are computed as

$$\mathbf{s}_M^- = \sum_{k=1}^{N} \mathbf{h}_{m,k}^- u_{m,ik}^-, \quad \mathbf{s}_M^+ = \sum_{k=1}^{N} \mathbf{h}_{m,k}^+ u_{m,ik}^+ \tag{6.36}$$

where $u_{m,ik}^-$ and $u_{m,ik}^+$ are the membership grades, $\mathbf{h}_{m,k}^-$ and $\mathbf{h}_{m,k}^+$ are $c \times 1$ column vector of the tag result (e.g., $[0, 1, 0, \ldots, 0]$ which has the maximal membership toward second cluster), $i = 1, 2, \ldots, c$. With centroid defuzzification techniques, the multi-output long-term prediction intervals can be obtained as

$$\widehat{\mathbf{X}}_M = \left[\widehat{\mathbf{x}_M^-}, \widehat{\mathbf{x}_M^+}\right] = \left[\mathbf{s}_M^{-T} \mathbf{V}_M^-, \mathbf{s}_M^{+T} \mathbf{V}_M^+\right] \tag{6.37}$$

However, for the lack of optimization, they cannot be deemed as the final results.

Besides, to maximize the data amount covered by the granulation area, the optimization model can be formulated by

$$\begin{aligned} &\max\ \operatorname{card}\left\{s_{m,kj} \in \left[\mathbf{s}_M^{-T}\mathbf{V}_M^-, \mathbf{s}_M^{+T}\mathbf{V}_M^+\right]\right\} \\ &\text{s.t.}\ \sum_{j=1}^{n} \frac{\varepsilon_{m,j}}{n} = \varepsilon_{m,0} \end{aligned} \tag{6.38}$$

where $s_{m,\ kj}$ is the elements of testing dataset. $k = 1, 2, \ldots, N; j = 1, 2, \ldots, n; m = 1, 2$. The constraint condition is to constrain the boundaries of the vertical extension on the prototypes, i.e., the average value of the cumulative granular levels $\varepsilon_{m,\ j}$ is restricted to the given values $\varepsilon_{m,\ 0}$. Considering the convergence efficiency, the parameter to be determined, and the optimization accuracy, the PSO algorithm is employed for optimal allocation of information granularity. Here, the fitness function of the optimization can be formulated by

$$\min n \times m\text{-card}\left\{s_{m,kj} \in \left[\mathbf{s}_M^{-T}\mathbf{V}_M^-, \mathbf{s}_M^{+T}\mathbf{V}_M^+\right]\right\} \tag{6.39}$$

where n refers to the length of the testing dataset, i.e., the predicting horizon. m is the number of the output. M denotes the two gasholder levels $L^{(1)}$ and $L^{(2)}$.

To clarify the detailed computing procedures of the multi-dimension granular-based long-term prediction intervals (M-GrC-PIs) approach, it can be listed as follows:

Algorithm 6.5: Multi-dimension Granular-Based Long-Term Prediction Intervals (M-GrC-PIs) Approach

Step 1: *The multi-output model of hybrid fuzzy clustering should be summarized into two phases. Firstly, the FCM is individually utilized for each set to obtain the initial memberships and the prototypes. The number of the clusters has to be the same to all five datasets. Secondly, the collaborative prototypes along with fuzzy membership grades are determined until a termination criterion is satisfied.*

Step 2: *Extend prototypes as interval values by* (6.34) *and compute related membership grades. Based on such results, deploy fuzzy inference and defuzzification techniques to obtain initial PIs.*

Step 3: *Regard the optimization model* (6.35) *to optimize the PIs and complete the multi-output long-term PIs prediction for the gas tank levels.*

6.5.1 Case Study

The comparative long-term PIs results of the holder levels in #1 subsystem are presented in Fig. 6.24, where two holders are involved. Since the delta and the MVE deployed iteration mechanism, they can only give accurate results in about first 60 min. And they gradually behave worse along with the prediction goes on. Similarly, the ESN failed to give satisfied results especially for the period from 70th to 230th points. Besides, for the ignorance of the mutual impacts between two gasholders, these single-output models are out of the reality and the outcomes are consequently unsatisfactory. However, the M-GrC-PIs method presents the better

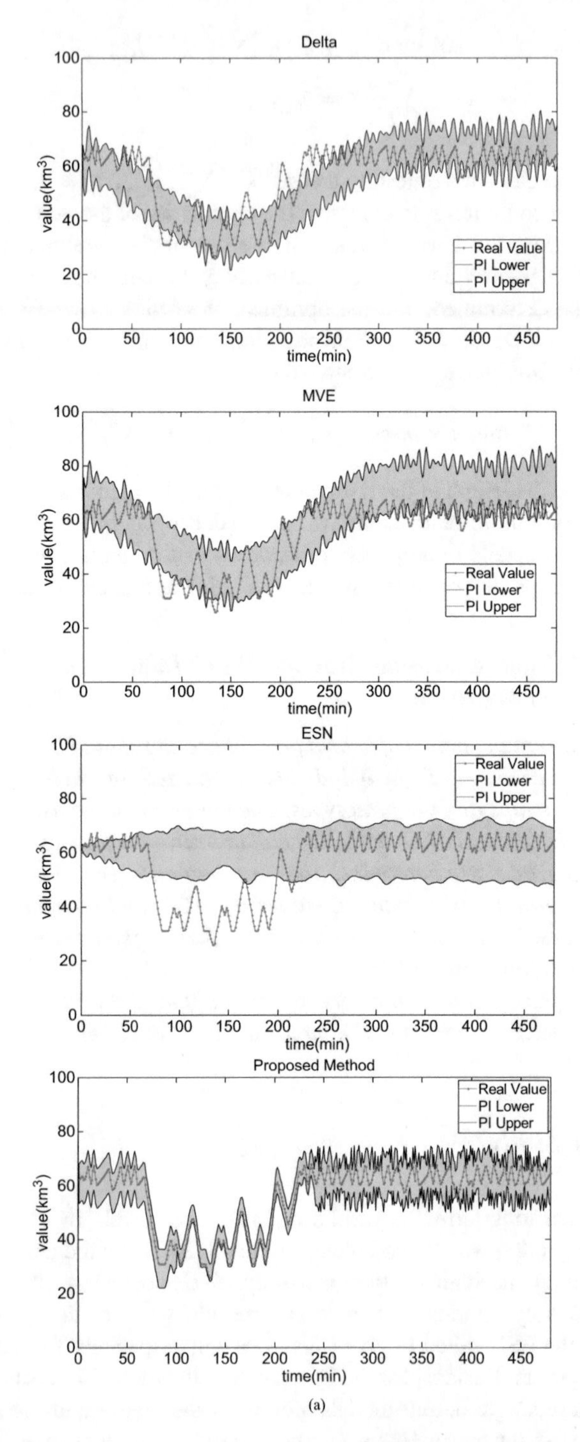

Fig. 6.24 Comparison of the long-term PIs results on #1 subsystem (**a**) #1 holder, (**b**) #2 holder

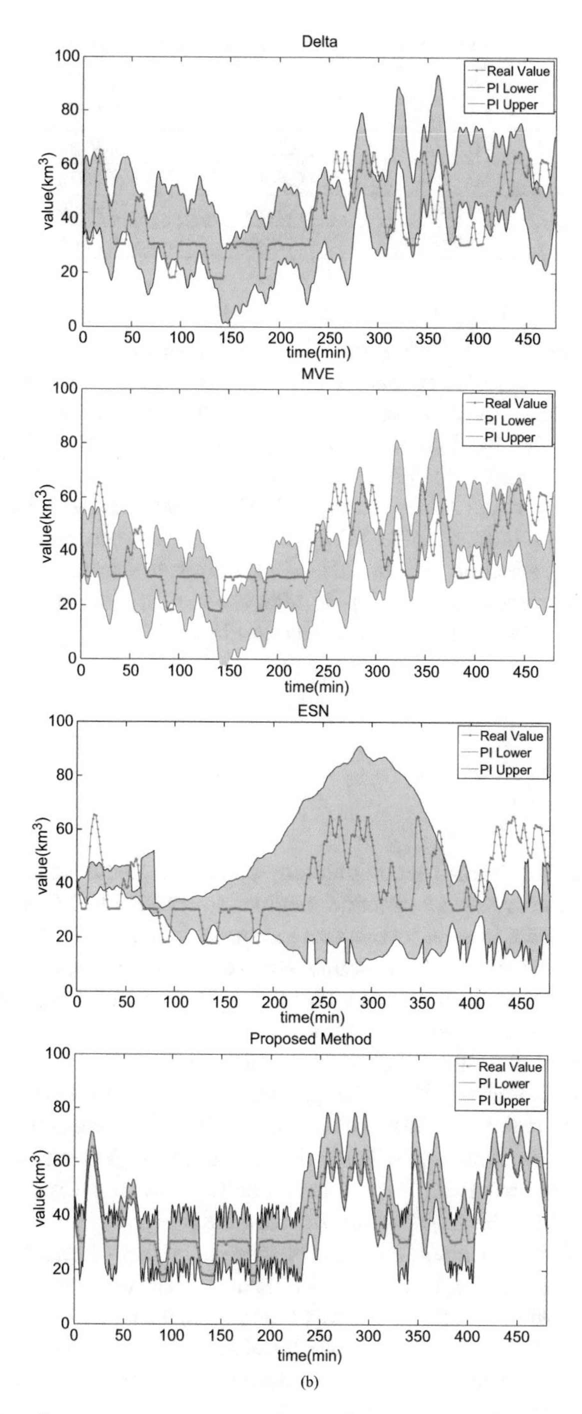

(b)

Fig. 6.24 (continued)

Table 6.20 Error statistics of different long-term PIs construction methods on #1 subsystem

	PICP		MPIW	
Predict methods	#1 holder	#2 holder	#1 holder	#2 holder
Delta	0.6958	0.3938	14.9481	16.0835
MVE	0.7604	0.6083	16.3878	25.6086
ESN	0.5433	0.5995	11.6589	28.4652
M-GrC-PIs	0.9521	0.9088	10.3365	12.2555

ones, indicating that it enables to successfully estimate the future trends of the levels for two gas holders simultaneously. Table 6.20 illustrates the quantitative error statistics. The MVE performs well at PICP, whereas poor at the MPIW. Although the MPIW of delta is acceptable, the PICP of results on holder #2 is rather low. On the contrary, the M-GrC-PIs have an excellent performance on both indices.

Another set of long-term PIs results for #2 subsystem are shown in Fig. 6.25, along with the error statistics given in Table 6.21. Compared to the experiments on #1 subsystem, the delta and the MVE still cannot give accurate results. The PIs obtained by ESN are either overbroad (see #3) or fails to accurately predict the trend (see #4). While the results of the M-GrC-PIs successfully forecast the holder levels in terms of PIs, which has high coverage and low width with respect to the error statistics.

6.6 Discussion

In this chapter, we present a set of long-term prediction methods as well as the long-term prediction intervals construction methods for the industrial process data, and take the energy data in steel industry as instances, which are all based on granular computing and the industrial semantics feature. The experimental results indicate that these methods are superior to the others in terms of long-term prediction, prediction intervals coverage, interval width, and computing cost.

With regard to the future work and research directions, there are a number of unexplored topics which deserve further attention. First, the length of information granules corresponding to a certain of industrial semantics were established based on some experiential (expert) knowledge or obtained through intensive experimentation. It would be beneficial to perform theoretical analysis that may lead to the construction of granular relationships between the equipment operational mode and the data variation tendency. Second, a precise estimation of real-world data noise characteristics becomes crucial to further enhancements of accuracy, especially for long-term estimation procedures. In this sense, further improvements of the quantification of the data trend uncertainty deserve future studying. Besides, in the perspective of industrial application, the levels of the gasholders play a vital role for the energy scheduling optimization. Therefore, the method constructing prediction intervals for gasholder levels is worth exploring, of which the analysis on theirs

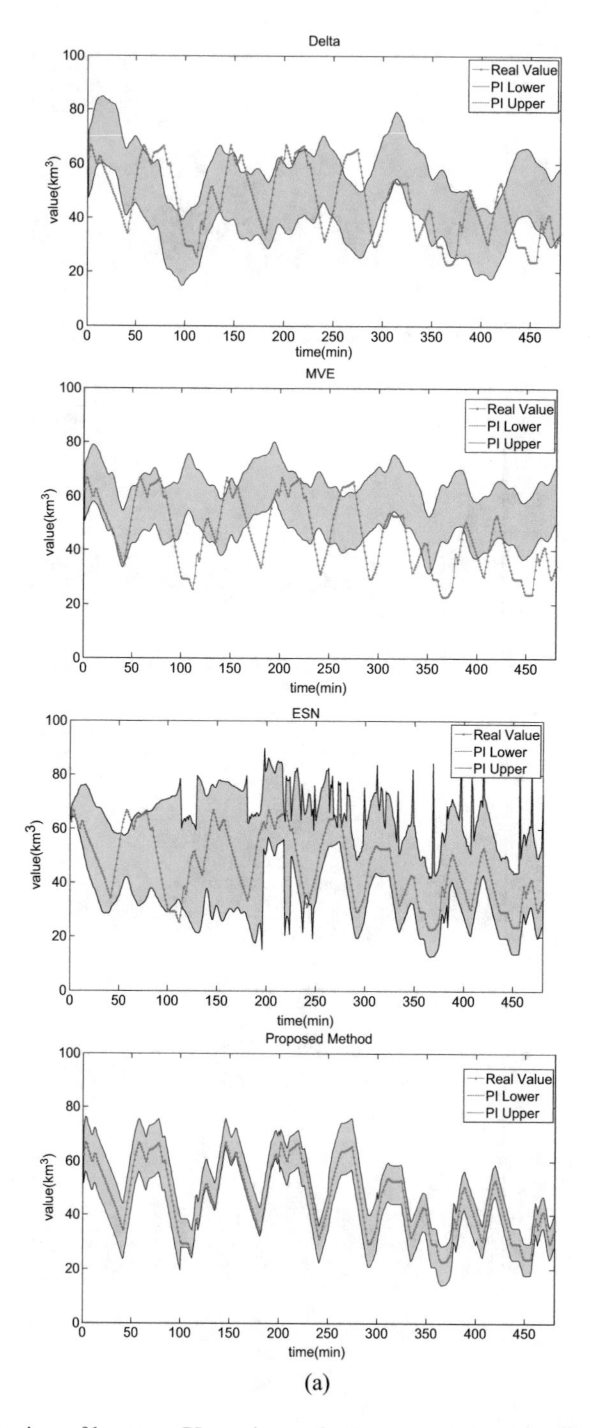

Fig. 6.25 Comparison of long-term PIs results on #2 subsystem (**a**) #3 holder, (**b**) #4 holder

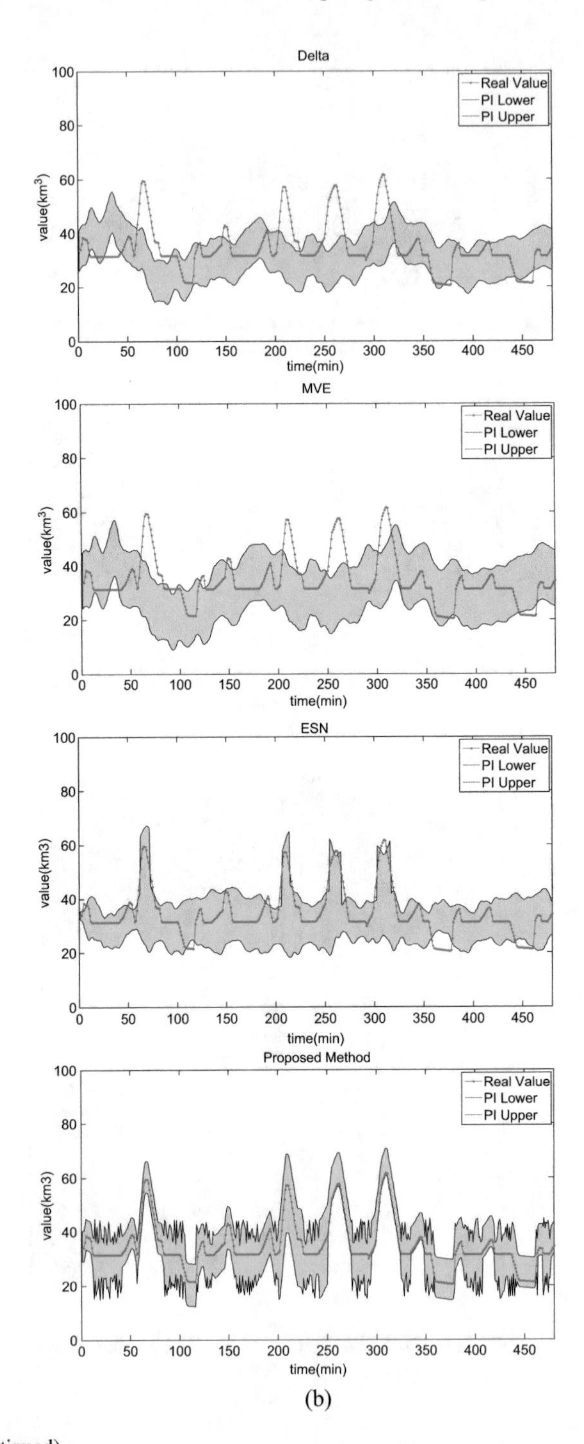

Fig. 6.25 (continued)

Table 6.21 Error statistics of different long-term PIs construction methods on #1 subsystem

	PICP		MPIW	
Prediction methods	#3 holder	#4 holder	#3 holder	#4 holder
Delta	0.5021	0.6563	24.6017	19.1564
MVE	0.4917	0.7354	20.9931	20.6457
ESN	0.8965	0.8452	30.8512	21.3645
M-GrC-PIs	0.9755	0.9152	14.6972	17.9525

influence factors is also needed. As for the theoretical aspects, combining granular computing with statistics is of great research value, in which the mean and the variance should be calculated. And the granular neural network would be more appropriate in such case compared with the FCM clustering method.

References

1. Zhao, J., Wang, W., Pedrycz, W., et al. (2012). Online parameter optimization-based prediction for converter gas system by parallel strategies. *IEEE Transactions on Control Systems Technology, 20*(3), 835–845.
2. Zhao, J., Liu, Y., Zhang, X., et al. (2012). A MKL based on-line prediction for gasholder level in steel industry. *Control Engineering Practice, 20*(6), 629–641.
3. Han, Z., Liu, Y., Zhao, J., et al. (2012). Real time prediction for converter gas tank levels based on multi-output least square support vector regressor. *Control Engineering Practice, 20*(12), 1400–1409.
4. Wang, W., Pedrycz, W., & Liu, X. (2015). Time series long-term forecasting model based on information granules and fuzzy clustering. *Engineering Applications of Artificial Intelligence, 41*, 17–24.
5. Zhao, J., Han, Z., Pedrycz, W., et al. (2016). Granular model of long-term prediction for energy system in steel industry. *IEEE Transactions on Cybernetics, 46*(2), 388.
6. Han, Z., Zhao, J., Liu, Q., et al. (2016). Granular-computing based hybrid collaborative fuzzy clustering for long-term prediction of multiple gas holders levels. *Information Sciences, 330*, 175–185.

Chapter 7
Parameter Estimation and Optimization

Abstract The selection of parameters or hyper-parameters gives great impact on the performance of a data-driven model. This chapter introduces some commonly used parameter optimization and estimation methods, such as the gradient-based methods (e.g., gradient descend, Newton method, and conjugate gradient method) and the intelligent optimization ones (e.g., genetic algorithm, differential evolution algorithm, and particle swarm optimization). In particular, in this chapter, the conjugate gradient method is employed to optimize the hyper-parameters in a LSSVM model based on noise estimation, which enable to alleviate the impact of noise on the performance of the LSSVM. As for dynamic models, this chapter introduces nonlinear Kalman-filter methods for parameter estimation. The well-known ones include the extended Kalman-filter, the unscented Kalman-filter, and the cubature Kalman-filter. Here, a dual estimation model based on two Kalman-filters is illustrated, which simultaneously estimates the uncertainties of internal state and the output. Besides, the probabilistic methods for parameter estimation are also introduced, where a Bayesian model, especially a variational inference framework, is elaborated in details. In such a framework, a particular variational relevance vector machine (RVM) model based on automatic relevance determination kernel is introduced, which provides the approximated posterior distributions over the kernel parameters. Finally, we give some case studies by employing a number of industrial data.

7.1 Introduction

In machine learning community, there are usually various parameters involved in prediction models, which extremely affects their prediction performance. For example, in a kernel-based model, e.g., support vector machine (SVM), the kernel parameter, and the penalty factor are two important parameters. With their suitable values, the SVM model can exhibit excellent prediction performance. Besides, in a Gaussian process (GP) regression model the kernel parameters can also play a crucial role in the prediction.

J. Zhao et al., *Data-Driven Prediction for Industrial Processes and Their Applications*, Information Fusion and Data Science,
https://doi.org/10.1007/978-3-319-94051-9_7

Currently, various parameter estimation and optimization methods are developed for these important parameters, including conventional gradient-based methods, intelligent optimization methods, probabilistic methods, etc.

Gradient-based methods employ the gradient information to define a search direction for reaching to the optimal solution in unconstraint optimization problems, including gradient descent, batch gradient descent, stochastic gradient descent, mini-batch gradient descent, Newton methods, and conjugate gradient method, etc. The class of algorithm has specific search directions, but they are vulnerable to the worse initial values of the parameters when the objective function is multi-modal. Moreover, it requires that the objective function has specific mathematical form and is differentiable.

Intelligent optimization algorithms, such as genetic algorithm, differential evolution algorithm, particle swarm optimization, and simulated annealing algorithm, can deal with the multi-modal problems, and can work well when the objective function and constraints are not continuous and convex. However, these algorithms are usually unstable and can produce different solutions because they involve random searching procedures.

Besides, probabilistic methods take the uncertainty of data into consideration. The model parameters are assigned with prior distributions. Maximum likelihood estimation method only considers the variance of output noise, which is equivalent to least square method. While Bayesian methods add prior distributions to the model parameters, thus their posterior distributions can be calculated. In addition, when the analytical solutions of the posterior distribution are not tractable, variational inference method can provide approximated solutions of the posterior in simple form.

Particularly, for dynamic model of time series, Kalman-filter estimation method can give a procedure for estimating the parameters of the dynamic model. Besides, various extensions based on it are developed through these years, including nonlinear Kalman-filter estimation, unscented Kalman-filter estimation, and cubature Kalman-filter estimation. Moreover, dual estimation methods of linear/nonlinear Kalman-filter are proposed to estimate the model parameters and the internal states simultaneously.

As for noisy industrial data, the noise could largely affect the performance of prediction model. Thereby, it is necessary to take the data noise into consideration in a modeling process, particularly, the noise estimation-based parameter optimization method improves the model prediction accuracy by considering the noise variance in training process.

In this chapter, we will introduce these aforementioned parameter estimation and optimization methods and give a number of case studies for industrial applications.

7.2 Gradient-Based Methods

Gradient-based method is a class of optimization algorithms to solve unconstrained optimization problems

$$\min_{\mathbf{x}\in R^p} F(\mathbf{x}) \tag{7.1}$$

where $\mathbf{x} = [x_1, x_2, \ldots, x_p]^{\mathrm{T}}$ is a p-dimensional parameter vector of the function $F(\mathbf{x})$ which is to be minimized. The search directions are related to the gradient of the function at the current point, based on which gradient-based methods minimize $F(\mathbf{x})$ respect to $\mathbf{x}$.

Gradient is a multi-variable version of derivatives. For a multi-variable function, its gradient is a vector-valued function, as opposed to a derivative, which is scalar valued. For example, as for a function $F(x, y, z)$, its gradient is represented as $\nabla F(x, y, z) = \left(\frac{\partial F}{\partial x}, \frac{\partial F}{\partial y}, \frac{\partial F}{\partial z}\right)^{\mathrm{T}}$, and as for a certain point (x_0, y_0, z_0), its gradient is $\left(\frac{\partial F}{\partial x_0}, \frac{\partial F}{\partial y_0}, \frac{\partial F}{\partial z_0}\right)^{\mathrm{T}}$.

Gradient-based methods contain gradient descent, Newton method, and conjugate gradient. Besides, gradient descent also includes batch gradient descent (BGD), stochastic gradient descent (SGD), and mini-batch gradient descent (MBGD) algorithms.

7.2.1 Gradient Descent

Gradient descent (GD) is an iterative optimization algorithm for minimizing the objective function in an unconstrained minimization problem, in which one updates the current solution at the direction proportional to the negative of the gradient in each iterative step.

We firstly describe some concepts before introducing the GD method, including the learning rate, the dataset, the underlying function, and the loss function. The learning rate means the length of the parameters moving at the direction of the negative of the gradient. The dataset contains the input and output data, where the input matrix is denoted by $\mathbf{X} = [\mathbf{x}_1, \mathbf{x}_2, \ldots, \mathbf{x}_i \ldots, \mathbf{x}_N]^{\mathrm{T}}$, $\mathbf{x}_i$ is the ith input sample with its corresponding output t_i, and the output vector is $\mathbf{t} = [t_1, t_2, \ldots, t_i, \ldots, t_N]^{\mathrm{T}}$. Thus, we assume that there is an underlying function from the input to the output, fitting the dataset, i.e., $t_i = F(\mathbf{x}_i; \boldsymbol{\theta})$, where $\boldsymbol{\theta}$ denotes the set of parameters of the function $F(.)$. To find the optimal parameters $\widehat{\boldsymbol{\theta}}$ which can fit the dataset perfectly, we define a loss function, i.e.,

$$J(\boldsymbol{\theta}) = \frac{1}{N}\sum_{i=1}^{N} (F(\mathbf{x}_i; \boldsymbol{\theta}) - t_i)^2 \tag{7.2}$$

which is to be minimized.

Supposing that the loss function $J(\boldsymbol{\theta})$ is differentiable in a neighborhood of a point $\boldsymbol{\theta}$, and the value of $J(\boldsymbol{\theta})$ reduces fastest at the direction of the negative of the gradient of $J(\boldsymbol{\theta})$ at $\boldsymbol{\theta}$, $-\nabla J(\boldsymbol{\theta})$, the iteration rule of $\boldsymbol{\theta}$ is given by

$$\boldsymbol{\theta}_{n+1} = \boldsymbol{\theta}_n - \gamma \nabla J(\boldsymbol{\theta}_n) \tag{7.3}$$

where γ denotes the learning rate. With γ small enough, one has $J(\boldsymbol{\theta}_n) \geq J(\boldsymbol{\theta}_{n+1})$. As such, one starts from an initial point $\boldsymbol{\theta}_0$ for a local minimum of $J(\boldsymbol{\theta})$ so as to produce a sequence $\{\boldsymbol{\theta}_i\}$. We have $J(\boldsymbol{\theta}_i) \geq J(\boldsymbol{\theta}_{i+1})$, with $i = 0, 1, 2, \ldots$, desiring to converge to a local minimum with the sequence $\{\boldsymbol{\theta}_i\}$.

If the value of the learning rate γ is too small, the convergence speed of the iteration process will be too slow; otherwise, too large learning rate can lead to quick iteration as well as the fact that the process will not converge to a local minimum with fluctuation. Note that the value of γ is allowed to change at each iteration. We give an iteration rule of γ, where ∇J holds the Lipschitz continuity condition [1] and is chosen via a line search which satisfies the Barzilai-Borwein method [2].

$$\gamma_n = \frac{(\boldsymbol{\theta}_n - \boldsymbol{\theta}_{n-1})^{\mathrm{T}}[\nabla J(\boldsymbol{\theta}_n) - \nabla J(\boldsymbol{\theta}_{n-1})]}{\|\nabla J(\boldsymbol{\theta}_n) - \nabla J(\boldsymbol{\theta}_{n-1})\|^2} \tag{7.4}$$

of which its convergence to a local minimum can be guaranteed.

Besides, the learning process will also be affected by initial selection of the parameters $\boldsymbol{\theta}_0$. When the function J is convex, all local minima are also global minima with different initial parameters, thus in this case the GD can converge to the global solution. While the initial value of $\boldsymbol{\theta}_0$ must be carefully chosen with the non-convex function J (namely multi-modal), in order to avoid getting stuck into the local minimum. Or for this case, one can run the algorithm for different $\boldsymbol{\theta}_0$s, then choose the best one which produces least value of the objective function.

Here, we illustrate an example of the GD algorithm. Considering a linear regression model of the form,

$$h(\mathbf{x}|\boldsymbol{\theta}) = \theta_0 + \theta_1 x_1 + \theta_2 x_2 + \theta_3 x_3 = \boldsymbol{\theta}^{\mathrm{T}}\mathbf{x} \tag{7.5}$$

where $\boldsymbol{\theta} = [\theta_0, \theta_1, \theta_2, \theta_3]^{\mathrm{T}}$ is the parameter vector of function $h(\mathbf{x}|\boldsymbol{\theta})$. Besides, given a dataset $\mathbf{X} = [\mathbf{x}_1, \mathbf{x}_2, \ldots, \mathbf{x}_N]^{\mathrm{T}}$, where $\mathbf{x}_i = [x_{i1}, x_{i2}, x_{i3}]^{\mathrm{T}}$, and its corresponding output dataset $\mathbf{t} = [t_1, t_2, \ldots, t_N]^{\mathrm{T}}$, we intend to find a suitable $\boldsymbol{\theta}^*$ that fits the dataset well. The loss function is defined by.

$$J(\boldsymbol{\theta}) = \frac{1}{N}\sum_{i=1}^{N}(h(\mathbf{x}_i|\boldsymbol{\theta}) - t_i)^2 \tag{7.6}$$

This objective function indicates that we aim to minimize the sum of all of the prediction errors with respect to $\boldsymbol{\theta}$.

$$\frac{\partial J(\boldsymbol{\theta})}{\partial \theta_j} = \frac{2}{N}\sum_{i=1}^{N}\theta_j(\boldsymbol{\theta}^{\mathrm{T}}\mathbf{x}_i - t_i), \quad j = 1, 2, 3 \tag{7.7}$$

Therefore, the update rule of $\boldsymbol{\theta}$ is given by

$$\theta_j^{\text{new}} = \theta_j - \gamma \frac{\partial J(\boldsymbol{\theta})}{\partial \theta_j}, \quad j = 1, 2, 3 \tag{7.8}$$

Batch Gradient Descent

Batch gradient descent (BGD) algorithm is a commonly used GD one. Specifically, assuming that we have M samples and the BGD uses the M data points to optimize the parameters $\boldsymbol{\theta}$ at one iteration, the loss function of $\boldsymbol{\theta}$ is defined as

$$J(\boldsymbol{\theta}) = \frac{1}{M} \sum_{i=1}^{M} (F(\mathbf{x}_i; \boldsymbol{\theta}) - t_i)^2 \tag{7.9}$$

Then, we use (7.3) to update the parameters.

The BGD algorithm uses all the samples to optimize the parameters of the model. While if the number of samples is too large, the computing cost of this algorithm will be very high.

Stochastic Gradient Descent

Since the BGD often suffers from high computational load, especially with very large number of the training samples, stochastic gradient descent (SGD), also known as incremental gradient descent, is proposed to tackle the above computational problem. It is a stochastic approximation of the GD method for minimizing an objective function, written as a sum of differentiable functions.

Considering a general regression model $y = g(\mathbf{x}|\boldsymbol{\theta})$, where $\boldsymbol{\theta} = [\theta_1, \theta_2, \ldots, \theta_m]^{\mathrm{T}}$ is an m-dimensional parameter vector, the objective function is rewritten as

$$J(\boldsymbol{\theta}) = \frac{1}{N} \sum_{i=1}^{N} (g(\mathbf{x}_i|\boldsymbol{\theta}) - t_i)^2 = \frac{1}{N} \sum_{i=1}^{N} J_i(\boldsymbol{\theta}) \tag{7.10}$$

where $J_i(\boldsymbol{\theta}) = (g(\mathbf{x}_i|\boldsymbol{\theta}) - t_i)^2$ is the sub-objective function with ith training point. In SGD, the true gradient of $J(\boldsymbol{\theta})$ is approximated by a gradient for a single example,

$$\boldsymbol{\theta}^{\text{new}} = \boldsymbol{\theta} - \gamma \nabla J_i(\boldsymbol{\theta}), \quad i = 1, 2, \ldots, N \tag{7.11}$$

As the algorithm sweeps through the whole training set, it performs the above update for each training example. Several passes can be made over the training set until the algorithm converges. If this is done, the data can be shuffled for each pass to prevent cycles. Therefore, SGD can be described as

Algorithm 7.1: Stochastic Gradient Descent

Initialization: *Choose an initial vector of parameters* $\boldsymbol{\theta}_0$*and the learning rate* γ.

Repeat until a minimum is reached.

Randomly shuffle the training examples.

For i=1,2,...,N

$$\boldsymbol{\theta}^{\text{new}} = \boldsymbol{\theta} - \gamma \nabla J_i(\boldsymbol{\theta})$$

End

End

End

For the convergence of SGD, when the learning rates γ decrease with an appropriate rate, SGD converges almost surely to a global minimum when the objective function is convex; otherwise, it converges almost surely to a localminimum [3, 4].

Mini-batch Gradient Descent

Mini-batch gradient descent realizes a compromise between computing the true gradient and the gradient at a single example, which is to compute the gradient against more than one training example at each step. For M samples in a dataset, we firstly split these samples into p parts randomly, and the loss function is given by

$$J(\boldsymbol{\theta}) = \frac{1}{p} \sum_{k=1}^{p} Jm_k(\boldsymbol{\theta}) \tag{7.12}$$

where $Jm_k(\boldsymbol{\theta})$ is the loss function for the kth part of the dataset.

$$Jm_k(\boldsymbol{\theta}) = \frac{1}{N_k} \sum_{i=1}^{N_k} \left(g\left(\mathbf{x}_i^{(k)} | \boldsymbol{\theta}\right) - t_i^{(k)} \right)^2 \tag{7.13}$$

where $\mathbf{x}_i^{(k)}$ is the ith input sample in the kth part dataset, $t_i^{(k)}$ is its corresponding output, and N_k is the number of samples in kth part dataset. Thus, in many situations one prefers to use mini-batch gradient descent: each update to the parameters $\boldsymbol{\theta}$ is done using a small batch of the data $\left\{\mathbf{x}_i^{(k)}\right\}_{i=1}^{N_k}$. Using this method, the direction of the updates is somewhat corrected in comparison with the stochastic updates, but is updated much more regularly than in the case of the original gradient descent.

7.2.2 *Newton Method*

We firstly consider the minimization of a one-dimensional nonlinear function $f(x)$. Suppose that the function $f(x)$ is a convex function and second-order continuously differentiable. The basic idea of Newton method is to use a second-order Taylor expansion at the current estimated minima point x_k to approximate the function $f(x)$, i.e.,

$$\varphi(x) = f(x_k) + f'(x_k)(x - x_k) + \frac{1}{2} f''(x_k)(x - x_k)^2 \tag{7.14}$$

ignoring the higher order terms of $(x - x_k)$. Since we shall find the minima of $f(x)$, $\varphi(x)$ should satisfy the following condition:

$$\varphi'(x) = 0 \tag{7.15}$$

Therefore, one has

$$f'(x_k) + f''(x_k)(x - x_k) = 0 \tag{7.16}$$

which produces

$$x = x_k - \frac{f'(x_k)}{f''(x_k)} \tag{7.17}$$

Thus, given the initial point x_0, one can obtain the following iteration rule:

$$x_{k+1} = x_k - \frac{f'(x_k)}{f''(x_k)}, \quad k = 0, 1, 2, \ldots \tag{7.18}$$

This rule can produce a sequence $\{x_k\}$ in order to approach the minima of the function $f(x)$. Similarly, for multi-variable function $f(\mathbf{x})$, we also have

$$\varphi(\mathbf{x}) = f(\mathbf{x}_k) + \nabla f(\mathbf{x}_k)^{\mathrm{T}}(\mathbf{x} - \mathbf{x}_k) + \frac{1}{2}(\mathbf{x} - \mathbf{x}_k)^T \nabla^2 f(\mathbf{x}_k)(\mathbf{x} - \mathbf{x}_k) \tag{7.19}$$

where ∇f is the gradient vector of $f(\mathbf{x})$, and $\nabla^2 f$ is the Hessian matrix, with their expression as

$$\nabla f = \begin{bmatrix} \frac{\partial f}{\partial x_1} \\ \frac{\partial f}{\partial x_2} \\ \cdots \\ \frac{\partial f}{\partial x_p} \end{bmatrix}, \nabla^2 f = \begin{bmatrix} \frac{\partial^2 f}{\partial x_1^2} & \frac{\partial^2 f}{\partial x_1 x_2} & \cdots & \frac{\partial^2 f}{\partial x_1 x_p} \\ \frac{\partial^2 f}{\partial x_2 x_1} & \frac{\partial^2 f}{\partial x_2^2} & \cdots & \frac{\partial^2 f}{\partial x_2 x_p} \\ \vdots & \vdots & \cdots & \vdots \\ \frac{\partial^2 f}{\partial x_p x_1} & \frac{\partial^2 f}{\partial x_p x_2} & \cdots & \frac{\partial^2 f}{\partial x_p^2} \end{bmatrix} \tag{7.20}$$

For simplicity, we denote ∇f and $\nabla^2 f$ by $\mathbf{g}$ and $\mathbf{H}$, and $\mathbf{H}$ must be positive definite. Because $\mathbf{g}$ and $\mathbf{H}$ are both function of $\mathbf{x}$, we can denote them by $\mathbf{g}_k$ and $\mathbf{H}_k$ at point $\mathbf{x}_k$. Similarly, to minimize the function f, we compute the derivative,

$$\nabla \varphi(\mathbf{x}) = 0 \tag{7.21}$$

which gives the equation

$$\mathbf{g}_k + \mathbf{H}_k(\mathbf{x} - \mathbf{x}_k) = 0 \tag{7.22}$$

If the matrix $\mathbf{H}_k$ is positive definite, then

$$\mathbf{x} = \mathbf{x}_k - \mathbf{H}_k^{-1}\mathbf{g}_k \tag{7.23}$$

Therefore, given the initial solution $\mathbf{x}_0$, the iteration rule is given by

$$\mathbf{x}_{k+1} = \mathbf{x}_k - \mathbf{H}_k^{-1}\mathbf{g}_k, \quad k = 0, 1, 2, \ldots \tag{7.24}$$

where the search direction is defined by $\mathbf{d}_k = -\mathbf{H}_k^{-1}\mathbf{g}_k$. In the following, the procedure of the Newton is presented.

Algorithm 7.2: Newton Method
Initialization: *Set an initial solution* $\mathbf{x}_0$*and the threshold* ε, *and let* $k = 0$.

Step 1: *Compute* $\mathbf{g}_k$*and* $\mathbf{H}_k$

Step 2: *If* $\|\mathbf{g}_k\| < \varepsilon$, *terminate the iteration; otherwise, compute the search direction* $\mathbf{d}_k = -\mathbf{H}_k^{-1}\mathbf{g}_k$.

Step 3: *Compute the new point* $\mathbf{x}_{k+1} = \mathbf{x}_k + \mathbf{d}_k$

Step 4: *Let* $k = k + 1$, *and return to Step 1.*

When the objective function is a quadratic one, the Hessian matrix becomes a constant matrix because the second-order Taylor expansion approximation is the same as the original function. We can obtain the true minima from arbitrary initial point. For a non-quadratic objective function that is nearer to a quadratic one, the convergence speed will be very fast. This is an advantage of the Newton method. However, there is no learning rate in the iteration rule of the original Newton method. For a non-quadratic objective function, we cannot guarantee the convergence of the algorithm. To eliminate such a drawback, researchers proposed the damped Newton method, which still used the search direction $\mathbf{d}_k$, but in each iteration the line search performed to search for the optimal learning rate, i.e.,

$$\lambda_k = \arg\min_{\lambda \in R} f(\mathbf{x}_k + \lambda \mathbf{d}_k) \tag{7.25}$$

Then, we present the procedure of the damped Newton method.

Algorithm 7.3: Damped Newton Method
Initialization: *Set an initial solution* $\mathbf{x}_0$*and the threshold* ε, *and let* $k = 0$.

Step 1: *Compute* $\mathbf{g}_k$*and* $\mathbf{H}_k$

Step 2: *If* $\|\mathbf{g}_k\| < \varepsilon$, *terminate the iteration; otherwise, compute the search direction* $\mathbf{d}_k = -\mathbf{H}_k^{-1}\mathbf{g}_k$.

Step 3: *Compute the learning rate* λ_k*by using (7.25), and compute the new point* $\mathbf{x}_{k+1} = \mathbf{x}_k + \lambda_k\mathbf{d}_k$

Step 4: *Let* $k = k + 1$, *and return to Step 1.*

The Newton method is the development of the GD algorithm. The GD algorithm, only using the local property of the objective function, is an algorithm which utilizes the first-order gradient information and considers the negative gradient as the search direction. But the Newton method uses not only the first-order gradient but also the second-order one. Thereby, it can get faster convergence than that of the GD algorithm. However, it still has several drawbacks. 1) The objective function must be continuous and exist the first-order and second-order partial derivatives, with its Hessian matrix being positive definite. 2) The computing procedure is rather complex because of the requirement of computation of the Hessian matrix and its inversion.

7.2.3 *Quasi-Newton Method*

Although the Newton method has fast convergence speed, the computation of the Hessian matrix is very complex. Besides, it is very hard to guarantee that it is always positive definite. To tackle these two problems, researchers proposed the quasi-Newton method [5], of which its basic idea is to construct a positive-definite matrix to approximate the Hessian matrix for avoiding to compute the second-order partial derivatives.

Before introducing the specific quasi-Newton methods, we firstly derive the quasi-Newton condition, also known as the secant condition. We use the matrix $\mathbf{B}$ to approximate the Hessian matrix $\mathbf{H}$, and the matrix $\mathbf{D}$ to approximate $\mathbf{H}^{-1}$, i.e., $\mathbf{B} \approx \mathbf{H}, \mathbf{D} \approx \mathbf{H}^{-1}$.

Supposing that after $k+1$ iterations the point $\mathbf{x}_{k+1}$ is obtained, one can perform the Taylor expansion with respect to the objective function f at $\mathbf{x}_{k+1}$, i.e.,

$$f(\mathbf{x}) \approx f(\mathbf{x}_{k+1}) + \nabla f(\mathbf{x}_{k+1})^{\mathrm{T}}(\mathbf{x} - \mathbf{x}_{k+1}) + \frac{1}{2}(\mathbf{x} - \mathbf{x}_{k+1})^{\mathrm{T}} \nabla^2 f(\mathbf{x}_{k+1})(\mathbf{x} - \mathbf{x}_{k+1}) \tag{7.26}$$

Performing the gradient operator ∇ on both sides of (7.26) gives

$$\nabla f(\mathbf{x}) \approx \nabla f(\mathbf{x}_{k+1}) + \mathbf{H}_{k+1} \cdot (\mathbf{x} - \mathbf{x}_{k+1}) \tag{7.27}$$

Let $\mathbf{x} = \mathbf{x}_k$ in (7.27), one has

$$\mathbf{g}_{k+1} - \mathbf{g}_k \approx \mathbf{H}_{k+1} \cdot (\mathbf{x}_{k+1} - \mathbf{x}_k) \tag{7.28}$$

Next we define the symbols

$$\mathbf{s}_k = \mathbf{x}_{k+1} - \mathbf{x}_k, \quad \mathbf{y}_k = \mathbf{g}_{k+1} - \mathbf{g}_k \tag{7.29}$$

Therefore, one can rewrite (7.28) as

$$\mathbf{y}_k \approx \mathbf{H}_{k+1} \cdot \mathbf{s}_k \tag{7.30}$$

or

$$\mathbf{s}_k \approx \mathbf{H}_{k+1}^{-1} \cdot \mathbf{y}_k \tag{7.31}$$

This is the quasi-Newton condition, which keeps a rein on the Hessian matrix $\mathbf{H}_{k+1}$. Then, one has

$$\mathbf{y}_k \approx \mathbf{B}_{k+1} \cdot \mathbf{s}_k \tag{7.32}$$

or

$$\mathbf{s}_k \approx \mathbf{D}_{k+1} \cdot \mathbf{y}_k \tag{7.33}$$

Broyden–Fletcher–Goldfarb–Shanno (BFGS) Algorithm

Broyden–Fletcher–Goldfarb–Shanno (BFGS) method is one of the mostly used quasi-Newton methods to solve unconstrained nonlinear optimization problems [6]. This method approximates the Hessian matrix, i.e., $\mathbf{B}_k \approx \mathbf{H}_k$, in an iterative way. The iteration rule is given by

$$\mathbf{B}_{k+1} = \mathbf{B}_k + \Delta\mathbf{B}_k, \quad k = 0, 1, 2, \ldots \tag{7.34}$$

where $\mathbf{B}_0$ is often fixed to the identity matrix $\mathbf{I}$. Next we construct the corrected matrix $\Delta\mathbf{B}_k$, i.e.,

$$\Delta\mathbf{B}_k = \alpha\mathbf{u}\mathbf{u}^{\mathrm{T}} + \beta\mathbf{v}\mathbf{v}^{\mathrm{T}} \tag{7.35}$$

Substituting (7.35) into (7.34) and using the quasi-Newton condition (7.30), one has

$$\mathbf{y}_k = \mathbf{B}_k\mathbf{s}_k + \left(\alpha\mathbf{u}^{\mathrm{T}}\mathbf{s}_k\right)\mathbf{u} + \left(\beta\mathbf{v}^{\mathrm{T}}\mathbf{s}_k\right)\mathbf{v} \tag{7.36}$$

Let $\alpha\mathbf{u}^{\mathrm{T}}\mathbf{s}_k = 1$, $\beta\mathbf{v}^{\mathrm{T}}\mathbf{s}_k = -1$, and

$$\mathbf{u} = \mathbf{y}_k, \quad \mathbf{v} = \mathbf{B}_k\mathbf{s}_k \tag{7.37}$$

which gives

$$\alpha = \frac{1}{\mathbf{y}_k^{\mathrm{T}}\mathbf{s}_k}, \quad \beta = -\frac{1}{\mathbf{s}_k^{\mathrm{T}}\mathbf{B}_k\mathbf{s}_k} \tag{7.38}$$

Therefore, the corrected matrix $\Delta\mathbf{B}_k$ is given by

$$\Delta\mathbf{B}_k = \frac{\mathbf{y}_k\mathbf{y}_k^{\mathrm{T}}}{\mathbf{y}_k^{\mathrm{T}}\mathbf{s}_k} - \frac{\mathbf{B}_k\mathbf{s}_k\mathbf{s}_k^{\mathrm{T}}\mathbf{B}_k}{\mathbf{s}_k^{\mathrm{T}}\mathbf{B}_k\mathbf{s}_k} \tag{7.39}$$

Next we present the whole procedure of the BFGS method.

Algorithm 7.4: BFGS-1 Method

Initialization: *Set an initial solution* $\mathbf{x}_0$*and the threshold* ε*, and let* $\mathbf{B}_0 = \mathbf{I}$ *and* $k = 0$.

Step 1: *Compute the search direction* $\mathbf{d}_k = -\mathbf{B}_k^{-1} \cdot \mathbf{g}_k$.

Step 2: *Compute the learning rate* λ_k*using (7.25), and let* $\mathbf{s}_k = \lambda_k\mathbf{d}_k$, $\mathbf{x}_{k+1} = \mathbf{x}_k + \mathbf{s}_k$.

Step 3: *If* $\|\mathbf{g}_{k+1}\| < \varepsilon$*, terminate the iteration.*

Step 4: *Compute* $\mathbf{y}_k = \mathbf{g}_{k+1} - \mathbf{g}_k$.

Step 5: *Compute* $\mathbf{B}_{k+1} = \mathbf{B}_k + \frac{\mathbf{y}_k\mathbf{y}_k^{\mathrm{T}}}{\mathbf{y}_k^{T}\mathbf{s}_k} - \frac{\mathbf{B}_k\mathbf{s}_k\mathbf{s}_k^{\mathrm{T}}\mathbf{B}_k}{\mathbf{s}_k^{T}\mathbf{B}_k\mathbf{s}_k}$.

Step 6: *Let* $k = k + 1$*, and return to Step 1.*

In Algorithm 7.4, the search direction $\mathbf{d}_k = -\mathbf{B}_k^{-1} \cdot \mathbf{g}_k$ can be computed by solving a linear equation system $\mathbf{B}_k\mathbf{d}_k = -\mathbf{g}_k$ in order to avoid the inverse of $\mathbf{B}_k$. Besides, in Step 5, the recursion formula of $\mathbf{B}_{k+1}$ is generally substituted by the Sherman-Morrison formula, which gives the relationship of $\mathbf{B}_{k+1}^{-1}$ and $\mathbf{B}_k^{-1}$ [7].

$$\mathbf{B}_{k+1}^{-1} = \left(\mathbf{I} - \frac{\mathbf{s}_k\mathbf{y}_k^{\mathrm{T}}}{\mathbf{y}_k^{\mathrm{T}}\mathbf{s}_k}\right)\mathbf{B}_k^{-1}\left(\mathbf{I} - \frac{\mathbf{y}_k\mathbf{s}_k^{\mathrm{T}}}{\mathbf{y}_k^{\mathrm{T}}\mathbf{s}_k}\right) + \frac{\mathbf{s}_k\mathbf{s}_k^{\mathrm{T}}}{\mathbf{y}_k^{\mathrm{T}}\mathbf{s}_k} \tag{7.40}$$

And this can be written in another form

$$\mathbf{B}_{k+1}^{-1} = \mathbf{B}_k^{-1} + \left(\frac{1}{\mathbf{s}_k^{T}\mathbf{y}_k} + \frac{\mathbf{y}_k^{\mathrm{T}}\mathbf{B}_k^{-1}\mathbf{y}_k}{\left(\mathbf{s}_k^{\mathrm{T}}\mathbf{y}_k\right)^2}\right)\mathbf{s}_k\mathbf{s}_k^{\mathrm{T}} - \frac{1}{\mathbf{s}_k^{\mathrm{T}}\mathbf{y}_k}\left(\mathbf{B}_k^{-1}\mathbf{y}_k\mathbf{s}_k^{\mathrm{T}} + \mathbf{s}_k\mathbf{y}_k^{\mathrm{T}}\mathbf{B}_k^{-1}\right) \tag{7.41}$$

In addition, for completeness, the Sherman-Morrison formula is given as follows. Supposing that $\mathbf{A} \in \mathbf{R}^n$ is a nonsingular matrix, and $\mathbf{u}, \mathbf{v} \in \mathbf{R}^n$; if $1 + \mathbf{v}^{\mathrm{T}}\mathbf{A}^{-1}\mathbf{u} \neq 0$, one has

$$\left(\mathbf{A} + \mathbf{u}\mathbf{v}^{\mathrm{T}}\right)^{-1} = \mathbf{A}^{-1} - \frac{\mathbf{A}^{-1}\mathbf{u}\mathbf{v}^{\mathrm{T}}\mathbf{A}^{-1}}{1 + \mathbf{v}^{\mathrm{T}}\mathbf{A}^{-1}\mathbf{u}} \tag{7.42}$$

Using (7.40), we get another BFGS algorithm which can avoid the matrix inverse or solving the linear equation system.

Algorithm 7.5: BFGS-2 Method

Initialization: *Set an initial solution* $\mathbf{x}_0$*and the threshold* ε*, and let* $\mathbf{B}_0 = \mathbf{I}$ *and* $k = 0$.

Step 1: *Compute the search direction* $\mathbf{d}_k = -\mathbf{B}_k^{-1} \cdot \mathbf{g}_k$

Step 2: *Compute the learning rate* λ_k*using (7.25), and let* $\mathbf{s}_k = \lambda_k\mathbf{d}_k$, $\mathbf{x}_{k+1} = \mathbf{x}_k + \mathbf{s}_k$.

Step 3: *If* $\|\mathbf{g}_{k+1}\| < \varepsilon$*, terminate the iteration;*

Step 4: *Compute* $\mathbf{y}_k = \mathbf{g}_{k+1} - \mathbf{g}_k$

***Step 5**: Compute* $\mathbf{B}_{k+1}^{-1} = \left(\mathbf{I} - \frac{\mathbf{s}_k\mathbf{y}_k^{\mathrm{T}}}{\mathbf{y}_k^{\mathrm{T}}\mathbf{s}_k}\right)\mathbf{B}_k^{-1}\left(\mathbf{I} - \frac{\mathbf{y}_k\mathbf{s}_k^{\mathrm{T}}}{\mathbf{y}_k^{\mathrm{T}}\mathbf{s}_k}\right) + \frac{\mathbf{s}_k\mathbf{s}_k^{\mathrm{T}}}{\mathbf{y}_k^{\mathrm{T}}\mathbf{s}_k}$

***Step 6**: Let* $k = k + 1$, *and return to Step 1.*

L-BFGS Method

Limited-memory BFGS (L-BFGS or LM-BFGS) is an optimization algorithm, which approximates the BFGS algorithm using a limited amount of computer memory. It is a popular algorithm for parameter optimization in machine learning [8, 9]. Similar to the original BFGS, the L-BFGS uses an estimation way for the inverse of Hessian matrix to steer its search through variable space. The BFGS stores a dense $n \times n$ approximation to the inverse of Hessian (n is the number of variables in the problem), while the L-BFGS stores only a few vectors. Since the L-BFGS method requires limited computer memory, it is particularly suitable for the optimization problems with a large number of variables.

In the L-BFGS, instead of performing the inverse of the Hessian $\mathbf{H}_k$, a history of the past m updates of the position $\mathbf{x}$ and gradient $\nabla f(\mathbf{x})$ is saved, where the history size m can be small (often $m < 10$). These updates are used to implicitly do operations requiring the $\mathbf{H}_k$-vector product.

We define $\rho_k = \frac{1}{\mathbf{y}_k^{\mathrm{T}}\mathbf{s}_k}$ and $\mathbf{V}_k = \mathbf{I} - \rho_k\mathbf{y}_k\mathbf{s}_k^{\mathrm{T}}$. And $\mathbf{D}_0$ will be the initial approximation of the inverse of the Hessian matrix. This algorithm is based on the recursion operation for the inverse of the Hessian defined by (7.40). Thus, (7.40) is rewritten as

$$\mathbf{D}_{k+1} = \mathbf{V}_k^{\mathrm{T}}\mathbf{D}_k\mathbf{V}_k + \rho_k\mathbf{s}_k\mathbf{s}_k^{\mathrm{T}} \tag{7.43}$$

Then, one can compute $\mathbf{D}_1, \mathbf{D}_2, \mathbf{D}_3, \ldots$ recursively, i.e.,

$$\mathbf{D}_1 = \mathbf{V}_0^{\mathrm{T}}\mathbf{D}_0\mathbf{V}_0 + \rho_0\mathbf{s}_0\mathbf{s}_0^{\mathrm{T}} \tag{7.44}$$

$$\begin{aligned}\mathbf{D}_2 &= \mathbf{V}_1^{\mathrm{T}}\mathbf{D}_1\mathbf{V}_1 + \rho_1\mathbf{s}_1\mathbf{s}_1^{\mathrm{T}} \\ &= \mathbf{V}_1^{\mathrm{T}}\left(\mathbf{V}_0^{\mathrm{T}}\mathbf{D}_0\mathbf{V}_0 + \rho_0\mathbf{s}_0\mathbf{s}_0^{\mathrm{T}}\right)\mathbf{V}_1 + \rho_1\mathbf{s}_1\mathbf{s}_1^{\mathrm{T}} \\ &= \mathbf{V}_1^{\mathrm{T}}\mathbf{V}_0^{\mathrm{T}}\mathbf{D}_0\mathbf{V}_0\mathbf{V}_1 + \mathbf{V}_1^{\mathrm{T}}\rho_0\mathbf{s}_0\mathbf{s}_0^{\mathrm{T}}\mathbf{V}_1 + \rho_1\mathbf{s}_1\mathbf{s}_1^{\mathrm{T}}\end{aligned} \tag{7.45}$$

$$\begin{aligned}\mathbf{D}_3 &= \mathbf{V}_2^{\mathrm{T}}\mathbf{D}_2\mathbf{V}_2 + \rho_2\mathbf{s}_2\mathbf{s}_2^{\mathrm{T}} \\ &= \mathbf{V}_2^{\mathrm{T}}\left(\mathbf{V}_1^{\mathrm{T}}\mathbf{V}_0^{\mathrm{T}}\mathbf{D}_0\mathbf{V}_0\mathbf{V}_1 + \mathbf{V}_1^{\mathrm{T}}\rho_0\mathbf{s}_0\mathbf{s}_0^{\mathrm{T}}\mathbf{V}_1 + \rho_1\mathbf{s}_1\mathbf{s}_1^{\mathrm{T}}\right)\mathbf{V}_2 + \rho_2\mathbf{s}_2\mathbf{s}_2^{\mathrm{T}} \\ &= \mathbf{V}_2^{\mathrm{T}}\mathbf{V}_1^{\mathrm{T}}\mathbf{V}_0^{\mathrm{T}}\mathbf{D}_0\mathbf{V}_0\mathbf{V}_1\mathbf{V}_2 + \mathbf{V}_2^{\mathrm{T}}\mathbf{V}_1^{\mathrm{T}}\rho_0\mathbf{s}_0\mathbf{s}_0^{\mathrm{T}}\mathbf{V}_1\mathbf{V}_2 + \mathbf{V}_2^{\mathrm{T}}\rho_1\mathbf{s}_1\mathbf{s}_1^{\mathrm{T}}\mathbf{V}_2 + \rho_2\mathbf{s}_2\mathbf{s}_2^{\mathrm{T}}\end{aligned} \tag{7.46}$$

.... Generally, one has

$$
\begin{aligned}
\mathbf{D}_{k+1} = & \left(\mathbf{V}_k^{\mathrm{T}}\mathbf{V}_{k-1}^{\mathrm{T}}\ldots\mathbf{V}_1^{\mathrm{T}}\mathbf{V}_0^{\mathrm{T}}\right)\mathbf{D}_0\left(\mathbf{V}_0\mathbf{V}_1\ldots\mathbf{V}_{k-1}\mathbf{V}_k\right) \\
& +\left(\mathbf{V}_k^{\mathrm{T}}\mathbf{V}_{k-1}^{\mathrm{T}}\ldots\mathbf{V}_2^{\mathrm{T}}\mathbf{V}_1^{\mathrm{T}}\right)\left(\boldsymbol{\rho}_0\mathbf{s}_0\mathbf{s}_0^{\mathrm{T}}\right)\left(\mathbf{V}_1\mathbf{V}_2\ldots\mathbf{V}_{k-1}\mathbf{V}_k\right) \\
& +\left(\mathbf{V}_k^{\mathrm{T}}\mathbf{V}_{k-1}^{\mathrm{T}}\ldots\mathbf{V}_3^{\mathrm{T}}\mathbf{V}_2^{\mathrm{T}}\right)\left(\boldsymbol{\rho}_1\mathbf{s}_1\mathbf{s}_1^{\mathrm{T}}\right)\left(\mathbf{V}_2\mathbf{V}_3\ldots\mathbf{V}_{k-1}\mathbf{V}_k\right) \\
& +\ldots \\
& +\left(\mathbf{V}_k^{\mathrm{T}}\mathbf{V}_{k-1}^{\mathrm{T}}\right)\left(\boldsymbol{\rho}_{k-2}\mathbf{s}_{k-2}\mathbf{s}_{k-2}^{\mathrm{T}}\right)\left(\mathbf{V}_{k-1}\mathbf{V}_k\right) \\
& +\mathbf{V}_k^{\mathrm{T}}\left(\boldsymbol{\rho}_{k-1}\mathbf{s}_{k-1}\mathbf{s}_{k-1}^{\mathrm{T}}\right)\mathbf{V}_k \\
& +\boldsymbol{\rho}_k\mathbf{s}_k\mathbf{s}_k^{\mathrm{T}}
\end{aligned}
\tag{7.47}
$$

From (7.47), the computation of $\mathbf{D}_{k+1}$ needs the terms $\{\mathbf{s}_i, \mathbf{y}_i\}_{i=0}^{k}$. If we store m terms $\{\mathbf{s}_i, \mathbf{y}_i\}_{i=0}^{m}$, we can compute $\mathbf{D}_{m+1}$. Thus, when computing $\mathbf{D}_{m+2}$, we can abandon some variables, i.e., $\{\mathbf{s}_0, \mathbf{y}_0\}$. We write the general expression with $\widehat{m} = \min\{m, k-1\}$,

$$
\begin{aligned}
\mathbf{D}_{k+1} = & \left(\mathbf{V}_k^{\mathrm{T}}\mathbf{V}_{k-1}^{\mathrm{T}}\cdots\mathbf{V}_{k-\widehat{m}+1}^{\mathrm{T}}\mathbf{V}_{k-\widehat{m}}^{\mathrm{T}}\right)\left(\boldsymbol{\rho}_{k-\widehat{m}-1}\mathbf{s}_{k-\widehat{m}-1}\mathbf{s}_{k-\widehat{m}-1}^{\mathrm{T}}\right)\left(\mathbf{V}_{k-\widehat{m}}\mathbf{V}_{k-\widehat{m}+1}\cdots\mathbf{V}_{k-1}\mathbf{V}_k\right) \\
& +\left(\mathbf{V}_k^{\mathrm{T}}\mathbf{V}_{k-1}^{\mathrm{T}}\cdots\mathbf{V}_{k-\widehat{m}+2}^{\mathrm{T}}\mathbf{V}_{k-\widehat{m}+1}^{\mathrm{T}}\right)\left(\boldsymbol{\rho}_{k-\widehat{m}}\mathbf{s}_{k-\widehat{m}}\mathbf{s}_{k-\widehat{m}}^{\mathrm{T}}\right)\left(\mathbf{V}_{k-\widehat{m}+1}\mathbf{V}_{k-\widehat{m}+2}\cdots\mathbf{V}_{k-1}\mathbf{V}_k\right) \\
& +\left(\mathbf{V}_k^{\mathrm{T}}\mathbf{V}_{k-1}^{\mathrm{T}}\cdots\mathbf{V}_{k-\widehat{m}+3}^{\mathrm{T}}\mathbf{V}_{k-\widehat{m}+2}^{\mathrm{T}}\right)\left(\boldsymbol{\rho}_{k-\widehat{m}+1}\mathbf{s}_{k-\widehat{m}+1}\mathbf{s}_{k-\widehat{m}+1}^{\mathrm{T}}\right)\left(\mathbf{V}_{k-\widehat{m}+2}\mathbf{V}_{k-\widehat{m}+3}\cdots\mathbf{V}_{k-1}\mathbf{V}_k\right) \\
& +\ldots \\
& +\left(\mathbf{V}_k^{\mathrm{T}}\mathbf{V}_{k-1}^{\mathrm{T}}\right)\left(\boldsymbol{\rho}_{k-2}\mathbf{s}_{k-2}\mathbf{s}_{k-2}^{\mathrm{T}}\right)\left(\mathbf{V}_{k-1}\mathbf{V}_k\right) \\
& +\mathbf{V}_k^{\mathrm{T}}\left(\boldsymbol{\rho}_{k-1}\mathbf{s}_{k-1}\mathbf{s}_{k-1}^{\mathrm{T}}\right)\mathbf{V}_k \\
& +\boldsymbol{\rho}_k\mathbf{s}_k\mathbf{s}_k^{\mathrm{T}}
\end{aligned}
\tag{7.48}
$$

Note that $\mathbf{D}_{k+1} \approx \mathbf{H}_{k+1}^{-1}$ as defined in Sect. 7.2.3. Therefore, the overall procedure of the L-BFGS is given by

Algorithm 7.6: L-BFGS Method

Initialization: *Let* $\delta = \begin{cases} 0, & \text{if } k \le m \\ k-m, & \text{if } k > m \end{cases}$, $L = \begin{cases} k, & \text{if } k \le m \\ m, & \text{if } k > m \end{cases}$, *and* $\mathbf{q}_L = \mathbf{g}_k$.

Step 1: *Backward loop*

For *i=L − 1 to 0* ***Do***

$j = i + \delta$

$\alpha_i = \boldsymbol{\rho}_j\mathbf{s}_j^{\mathrm{T}}\mathbf{q}_{i+1}$

$\mathbf{q}_i = \mathbf{q}_{i+1} - \alpha_i\mathbf{y}_j$

End

Step 2: *Forward loop*

Compute $\mathbf{r}_0 = \mathbf{D}_0 \cdot \mathbf{q}_0$.

For *i = 0 to L − 1* ***Do***

$j = i + \delta$

$\beta_i = \boldsymbol{\rho}_j\mathbf{y}_j^{\mathrm{T}}\mathbf{r}_i$

$\mathbf{r}_{i+1} = \mathbf{r}_i - (\alpha_i - \beta_i)\mathbf{s}_j$

End

Step 3: *finally, we have the value of* $\mathbf{r}_L$*which is equal to* $\mathbf{H}_k\mathbf{g}_k$.

7.2.4 Conjugate Gradient Method

Conjugate gradient (CG) method is to find the numerical solution of symmetric and positive-definite systems of linear equations [10]. It is often implemented as an iterative algorithm, which is suitable for the problem with huge size.

Considering the following system of linear equations,

$$\mathbf{A}\mathbf{x} = \mathbf{b} \tag{7.49}$$

where the known $n \times n$ matrix $\mathbf{A}$ is symmetric, positive definite, and real, and the vector $\mathbf{x}$ is to be solved. The unique solution of this system is denoted by $\mathbf{x}_*$.

First, we introduce the concept of conjugacy. Two non-zero vectors $\mathbf{u}$ and $\mathbf{v}$ are conjugate with respect to $\mathbf{A}$ if

$$\mathbf{u}^{\mathrm{T}}\mathbf{A}\mathbf{v} = 0 \tag{7.50}$$

The left hand side of (7.50) defines an inner product since $\mathbf{A}$ is symmetric and positive definite. Since the conjugate property is symmetric, one can see that if $\mathbf{u}$ is conjugate to $\mathbf{v}$, then $\mathbf{v}$ is conjugate to $\mathbf{u}$. Supposing that

$$\mathbf{P} = [\mathbf{p}_1, \mathbf{p}_2, \ldots, \mathbf{p}_n] \tag{7.51}$$

which is a set of n mutually conjugate vectors, $\mathbf{P}$ forms a basis for $\mathbf{R}^n$. One can describe the solution $\mathbf{x}_*$ of $\mathbf{A}\mathbf{x} = \mathbf{b}$ in the following basis:

$$\mathbf{x}_* = \sum_{i=1}^{n} \alpha_i \mathbf{p}_i \tag{7.52}$$

Multiplying $\mathbf{A}$ in both sides gives

$$\mathbf{A}\mathbf{x}_* = \sum_{i=1}^{n} \alpha_i \mathbf{A}\mathbf{p}_i \tag{7.53}$$

Then, one has (7.54) by multiplying $\mathbf{p}_k^T$ in both sides of (7.53), i.e.,

$$\mathbf{p}_k^{\mathrm{T}}\mathbf{A}\mathbf{x}_* = \sum_{i=1}^{n} \alpha_i \mathbf{p}_k^{\mathrm{T}}\mathbf{A}\mathbf{p}_i \tag{7.54}$$

which implies:

$$\alpha_k = \frac{\mathbf{p}_k^{\mathrm{T}}\mathbf{b}}{\mathbf{p}_k^{\mathrm{T}}\mathbf{A}\mathbf{p}_k} \tag{7.55}$$

In the following, we give an iterative procedure to find n conjugate vectors $\{\mathbf{p}_k\}_{k=1}^{n}$, then the coefficients α_k will be computed. We minimize the following quadratic function with respect to $\mathbf{x}$, instead of computing $\mathbf{x}_*$ for (7.49)

$$f(\mathbf{x}) = \frac{1}{2}\mathbf{x}^{\mathrm{T}}\mathbf{A}\mathbf{x} - \mathbf{x}^{\mathrm{T}}\mathbf{b}, \quad \mathbf{x} \in \mathbf{R}^n \tag{7.56}$$

The gradient of $f(\mathbf{x})$ is equal to $\mathbf{Ax} - \mathbf{b}$. First, we give an initial $\mathbf{x}_0$ (e.g., $\mathbf{x}_0 = 0$) for $\mathbf{x}_*$, and the first basis vector $\mathbf{p}_0$ is denoted by the negative of the gradient of $f(\mathbf{x})$ at $\mathbf{x} = \mathbf{x}_0$, meaning that we take $\mathbf{p}_0 = \mathbf{b} - \mathbf{Ax}_0$. The other vectors in the basis will be conjugate to the gradient. Next, in an iterative process, let $\mathbf{r}_k$ denotes the residual at the kth step, i.e.,

$$\mathbf{r}_k = \mathbf{b} - \mathbf{Ax}_k \tag{7.57}$$

Note that $\mathbf{r}_k$ is the negative gradient of $f(\mathbf{x})$ at $\mathbf{x} = \mathbf{x}_k$, so the CG would be to move in the direction $\mathbf{r}_k$. Here, we assume that the directions $\mathbf{p}_k$ be conjugate to each other. Requiring that the next search direction be built out of the current residue and all previous search directions, which is reasonable enough in practice, we give the expression of $\mathbf{p}_k$, i.e.,

$$\mathbf{p}_k = \mathbf{r}_k - \sum_{i<k} \frac{\mathbf{p}_i^{\mathrm{T}}\mathbf{A}\mathbf{r}_k}{\mathbf{p}_i^{T}\mathbf{A}\mathbf{p}_i} \tag{7.58}$$

Guided by this direction, the next optimal location is given by

$$\mathbf{x}_{k+1} = \mathbf{x}_k + \alpha_k \mathbf{p}_k \tag{7.59}$$

with

$$\alpha_k = \frac{\mathbf{p}_k^{\mathrm{T}}\mathbf{b}}{\mathbf{p}_k^{\mathrm{T}}\mathbf{A}\mathbf{p}_k} = \frac{\mathbf{p}_k^{\mathrm{T}}(\mathbf{r}_k + \mathbf{A}\mathbf{x}_k)}{\mathbf{p}_k^{\mathrm{T}}\mathbf{A}\mathbf{p}_k} = \frac{\mathbf{p}_k^{\mathrm{T}}\mathbf{r}_k}{\mathbf{p}_k^{\mathrm{T}}\mathbf{A}\mathbf{p}_k} \tag{7.60}$$

where the last equality holds as $\mathbf{p}_k$ and $\mathbf{x}_k$ are conjugate.

The above equation gives a straightforward explanation of the CG. From these equations, the algorithm requires storage of all previous searching directions and residue vectors, as well as many matrix-vector multiplications. However, we can see that $\mathbf{r}_k$ is orthogonal to $\mathbf{p}_i$ for all $i < k$, $\mathbf{r}_i$ is orthogonal to $\mathbf{r}_j$ for $i \neq j$, therefore, for $i \neq j$ and $i \neq j + 1$, $\mathbf{r}_i$, and $\mathbf{p}_j$ are $\mathbf{A}$-orthogonal. And only $\mathbf{r}_k$, $\mathbf{p}_k$, and $\mathbf{x}_k$ are needed to construct $\mathbf{r}_{k+1}$, $\mathbf{p}_{k+1}$, and $\mathbf{x}_{k+1}$. Furthermore, one needs to perform only one matrix-vector multiplication in each iteration.

The iterative algorithm is described below for solving $\mathbf{Ax} = \mathbf{b}$. The initial input vector $\mathbf{x}_0$ needs to be a suitable initial solution.

Algorithm 7.7: Conjugate Gradient Method

Initialization: *Let* $\mathbf{r}_0 = \mathbf{b} - \mathbf{A}\mathbf{x}_0$, $\mathbf{p}_0 = \mathbf{r}_0$, $k = 0$

Repeat $\alpha_k = \frac{\mathbf{r}_k^{\mathrm{T}}\mathbf{r}_k}{\mathbf{p}_k^{\mathrm{T}}\mathbf{A}\mathbf{p}_k}$

$\mathbf{x}_{k+1} = \mathbf{x}_k + \alpha_k \mathbf{p}_k$

$\mathbf{r}_{k+1} = \mathbf{r}_k - \alpha_k \mathbf{A}\mathbf{p}_k$

If $\mathbf{r}_{k+1}$ *is small enough,* ***then*** *exit loop* $\beta_k = \frac{\mathbf{r}_{k+1}^{\mathrm{T}}\mathbf{r}_{k+1}}{\mathbf{r}_k^{T}\mathbf{r}_k}$

$\mathbf{p}_{k+1} = \mathbf{r}_{k+1} + \beta_k \mathbf{p}_k$

$k = k + 1$

End

The result is $\mathbf{x}_{k+1}$. This is the most commonly used CG algorithm for linear optimization.

Nonlinear Conjugate Gradient Method

Nonlinear conjugate gradient method is a generalization of the CG method for nonlinear optimization. Supposing that a nonlinear function $f(\mathbf{x})$: $R^n \to R$ is continuous and differentiable, and $\mathbf{g}(\mathbf{x}) = [g_1(\mathbf{x}), \ldots, g_i(\mathbf{x}), \ldots, g_p(\mathbf{x})]^{\mathrm{T}}$ is the gradient of $f(\mathbf{x})$ at point $\mathbf{x}$, where $g_i(\mathbf{x}) = \partial f(\mathbf{x})/\partial x_i$, we minimize the nonlinear function $f(\mathbf{x})$ with respect to $\mathbf{x}$, i.e.,

$$\min_{\mathbf{x}} f(\mathbf{x}), \quad \mathbf{x} \in R^n \tag{7.61}$$

Thus, the iteration rule is expressed as

$$\mathbf{x}_{k+1} = \mathbf{x}_k + \alpha_k \mathbf{d}_k \tag{7.62}$$

where the scalar α_k is computed by some line search method. The search direction $\mathbf{d}_k$ is defined as

$$\mathbf{d}_k = \begin{cases} -\mathbf{g}_k, & \mathit{if} \quad k = 0 \\ -\mathbf{g}_k + \beta_k \mathbf{d}_{k-1}, & \mathit{if} \quad k > 0 \end{cases} \tag{7.63}$$

where $\mathbf{g}_k = \mathbf{g}(\mathbf{x}_k)$, and β_k is a parameter.

Different choices of β_k correspond to different nonlinear CG methods. The well-known ones contain the HS method proposed by Hestenes and Stiefel [11], the FR method proposed by Fletcher and Reeves [12], the PRP method reported in [13, 14], the CD method proposed by Fletcher [15], the LS method reported in [16], and the DY method proposed by Dai and Yuan [17]. Their computations of β_k are specified in the following, i.e.,

$$\beta_k^{FR} = \frac{\|\mathbf{g}_k\|^2}{\|\mathbf{g}_{k-1}\|^2} \tag{7.64}$$

$$\beta_k^{PRP} = \frac{\mathbf{g}_k^{\mathrm{T}}\mathbf{y}_{k-1}}{\|\mathbf{g}_{k-1}\|^2} \tag{7.65}$$

$$\beta_k^{HS} = \frac{\mathbf{g}_k^{\mathrm{T}}\mathbf{y}_{k-1}}{\mathbf{d}_{k-1}^{\mathrm{T}}\mathbf{y}_{k-1}} \tag{7.66}$$

$$\beta_k^{CD} = -\frac{\|\mathbf{g}_k\|^2}{\mathbf{d}_{k-1}^{\mathrm{T}}\mathbf{g}_{k-1}} \tag{7.67}$$

$$\beta_k^{DY} = \frac{\|\mathbf{g}_k\|^2}{\mathbf{d}_{k-1}^{\mathrm{T}}\mathbf{y}_{k-1}} \tag{7.68}$$

$$\beta_k^{PRP} = -\frac{\mathbf{g}_k^{\mathrm{T}}\mathbf{y}_{k-1}}{\mathbf{d}_{k-1}^{\mathrm{T}}\mathbf{g}_{k-1}} \tag{7.69}$$

where $\mathbf{y}_{k-1} = \mathbf{g}_k - \mathbf{g}_{k-1}$.

In order to ensure the convergence of the algorithm, we need a line search strategy. Here, some rules for line search are introduced.

1. Accurate line search: Compute α_k that satisfies the following condition:

$$f(\mathbf{x}_k + \alpha_k\mathbf{d}_k) = \min_{\alpha>0} f(\mathbf{x}_k + \alpha\mathbf{d}_k) \tag{7.70}$$

2. Armijo line search: Given $\rho \in (0, 1)$, compute $\alpha_k = \max\ \{\rho^j, j = 0, 1, 2, \ldots\}$ that satisfies

$$f(\mathbf{x}_k + \alpha_k\mathbf{d}_k) \le f(\mathbf{x}_k) + \delta\alpha_k\mathbf{g}_k^{\mathrm{T}}\mathbf{d}_k \tag{7.71}$$

3. Wolfe line search: Compute α_k which satisfies (7.71) and

$$\mathbf{d}_k^{\mathrm{T}}\mathbf{g}(\mathbf{x}_k + \alpha_k\mathbf{d}_k) \ge \sigma\mathbf{d}_k^{\mathrm{T}}\mathbf{g}_k \tag{7.72}$$

   with $0 < \delta \le \sigma < 1$.

4. Strong Wolfe line search: Compute α_k that satisfies (7.72) and

$$\left|\mathbf{d}_k^{\mathrm{T}}\mathbf{g}(\mathbf{x}_k + \alpha_k\mathbf{d}_k)\right| \ge \sigma\left|\mathbf{d}_k^{\mathrm{T}}\mathbf{g}_k\right| \tag{7.73}$$

   with $0 < \delta \le \sigma < 1$.

For theoretical analysis, the accurate line search is often used. However, since it suffers from heavy computational load, in practice one usually uses the Armijo and Wolfe search. One can define that the search direction $\mathbf{d}_k$ is a descent direction when $\mathbf{d}_k^{\mathrm{T}}\mathbf{g}_k < 0$. From the definition of $\mathbf{d}_k$ one has $\mathbf{d}_k^{\mathrm{T}}\mathbf{g}_k = -\|\mathbf{g}_k\|^2 < 0$ using the accurate line search. Therefore, the accurate line search-based nonlinear CG algorithms are all descent algorithms. However, we cannot guarantee that it uses a descent direction in each iteration when employing the inaccurate line search strategies, thus one can treat the negative gradient as the search direction.

7.2.5 Illustration: A Gradient Grid Search Algorithm

To make illustrate of the gradient-based algorithm more clearly, we present a gradient grid search (GGS) algorithm [18] for optimizing the parameters of the LSSVM regression model.

In a LSSVM model, the parameters, including the penalty factor and the parameter of the Gaussian kernel, have to be optimized to produce better prediction performance. Generally, the cross-validation and grid research are usually employed. However, these optimization methods often suffer from high computational load since one has to perform the repeated inverse of matrix and these methods do not have a certain search direction. Thus, we present a gradient grid search algorithm to tackle the aforementioned problems, where the grid search method is combined with the gradient to avoid the blindness of the search.

In Sect. 3.6.3, we introduced the LSSVM regression model in details. Note that the Gaussian kernel function is defined as $k(x_i, x_j) = \exp\left(-\|x_i - x_j\|^2/\sigma\right)$. Before introducing the GGS algorithm, the fast leave-one-out and the gradient of the grid search are introduced as follows. We define

$$\mathbf{A} = \begin{bmatrix} \mathbf{K} + \gamma^{-1}\mathbf{I} & \vec{1} \\ \vec{1}^T & 0 \end{bmatrix}, \quad \mathbf{B} = \begin{bmatrix} \mathbf{y} \\ 0 \end{bmatrix}, \quad \mathbf{s} = \begin{bmatrix} \boldsymbol{\alpha} \\ b \end{bmatrix} \tag{7.74}$$

The coefficients of the LSSVM model for the ith leave-one-out cross-validation are written as

$$\mathbf{s}_i = \mathbf{s}_{i^-} - \frac{s(i)}{\mathbf{A}^{-1}(i,i)} \mathbf{A}^{-1}(i^-, i) \tag{7.75}$$

where $\mathbf{s}_i$ is a vector with length N, $\mathbf{s}_{i^-}$ is the coefficient vector with its ith element omitted in $\mathbf{s}$ for all the training set, $s(i)$ denotes the ith element of $\mathbf{s}$, $\mathbf{A}^{-1}(i^-, i)$ is the vector of ith column of $\mathbf{A}^{-1}$ with its ith element omitted, and $\mathbf{A}^{-1}(i, i)$ denotes the ith element of the diagonal of $\mathbf{A}^{-1}$. $\mathbf{A}^{-1}$ denotes the inverse of $\mathbf{A}$. Defining $\tilde{\mathbf{K}} = [\mathbf{K},\ \ 1]$, the predicted residual sum of squares (PRSEE) is defined by

$$\mathrm{PRSEE} = \sum_{i=1}^{N} \left(\tilde{\mathbf{K}}(i, i^-)\mathbf{s}_i - \mathbf{y}_i\right)^2 \tag{7.76}$$

From (7.76), as for N times running of cross-validation, only one inverse operation of $\mathbf{A}$ is required. This section derives the partial derivatives of PRSEE with respect to the penalty factor γ and the parameter of Gaussian kernel σ, treated as the direction of the GGS. The computations of the derivations are as follows:

Step 1: Compute the partial derivative of $\tilde{\mathbf{K}}$ and $\mathbf{K}_{i,j}$ with respect to σ, i.e.,

$$\frac{\partial \tilde{\mathbf{K}}}{\partial \sigma} = \left[\frac{\partial \mathbf{K}}{\partial \sigma}, \quad 0\right], \quad \frac{\partial \mathbf{K}_{i,j}}{\partial \sigma} = -2\sigma^{-3}\mathbf{K}_{i,j}\left\|\mathbf{x}_i - \mathbf{x}_j\right\|^2 \tag{7.77}$$

Step 2: Compute the partial derivative of $\mathbf{A}^{-1}$ with respect to γ and σ, i.e.,

$$\frac{\partial \mathbf{A}^{-1}}{\partial \gamma} = -\mathbf{A}^{-1}\begin{bmatrix} -\gamma^{-2}\mathbf{I} & 0 \\ 0^{\mathrm{T}} & 0 \end{bmatrix}\mathbf{A}^{-1}, \quad \frac{\partial \mathbf{A}^{-1}}{\partial \sigma} = -\mathbf{A}^{-1}\begin{bmatrix} \partial \mathbf{K}/\partial \sigma & 0 \\ 0^{\mathrm{T}} & 0 \end{bmatrix}\mathbf{A}^{-1} \tag{7.78}$$

Step 3: Compute the partial derivative of $\mathbf{s}$ with respect to γ and σ, i.e.,

$$\frac{\partial \mathbf{s}}{\partial \gamma} = -\mathbf{A}^{-1}\begin{bmatrix} -\gamma^{-2}\mathbf{I} & 0 \\ 0^{\mathrm{T}} & 0 \end{bmatrix}\mathbf{A}^{-1}\mathbf{B}, \quad \frac{\partial \mathbf{s}}{\partial \sigma} = -\mathbf{A}^{-1}\begin{bmatrix} \partial \mathbf{K}/\partial \sigma & 0 \\ 0^{\mathrm{T}} & 0 \end{bmatrix}\mathbf{A}^{-1}\mathbf{B} \tag{7.79}$$

Step 4: Compute the partial derivative of $\mathbf{s}_i$ with respect to γ and σ, i.e.,

$$\frac{\partial \mathbf{s}_i}{\partial \gamma} = \frac{\partial \mathbf{s}}{\partial \gamma}(i^-) - \frac{s(i)\frac{\partial \mathbf{A}^{-1}}{\partial \gamma}(i^-, i) + \frac{\partial \mathbf{s}}{\partial \gamma}(i)\mathbf{A}^{-1}(i^-, i)}{\mathbf{A}^{-1}(i,i)} + \frac{s(i)\mathbf{A}^{-1}(i^-, i)}{\left[\mathbf{A}^{-1}(i,i)\right]^2}\frac{\partial \mathbf{A}^{-1}}{\partial \gamma}(i,i)$$

$$\frac{\partial \mathbf{s}_i}{\partial \sigma} = \frac{\partial \mathbf{s}}{\partial \sigma}(i^-) - \frac{s(i)\frac{\partial \mathbf{A}^{-1}}{\partial \sigma}(i^-, i) + \frac{\partial \mathbf{s}}{\partial \sigma}(i)\mathbf{A}^{-1}(i,i)}{\mathbf{A}^{-1}(i,i)} + \frac{s(i)\mathbf{A}^{-1}(i^-, i)}{\left[\mathbf{A}^{-1}(i,i)\right]^2} \times \frac{\partial \mathbf{A}^{-1}}{\partial \sigma}(i,i) \tag{7.80}$$

Step 5: Compute the partial derivative of PRSEE with respect to γ and σ, i.e.,

$$\frac{\partial \mathrm{PRSEE}}{\partial \gamma} = \sum_{i=1}^{N} 2\left(\tilde{\mathbf{K}}(i, i^-)\mathbf{s}_i - \mathbf{y}_i\right)\frac{\partial \mathbf{s}_i}{\partial \gamma}, \quad \frac{\partial \mathrm{PRSEE}}{\partial \sigma} = \sum_{i=1}^{N} 2\left(\tilde{\mathbf{K}}(i, i^-)\mathbf{s}_i - \mathbf{y}_i\right) \times \left(\frac{\partial \tilde{\mathbf{K}}}{\partial \sigma}(i, i^-) + \tilde{\mathbf{K}}(i, i^-)\frac{\partial \mathbf{s}_i}{\partial \sigma}\right) \tag{7.81}$$

Step 6: Substituting (7.77)–(7.80) into (7.81) produces the gradient direction of the grid search.

Based on the above derivatives, we give the procedure of optimizing the parameters of LSSVM by the GGS algorithm.

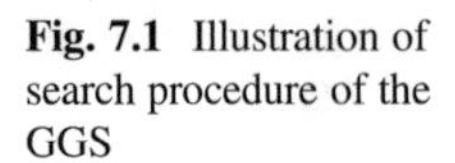

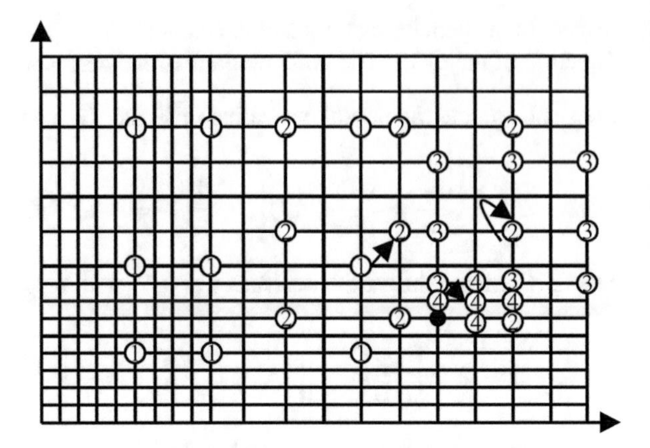

Fig. 7.1 Illustration of search procedure of the GGS

Algorithm 7.8: The GGS Algorithm for Optimizing the Parameters of LSSVM

Step 1: *Make 100 grids with unequal lengths for the parameters γ and σ, treated as the x axil and y axil, respectively, as shown in Fig. 7.1.*

Step 2: *Find out the center point in a two-dimensional plane $(\gamma,\ \sigma)$ constructed in Step 1, as well as find 8 points (marked by ① in Fig. 7.1) around the center point of step length 10.*

Step 3: *Calculate the values of PRSEE for the 9 points mentioned in Step 2 and select the point $(\gamma,\ \sigma)_{\mathrm{minPRSEE}}$ with least value of PRSEE.*

Step 4: *Compute the gradient defined by (7.80) for the point $(\gamma,\ \sigma)_{\mathrm{minPRSEE}}$, which is treated as the gradient direction of the grid. Update the value of $(\gamma,\ \sigma)$ by using the GD algorithm, and choose the nearest points around this updated $(\gamma,\ \sigma)$ as new center points (as illustrated by ② in Fig. 7.1), meanwhile, the step length is subtracted by 1. Then, 8 points are selected around these new center points.*

Step 5: *Repeat Step 3 and 5 until the step length is subtracted to zero. If PRSEE is not small enough, repeat these above steps until the optimal point is found out.*

7.3 Intelligent Optimization Algorithms

Intelligent optimization algorithms are a class of random search algorithms inspired by the biological evolution or the physical phenomena. The theoretical analysis of these algorithms is not developed perfectly, and they cannot ensure that the obtained solutions are always optimal. Hence they are called the "meta-heuristic" methods. While, in practical applications, this class of algorithms can work well when the objective function and constraints are not continuous and convex, even when there is no explicit analytical objective function.

Specially, evolutionary algorithm (EA) is a class of well-known intelligent optimization algorithm, which typically includes genetic algorithm, differential evolution algorithm, etc. These algorithms use mechanisms inspired by biological evolution, such as reproduction, mutation, recombination, and selection, where the candidate solutions to the optimization problem play the role of individuals in a population, and the fitness function determines the quality of the solutions (see also loss function). Evolution of the population is performed after using the above operators repeatedly.

7.3.1 Genetic Algorithm

Genetic algorithm (GA), motivated by evolutionary biology, is an adaptive heuristic search algorithm based on the evolutionary ideas of natural selection and genetics [19]. As such, an intelligent exploitation of a random search is performed to solve optimization problems.

Specifically, in a GA, there are some important concepts, including "*population,*" "*individual,*" "*chromosome,*" "*gene,*" and "*fitness function.*" A population is a collection of individuals which are characterized by their chromosomes. A chromosome consists of genes. The goodness of an individual is measured by its fitness function for the optimization criterion. We consider a general optimization problem,

$$\min f(\mathbf{x}) \tag{7.82}$$

where $\mathbf{x} = [x_1, x_2, \ldots, x_i, \ldots, x_p]$ is a p-dimensional input variables, and $f(\mathbf{x})$ is the objective function to be optimized. Supposing that there are N solutions $\{\mathbf{x}_i\}_{i=1}^N$ for $f(\mathbf{x})$, each solution $\mathbf{x}_i = [x_{i1}, x_{i2}, \ldots, x_{ij} \ldots, x_{ip}]$ must be converted to a binary normal number that only contains 0 and 1. For example, one can suppose that the jth element x_{ij} in $\mathbf{x}_i$ is equal to 10, which will be converted to a binary number with its value "00001010" of eight bits. Thus, a binary form of $\mathbf{x}_i$ will have $8 \times p$ bits, and this binary number serves as a chromosome, in which each bit is a gene. Therefore, the binary form of $\{\mathbf{x}_i\}_{i=1}^N$ constitutes a population. Actually, except for the binary coding scheme, the floating point representation of a chromosome can also be employed, in which each chromosome vector is coded as a vector of floating point numbers, of the same length as the solution vector. Each element is initialized within the desired range, and the operators are carefully designed to preserve this requirement. Figure 7.2 shows an exhibition for gene, chromosome, and population.

A GA starts by generating a random population (the original population). After an initial population is randomly generated, the algorithm proceeds through the following three operators: including the selection operator which chooses the individuals according to their fitness values, the crossover operator which represents mating between individuals, and the mutation operator which introduces random modifications.

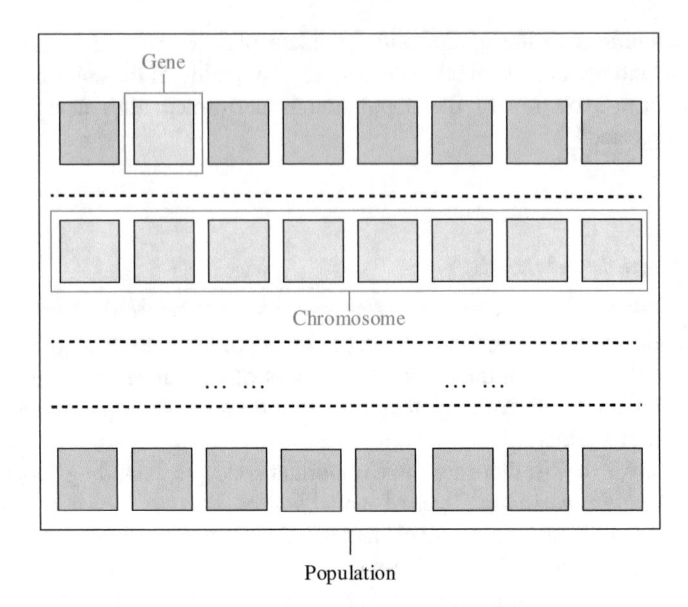

Fig. 7.2 Exhibition for gene, chromosome, and population

Selection Operator

Better individuals have higher opportunity to be selected by the selection operator which allows them to pass on their genes to the next generation. The goodness of each individual depends on the value of its fitness function. The well-known selection schemes include the roulette wheel selection, the stochastic universal selection, the tournament selection, and so forth.

Here, we introduce the roulette wheel selection, whose basic idea is to select an individual with a probability proportional to the value of its fitness function. The detailed procedure is as follows:

Step 1: Compute the value of the fitness function for each individual $\mathbf{x}_i$, $i = 1, 2, \ldots, N$. (For example, one can treat the negative of the objective function $g(\mathbf{x}) = -f(\mathbf{x})$ as the fitness function.)

Step 2: Compute the probability of every individual passing on to the next generation.

$$P(\mathbf{x}_i) = \frac{g(\mathbf{x}_i)}{\sum_{i=1}^{N} g(\mathbf{x}_i)} \tag{7.83}$$

Step 3: Compute the cumulate probability of the ith individual $\mathbf{x}_i$, $i = 1, 2, \ldots, N$ denoted by q_i.

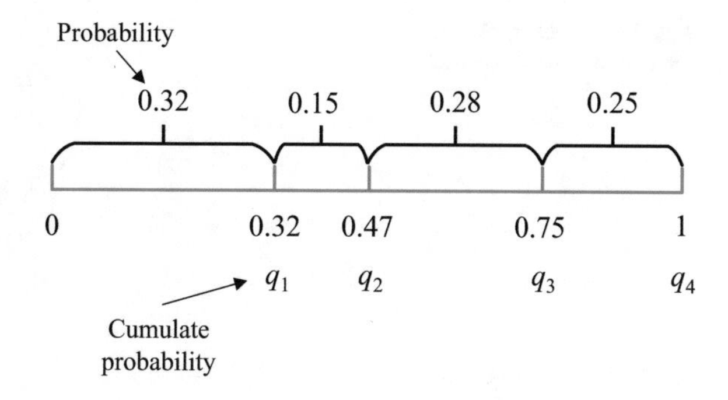

Fig. 7.3 Illustration of the cumulate probability

$$q_i = \sum_{j=1}^{i} P(\mathbf{x}_j) \tag{7.84}$$

For illustration, there is a specific example for the cumulate probability in Fig. 7.3.

Step 4: Generate a random value r ranged in the unit interval with a unique probability.
Step 5: If $r > q_1$, we select $\mathbf{x}_1$ as the individual for the next crossover operator; otherwise, select $\mathbf{x}_k$, if $q_{k-1} \leq r \leq q_k$.
Step 6: Repeat Step 4 and Step 5 until M individuals are selected.

Crossover Operator

Crossover occurs with probability P_{c} when individuals mate, in which the chromosomes of children are generated stochastically from their parents' chromosomes. Note that individuals with higher fitness values are more likely to mate and produce better children. Based on this idea, the commonly used binary valued crossovers contain the single-point crossover, the multiple-point crossover, the uniform crossover, and so forth. Here, for simplicity, Fig. 7.4 illustrates an example for the single-point crossover operator.

Mutation Operator

Mutation occurs when a few genes are randomly changed, e.g., from on to off or vice versa. With a low probability P_{m}, a portion of the new individuals will have some of their bits flipped, whose purpose is to maintain diversity within the population and avoid premature convergence. Mutation induces a random walk through the search space. Figure 7.5 shows an example of the mutation operator.

Fig. 7.4 An example for the single-point crossover

Before performing crossover:

01000|01110000000010000

11100|00000111111000101

After performing crossover:

01000|00000111111000101

11100|01110000000010000

Fig. 7.5 An example for the mutation operator

Before performing mutation:

0100001110000 000010000

After performing mutation:

0100001110000 100010000

The size of a population (number of chromosomes) in a GA does not change during the iterations. To ensure this, some individuals in the parents must be replaced by their offspring so as to create a new generation, where the individuals with higher value of the fitness function have higher opportunity to survive. Thus, we hope that through a number of iterations better solutions will be obtained while the least fit solutions will die out. Finally, an overall procedure for performing a GA is presented in Fig. 7.6.

Although the GA can give the global optimum through a number of iterations of selection, crossover, and mutation, there are still some limitations of this algorithm. Those are: 1) For complex optimization problem (e.g., high-dimensional, multi-modal), the evaluation of the objective function for a population may be very expensive in the computational cost. 2) In a GA, the stop criterion is very hard to be created, and a maximum setting for the number of iterations is often used to stop the algorithm. 3) In many cases, the GA may have a tendency that the local optimum, rather than a global one, is found. That is, it does not "know how" to sacrifice short-term fitness to gain longer term fitness. The possibility of this depends on the shape of the fitness landscape.

To encounter the above problems, adaptive genetic algorithm (AGA) was developed. The AGA is a variant of GA equipped adaptive parameters. The solution accuracy and the convergence speed are greatly determined by the probabilities of crossover (P_c) and mutation (P_m). Instead of using fixed values of P_c and P_m, the AGA utilize the population information in each generation and adaptively adjust the P_c and P_m, in order to maintain the diversity of the population as well as to sustain the convergence capacity. For example, in [20], the adjustment of P_c and P_m depends on the fitness values of the solutions. And in [21], a clustering-based AGA was reported, where through the use of clustering analysis to judge the optimization states of the population, the adjustment of P_c and P_m depends on these optimization states.

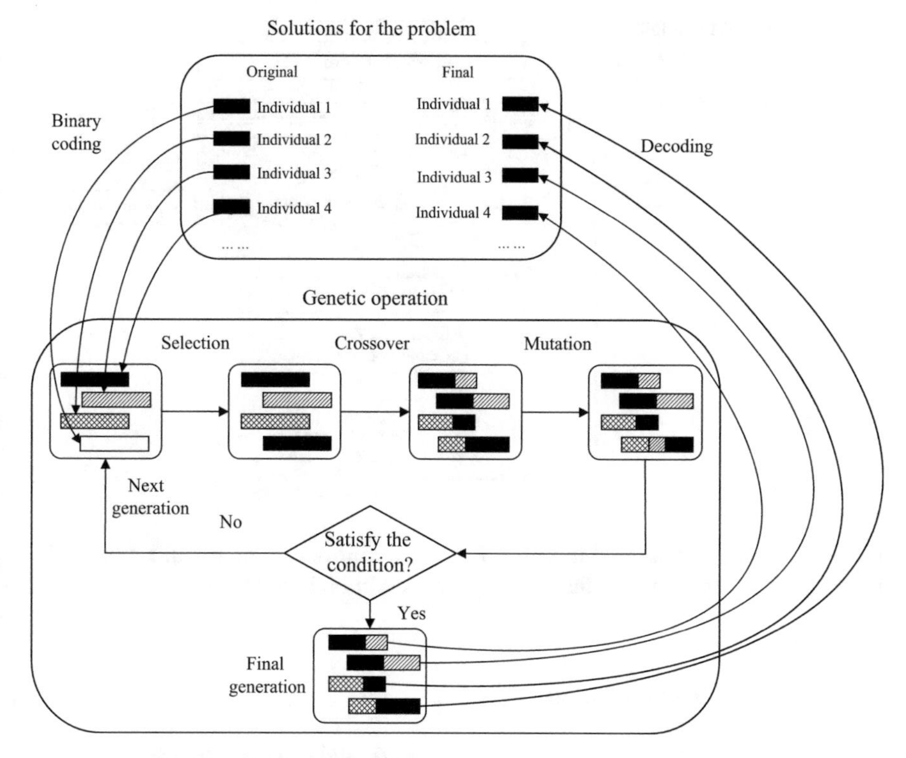

Fig. 7.6 A general description of GA

7.3.2 Differential Evolution Algorithm

Differential evolution (DE) algorithm, proposed by Storn in 1995 [22], is a stochastic optimization algorithm simulating natural evolution, which employs alternative differential operators to replace the classical crossover and mutation schemes of the GA. While it retains the population-based global search procedure, and adopts the float number encoding and difference-based mutation operation, in order to reduce the complexity of the genetic operation. Meanwhile, the search condition can be traced in real time by its setting of particular memory in order to adjust its search strategy; therefore, the DE has higher ability of convergence and robustness than other algorithms. Besides, the DE can solve the complex optimization problems which cannot be handled by the conventional mathematical programming methods without using the characteristic information of the problem. Based on these advantages of the DE, it has been applied in many fields, e.g., signal processing, mechanical design.

In a DE, the notion of the *parameter vectors* denotes the individuals which constitute a population. The differences of the parameter vectors are used to search the objective function space and a cycle of three operators containing crossover, mutation, and selection as the GA does is used to find the optimal solution of the

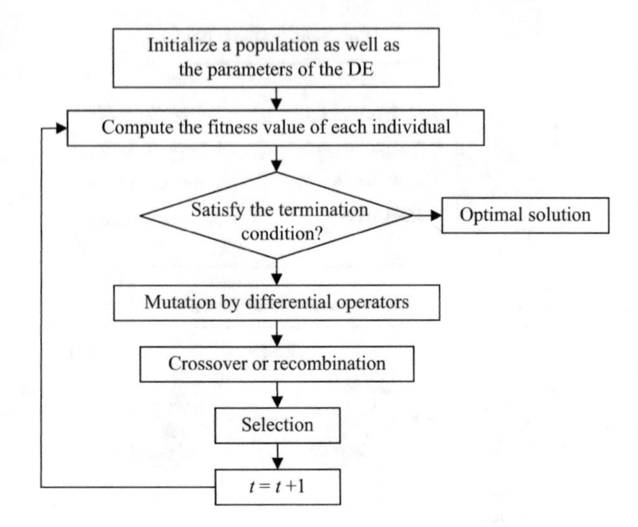

Fig. 7.7 A chart flow of DE

problem, which is presented in Fig. 7.7. First, we introduce the initialization of the parameter vectors, and next the three operators will be introduced.

Initialization of the Population

Now we also consider the optimization problem formulated by (7.82). Similar to the GA, the DE starts the algorithm with a randomly initialized population of N p-dimensional real-valued parameter vectors. Each vector, also known as chromosome, is treated as a candidate solution to the p-dimensional optimization problem.

Since these vector may be changed through the successive iterations, we adopt the following notation for representing the ith vector of the population at iteration t ($t = 0, 1, 2, ..., t, t + 1, ...$).

$$\mathbf{x}_i^t = \left[x_{i1}^t, x_{i2}^t, \ldots, x_{ip}^t\right]^{\mathrm{T}} \tag{7.85}$$

where $i = 1, 2, \ldots, N$, and N is the size of the population. Through the successive iterations these parameter vectors may be changed. Thus, the population is initialized within certain ranges for each parameter randomly. Then, we have a population $\left\{\mathbf{x}_i^t\right\}_{i=1}^N$.

Mutation with Differential Operators

In the evolutionary computing framework, mutation means a sudden change in the gene characteristics of a chromosome, or specifically indicates a random change in a random position of a chromosome. In a DE, this is generated by a predetermined probability density function (PDF) as the GA dose. Particularly, being real-coded, the DE performs the mutation by a differential operator. This differential operator produces mutant vectors for mutation by computing difference of two chromosomes, with the details described in the following. In the mutation stage of a DE, the mutant vector $\mathbf{v}_i^t$ is firstly created for changing each population member $\mathbf{x}_i^t$ in each iteration. To create $\mathbf{v}_i^t$ at iteration t, three parameter vectors, denoted by $\mathbf{x}_{r_1^i}^t$, $\mathbf{x}_{r_2^i}^t$, and $\mathbf{x}_{r_3^i}^t$, are randomly selected from the current population. Different from the base vector index i, the indices r_1^i, r_2^i, and r_3^i are randomly chosen from the range of $[1, N]$, which are also mutually different. Then, we compute the difference between any two of these three parameter vectors, and this difference is multiplied by a scale factor. The mutant vector $\mathbf{v}_i^t$ is obtained by adding the third vector and the scaled difference. Thus, we give the expression of $\mathbf{v}_i^t$,

$$\mathbf{v}_i^t = \mathbf{x}_{r_1^i}^t + F \cdot \left(\mathbf{x}_{r_2^i}^t - \mathbf{x}_{r_3^i}^t\right) \tag{7.86}$$

where F is a real scaling factor which is selected in the range of [0,2] empirically.

Crossover

To increase the diversity of the population, a crossover operation is performed generating the mutant vector. In this operation, a new vector, called the trial vector, is created by exchanging some elements of the mutant vector $\mathbf{v}_i^t$ with its corresponding target vector $\mathbf{x}_i^t$. This trial vector can be denoted by $\mathbf{u}_i^t = \left[u_{i1}^t, u_{i2}^t, \ldots, u_{ip}^t\right]^{\mathrm{T}}$, which is formulated by

$$u_{ij}^t = \begin{cases} v_{ij}^t, & \textit{if}\ (\mathrm{randb}(j) \le \mathrm{CR}) \quad \text{or} \quad j = \mathrm{rnbr}(i) \\ x_{ij}^t, & \textit{if}\ (\mathrm{randb}(j) > \mathrm{CR}) \quad \text{and} \quad j \ne \mathrm{rnbr}(i) \end{cases}, \quad j = 1, 2, \ldots, p \tag{7.87}$$

where $\mathrm{randb}(j)$ is a random number drawn from a uniform distribution for the jth position of the chromosome, $\mathrm{CR} \in (0, 1)$ is the crossover rate which is predetermined by the user (the better choice of CR is a small value [23], since if the value of CR is much larger, the speed of convergence becomes higher, which may lead to the prematurity of the solutions), and $\mathrm{rnbr}(i)$ is an index randomly chosen from $\{1, 2, \ldots, p\}$ which ensures that $\mathbf{u}_i^t$ gets at least one element from $\mathbf{v}_i^t$.

Selection

Based on the results produced by the mutation and crossover operations, the DE proceeds the iteration by the last operation called selection which decides who (the target vector $\mathbf{x}_i^t$ and the trial vector $\mathbf{u}_i^t$) will survive to the next generation. This decision will depend on the value of the fitness function of $\mathbf{x}_i^t$ and $\mathbf{u}_i^t$. If the trial vector $\mathbf{u}_i^t$ yields a better fitness value, it will replace the target vector; otherwise, the target vector $\mathbf{x}_i^t$ will be passed to the next iteration. The selection operation is outlined as

$$\mathbf{x}_i^{t+1} = \begin{cases} \mathbf{u}_i^t, & \text{if} \quad f(\mathbf{u}_i^t) \leq f(\mathbf{x}_i^t) \\ \mathbf{x}_i^t, & \text{if} \quad f(\mathbf{u}_i^t) > f(\mathbf{x}_i^t) \end{cases} \tag{7.88}$$

where $f(\mathbf{x})$ is the objective function defined by (7.82). This operation can ensure that the population size remains fixed through generations since it employs a binary selection, i.e., only one between the target vector and its offspring survives. Moreover, through this operation the quality of the population either improves over generations or remains unchanged, but never degrades. An overall procedure for the DE algorithm is listed as below.

Algorithm 7.9: DE Algorithm

Input: Population: N; Dimension of the variables: p; Maximum generation: T

Output: The best solution: $\mathbf{x}_{\text{best}}$

Initialization: $t \leftarrow 1$

For $i = 1$ to N ***do***

For $j = 1$ to p ***do***

$x_{ij}^0 = x_{j,\min} + \text{rand}(0,1) \cdot (x_{j,\max} - x_{j,\min})$

End

End

Main loop:

While $(|f(\mathbf{x}_{\text{best}})| \geq \varepsilon)\text{or}(t \leq T)$ ***do***

For $i = 1$ to N ***do***

(Mutation and Crossover)

For $j = 1$ to p ***do***

v_{ij}^t=*Mutation* (x_{ij}^t)

u_{ij}^t=*Crossover* (x_{ij}^t, v_{ij}^t)

End

(Greedy Selection)

If $f(\mathbf{u}_i^t) \leq f(\mathbf{x}_i^t)$ ***then***

$\mathbf{x}_i^t \leftarrow \mathbf{u}_i^t$

If $f(\mathbf{x}_i^t) \leq f(\mathbf{x}_{\text{best}})$ ***then***

$\mathbf{x}_{\text{best}} \leftarrow \mathbf{x}_i^t$

End

Else

$\mathbf{x}_i^{t+1} \leftarrow \mathbf{x}_i^t$

End

End

$t \leftarrow t + 1$

End

In the DE, the terminating condition can be defined in two ways: 1) Define a fixed number of iterations T whose value depends on the complexity of the objective function. 2) When the value of best fitness does not change appreciably over successive iterations.

In addition, to improve the performance of the standard DE, there are also many variants based on this algorithm, see [23]. The popular way is to adapt the DE parameters (including the scale parameter F and the crossover rate (CR)) during optimization.

As for the selected three parameters $\mathbf{x}_{r_1^i}^t$, $\mathbf{x}_{r_2^i}^t$, and $\mathbf{x}_{r_3^i}^t$, we sort them according to their quality from the best to worst, denoted by $\mathbf{x}_{\mathrm{b}}^t$, $\mathbf{x}_{\mathrm{m}}^t$, and $\mathbf{x}_{\mathrm{w}}^t$. Then, the mutation operator is changed into

$$\mathbf{v}_i^t = \mathbf{x}_{\mathrm{b}}^t + F_i \cdot \left(\mathbf{x}_{\mathrm{m}}^t - \mathbf{x}_{\mathrm{w}}^t\right) \tag{7.89}$$

The adaptive factor F_i changes according to difference of the above three selected solutions, which balances the global search and the local search.

$$F_i = F_{\mathrm{l}} + (F_{\mathrm{u}} - F_{\mathrm{l}})\frac{f\left(\mathbf{x}_{\mathrm{m}}^t\right) - f\left(\mathbf{x}_{\mathrm{b}}^t\right)}{f\left(\mathbf{x}_{\mathrm{w}}^t\right) - f\left(\mathbf{x}_{\mathrm{b}}^t\right)} \tag{7.90}$$

where F_{l} is equal to 0.1 and F_{u} is equal to 0.9.

Besides, for the selection of CR, it can also be updated during each iteration. The following is the updating formula for CR.

$$\mathrm{CR}_i = \begin{cases} \mathrm{CR}_{\mathrm{l}} + (\mathrm{CR}_{\mathrm{u}} - \mathrm{CR}_{\mathrm{l}})\dfrac{f\left(\mathbf{x}_i^t\right) - f_{\min}}{f_{\max} - f_{\min}}, & \text{if } f\left(\mathbf{x}_i^t\right) > \bar{f} \\ \mathrm{CR}_{\mathrm{l}} & \text{if } f\left(\mathbf{x}_i^t\right) < \bar{f} \end{cases} \tag{7.91}$$

where $f_{\min}$ and $f_{\max}$ are the least and the highest fitness in the current population, respectively, and $\bar{f}$ is the mean value of the fitness among the population. Besides, CR_{l} and CR_{u} are the lower and upper bound of CR.

7.3.3 *Particle Swarm Optimization Algorithm*

Particle swarm optimization (PSO) is an evolutionary computation algorithm proposed by Kennedy and Eberhart [24]. The concept of the particle swarm was used as a simulation of a simplified social system originally; whose original purpose was to graphically simulate the unpredictable choreography of a bird flock. Preliminary simulations were changed to incorporate nearest-neighbor velocity matching, eliminate the ancillary variables, and combine with multi-dimensional search and acceleration by distance [24, 25]. With the development of this algorithm, it became an optimizer in fact. Through a process of trial and error, many parameters extraneous to optimization were eliminated from the algorithm, resulting in the very simple original implementation. PSO is similar to a GA in that it is initialized with a population of random solutions. However, the difference from a GA lies in that each potential solution is also assigned a randomized velocity, and these potential solutions, called particles, can "move" through the problem space.

Assume that the optimization problem has n variables. PSO begins with a random initialization of a population (swarm) of individuals (particles) in the n-dimensional search space. The particles move through the search space with adaptive velocities, and each particle keeps two values in its memory, including its own best experience and the best experience of the whole swarm. The former is the one with the best fitness value (best fitness value corresponds to least objective value since fitness function is conversely proportional to objective function), whose position and objective value are called $\mathbf{p}_i = [p_{i1}, p_{i2}, \ldots, p_{id}, \ldots, p_{in}]$ and p_{best}, respectively, and the position and objective value of the best experience of the whole swarm are denoted by $\mathbf{p}_g = [p_{g1}, p_{g2}, \ldots, p_{gd}, \ldots, p_{gn}]$ and g_{best}, respectively. The position and velocity of particle i are given by the following vectors:

$$\mathbf{x}_i = [x_{i1}, x_{i2}, \ldots, x_{id}, \ldots, x_{in}] \tag{7.92}$$

$$\mathbf{v}_i = [v_{i1}, v_{i2}, \ldots, v_{id}, \ldots, v_{in}] \tag{7.93}$$

One can update the velocities and positions of particles in each time step by the following equations:

$$v_{id}(t+1) = v_{id}(t) + c_1 r_{1d}(p_{id} - x_{id}) + c_2 r_{2d}(p_{gd} - x_{id}) \tag{7.94}$$

$$x_{id}(t+1) = x_{id}(t) + v_{id}(t+1) \tag{7.95}$$

where c_1 and c_2 are two positive numbers, and r_{1d} and r_{2d} are two random numbers with uniform distribution in the interval of [0,1]. Besides, as for the right hand side of the velocity update (7.94), the first term $v_{id}(t)$ denotes the tendency of a particle to remain in the same direction it has traversing and it is called "inertia," "habit," or "momentum." The second term $c_1 r_{1d}(p_{id} - x_{id})$ is a linear attraction toward the particle's own best experience scaled by a random weight $c_1 r_{1d}$. This term is called "memory," "nostalgia," or "self-knowledge." The third term $c_2 r_{2d}(p_{gd} - x_{id})$ is called "cooperation," "shared information," or "social knowledge," which is a linear attraction toward the best experience of the all particles in the swarm, scaled by a random weight $c_2 r_{2d}$. Next, we give the procedure for implementation of PSO.

Algorithm 7.10: PSO Algorithm

Initialization: *Initialize the particles' velocities and positions randomly and calculate the objective value of all particles. The position and objective of each particle are set as its* $\mathbf{p}_i$ *and* p_{best}, *respectively, and also the position and objective of the particle with the best fitness (least objective) are set as* $\mathbf{p}_g$ *and* g_{best}, *respectively.*

Step 1: *Update the particles' velocities and positions according to (7.94) and (7.95).*

Step 2: *Update each particle's* $\mathbf{p}_i$ *and* p_{best}. *Specifically, if the current fitness of the particle is better than its* p_{best}, p_{best} *and* $\mathbf{p}_i$ *are replaced with current objective value and position vector, respectively.*

Step 3: *Update* $\mathbf{p}_g$ *and* g_{best}. *Specifically, if the current best fitness of the whole swarm is fitter than* g_{best}, g_{best} *and* $\mathbf{p}_g$ *are replaced with current best objective and its corresponding position vector, respectively.*

Step 4: *Repeat Step 1 to Step 3 until stopping criterion (usually a pre-specified number of iterations or a quality threshold for objective value) reaches.*

It is noticeable that the particles may become uncontrollable or exceed search space since the velocity update equations are stochastic and the velocities may become too high. Therefore, velocities are limited to a maximum value $v_{\max}$ [26], i.e.,

$$\textit{if} \quad |v_{id}| > v_{\max} \quad \text{then} \quad v_{id} = \text{sign}(v_{id})v_{\max} \tag{7.96}$$

where *sign* represents sign function. $v_{\max}$ is an important parameter in PSO. It determines the resolution, or fineness, with which regions between the present position and the target (best-so-far) position are searched. If $v_{\max}$ is too high, particles might move past good solutions. If $v_{\max}$ is too small, on the contrast, particles may not explore sufficiently beyond locally good regions. In fact, they could become trapped in local optima, unable to move far enough to reach a better position in the problem space. We often set it at about 10–20% of the dynamic range of the variable on each dimension.

However, this primary PSO, characterized by (7.94) and (7.95), often does not work well, since there is no strategy of adjusting the trade-off between explorative and exploitative capabilities. Thereby, the inertia weight PSO is introduced to alleviate this drawback [27, 28]. In the inertia weight PSO, which is the most commonly used variant for PSO, the velocities of particles in previous time step is multiplied by a parameter called inertia weight. The corresponding velocity update equations are given by,

$$v_{id}(t+1) = w \cdot v_{id}(t) + c_1 r_{1d}(p_{id} - x_{id}) + c_2 r_{2d}(p_{gd} - x_{id}) \tag{7.97}$$

where inertia weight w adjusts the trade-off between exploration and exploitation capabilities of PSO. The less the inertia weight is, the more the exploration capability of PSO will be and vice versa. Commonly, it is decreased linearly during the algorithm, thus, at initial stages the search procedure is mainly focused on exploration and at latter stages it is focused more on exploitation.

The generic PSO is only designed for continuous optimization problems. Here, we introduce some variants of PSO for addressing discrete optimization ones.

The round off-based PSO approach is the most commonly used method for discrete optimization [23]. The discrete variables in this approach are treated the same as continuous variables during the optimization process, while at the end of this optimization they are rounded off to the nearest discrete value. In practice, the experimental results show that rounding off does not affect the accuracy of the optimization so much. The main advantages of this approach lies in its simplicity and low computational load.

The penalty approach, reported in [29–31], is another method for discrete optimization. In this approach, discrete variables are regarded as continuous ones by penalizing points existent at intervals. Here, we give the penalty function [30, 31].

$$P(\mathbf{x}) = \sum_{i=1}^{n_d} \frac{1}{2}\left[\sin \frac{2\pi\{c_i - 0.25(d_{i,j+1} + 3d_{i,j})\}}{d_{i,j+1} - d_{i,j}} + 1\right] \tag{7.98}$$

where $d_{i,j}$ and $d_{i,j+1}$ are the jth and $(j+1)$th discrete values of ith discrete variable, c_i denotes the continuous value between them and n_{d} represents the number of discrete variables. From (7.98), one can see that around the discrete values the penalty value is small, but it becomes large at regions away from discrete values. Thus, the augmented objective function is constructed as follows:

$$\varphi(\mathbf{x}) = f(\mathbf{x}) + r \cdot P(\mathbf{x}) \tag{7.99}$$

where r represents the penalty parameter, and $f(\mathbf{x})$ is the objective function to be minimized in a generic PSO. First, the penalty parameter r for particle i ($i = 1,2,\ldots, N$) is computed as

$$r_i = 1 + P(\mathbf{x}_i) \tag{7.100}$$

Then, initial value of r denoted by r_{initial} is determined as:

$$r_{\mathrm{initial}} = \min_i \{r_1, r_2, \ldots, r_i, \ldots, r_N\} \tag{7.101}$$

During the optimization process, at each iteration t, r is updated by

$$r^{t+1} = r^t \cdot \exp\left(1 + P\left(\mathbf{p}_g^t\right)\right) \tag{7.102}$$

where $\mathbf{p}_g^t$ denotes the best location of the whole swarm at iteration t. However, the penalty approach suffers from heavy computational load which extremely limits its application in complex optimization problems.

Besides, the set-based approach [32, 33], the binary coding-based approach [34, 35], and the approaches based on modification of flight equations [36] are other approaches for tackling discrete optimization problems.

7.3.4 Simulated Annealing Algorithm

Simulated annealing (SA) algorithm, proposed by S. Kirkpatrick in 1983 [37] and V. Černý in 1985 [38], is a probability-based optimization algorithm to search for the optimal solution of an optimization problem. The idea of SA originates from the terminology of annealing in metallurgy, a technique involving heating and controlled slow cooling of a solid material in order to increase the size of its crystals and reduce their defects, where heating and cooling the material will affect both its temperature and thermodynamic free energy. In the annealing process, the temperature and the free energy will be reduced to a certain stable value. For more information, one can refer to survey articles that provide a good overview of simulated annealing's theoretical development and domains of application [39–41].

The SA algorithm simulates the annealing process in which the objective function is treated as the free energy function. The slow cooling scheme implemented in the SA is interpreted as a slow decrease of the temperature with a probability of accepting worse solutions when the solution space is explored (accepting worse solutions is a necessary strategy since it allows for a more extensive search for the optimal solution). Simple heuristics-based algorithms like hill climbing, which move by finding better neighbor after better neighbor and stop when they have reached a solution having no neighbors that are better solutions, cannot guarantee that the obtained best solution is a global optimum. However, meta-heuristics like the SA explore the solutions space by means of producing the neighbors of a solution, and although they prefer better neighbors, they also accept worse neighbors with some probability in order to avoid getting stuck in a local optimum. Thus, the meta-heuristics-based methods have higher possibility of finding the global optimum if run for large number of iterations.

For specific optimization, let us suppose that the optimization problem to be solved is that minimization of $f(\mathbf{x})$ with respect to parameter vector $\mathbf{x}$, and p is the dimensionality of $\mathbf{x}$. Each possible solution of $f(\mathbf{x})$ is denoted by a corresponding state $\mathbf{s}$ with dimensionality p.

For the initialization of the SA, an initial solution (namely state $\mathbf{s}$) is produced randomly. Then, at each step of the SA, one produces some neighboring state $\mathbf{s}'$ of the current state $\mathbf{s}$, and probabilistically determines between moving the system to state $\mathbf{s}'$ and staying in state $\mathbf{s}$. These actions ultimately lead the system to move to states of lower energy (lower value of the objective function). In practice, this step is repeated many times until the system reaches a state that is good enough for the problem, or until a limit of the number of the iterations reaches.

Actually, the above step for the state transition is involved in the annealing scheme. This scheme makes a gradual reduction of the temperature as the simulation proceeds. The initial temperature T is set to a very large value with a lower bound of the temperature $T_{\min}$ being defined initially at the same time. Then at each iteration, it is decreased following some annealing schedule specified by the user, which must end when $T < T_{\min}$ holds. In this way, the system is expected to wander initially toward a broad region of the search space containing good solutions; then drift toward low-energy regions that become narrower and narrower; and finally move to the global optimum. To make the readers more clearly, we give the specific procedure of the SA algorithm in the following:

Algorithm 7.11: SA Algorithm

Initialization: *Initialize the temperature T with a large value and a lower bound of T ($T_{\min}$) with a very small value, initialize a solution vector* **s**, *and fix the number of iteration L.*

While $T > T_{\min}$ ***do***

For $i = 1$ to L ***do***

Step 1: *Produce a neighboring solution* $\mathbf{s}'$ *from* **s**.

Step 2: *Compute the difference* $\Delta f = f(\mathbf{s}') - f(\mathbf{s})$.

Step 3: *If* $\Delta f < 0$, *accept* $\mathbf{s}'$ *as the new current solution* **s**; *otherwise, accept* $\mathbf{s}'$ *with probability* $P(\mathbf{s}', \mathbf{s}, T)$.

End

Step 4: *Terminate the process if the termination condition satisfies, and produce the current solution as the optimum; otherwise, decrease the temperature T and continue the 'While' loop.*

End

To apply the SA to a specific optimization problem, one must specify the following parameters: the state-space, the objective function, the candidate generator procedure, the acceptance probability function, and the annealing schedule and the initial temperature. The choices of these parameters will impose significant impacts on the method's performance. In the following, we will introduce more details of the SA algorithm, including the settings of the parameters of the SA, which makes this algorithm more clearly.

In the above SA, the new state $\mathbf{s}'$ is produced from the current state **s**. The well-defined way to produce neighboring state is called a "move" ($\mathbf{s}' = \mathbf{s} + \Delta\mathbf{s}$), and different moves give different sets of neighboring states. These moves usually result in minimal alterations of the last state, in an attempt to progressively improve the solution through iteratively improving its parts as described in the SA algorithm.

As for the *acceptance probability function* $P(\mathbf{s}', \mathbf{s}, T)$, it indicates the probability of making a transition from the current state **s** to a candidate new state $\mathbf{s}'$. State with a smaller value of the objective function is better than those with a greater value. The probability function $P(\mathbf{s}', \mathbf{s}, T)$ must be positive even when $f(\mathbf{s}')$ is greater than $f(\mathbf{s})$. This feature prevents the method from becoming stuck at a local minimum. One must choose the $P(\mathbf{s}', \mathbf{s}, T)$ following the characteristic that the probability of accepting a transition decreases when the difference Δf increases. Besides, given these properties, the temperature T plays an important role in steering the evolution of the state **s** of the system with regard to its sensitivity to the variations of objective function value. Specifically, for a large T, the evolution of **s** is sensitive to coarse variations, while it is sensitive to finer variations when T is small. Thus, one can see that the probability function $P(\mathbf{s}', \mathbf{s}, T)$ depends on the objective function $f(\mathbf{s})$ and $f(\mathbf{s}')$ of the two states, and on the global time-varying temperature T. Here, we give a commonly used definition of $P(\mathbf{s}', \mathbf{s}, T)$,

$$P(\mathbf{s}', \mathbf{s}, T) = \exp\left(-\frac{f(\mathbf{s}') - f(\mathbf{s})}{k \cdot T}\right) \tag{7.103}$$

where k is a predefined positive integer.

The cooling scheme of the temperature T is also very significant to the SA algorithm, which has great impact on the effectiveness of the SA, thus, it must be carefully designed. In practice, the specific form of the cooling is often given by $T_{\text{new}} = \alpha \cdot T$, where $\alpha \in (0, 1)$, and for a larger search space, α can be set to a value around 1, e.g., 0.9 or 0.95.

7.4 Nonlinear Kalman-Filter Estimation

For a dynamical system, one not only considers the relationship between the input and output variables at the same time slice, but also is interested in the relationship between different time slices. Besides, there are several latent state variables involved in the dynamical system. To estimate these latent state variables, the Kalman-filter estimation method is employed.

To illustrate nonlinear Kalman-filter, one can consider a typical nonlinear system as the state-space model whose equations can be written as

$$\begin{aligned} \mathbf{x}_k &= f(\mathbf{x}_{k-1}, \mathbf{u}_{k-1}, \mathbf{w}) + \mathbf{q}_{k-1} \\ \mathbf{t}_k &= h(\mathbf{x}_k, \mathbf{u}_k, \mathbf{w}) + \boldsymbol{\gamma}_k \end{aligned} \tag{7.104}$$

where f and h are known nonlinear functions. $\mathbf{u}_k \in \mathbb{R}^m$ is a known input at time k. $\mathbf{x}_k \in \mathbb{R}^N$ is the states to be estimated. $\mathbf{t}_k \in \mathbb{R}^l$ is the observed targets. $\mathbf{q}_{k-1}$ and $\boldsymbol{\gamma}_k$ are uncorrelated process and measurement Gaussian noise with mean zero and covariance $\mathbf{R}_{k-1}^q$ and $\mathbf{R}_k^\gamma$, respectively.

The state-space model of the Kalman-filtering problem embodies two equations, namely the process equation and the measurement equation. The process equation describes the temporal evolution of the hidden state, whereas the measurement equation maps the hidden states to the measurements. Kalman-filter devotes itself to using the feedback control technique to estimate the process states. The objective of nonlinear optimal filter is to recursively compute the posterior density function $p(\widehat{\mathbf{x}}_k|D_k)$ when the history of the inputs and measurements $D_k = \{(\mathbf{t}_i, \mathbf{u}_i)\}_{i=1}^k$ are available before time k. The posterior density provides a complete statistical description of the states at that time and is computed by two basic update steps, namely the time update and the measurement update. The main task of the time-update equations is to construct the priori distribution of the process state in next moment, i.e., using the current state and its covariance matrix to forward calculate the priori distribution of the next state, of the form

$$\begin{aligned} p(\widehat{\mathbf{x}}_{k|k-1}|D_{k-1}) &= \int_{\mathbb{R}^{n_{\mathbf{x}}}} p(\mathbf{x}_{k|k-1}, \mathbf{x}_{k-1}|D_{k-1}) d\mathbf{x}_{k-1} \\ &= \int_{\mathbb{R}^{n_{\mathbf{x}}}} p(\mathbf{x}_{k-1}|D_{k-1}) p(\mathbf{x}_{k|k-1}|\mathbf{x}_{k-1}, \mathbf{u}_{k-1}) d\mathbf{x}_{k-1} \end{aligned} \tag{7.105}$$

where $p(\mathbf{x}_{k-1}|D_{k-1})$ is the old posterior density at time (k − 1), and $p(\mathbf{x}_{k|k-1}|\mathbf{x}_{k-1}, \mathbf{u}_{k-1})$ is the state transition density function.

The measurement update is to construct the posterior distribution over the states at time k based on the priori estimation and the feedback information coming from the noisy measurements. Theoretically, the Kalman-filter is one special case of Bayesian rule, so the nonlinear Kalman-filter-based parameters estimation is a kind of optimal Bayesian filter. According to Bayes' rule, the predictive density is updated and the posterior density function of the current state is obtained as follows:

$$\begin{aligned} p(\widehat{\mathbf{x}}_k|D_k) &= p(\widehat{\mathbf{x}}_k|D_{k-1}, \mathbf{u}_k, t_k) \\ &= \frac{1}{c_k} p(\widehat{\mathbf{x}}_{k|k-1}|D_{k-1}, \mathbf{u}_k) p(t_k|\mathbf{x}_{k|k-1}, \mathbf{u}_k) \end{aligned} \tag{7.106}$$

where $p(t_k|\mathbf{x}_{k|k-1}, \mathbf{u}_k)$ is the likelihood function and c_k is the normalizing constant that can be written as

$$\begin{aligned} c_k &= p(t_k|D_{k-1}, \mathbf{u}_k) \\ &= \int p(\widehat{\mathbf{x}}_{k|k-1}|D_{k-1}, \mathbf{u}_k) p(t_k|\mathbf{x}_{k|k-1}, \mathbf{u}_k) d\mathbf{x}_{k|k-1} \end{aligned} \tag{7.107}$$

The typical cases of nonlinear Bayesian filter include extended Kalman-filter (EKF), unscented Kalman-filter (UKF) and cubature Kalman-filter (CKF). As follows, we will introduce the principles and specific implementation of these three filters.

7.4.1 Extended Kalman-Filter

For nonlinear system, EKF provides approximate maximum likelihood estimates [42]. The mean and covariance of the state are again recursively updated; however, a first-order linearization of the dynamics is necessary in order to analytically propagate the Gaussian random-variable representation. Effectively, the nonlinear dynamics are approximated by a time-varying linear system, and the linear Kalman-filter equations are applied. The standard EKF remains the most popular approach owing to its simplicity. As follows, we will introduce the standard EKF for state estimation and weights estimation.

Assuming that the parameters in (7.104) are known, when estimating the internal states with EKF, (7.104) has to be linearized by using the first-order Taylor series expansion, that is

$$\mathbf{x}_k \approx \tilde{\mathbf{x}}_k + \mathbf{F}_{\mathbf{x}}(\tilde{\mathbf{x}}_{k-1}, \mathbf{u}_k)(\mathbf{x}_{k-1} - \widehat{\mathbf{x}}_{k-1}) \tag{7.108}$$

where $\widehat{\mathbf{x}}_{k-1}$ is the posterior estimation in time step $k-1$. $\tilde{\mathbf{x}}_k$ can be obtained from (7.109). $\mathbf{F}_{\mathbf{x}}(\tilde{\mathbf{x}}_{k-1}, \mathbf{u}_k)$ is the Jacobian matrix of the partial derivative of f with respect to $\mathbf{x}$, i.e.,

$$\tilde{\mathbf{x}}_k = f\left(\mathbf{u}_k, \widehat{\mathbf{x}}_{k-1}, \mathbf{w}\right) \tag{7.109}$$

$$\left[\mathbf{F}_{\mathbf{x}}\left(\widehat{\mathbf{x}}_{k-1}, \mathbf{u}_k\right)\right]_{i,j} = \left.\frac{\partial f_i(\mathbf{x}, \mathbf{u}_k)}{\partial \mathbf{x}_j}\right|_{\mathbf{x}=\widehat{\mathbf{x}}_{k-1}} \tag{7.110}$$

As such, the nature of EKF can also be viewed as a kind of feedback control. And, the equations in EKF are divided into two parts, including the time-update equations and the measurement-update equations. The time-update equations are responsible for obtaining the priori estimates in the next time step, in which the priori $\widehat{\mathbf{x}}_{k|k-1} = \tilde{\mathbf{x}}_k$ is assigned with covariance $\mathbf{P}_{\mathbf{x}_{k|k-1}}$. Assume that the original conditions of this estimation process are $\widehat{\mathbf{x}}_0 = E[\mathbf{x}_0]$ and $\mathbf{P}_{\mathbf{x}_0} = E\left[\left(\mathbf{x}_0 - \widehat{\mathbf{x}}_0\right)\left(\mathbf{x}_0 - \widehat{\mathbf{x}}_0\right)^{\mathrm{T}}\right]$, and then the time update equations in EKF are

$$\begin{aligned} \widehat{\mathbf{x}}_{k|k-1} &= f\left(\widehat{\mathbf{x}}_{k-1}, \mathbf{u}_k, \mathbf{w}\right) \\ \mathbf{P}_{\mathbf{x}_{k|k-1}} &= \mathbf{A}_{k-1}\mathbf{P}_{\mathbf{x}_{k-1}}\mathbf{A}_{k-1}^T + \mathbf{R}_{k-1}^q \end{aligned} \tag{7.111}$$

where $k \in \{1, 2, \cdots, \infty\}$, and $\mathbf{A}_k \triangleq \left.\frac{\partial f(\mathbf{x}, \mathbf{u}_k, \mathbf{w})}{\partial \mathbf{x}}\right|_{\widehat{\mathbf{x}}_k}$.

The measurement-update equations are responsible for estimating the posterior distributions through correcting the priori based on the current noisy measurement $\mathbf{t}_k$.

$$\begin{aligned} \widehat{\mathbf{x}}_k &= \widehat{\mathbf{x}}_{k|k-1} + \mathcal{K}_k^{\mathbf{x}}\left(\mathbf{t}_k - h\left[\mathbf{u}_k, \widehat{\mathbf{x}}_{k|k-1}\right]\right) \\ \mathbf{P}_{\mathbf{x}_k} &= \left(\mathbf{I} - \mathcal{K}_k^{\mathbf{x}}\mathbf{H}_k\right)\mathbf{P}_{\mathbf{x}_{k|k-1}} \end{aligned} \tag{7.112}$$

where $\mathcal{K}_k^{\mathbf{x}} = \mathbf{P}_{\mathbf{x}_{k|k-1}}\mathbf{H}_k^{\mathrm{T}}\left(\mathbf{H}_k\mathbf{P}_{\mathbf{x}_{k|k-1}}\mathbf{H}_k^{\mathrm{T}} + \mathbf{R}_k^{\gamma}\right)^{-1}$ is the Kalman gain, and $\mathbf{H}_k \triangleq \left.\frac{\partial h(\mathbf{x}, \mathbf{u}_k, \mathbf{w})}{\partial \mathbf{x}}\right|_{\widehat{\mathbf{x}}_k}$. $\mathbf{R}_k^q$ and $\mathbf{R}_k^{\gamma}$ are the covariance of q_k and γ_k, respectively.

The first-order EKF may be biased in highly nonlinear system, which is its obvious disadvantage. As a solution to alleviate the biased error, the second-order EKF was reported in [43], however, it requires the computation of Jacobians and Hessians, which is the procedure that are often numerically unstable, and computationally intensive. Particularly, the Hessian turns out to be a three dimensional matrix with its entries being second-order partial derivatives. In some systems, they do not even exist, e.g., models representing abruptly changing behavior. In practice, the second-order EKF is not commonly used, and higher order approximations are almost never used. Due to the obvious drawbacks of the second-order EKF, a kind of central difference Kalman-filter (CDKF) is developed [44], in which the nonlinear functions of the dynamic system are also approximated by the second-order Taylor series expansion. Differently, its Jacobians and Hessians are replaced by central difference approximations. Totally, the basic principles of the first-order EKF, the second-order EKF and the CDKF are very similar, so the specific implementation process of the second-order EKF and the CDKF will not be described here.

7.4.2 Unscented Kalman-Filter

In this subsection, we first summarize the unscented transformation. The intuition of the unscented transformation is that it should be easier to approximate a Gaussian distribution than any other nonlinear function/transformation with a fixed number of parameters. Following this intuition, we wish to generate a discrete distribution having the same first and second moments as the original distribution, where each point in the discrete approximation can be directly transformed. The mean and covariance of the transformed ensemble can then be computed as the estimate of the nonlinear transformation of the original distribution. Based on the above intuition, the unscented transformation algorithm emerges. Given an n-dimensional Gaussian distribution with covariance $\mathbf{P}$, we can generate a set of points having the same sample covariance from the columns (or rows) of the matrix $\pm\sqrt{n\mathbf{P}}$ (the positive and negative roots). The mean of this set of points is zero, but if the original distribution has mean $\bar{\mathbf{x}}$, then simply adding to each of the points yields a symmetric set of $2n$ points having the desired mean and covariance [45]. Since the set is symmetric, its odd central moments are zero. Therefore, its first three moments are the same as the original Gaussian distribution.

According to the principle stated above, the process of the unscented transformation is listed as follows. First, a set of translated sigma points is computed from the $n \times n$ matrix $\mathbf{P_{xx}}$ as

$$\begin{aligned} \sigma &\leftarrow 2n \text{ rows or columns from} \pm \sqrt{(n+\kappa)\mathbf{P_{xx}}} \\ \chi_0 &= \bar{\mathbf{x}}, \quad \chi_i = \sigma_i + \bar{\mathbf{x}} \end{aligned} \tag{7.113}$$

which assures that $\mathbf{P_{xx}} = \dfrac{1}{2(n+\kappa)} \sum_{i=1}^{2n} [\chi_i - \bar{\mathbf{x}}][\chi_i - \bar{\mathbf{x}}]^{\mathrm{T}}$.

Second, another set of the transformed sigma points are evaluated for each of the $0 - 2n$ points by $\mathcal{Y}_i = g[\chi_i]$. Based on the set of the transformed sigma points, the predicted mean is computed as

$$\bar{\mathbf{y}} = \frac{1}{n+\kappa}\left\{\kappa\mathcal{Y}_0 + \frac{1}{2}\sum_{i=1}^{2n}\chi_i\right\} \tag{7.114}$$

Finally, the predicted covariance is given by

$$\mathbf{P_{yy}} = \frac{1}{n+\kappa}\left\{\kappa[\mathcal{Y}_0 - \bar{\mathbf{y}}][\mathcal{Y}_0 - \bar{\mathbf{y}}]^{\mathrm{T}} + \frac{1}{2}\sum_{i=1}^{2n}[\mathcal{Y}_i - \bar{\mathbf{y}}][\mathcal{Y}_i - \bar{\mathbf{y}}]^{\mathrm{T}}\right\} \tag{7.115}$$

Comparing this algorithm with the linearization algorithm, we see a number of significant advantages: It is not necessary to calculate the Jacobian matrix or make any other approximations of the nonlinear function of the dynamic system. The prediction stage only consists of standard linear algebra operations (matrix square roots, outer products, matrix and vector summation). The number of computations (including an efficient matrix square-root algorithm) scales with dimensions at the same rate as linearization.

Based on the unscented transformation, UKF is not necessarily limited to the usage of differentiable functions. Using UKF, the state random variable has to be redefined as the concatenation of internal states and noise variable, $\mathbf{x}^a = [\mathbf{x}^{\mathrm{T}}, \mathbf{q}^{\mathrm{T}}, \boldsymbol{\gamma}^{\mathrm{T}}]^{\mathrm{T}}$, namely an augmented variable, whose covariance $\mathbf{P}^a$ can be represented as diag($\mathbf{P}_{\mathbf{x}}$, $\mathbf{R}^q, \mathbf{R}^\gamma$). That is,

$$\chi_{k-1}^a = \left[\hat{\mathbf{x}}_{k-1}^a \quad \hat{\mathbf{x}}_{k-1}^a + \sigma_{\mathbf{x}} \quad \hat{\mathbf{x}}_{k-1}^a - \sigma_{\mathbf{x}}\right] \tag{7.116}$$

where $\sigma_{\mathbf{x}} = \sqrt{(n_{\mathbf{x}}^a + \lambda)\mathbf{P}_{k-1}^a}$, and $n_{\mathbf{x}}^a$ is the dimensionality of $\mathbf{x}^a$ that is the sum of the dimensionality of $\mathbf{x}$, ν and n. In this way, the mean and covariance of these points, denoted by $\hat{\mathbf{x}}_{k-1}^a$ and $\mathbf{P}_{k-1}^a$, respectively, are equal to those of the augmented variables. Each sigma point is instantiated through the nonlinear function f to yield a set of alternatives, of the form

$$\chi_{k|k-1}^{\mathbf{x}} = f\left(\chi_{k-1}^{\mathbf{x}}, \mathbf{u}_k, \mathbf{w}\right) + \chi_{k-1}^q \tag{7.117}$$

Then, the prior mean and covariance of internal states exhibit as

$$\hat{\mathbf{x}}_{k|k-1} = \frac{1}{2\left(n_{\mathbf{x}}^a + \lambda\right)} \sum_{i=1}^{2n_{\mathbf{x}}^a} \chi_{i,k|k-1}^{\mathbf{x}} \tag{7.118}$$

$$\mathbf{P}_{\mathbf{x}_{k|k-1}} = \frac{1}{2\left(n_{\mathbf{x}}^a + \lambda\right)} \sum_{i=1}^{2n_{\mathbf{x}}^a} \left[\chi_{i,k|k-1}^{\mathbf{x}} - \hat{\mathbf{x}}_{k|k-1}\right] \left[\chi_{i,k|k-1}^{\mathbf{x}} - \hat{\mathbf{x}}_{k|k-1}\right]^{\mathrm{T}} \tag{7.119}$$

To complete the measurement update process, another transformation of the output equation is also required

$$\mathcal{Y}_{k|k-1} = h\left(\chi_{k|k-1}^{\mathbf{x}}, \mathbf{u}_k, \mathbf{w}\right) + \chi_k^\gamma \tag{7.120}$$

$$\hat{\mathbf{y}}_{k|k-1} = \frac{1}{2\left(n_{\mathbf{x}}^a + \lambda\right)} \sum_{i=1}^{2n_{\mathbf{x}}^a} \mathcal{Y}_{i,k|k-1} \tag{7.121}$$

Then, the measurement update equations are

$$\mathbf{P}_{\mathbf{y}\mathbf{y}_{k|k-1}} = \frac{1}{2\left(n_{\mathbf{x}}^a + \lambda\right)} \sum_{i=1}^{2n_{\mathbf{x}}^a} \left[\mathcal{Y}_{i,k|k-1}^{\mathbf{x}} - \hat{\mathbf{y}}_{k|k-1}\right] \left[\mathcal{Y}_{i,k|k-1}^{\mathbf{x}} - \hat{\mathbf{y}}_{k|k-1}\right]^{\mathrm{T}} \tag{7.122}$$

$$\mathbf{P}_{\mathbf{xy}_{k|k-1}} = \frac{1}{2(n_{\mathbf{x}}^{a} + \lambda)} \sum_{i=1}^{2n_{\mathbf{x}}^{a}} \left[\mathcal{X}_{i,k|k-1}^{\mathbf{x}} - \widehat{\mathbf{x}}_{k|k-1}\right]\left[\mathcal{Y}_{i,k|k-1}^{\mathbf{x}} - \widehat{\mathbf{y}}_{k|k-1}\right]^{\mathrm{T}} \tag{7.123}$$

$$\mathcal{K}_{k}^{\mathbf{x}} = \mathbf{P}_{\mathbf{x}_k\mathbf{y}_k}\left(\mathbf{P}_{\tilde{\mathbf{y}}_k\tilde{\mathbf{y}}_k}\right)^{-1} \tag{7.124}$$

$$\widehat{\mathbf{x}}_{k} = \widehat{\mathbf{x}}_{k|k-1} + \mathcal{K}_{k}\left(\mathbf{t}_{k} - \widehat{\mathbf{y}}_{k|k-1}\right) \tag{7.125}$$

$$\mathbf{P}_{\mathbf{x}_k} = \mathbf{P}_{\mathbf{x}_{k|k-1}} - \mathcal{K}_{k}\mathbf{P}_{\mathbf{yy}_{k|k-1}}\mathcal{K}_{k}^{\mathrm{T}} \tag{7.126}$$

where λ is the composite scaling parameter, $\mathbf{R}^{q}$ is the process-noise covariance, and $\mathbf{R}^{\gamma}$is the measurement-noise covariance.

With respect to the accuracy, UKF is usually superior to that of EKF, but the computational efficiency is far from satisfactory. In addition, to generate a set of sigma points, the Cholesky decomposition needs to compute the square root covariance matrix. However, the UKF-computed covariance matrix cannot always be positive definite. Thus, the implementation process of the filter might be instable.

7.4.3 Cubature Kalman-Filter

The principle of the CKF has many similarities to that of the UKF. Differently, the CKF is built on the basis of the numerical integration [46]. Now, consider a multidimensional weighted integral of the form

$$I(f) = \int_{D} f(\mathbf{x})w(\mathbf{x})d\mathbf{x} \tag{7.127}$$

where $f(\cdot)$ is some arbitrary function, $D \subseteq \mathbb{R}^{n}$ is the region of integration, and the known weighting function $w(\mathbf{x}) \geq 0$ for all $\mathbf{x} \in D$. For example, in a Gaussian weighted integral, $w(\mathbf{x})$ is a Gaussian density function and satisfies the non-negativity condition in the entire region. To compute it approximately, we seek numerical integration methods. The basic task of numerically computing the integral (7.127) is to find a set of points $\mathbf{x}_i$, and weights w_i that approximates the integral $I(f)$ using a weighted sum of function evaluations:

$$I(f) \approx \sum_{i=1}^{m} w_{i} f(\mathbf{x}_{i}) \tag{7.128}$$

The methods used to find the weighted point set $\{\mathrm{x}_i, w_i\}$ are often described as cubature rules, and CKF is designed based on them. Consider another more commonly used numerical integration as follows:

$$\int_{\mathbb{R}^n} f(\mathbf{x})\mathcal{N}(\mathbf{x}|\mu,\Sigma)d\mathbf{x} = \int_{\mathbb{R}^n} f\left(\sqrt{\Sigma}\mathbf{x}+\mu\right)\mathcal{N}(\mathbf{x}|0,\mathbf{I})d\mathbf{x} \tag{7.129}$$

The cubature rules generated for (7.129) can be used to design a CKF model, in which the states are estimated in two update steps. Firstly, the priori statistic features of the states can be computed by using the priori distribution and the historical states, i.e.,

$$\widehat{\mathbf{x}}_{k|k-1} = \int_{\mathbb{R}^{n_\mathbf{x}}} f(\mathbf{u}_k,\mathbf{x}_{k-1},\mathbf{w})p\left(\mathbf{x}_{k-1}|\widehat{\mathbf{x}}_{k-1},\mathbf{P}_{\mathbf{x}_{k-1}}\right)d\mathbf{x}_{k-1} \tag{7.130}$$

$$\begin{aligned}\mathbf{P}_{\mathbf{x}_{k|k-1}} = &\int_{\mathbb{R}^{n_\mathbf{x}}} f(\mathbf{u}_k,\mathbf{x}_{k-1},\mathbf{w})f^{\mathrm{T}}(\mathbf{u}_k,\mathbf{x}_{k-1},\mathbf{w})p\left(\mathbf{x}_{k-1}|\widehat{\mathbf{x}}_{k-1},\mathbf{P}_{\mathbf{x}_{k-1}}\right)d\mathbf{x}_{k-1}\\ &-\widehat{\mathbf{x}}_{k|k-1}\left(\widehat{\mathbf{x}}_{k|k-1}\right)^{\mathrm{T}}+\mathbf{R}^{\nu}\end{aligned} \tag{7.131}$$

where $p\left(\mathbf{x}_{k-1}|\widehat{\mathbf{x}}_{k-1},\mathbf{P}_{\mathbf{x}_{k-1}}\right)$ is the probability density function over $\mathbf{x}_{k-1}$. Secondly, the corresponding posterior estimation is computed by the current noisy measurement $\mathbf{t}_k$.

$$\widehat{\mathbf{x}}_k = \widehat{\mathbf{x}}_{k|k-1} + \mathcal{K}_k\left(\mathbf{t}_k - \widehat{\mathbf{y}}_{k|k-1}\right) \tag{7.132}$$

where $\widehat{\mathbf{y}}_{k|k-1}$ is the output of the nonlinear model given the prior estimation $\widehat{\mathbf{x}}_{k|k-1}$ of the states

$$\widehat{\mathbf{y}}_{k|k-1} = \int_{\mathbb{R}^{n_x}} h(\mathbf{x}_k,\mathbf{u}_k)\mathcal{N}\left(\mathbf{x}_k|\widehat{\mathbf{x}}_{k|k-1},\mathbf{P}_{\mathbf{x}_{k|k-1}}\right)d\mathbf{x}_k \tag{7.133}$$

And $\mathcal{K}_k$ is the Kalman gain which is defined as

$$\mathcal{K}_k = \mathbf{P}_{\mathbf{xy}_{k|k-1}}\mathbf{P}^{-1}_{\mathbf{yy}_{k|k-1}} \tag{7.134}$$

where $\mathbf{P}_{\mathbf{yy}_{k|k-1}}$ is the innovation covariance and $\mathbf{P}_{\mathbf{xy}_{k|k-1}}$ is the cross-covariance. $\mathbf{P}_{\mathbf{yy}_{k|k-1}}$ and $\mathbf{P}_{\mathbf{xy}_{k|k-1}}$ can be calculated by (7.135) and (7.136)

$$\mathbf{P}_{\mathbf{yy}_{k|k-1}} = \int_{\mathbb{R}^{n_x}} h(\mathbf{x}_k,\mathbf{u}_k)h^{\mathrm{T}}(\mathbf{x}_k,\mathbf{u}_k)\mathcal{N}\left(\mathbf{x}_k|\widehat{\mathbf{x}}_{k|k-1},\mathbf{P}_{\mathbf{x}_{k|k-1}}\right)d\mathbf{x}_k - \widehat{\mathbf{y}}_{k|k-1}\widehat{\mathbf{y}}^{\mathrm{T}}_{k|k-1} + \mathbf{R}^{\gamma}_k \tag{7.135}$$

$$\mathbf{P}_{\mathbf{xy}_{k|k-1}} = \int_{\mathbb{R}^{n_x}} \mathbf{x}_k h^{\mathrm{T}}(\mathbf{x}_k,\mathbf{u}_k)\mathcal{N}\left(\mathbf{x}_k|\widehat{\mathbf{x}}_{k|k-1},\mathbf{P}_{\mathbf{x}_{k|k-1}}\right)d\mathbf{x}_k - \widehat{\mathbf{x}}_{k|k-1}\widehat{\mathbf{y}}^{\mathrm{T}}_{k|k-1} \tag{7.136}$$

Finally, the posterior covariance of the states is updated by (7.137)

$$\mathbf{P}_{\mathbf{x}_k} = \mathbf{P}_{\mathbf{x}_{k|k-1}} - \mathcal{K}_k \mathbf{P}_{\mathbf{y}\mathbf{y}_{k|k-1}} \mathcal{K}_k^{\mathrm{T}} \tag{7.137}$$

The key of CKF is to compute the integrals. Here, we assume the probability density in the formulas satisfy the Gaussian density $\mathcal{N}\left(\mathbf{x}_{k-1}|\widehat{\mathbf{x}}_{k-1}, \mathbf{P}_{\mathbf{x}_{k-1}}\right)$, i.e., to compute its mean and covariance. Then, the Gaussian weighted integrals can be computed by a third-degree cubature rule. For instance, approximated by a Gauss-Laguerre formula, we have

$$\int_{\mathbb{R}^{n_\mathbf{x}}} f(\mathbf{x})\mathcal{N}(\mathbf{x};\mu,\Sigma)d\mathbf{x} \approx \frac{1}{2n_\mathbf{x}} \sum_{i=1}^{2n_\mathbf{x}} f\left(\mu + \sqrt{\Sigma}\xi_i\right) \tag{7.138}$$

where a square root factor of the covariance Σ satisfies $\Sigma = \sqrt{\Sigma}\sqrt{\Sigma}^{\mathrm{T}}$, and the set of $2n_\mathbf{x}$ cubature points are given by

$$\xi_i = \begin{cases} \sqrt{n_\mathbf{x}}\mathbf{e}_i, & i = 1, 2, \cdots, n_x \\ -\sqrt{n_\mathbf{x}}\mathbf{e}_{i-n_\mathbf{x}}, & i = n_x + 1, n_x + 2, \cdots, 2n_x \end{cases} \tag{7.139}$$

with $\mathbf{e}_i \in \mathbb{R}^n$ being the ith elementary column vector.

Since the probability density function of non-Gaussian process can be estimated by a Gaussian mixture density with a finite number of weighted sums of Gaussian densities $\left(p(\mathbf{x}) \sim \sum_{i=1}^{n} a_i \mathcal{N}\left(\mathbf{x}|\bar{\mathbf{x}}_i, \mathbf{P}_{\mathbf{x}_i}\right)\right)$, the CKF can also be extended to non-Gaussian problems.

7.4.4 *Nonlinear Kalman-Filters-Based Dual Estimation*

In Sect. 3.5.5, we introduced the dual estimation-based ESN, where the Kalman-filter is not elaborated. Then here we will describe the nonlinear Kalman-filters-based dual estimation in details, whose base model is the ESN. As we all know, the states are estimated above based on the fact that the parameters in (7.104) is known. However, for most of the nonlinear Kalman-filtering problems, the parameters are unknown and have to be estimated simultaneously when estimating the states. Thus, we can use two nonlinear Kalman-filters to construct one optimization problem that recursively determines the state and the weights in order to minimize the value of a cost function. It can be shown that the cost function consists of a weighted prediction error and estimation error components, of the form

$$\begin{aligned} J\left(\widehat{\mathbf{x}}_k, \widehat{\mathbf{w}}\right) = & \sum_{k=1}^{n} \left\{ \left[\mathbf{t}_k - h\left(\widehat{\mathbf{x}}_k, \widehat{\mathbf{w}}\right)\right]^{\mathrm{T}} \left(\mathbf{R}_k^{\gamma}\right)^{-1} \left[\mathbf{t}_k - h\left(\widehat{\mathbf{x}}_k, \widehat{\mathbf{w}}\right)\right] \right\} \\ & + \sum_{k=1}^{n} \left\{ \left(\widehat{\mathbf{x}}_k - \widehat{\mathbf{x}}_{k|k-1}\right) \left(\mathbf{R}_{k-1}^{q}\right)^{-1} \left(\widehat{\mathbf{x}}_k - \widehat{\mathbf{x}}_{k|k-1}\right) \right\} \end{aligned} \tag{7.140}$$

where $\mathbf{x}_{k|k-1} = F(\mathbf{x}_{k-1}, \mathbf{w})$ is the predicted state, and $\mathbf{R}_k^q$ and $\mathbf{R}_k^\gamma$ are the additive noise and innovations noise covariance, respectively.

From the above description, two nonlinear Kalman-filters-based dual estimation [47] is necessary to solve the objective function (7.140), where these two filters are used for the estimation of the states and the parameters alternatively. The basic principle of the dual estimation is already shown in Fig. 3.4. In this section, we take the dual estimation-based EFK (DEKF) as an example to illustrate the dual estimation-based parameters optimization.

As for the parameters estimation, another new state-space model is reconstructed on the basis of (7.104). In this state-space, the weight parameters are to be estimated.

$$\begin{aligned} \mathbf{w}_{k+1} &= \mathbf{w}_k + \boldsymbol{\varepsilon}_k \\ \mathbf{t}_k &= h(\mathbf{x}_k, \mathbf{u}_k, \mathbf{w}_{k+1}) + \boldsymbol{\nu}_k \end{aligned} \tag{7.141}$$

where $\boldsymbol{\varepsilon}_k$ and $\boldsymbol{\nu}_k$ are uncorrelated process and measurement Gaussian noise with mean zero and covariance $\mathbf{R}_k^\varepsilon$ and $\mathbf{R}_k^\nu$, respectively.

Based on the state-space models described in (7.104) and (7.141), we introduce the DEKF algorithm, which combines the Kalman state estimation and the weight estimation. Recall that the task is to estimate both the state and model from only noisy observations. Essentially, two nonlinear Kalman-filters are run concurrently. At each time step, a state filter estimates the state using the current model estimate $\widehat{\mathbf{w}}_k$, while the other weight filter estimates the weights using the current state estimate $\widehat{\mathbf{x}}_k$. The specific process of the DEKF is described here.

Assuming that the original values of the parameters and the states are initialized with $\widehat{\mathbf{w}}_0 = E[\mathbf{w}_0]$ and $\widehat{\mathbf{x}}_0 = E[\mathbf{x}_0]$, respectively. Meanwhile, the corresponding covariance of the original parameters and the states are

$$\mathbf{P}_{\mathbf{w}_0} = E\left[(\mathbf{w}_0 - \widehat{\mathbf{w}}_0)(\mathbf{w}_0 - \widehat{\mathbf{w}}_0)^{\mathrm{T}}\right], \quad \mathbf{P}_{\mathbf{x}_0} = E\left[(\mathbf{x}_0 - \widehat{\mathbf{x}}_0)(\mathbf{x}_0 - \widehat{\mathbf{x}}_0)^{\mathrm{T}}\right] \tag{7.142}$$

First, the time update equations of the weight filter are listed as follows for $\forall k \in \{1, 2, \cdots, \infty\}$. The prior estimation of the weight and the corresponding covariance are updated by

$$\widehat{\mathbf{w}}_{k|k-1} = \widehat{\mathbf{w}}_{k-1}, \quad \mathbf{P}_{\mathbf{w}_{k|k-1}} = \mathbf{P}_{\mathbf{w}_{k-1}} + \mathbf{R}_{k-1}^\varepsilon. \tag{7.143}$$

Meanwhile, as for the state filter, its time update step devotes itself to estimate the priori of the state and its covariance

$$\widehat{\mathbf{x}}_{k|k-1} = f(\widehat{\mathbf{x}}_{k-1}, \mathbf{u}_k, \widehat{\mathbf{w}}_{k|k-1}), \quad \mathbf{P}_{\mathbf{x}_{k|k-1}} = \mathbf{A}_{k-1}\mathbf{P}_{\mathbf{x}_{k-1}}\mathbf{A}_{k-1}^{\mathrm{T}} + \mathbf{R}_{k-1}^q \tag{7.144}$$

Second, the measurement-update equations for the state filter and the weight filter are used to estimate the posterior state and the posterior weight

$$\widehat{\mathbf{x}}_k = \widehat{\mathbf{x}}_{k|k-1} + \mathcal{K}_k^{\mathbf{x}}\left(\mathbf{t}_k - h\left[\mathbf{u}_k, \widehat{\mathbf{x}}_{k|k-1}, \widehat{\mathbf{w}}_{k|k-1}\right]\right) \tag{7.145}$$

$$\widehat{\mathbf{w}}_k = \widehat{\mathbf{w}}_{k|k-1} + \mathcal{K}_k^{\mathbf{w}}\left(\mathbf{t}_k - h\left[\mathbf{u}_k, \widehat{\mathbf{x}}_k, \widehat{\mathbf{w}}_{k|k-1}\right]\right) \tag{7.146}$$

where $\mathcal{K}_k^{\mathbf{x}}$ and $\mathcal{K}_k^{\mathbf{w}}$ are the Kalman gains for the state estimation and the weight estimation. The definition of $\mathcal{K}_k^{\mathbf{x}}$ and $\mathcal{K}_k^{\mathbf{w}}$ are

$$\mathcal{K}_k^{\mathbf{x}} = \mathbf{P}_{\mathbf{x}_{k|k-1}}\mathbf{H}_k^{\mathrm{T}}\left(\mathbf{H}_k\mathbf{P}_{\mathbf{x}_{k|k-1}}\mathbf{H}_k^{\mathrm{T}} + \mathbf{R}_k^{\gamma\mathrm{T}}\mathbf{R}_k^{\gamma}\right)^{-1} \tag{7.147}$$

$$\mathcal{K}_k^{\mathbf{w}} = \mathbf{P}_{\mathbf{w}_{k|k-1}}\mathbf{C}_k^{\mathrm{T}}\left(\mathbf{C}_k\mathbf{P}_{\mathbf{w}_{k|k-1}}\mathbf{C}_k^{\mathrm{T}} + \mathbf{R}_k^{n}\right)^{-1} \tag{7.148}$$

where $\mathbf{C}_k = (\partial h/\partial \mathbf{w})|_{\mathbf{w}=\widehat{\mathbf{w}}_{k|k-1}}$. To make another estimation for the next time step, the posterior covariance of the states and the weights are calculated by

$$\mathbf{P}_{\mathbf{x}_k} = \left(\mathbf{I} - \mathcal{K}_k^{\mathbf{x}}\mathbf{H}_k\right)\mathbf{P}_{\mathbf{x}_{k|k-1}} \tag{7.149}$$

$$\mathbf{P}_{\mathbf{w}_k} = \left(\mathbf{I} - \mathcal{K}_k^{\mathbf{w}}\mathbf{C}_k\right)\mathbf{P}_{\mathbf{w}_{k|k-1}} \tag{7.150}$$

The DEFK is one specific case of the nonlinear Bayesian filter-based dual estimation, so the strategy of training a DEFK is also suitable for training other nonlinear Bayesian filters, such as the UKF and the CKF. Moreover, as for implementation of the dual estimation, the filters for the state estimation and the weight estimation do not need to be the same. That is, we can choose one EKF and one UKF or one UKF and one CKF to complete the dual estimation.

7.4.5 *Dual Estimation of Linear/Nonlinear Kalman-Filter*

For a deeper understanding of the dual estimation, we study a more special model, namely the ESN-based prediction model. Considering an ESN model whose internal states and outputs are combining with the additive noises, its formulas are written as

$$\begin{aligned} \mathbf{x}_k &= f\left(\mathbf{W}^{\mathrm{in}}\mathbf{u}_k + \mathbf{W}\mathbf{x}_{k-1}\right) + \nu_{k-1} \\ y_k &= \mathbf{W}^{\mathrm{out}} \cdot \left[\mathbf{u}_k, \mathbf{x}_k\right] + n_k \end{aligned} \tag{7.151}$$

where y_k is a scalar quantity that shows the network is a single-output model. ν_{k-1} and n_k are independent white Gaussian noise sequence with covariance $\mathbf{R}^{\nu}$ and σ_n^2, respectively.

For the above model, the internal states and the output are combining with the uncertainties due to the interruption of the noise. To solve the weights of the model, the internal states are also required to be estimate simultaneously. Thus, the dual estimation is one effective technique for solving the parameters and the process states. However, the ESN model exhibits its own special characteristics, that is, its input weights $\mathbf{W}^{\mathrm{in}}$ and the internal weights $\mathbf{W}$ are given before training and fixed in the training process. The output weights are the unique unknown parameters to be solved. As a result, the ESN model described by (7.151) can be divided into one state-space and one weight space, where the state-space is nonlinear and the weight space is linear. The formula of the state-space can use the formula (7.151) and the formula of the weight space are rewritten as

$$\begin{aligned} \mathbf{W}_k^{\text{out}} &= \mathbf{W}_{k-1}^{\text{out}} + \mathbf{q}_{k-1} \\ y_k &= \mathbf{W}_k^{\text{out}} \cdot \left[\mathbf{u}_k, \widehat{\mathbf{x}}_k\right] + n_k \end{aligned} \tag{7.152}$$

From (7.152), one can see that the state transition matrix is simply an identity. Thus, one nonlinear/linear Bayesian filters-based dual estimation is required to design for the parameters estimation. Here, we consider to employ the CKF to estimate the internal states and the linear KF to estimate the weights. As follows, we will introduce the implementation of this model in details.

The priori statistic features of internal states can be computed by using the priori distribution and the historical internal states, i.e.,

$$\widehat{\mathbf{x}}_{k|k-1} = \int_{\mathbb{R}^{n_\mathbf{x}}} f\left(\mathbf{W}^{\text{in}}\mathbf{u}_k + \mathbf{W}\mathbf{x}_{k-1}\right) p\left(\mathbf{x}_{k-1}; \widehat{\mathbf{x}}_{k-1}, \mathbf{P}_{\mathbf{x}_{k-1}}\right) d\mathbf{x}_{k-1} \tag{7.153}$$

$$\begin{aligned} \mathbf{P}_{\mathbf{x}_{k|k-1}} = &\int_{\mathbb{R}^{n_\mathbf{x}}} f\left(\mathbf{W}^{\text{in}}\mathbf{u}_k + \mathbf{W}\mathbf{x}_{k-1}\right) f^{\mathrm{T}}\left(\mathbf{W}^{\text{in}}\mathbf{u}_k + \mathbf{W}\mathbf{x}_{k-1}\right) p\left(\mathbf{x}_{k-1}; \widehat{\mathbf{x}}_{k-1}, \mathbf{P}_{\mathbf{x}_{k-1}}\right) d\mathbf{x}_{k-1} \\ &- \widehat{\mathbf{x}}_{k|k-1}\left(\widehat{\mathbf{x}}_{k|k-1}\right)^{\mathrm{T}} + \mathbf{R}^{\nu} \end{aligned} \tag{7.154}$$

where $p\left(\mathbf{x}_{k-1}; \widehat{\mathbf{x}}_{k-1}, \mathbf{P}_{\mathbf{x}_{k-1}}\right)$ is the probability density function over $\mathbf{x}_{k-1}$.

As for the time-update equations of the weight filter, a priori of output weights at time k is estimated by using the output weight at time $k-1$ and its covariance estimation, of the form

$$\widehat{\mathbf{W}}_{k|k-1}^{\text{out}} = \widehat{\mathbf{W}}_{k-1}^{\text{out}}, \quad \mathbf{P}_{\mathbf{W}_{k|k-1}^{\text{out}}} = \mathbf{P}_{\mathbf{W}_{k-1}^{\text{out}}} + \mathbf{R}^{q} \tag{7.155}$$

Besides, the corresponding posterior estimation of the internal states is computed by the difference between the current noisy measurement t_k and the output $\widehat{y}_{k|k-1}$ of the ESN model

$$\widehat{\mathbf{x}}_k = \widehat{\mathbf{x}}_{k|k-1} + \mathcal{K}_k^{\mathbf{x}}\left(t_k - \widehat{y}_{k|k-1}\right) \tag{7.156}$$

where $\mathcal{K}_k^{\mathbf{x}}$ is the Kalman gain. $\widehat{y}_{k|k-1}$ is the expected output of ESN calculated by

$$\widehat{y}_{k|k-1} = \widehat{\mathbf{W}}_{k|k-1}^{\text{out}} \cdot \left[\mathbf{u}_k, \widehat{\mathbf{x}}_{k|k-1}\right] \tag{7.157}$$

The key of CKF is to compute the integrals in (7.153) and (7.154), and the concrete implementation can be seen in Sect. 7.4.3.

The posterior estimation of the weights is updated by

$$\widehat{\mathbf{W}}_k^{\text{out}} = \widehat{\mathbf{W}}_{k|k-1}^{\text{out}} + \mathcal{K}_k^{\mathbf{w}}\left(t_k - \widehat{y}_{k|k-1}\right) \tag{7.158}$$

where the Kalman gain $\mathcal{K}_k^{\mathbf{w}}$ and the expected output $\widehat{y}_{k|k-1}$ are determined by

$$\mathcal{K}_k^{\mathbf{w}} = \mathbf{P}_{\mathbf{W}_{k|k-1}^{\text{out}}}\left[\mathbf{u}_k, \widehat{\mathbf{x}}_k\right]\left(\left[\mathbf{u}_k, \widehat{\mathbf{x}}_k\right]^{\text{T}}\mathbf{P}_{\mathbf{W}_{k|k-1}^{\text{out}}}\left[\mathbf{u}_k, \widehat{\mathbf{x}}_k\right] + \sigma_n^2\right)^{-1} \tag{7.159}$$

$$\widehat{y}_{k|k-1} = \widehat{\mathbf{W}}_{k|k-1}^{\text{out}} \cdot \left[\mathbf{u}_k, \widehat{\mathbf{x}}_k\right] \tag{7.160}$$

As we all know, the parameters optimization of linear regression always suffers from the ill-condition problem, which is mainly because the parameters identification is related to the inverse of the singular matrix. As for the Kalman-filter-based parameters estimation, there is no need to compute the inverse of the singular matrix since the value of $\left(\left[\mathbf{u}_k, \widehat{\mathbf{x}}_k\right]^{\text{T}}\mathbf{P}_{\mathbf{W}_{k|k-1}^{\text{out}}}\left[\mathbf{u}_k, \widehat{\mathbf{x}}_k\right] + \sigma_n^2\right)^{-1}$ is a scalar which can be directly obtained, so that the ill-condition problem can be effectively avoided. All in all, the Kalman-filter can effectively estimates the output weights of the ESN accurately without ill-conditions, and the estimation process become very simple.

7.4.6 *Case Study*

In Chap. 3, we gave a case study of the dual estimation-based ESN with nonlinear KFs for the industrial data. Then, we will give experiments of the KF-based improved ESN, three nonlinear KF-based ESN, and the generic ESN for the noisy Mackey Glass time series.

The Mackey Glass time series, already introduced in Chap. 5, is added with the additive white Gaussian noise with variance 0.01. We obtain a noisy time series with the form of $\{u(1), u(2), \cdots u(n), \cdots\}$, where $n \geq 2000$. The sample $u(k + \Delta)$ can be estimated from a properly chosen time series $\{u(k), u(k - \Delta), \cdots, u(k - (d_{\text{E}} - 2)\Delta), u(k - (d_{\text{E}} - 1)\Delta)\}$, where d_E and Δ denote the embedding dimensionality and the delay, respectively.

For the prediction of the noisy time series, the input dimension of the improved ESN is empirically set to 50. The activation function of the internal neurons uses $\varphi(\nu) = \tanh(\nu)$, while a linear activation is set on output. The internal states and output weights are initialized with zero mean Gaussian with diagonal covariance of $0.005\mathbf{I}_{\mathbf{x}}$ and $0.005\mathbf{I}_{\mathbf{W}^{\text{out}}}$, respectively. And, $\mathbf{W}^{\text{in}}$ and $\mathbf{W}$ are randomly generated. According to issues of [48], the spectral radius of $\mathbf{W}$ is set as 0.8 and its sparse connectivity is equal to 2%.

To train the ESN model, we firstly construct a set of samples denoted as $\{(\mathbf{u}_i, y_i) | i = 1, 2, \cdots, 1000\}$, where $[\mathbf{u}_i] = [u(i - (d_{\text{E}} - 1)\Delta)\ u(i - (d_{\text{E}} - 2)\Delta)\ \cdots\ u(i)]$, $[y_i] = [u(i + \Delta)]$, $d_{\text{E}} = 50$, $\Delta = 1$. Besides training, the rest of data are used to test.

Here, we make $\mathbf{R}^{\nu}$ and $\mathbf{R}^{\mathbf{q}}$ decay such that $\mathbf{R}_0^{\mathbf{v}} = (1/\lambda - 1)\mathbf{P}_{\mathbf{x}_0}$ and $\mathbf{R}_0^{\mathbf{q}} = (1/\lambda - 1)$ $\mathbf{P}_{\mathbf{W}_0^{out}}$ with λ is equal to 0.9995, and the value of $\mathbf{R}^{\mathbf{n}}$ is determined by the intrinsic noise of the sample data.

To clarify the impact of the number of internal units on the prediction accuracy, a comparative experiment based on cross-validation is conducted, where the number of samples for cross-validation is 1000. Figure 7.8 shows the cross-validation results of the nonlinear/linear dual estimation-based ESN with different number of internal units. From this figure, when the number of internal units is equal to 40, the improved ESN based on CKF/KF dual estimation (ESN-DE) presents the best performance.

The training curves for the noisy Mackey Glass time series are shown in Fig. 7.9, where the relationship between the training epoch and the training error evaluated by RMSE is illustrated. It can be seen that the nonlinear/linear dual estimations are convergent to train the ESN-DE. The comparative results by the dual estimation-based ESN are shown in Fig. 7.10. From Fig. 7.10a, a 60-min prediction curve is comparatively presented by the three dual estimation-based networks, and their absolute prediction errors are shown as Fig. 7.10b. On the perspective of the average absolute error, the CKF/KF dual estimation-based ESN is the lowest.

To further represent the performance of the state estimation for the ESN, the prediction results produced by three methods, including the ESN based on CKF/KF dual estimation, the ESN based on KF without state estimation, and the generic ESN, are comparatively presented in Fig. 7.11. From this figure, the ESN based on CKF/KF dual estimation method presents the best performance on the respect of prediction accuracy, evaluated by the absolute error in Fig. 7.11b. For quantified

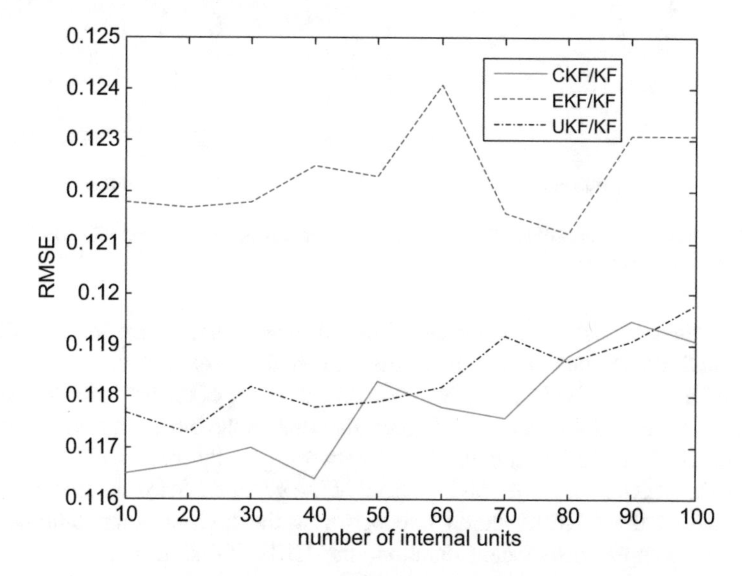

Fig. 7.8 Prediction accuracy of the dual estimation-based ESN with different number of internal units

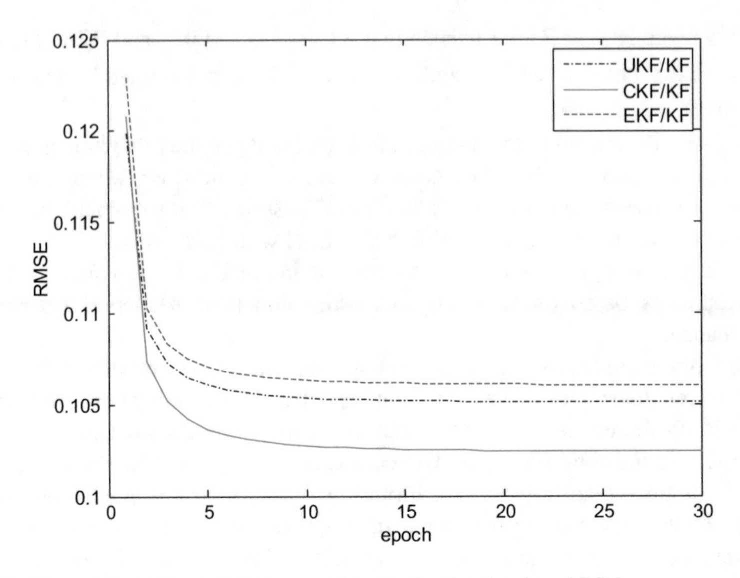

Fig. 7.9 The corresponding training curve of the dual estimation-based ESN

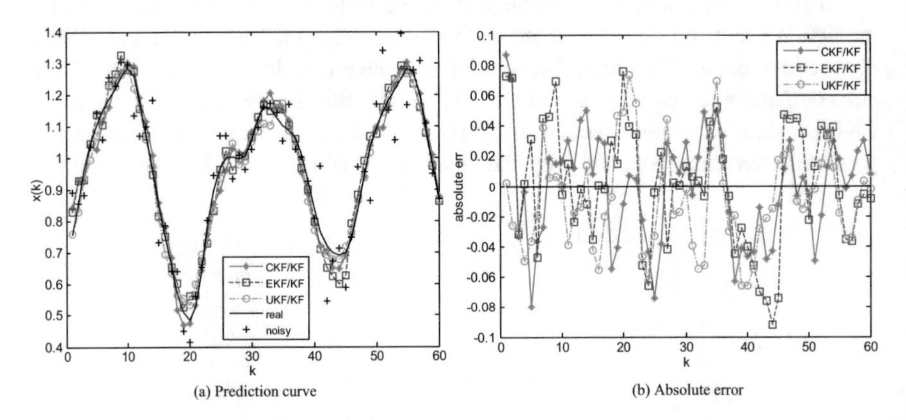

Fig. 7.10 Prediction results of the three nonlinear/linear dual estimation-based ESN. (**a**) Prediction curve. (**b**) Absolute error

statistics, the detailed results are also listed as Table 7.1, where the RMSE, the MAPE, and the computational time of these methods are exhibited.

From Table 7.1, it is clear that the prediction accuracy of the nonlinear/linear dual estimation-based ESN is definitely higher than that of the other methods. In particular, the CKF/KF dual estimation-based network gets the best result. As for the computational efficiency, the solving speed of the generic ESN is the faster than the others; while, the UKF/KF-based one performs the highest computational cost. Under the comprehensive consideration, the CKF/KF-based ESN is a suitable method with respect to the accuracy and the computing efficiency, which gives a sound outcome in the field of noisy nonlinear time series prediction.

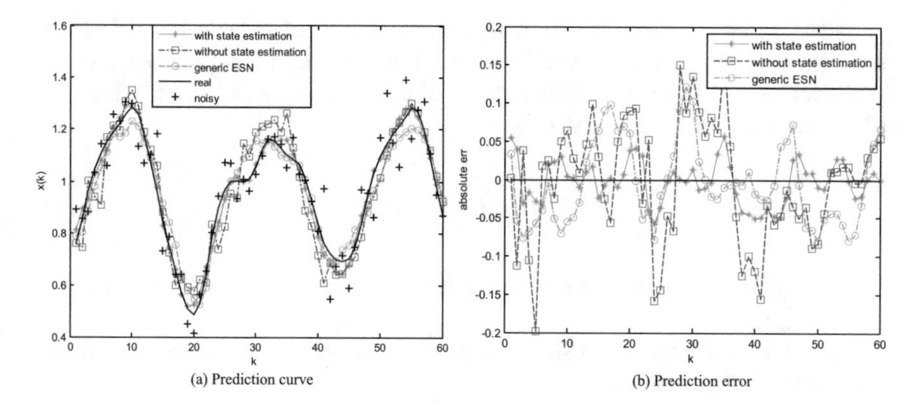

(a) Prediction curve

(b) Prediction error

Fig. 7.11 Comparison of prediction results produced by the ESN based on CKF/KF dual estimation, the ESN based on KF without internal states estimation and the generic ESN. (**a**) Prediction curve. (**b**) Prediction error

Table 7.1 The comparative results of prediction performance

Methods	RMSE	MAPE	Computational time(s)
EKF/KF-based improved ESN	0.04304	3.8283	39.606
UKF/KF-based improved ESN	0.04231	3.8386	249.237
CKF/KF-based improved ESN	0.04025	3.6515	89.159
KF-based improved ESN	0.06572	5.9925	22.628
Generic ESN	0.05111	4.7834	1.176

7.5 Probabilistic Methods

For parameter estimation and optimization, gradient-based methods and evolutionary algorithms can estimate or optimize the parameters of a prediction model. These methods can only find the point estimation of the parameters. However, a probabilistic model takes the data uncertainty into consideration, which can give probabilistic estimation. The probabilistic methods contain maximum likelihood method, Bayesian method, and variational inference methods.

7.5.1 *Maximum Likelihood Method*

Maximum likelihood estimation (MLE) is a method which can estimate the parameters of a statistical model given observations, by finding the parameter values that maximize the likelihood [49]. Intuitively, this lets the parameters agree with the dataset. MLE gives a unified approach to estimation, which is well defined in the case of the normal distribution and many other problems.

Without loss of generality, we firstly consider a general statistical model $p(\mathbf{x};\boldsymbol{\theta})$, where $\mathbf{x}$ denotes multivariate variables, and $\boldsymbol{\theta}$ denotes the set of parameters of this statistical model. And we have the samples of $\mathbf{x}$ denoted by $\mathbf{x}_1, \mathbf{x}_2, \ldots, \mathbf{x}_N$. Thus, the likelihood function is represented as

$$L = \prod_{i=1}^{N} p(\mathbf{x}_i;\boldsymbol{\theta}) \tag{7.161}$$

Maximizing L with respect to $\boldsymbol{\theta}$ will give the optimal $\boldsymbol{\theta}^*$ which most fits the data $\mathbf{x}_1, \mathbf{x}_2, \ldots, \mathbf{x}_N$.

Here, for simplicity, we consider a single real-valued variable x, and the Gaussian distribution is formulated by

$$p(x) = N\left(x|\mu,\sigma^2\right) = \frac{1}{\left(2\pi\sigma^2\right)^{1/2}} \exp\left\{-\frac{1}{2\sigma^2}(x-\mu)^2\right\} \tag{7.162}$$

with mean μ, and variance σ^2. Assume that there are N training samples of x, i.e., $D = \{x_1, x_2, \ldots, x_N\}$. We shall evaluate the parameters μ and σ^2 from the dataset D. Let us write down the likelihood function,

$$\begin{aligned} L &= \prod_{i=1}^{N} p(x_i) \\ &= \prod_{i=1}^{N} \frac{1}{\left(2\pi\sigma^2\right)^{1/2}} \exp\left\{-\frac{1}{2\sigma^2}(x_i-\mu)^2\right\} \\ &= \frac{1}{\left(2\pi\sigma^2\right)^{N/2}} \exp\left\{-\frac{1}{2\sigma^2}\sum_{i=1}^{N}(x_i-\mu)^2\right\} \end{aligned} \tag{7.163}$$

We consider the logarithm of this likelihood function, i.e.,

$$\log L = -\frac{N}{2}\log\left(2\pi\sigma^2\right) - \frac{1}{2\sigma^2}\sum_{i=1}^{N}(x_i-\mu)^2 \tag{7.164}$$

Computing the derivative of $\log L$ with respect to μ gives

$$\frac{\partial \log L}{\partial \mu} = \frac{1}{\sigma^2}\sum_{i=1}^{N}(x_i-\mu) \tag{7.165}$$

Then, setting it to zero gives the estimation formula for μ,

$$\widehat{\mu} = \frac{1}{N}\sum_{i=1}^{N} x_i \tag{7.166}$$

Similarly, maximizing (7.164) with respect to σ^2, we obtain the maximum likelihood solution for the variance of the form

$$\widehat{\sigma}^2 = \frac{1}{N}\sum_{i=1}^{N}(x_i - \widehat{\mu})^2 \tag{7.167}$$

MLE for Linear Regression Model

To further illustrate the MLE method for parameters estimation in a probabilistic model clearly, we specially consider a probabilistic linear regression model. Although the basis of the linear regression is already introduced in Chap. 3, we still introduce it in this section. The linear regression model is formulated by linear combinations of fixed nonlinear functions of the input variables, of the form,

$$y(\mathbf{x}, \mathbf{w}) = \sum_{i=1}^{M-1} w_i\phi_i(\mathbf{x}) + w_0 = \mathbf{w}^{\mathrm{T}}\boldsymbol{\phi}(\mathbf{x}) \tag{7.168}$$

where $\{\phi_i(\mathbf{x})\}$ are known as basis functions, and $\boldsymbol{\phi}(\mathbf{x}) = [\phi_1(\mathbf{x}), \phi_2(\mathbf{x}), \ldots, \phi_M(\mathbf{x})]^{\mathrm{T}}$. The total number of parameters in this model will be M. The latent function $y(\mathbf{x}, \mathbf{w})$ is corrupted by the noise ε which can be assigned by a zero mean Gaussian distribution $\varepsilon \sim N(\varepsilon| 0, \sigma^2)$. Then, the corresponding observed value t is written as

$$t = y(\mathbf{x}, \mathbf{w}) + \varepsilon \tag{7.169}$$

where t is a continuous variable whose density function is

$$f(t|\mathbf{x}, \mathbf{w}, \beta) = N\left(t|y(\mathbf{x}, \mathbf{w}), \beta^{-1}\right) \tag{7.170}$$

with $\beta \equiv \sigma^{-2}$. The conditional mean is formulated as

$$E[t|\mathbf{x}] = \int t \cdot f(t|\mathbf{x}, \mathbf{w}, \beta)dt = y(\mathbf{x}, \mathbf{w}) \tag{7.171}$$

One must note that this Gaussian noise assumption means that the conditional distribution of t given $\mathbf{x}$ is unimodal, which may be unsuitable for some applications.

Now consider a dataset of inputs $\mathbf{X} = [\mathbf{x}_1, \mathbf{x}_2, \ldots, \mathbf{x}_N]^{\mathrm{T}}$ with the corresponding target values $\mathbf{t} = [t_1, t_2, \ldots, t_N]^{\mathrm{T}}$. Assuming that these data points are the samples which are drawn independently from the distribution (7.170), we obtain the following likelihood function which is a function of the adjustable parameters $\mathbf{w}$ and β, i.e.,

$$L = p(\mathbf{t}|\mathbf{X}, \mathbf{w}, \beta) = \prod_{i=1}^{N} N\left(t_i|y(\mathbf{x}_i, \mathbf{w}), \beta^{-1}\right) \tag{7.172}$$

The logarithm of $p(\mathbf{t}|\mathbf{X}, \mathbf{w}, \beta)$ is given by

$$\begin{aligned}\log p(\mathbf{t}|\mathbf{X}, \mathbf{w}, \beta) &= \sum_{i=1}^{N} \log N\left(t_i|\mathbf{w}^{\mathrm{T}}\boldsymbol{\phi}(\mathbf{x}_i), \beta^{-1}\right) \\ &= \frac{N}{2}\log\beta - \frac{N}{2}\log(2\pi) - \frac{\beta}{2}\sum_{i=1}^{N}\{t_i - \mathbf{w}^{T}\boldsymbol{\phi}(\mathbf{x}_i)\}^2\end{aligned} \quad (7.173)$$

Thus, we can use maximum likelihood to determine $\mathbf{w}$ and β. First, we consider the maximization of (7.173) with respect to $\mathbf{w}$. The gradient of (7.173) is written as

$$\nabla \log p(\mathbf{t}|\mathbf{X}, \mathbf{w}, \beta) = -\frac{\beta}{2}\sum_{i=1}^{N}\{t_i - \mathbf{w}^{\mathrm{T}}\phi(\mathbf{x}_i)\}\phi(\mathbf{x}_i)^{\mathrm{T}} \quad (7.174)$$

Setting it to zero gives

$$0 = \sum_{i=1}^{N} t_i\phi(\mathbf{x}_i)^{\mathrm{T}} - \mathbf{w}^{\mathrm{T}}\left(\sum_{i-1}^{N}\phi(\mathbf{x}_i)\phi(\mathbf{x}_i)^{\mathrm{T}}\right) \quad (7.175)$$

Solving for $\mathbf{w}$ we obtain

$$\mathbf{w}_{\mathrm{ML}} = \left(\boldsymbol{\Phi}^{T}\boldsymbol{\Phi}\right)^{-1}\boldsymbol{\Phi}^{\mathrm{T}}\mathbf{t} \quad (7.176)$$

where $\boldsymbol{\Phi}$ is an $N \times M$ matrix, called the *design matrix*, whose elements are given by $\Phi_{ij} = \phi_j(\mathbf{x}_i)$, so that

$$\boldsymbol{\Phi} = \begin{pmatrix} \phi_0(\mathbf{x}_1) & \phi_1(\mathbf{x}_1) & \cdots & \phi_{M-1}(\mathbf{x}_1) \\ \phi_0(\mathbf{x}_2) & \phi_1(\mathbf{x}_2) & \cdots & \phi_{M-1}(\mathbf{x}_2) \\ \cdots & \cdots & \cdots & \cdots \\ \phi_0(\mathbf{x}_N) & \phi_1(\mathbf{x}_N) & \cdots & \phi_{M-1}(\mathbf{x}_N) \end{pmatrix} \quad (7.177)$$

The likelihood can be rewritten as

$$L = p(\mathbf{t}|\mathbf{X}, \mathbf{w}, \beta) = N\left(\mathbf{t}|\boldsymbol{\Phi}\mathbf{w}, \beta^{-1}\mathbf{I}\right) \quad (7.178)$$

We can also maximize the log likelihood function (7.173) with respect to the noise precision parameter β, then we obtain

$$\frac{1}{\beta_{\mathrm{ML}}} = \frac{1}{N}\sum_{i=1}^{N}\{t_i - \mathbf{w}_{\mathrm{ML}}^{\mathrm{T}}\phi(\mathbf{x}_i)\}^2 \quad (7.179)$$

and then we see that the inverse of the noise precision is given by the residual variance of the target values around the regression function.

7.5.2 Bayesian Method

The MLE method always leads to over-fitting due to outliers in data or sparsity of data. Therefore, we turn to Bayesian method, which can alleviate the over-fitting, and can automatically select the parameters using the training data alone [50]. The Bayesian method can provide posterior distributions over the parameters.

As introduced in Sect. 3.3.2, the linear regression is modeled by Bayesian method, where the weight vector $\mathbf{w}$, defined in (3.29), is assigned with a Gaussian prior distribution defined by (3.31). The posterior distribution over $\mathbf{w}$, also formulated by a Gaussian distribution, is computed by Bayesian rule, and then the weight vector is estimated through maximizing this posterior with respect to $\mathbf{w}$.

Alternatively, in this chapter, we resort to the evidence approximation method [51, 52] for the linear regression. This method is also known as type II maximum likelihood [53].

First of all, we recall that the weight vector $\mathbf{w}$, defined in (7.168), obeys a Gaussian prior distribution, i.e.,

$$p(\mathbf{w}) = N(\mathbf{w}|\boldsymbol{\mu}_0, \boldsymbol{\Sigma}_0) \tag{7.180}$$

where its mean is $\boldsymbol{\mu}_0$ and its covariance is $\boldsymbol{\Sigma}_0$. For simplicity, we consider a particular form of Gaussian prior. Specifically, a zero mean isotropic Gaussian governed by a single precision parameter α is used here as done in Sect. 3.3.2

$$p(\mathbf{w}|\alpha) = N\left(\mathbf{w}|\mathbf{0}, \alpha^{-1}\mathbf{I}\right) \tag{7.181}$$

where $\alpha > 0$. Then, the posterior distribution over $\mathbf{w}$ is given by [49]

$$p(\mathbf{w}|\mathbf{t}, \beta) = N(\mathbf{w}|\boldsymbol{\mu}_N, \boldsymbol{\Sigma}_N) \tag{7.182}$$

where

$$\boldsymbol{\mu}_N = \beta\boldsymbol{\Sigma}_N\boldsymbol{\Phi}^{\mathrm{T}}\mathbf{t} \tag{7.183}$$

and

$$\boldsymbol{\Sigma}_N = \left(\alpha\mathbf{I} + \beta\boldsymbol{\Phi}^{\mathrm{T}}\boldsymbol{\Phi}\right)^{-1} \tag{7.184}$$

The evidence approximation method considers a fully Bayesian viewpoint on this Bayesian linear regression model. Here, we introduce prior distributions over the hyper-parameters α and β and make predictions by marginalizing with respect to these hyper-parameters as well as with respect to the parameters $\mathbf{w}$. In this way, the posterior distributions over there unknown parameters become analytically intractable [49]. Therefore, we discuss an approximation in which we set the hyper-parameters to specific values determined by maximizing the marginal likelihood function. This marginal likelihood is obtained by integrating over the weight parameters $\mathbf{w}$, so that

$$\begin{aligned} p(\mathbf{t}|\alpha,\beta) &= \int p(\mathbf{t}|\mathbf{w},\beta)p(\mathbf{w}|\alpha)d\mathbf{w} \\ &= \int N\left(\mathbf{t}|\mathbf{\Phi}\mathbf{w},\beta^{-1}\mathbf{I}\right)N\left(\mathbf{w}|\mathbf{0},\alpha^{-1}\mathbf{I}\right)d\mathbf{w} \end{aligned} \tag{7.185}$$

Because this integral contains the conditional distribution in a linear Gaussian model. Thus, we can easily give this integral of the Gaussian form,

$$p(\mathbf{t}|\alpha,\beta) = N\left(\mathbf{t}|\mathbf{0},\beta^{-1}\mathbf{I} + \alpha^{-1}\mathbf{\Phi}\mathbf{\Phi}^{\mathrm{T}}\right) \tag{7.186}$$

The log of the marginal likelihood is given by

$$\log p(\mathbf{t}|\alpha,\beta) = \frac{M}{2}\log\alpha + \frac{N}{2}\log\beta - \frac{1}{2}\log|\mathbf{\Sigma}_N| - \frac{N}{2}\log(2\pi) - E(\boldsymbol{\mu}_N) \tag{7.187}$$

where $\mathbf{\Sigma}_N$ is defined by (7.184), $\boldsymbol{\mu}_N$ is defined by (7.183), and $E(\boldsymbol{\mu}_N) = \frac{\beta}{2}\|\mathbf{t} - \mathbf{\Phi}\boldsymbol{\mu}_N\|^2 + \frac{\alpha}{2}\boldsymbol{\mu}_N^{\mathrm{T}}\boldsymbol{\mu}_N$. This is called the *evidence function*.

Now, let us consider maximization of $p(\mathbf{t}|\alpha,\beta)$ with respect to α and β. Alternatively, we do this by maximizing $\log p(\mathbf{t}|\alpha,\beta)$ defined by (7.187). First, we maximize $\log p(\mathbf{t}|\alpha,\beta)$ with respect to α. Consider the following eigenvector equation, i.e.,

$$\left(\beta\mathbf{\Phi}^T\mathbf{\Phi}\right)\mathbf{u}_i = \lambda_i\mathbf{u}_i \tag{7.188}$$

where $\{\lambda_i\}$ denote the eigenvalues of $\beta\mathbf{\Phi}^{\mathrm{T}}\mathbf{\Phi}$, and from (7.184), it can be seen that $\mathbf{\Sigma}_N$ has eigenvalues $\alpha + \lambda_i$. Then, consider the derivative of $\log|\mathbf{\Sigma}_N|$ in (7.187) with respect to α, one has

$$\frac{\partial\log|\mathbf{\Sigma}_N|}{\partial\alpha} = \frac{\partial\log\prod_i(\lambda_i+\alpha)}{\partial\alpha} = \sum_i\frac{1}{\lambda_i+\alpha} \tag{7.189}$$

Thus, the stationary points of (7.187) with respect to α satisfy

$$0 = \frac{M}{2\alpha} - \frac{1}{2}\boldsymbol{\mu}_N^{\mathrm{T}}\boldsymbol{\mu}_N - \frac{1}{2}\sum_i\frac{1}{\lambda_i+\alpha} \tag{7.190}$$

Defining the quantity $\gamma = \sum_i\frac{\lambda_i}{\lambda_i+\alpha}$, one then has the iteration rule of α, i.e.,

$$\alpha = \frac{\gamma}{\boldsymbol{\mu}_N^T\boldsymbol{\mu}_N} \tag{7.191}$$

Besides, maximizing (7.187) with respect to β can be done in a similar way. To do this, we note that the eigenvalues λ_i defined by (7.188) are proportional to β, and then $\frac{\partial\lambda_i}{\partial\beta} = \frac{\lambda_i}{\beta}$. Thus, one has

$$\frac{\partial\log|\mathbf{\Sigma}_N|}{\partial\beta} = \frac{\partial\log\prod_i(\lambda_i+\alpha)}{\partial\beta} = \frac{1}{\beta}\sum_i\frac{\lambda_i}{\lambda_i+\alpha} = \frac{\gamma}{\beta} \tag{7.192}$$

The stationary point of the marginal likelihood therefore satisfies

$$0 = \frac{M}{2\beta} - \frac{1}{2}\sum_{i=1}^{N}\left\{t_i - \boldsymbol{\mu}_N^{\mathrm{T}}\boldsymbol{\phi}(\mathbf{x}_i)\right\}^2 - \frac{\gamma}{2\beta} \tag{7.193}$$

and one can obtain the iteration rule of β, i.e.,

$$\frac{1}{\beta} = \frac{1}{N-\gamma}\sum_{i=1}^{N}\left\{t_i - \boldsymbol{\mu}_N^{\mathrm{T}}\boldsymbol{\phi}(\mathbf{x}_i)\right\}^2 \tag{7.194}$$

Therefore, after obtaining the estimation of α, β, and the weights $\mathbf{w}$, one can make predictions for new input vectors, and since the corresponding formula is given in Sect. 3.3.2, here we do not present it.

7.5.3 *Variational Inference*

Variational inference is a kind of approximate method for computing the posterior distribution over parameters in a probabilistic model. It relates to the functional in mathematics. Generally, a function is treated as a mapping which considers the value of a variable as the input and produces the value of the function as the output. Similarly, a functional is also defined as a mapping that considers a function as the input and that produces the value of the functional as the output.

Let us consider in the inference problem how the variational optimization is carried out. Suppose there is a fully Bayesian model where all parameters are assigned with prior distributions. Latent variables may also be involved in parameters, and we shall denote the set of all latent variables and parameters by $\mathbf{Z}$. Similarly, we denote the set of all observed variables by $\mathbf{X}$. For example, we might have a set of N independent, identically distributed data, i.e., $\mathbf{X} = [\mathbf{x}_1, \mathbf{x}_2, \ldots, \mathbf{x}_N]^{\mathrm{T}}$ and $\mathbf{Z} = [\mathbf{z}_1, \mathbf{z}_2, \ldots, \mathbf{z}_N]^{\mathrm{T}}$. The probabilistic model is specified by the joint distribution $p(\mathbf{X}, \mathbf{Z})$, and our goal is to find an approximation for the posterior distribution $p(\mathbf{Z}|\mathbf{X})$ as well as for the marginal likelihood of the model $p(\mathbf{X})$. According to Bayesian rule, one can obtain

$$\begin{aligned}\log p(\mathbf{X}) &= \log\left(\frac{p(\mathbf{X}, \mathbf{Z})}{p(\mathbf{Z}|\mathbf{X})}\right) \\ &= \log p(\mathbf{X}, \mathbf{Z}) - \log p(\mathbf{Z}|\mathbf{X})\end{aligned} \tag{7.195}$$

Next, we introduce an arbitrary distribution over $\mathbf{Z}$, $q(\mathbf{Z})$, and it gives

$$\log p(\mathbf{X}) = [\log p(\mathbf{X}, \mathbf{Z}) - \log q(\mathbf{Z})] - [\log p(\mathbf{Z}|\mathbf{X}) - \log q(\mathbf{Z})] \tag{7.196}$$

Then, one has

$$\log p(\mathbf{X}) = \log\left(\frac{p(\mathbf{X},\mathbf{Z})}{q(\mathbf{Z})}\right) - \log\left(\frac{p(\mathbf{Z}|\mathbf{X})}{q(\mathbf{Z})}\right) \tag{7.197}$$

Taking the expectation given $q(\mathbf{Z})$ on both sides gives

$$\begin{aligned}\log p(\mathbf{X}) &= \int q(\mathbf{Z})\log\left(\frac{p(\mathbf{X},\mathbf{Z})}{q(\mathbf{Z})}\right)d\mathbf{Z} - \int q(\mathbf{Z})\log\left(\frac{p(\mathbf{Z}|\mathbf{X})}{q(\mathbf{Z})}\right)d\mathbf{Z}\\ &= \underbrace{\int q(\mathbf{Z})\log p(\mathbf{X},\mathbf{Z})d\mathbf{Z} - \int q(\mathbf{Z})\log q(\mathbf{Z})d\mathbf{Z}}_{L(q)} + \underbrace{\left(-\int q(\mathbf{Z})\log\left(\frac{p(\mathbf{Z}|\mathbf{X})}{q(\mathbf{Z})}\right)d\mathbf{Z}\right)}_{\mathrm{KL}(q\|p)}\\ &= L(q) + \mathrm{KL}(q\|p)\end{aligned} \tag{7.198}$$

where $L(q)$ is known as the lower bound of $\log p(\mathbf{X})$, and $\mathrm{KL}(q\|p)$ is the KL divergence between distribution q and p.

Besides, there is another way to achieve the above derivation using Jensen's inequality [49].

$$\begin{aligned}\log p(\mathbf{X}) &= \log\int_{\mathbf{Z}} p(\mathbf{X},\mathbf{Z})d\mathbf{Z}\\ &= \log\int_{\mathbf{Z}} p(\mathbf{X},\mathbf{Z})\frac{q(\mathbf{Z})}{q(\mathbf{Z})}d\mathbf{Z}\\ &= \log\left(E_q\left[\frac{p(\mathbf{X},\mathbf{Z})}{q(\mathbf{Z})}\right]\right)\\ &\geq E_q\left[\log\left(\frac{p(\mathbf{X},\mathbf{Z})}{q(\mathbf{Z})}\right)\right]\\ &= \int q(\mathbf{Z})\log p(\mathbf{X},\mathbf{Z})d\mathbf{Z} + \int q(\mathbf{Z})\log q(\mathbf{Z})d\mathbf{Z}\\ &\triangleq L(q)\end{aligned} \tag{7.199}$$

It can be proven easily that the "missing" part is $\log p(\mathbf{X}) - L(q) = \mathrm{KL}(q\|p)$.

We want to choose a $q(\mathbf{Z})$ function that minimizes the KL divergence, so that $q(\mathbf{Z})$ becomes closer to $p(\mathbf{Z}|\mathbf{X})$, and when $q(\mathbf{Z}) = p(\mathbf{Z}|\mathbf{X})$, one can obtain

$$\mathrm{KL}(q\|p) = -\int p(\mathbf{Z}|\mathbf{X})\log\left(\frac{p(\mathbf{Z}|\mathbf{X})}{p(\mathbf{Z}|\mathbf{X})}\right)d\mathbf{Z} = 0 \tag{7.200}$$

Minimizing $\mathrm{KL}(q\|p)$ is the same as maximizing the lower bound $L(q)$.

One way to do this is to restrict the family of $L(q)$, which can be done by using a parametric distribution $q(\mathbf{Z}|\mathbf{w})$ governed by a set of parameters $\mathbf{w}$. The lower bound $L(q)$ then becomes a function of $\mathbf{w}$, and the optimal values for the parameters can be determined by using well-known nonlinear optimization techniques. Suppose that let us choose $q(\mathbf{Z})$, of the form

$$q(\mathbf{Z}) = \prod_{i=1}^{M} q_i(\mathbf{Z}_i) \tag{7.201}$$

This factorized form of variational inference corresponds to an approximation framework developed in physics called mean field theory [54]. Substitute this choice into the lower bound $L(q)$, then dissect out the dependence on one of the factors $q_j(\mathbf{Z}_j)$. Denoting $q_j(\mathbf{Z}_j)$ by simply q_j to keep the notation uncluttered, we can obtain

$$\begin{aligned} L(q) &= \int q(\mathbf{Z})\log p(\mathbf{X},\mathbf{Z})d\mathbf{Z} - \int q(\mathbf{Z})\log q(\mathbf{Z})d\mathbf{Z} \\ &= \underbrace{\int \prod_{i=1}^{M} q_i(\mathbf{Z}_i)\log p(\mathbf{X},\mathbf{Z})d\mathbf{Z}}_{L_1(q)} - \underbrace{\int \prod_{i=1}^{M} q_i(\mathbf{Z}_i)\sum_{i=1}^{M}\log q_i(\mathbf{Z}_i)d\mathbf{Z}}_{L_2(q)} \end{aligned} \tag{7.202}$$

where

$$\begin{aligned} L_1(q) &= \int \prod_{i=1}^{M} q_i(\mathbf{Z}_i)\log p(\mathbf{X},\mathbf{Z})d\mathbf{Z} \\ &= \int_{\mathbf{Z}_1}\int_{\mathbf{Z}_2}\cdots\int_{\mathbf{Z}_M} \prod_{i=1}^{M} q_i(\mathbf{Z}_i)\log p(\mathbf{X},\mathbf{Z})d\mathbf{Z}_1 d\mathbf{Z}_2,\ \ldots, d\mathbf{Z}_M \end{aligned} \tag{7.203}$$

Rearrange the expression by taking a particular $q_j(\mathbf{Z}_j)$ out of the integral gives

$$L_1(q) = \int_{\mathbf{Z}_j} q_j(\mathbf{Z}_j)\left(\int_{\mathbf{Z}_{i\neq j}}\cdots\int \log p(\mathbf{X},\mathbf{Z})\prod_{i\neq j}^{M} q_i(\mathbf{Z}_i)d\mathbf{Z}_i\right)d\mathbf{Z}_j \tag{7.204}$$

or it can be converted into an expectation function, and since $\prod_{i\neq j}^{M} q_i(\mathbf{Z}_i)$ is a joint probability density

$$L_1(q) = \int_{\mathbf{Z}_j} q_j(\mathbf{Z}_j)\left(E_{i\neq j}[\log p(\mathbf{X},\mathbf{Z})]\right)d\mathbf{Z}_j \tag{7.205}$$

The second part $L_2(q)$ is expressed as

$$L_2(q) = \int \prod_{i=1}^{M} q_i(\mathbf{Z}_i)\sum_{i=1}^{M}\log q_i(\mathbf{Z}_i)d\mathbf{Z} \tag{7.206}$$

Note that the above needs to integrate out all $\mathbf{Z} = \{\mathbf{Z}_1, \mathbf{Z}_2, \ldots, \mathbf{Z}_M\}$. However, each term in the sum, $\sum_{i=1}^{M}\log q_i(\mathbf{Z}_i)$ involves only a single i; therefore, we are able to simplify the above into the following:

$$L_2(q) = \sum_{i=1}^{M}\left(\int_{\mathbf{Z}_i} q_i(\mathbf{Z}_i)\log q_i(\mathbf{Z}_i)d\mathbf{Z}_i\right) \tag{7.207}$$

For a particular $q_j(\mathbf{Z}_j)$, the rest of the sum can be treated like a constant, then $L_2(q)$ can be written as:

$$L_2(q) = \int_{\mathbf{Z}_j} q_j(\mathbf{Z}_j)\log q_j(\mathbf{Z}_j)d\mathbf{Z}_j + \text{const} \tag{7.208}$$

where const is the term that does not involve $\mathbf{Z}_j$.

Thereby, the entire lower bound $L(q)$ is rewritten as

$$\begin{aligned} L(q) &= L_1(q) - L_2(q) \\ &= \int_{\mathbf{Z}_j} q_j(\mathbf{Z}_j)\left(E_{i\neq j}[\log p(\mathbf{X},\mathbf{Z})]\right)d\mathbf{Z}_j - \int_{\mathbf{Z}_j} q_j(\mathbf{Z}_j)\log q_j(\mathbf{Z}_j)d\mathbf{Z}_j + \text{const} \end{aligned} \tag{7.209}$$

Since $E_{i\neq j}[\log p(\mathbf{X},\mathbf{Z})]$ would be some $\log p(\mathbf{Z}_j)$, we denote it by $\log\tilde{p}_j(\mathbf{X},\mathbf{Z}_j)$, i.e.,:

$$\log\tilde{p}_j(\mathbf{X},\mathbf{Z}_j) = E_{i\neq j}[\log p(\mathbf{X},\mathbf{Z})] \tag{7.210}$$

Or equivalently we can express lower bound in terms of:

$$L(q_j) = \int_{\mathbf{Z}_j} q_j(\mathbf{Z}_j)\log\left(\frac{\log\tilde{p}_j(\mathbf{X},\mathbf{Z}_j)}{q_j(\mathbf{Z}_j)}\right)d\mathbf{Z}_j + \text{const} \tag{7.211}$$

This is the same as $-\mathrm{KL}(E_{i\neq j}[\log p(\mathbf{X},\mathbf{Z})] \parallel q_j(\mathbf{Z}_j))$. We can maximize the lower bound $L(q)$, and then we find optimal $q_j^*(\mathbf{Z}_j)$, i.e.,

$$\log q_j^*(\mathbf{Z}_j) = E_{i\neq j}[\log p(\mathbf{X},\mathbf{Z})] \tag{7.212}$$

This is the iterative rule for maximizing the lower bound. It means that we can obtain the log of the optimal solution for factor q_j simply by considering the log of the joint distribution over all hidden and visible variables and then taking the expectation with respect to all of the other factors $\{q_j\}$ for $i \neq j$.

From (7.212), we see that for $j = 1, \ldots, M$ the M variables are coupled each other, and this equation does not represent an explicit solution; therefore, one must seek an optimal solution by first initializing all of the factors $q_i(\mathbf{Z}_i)$ appropriately and then iterating these equations for pursuing a convergence.

We now illustrate the factorized variational approximation using a Gaussian distribution over a single variable x. Our goal is to infer the posterior distribution for mean μ and precision β, given a dataset $D = \{x_1, x_2, \ldots, x_N\}$ of observed values of x which are assumed to be drawn independently from the Gaussian. The likelihood function is given by

$$\begin{aligned} p(D|\mu,\beta) &= \prod_{i=1}^{N}\left(\frac{\beta}{2\pi}\right)^{\frac{1}{2}}\exp\left(\frac{-\beta}{2}(x_i-\mu)^2\right) \\ &= \left(\frac{\beta}{2\pi}\right)^{\frac{N}{2}}\exp\left(\frac{-\beta}{2}\sum_{i=1}^{N}(x_i-\mu)^2\right) \end{aligned} \tag{7.213}$$

And the prior distributions over μ and β are given by

$$p(\mu|\beta) = N\left(\mu_0, (\lambda_0\beta)^{-1}\right) \propto \exp\left(\frac{-\lambda_0\beta}{2}(\mu-\mu_0)^2\right) \tag{7.214}$$

and

$$p(\beta) = \text{Gamma}(\beta|a_0, b_0) \propto \beta^{a_0-1}\mathrm{e}^{-b_0\beta} \tag{7.215}$$

Besides, the complete data-likelihood is expressed as

$$p(D,\mu,\beta) = p(D|\mu,\beta)p(\mu|\beta)p(\beta) \tag{7.216}$$

Due to conjugacy, the exact solution can be found, i.e.,

$$p(\mu,\beta|D) \propto p(D|\mu,\beta)p(\mu|\beta)p(\beta) = N\left(\mu_*, (\lambda_*\beta)^{-1}\right)\text{Gamma}(\beta|a_*, b_*) \tag{7.217}$$

where

$$\mu_* = \frac{\lambda_0\mu_0 + N\bar{x}}{\lambda_0 + N} \tag{7.218}$$

$$\lambda_* = \lambda_0 + N \tag{7.219}$$

$$a_* = a_0 + N/2 \tag{7.220}$$

$$b_* = b_0 + \frac{1}{2}\sum_{i=1}^{N}(x_i - \bar{x})^2 + \frac{\lambda_0 N(\bar{x}-\mu_0)}{2(\lambda_0+N)} \tag{7.221}$$

However, we use the variational inference to tackle this problem, assuming $q(\mu,\beta)$ is formulated in a factorized form, i.e.,

$$q(\mu,\beta) = q_\mu(\mu)q_\beta(\beta) \tag{7.222}$$

By using (7.212), $\log q_\mu^*(\mu)$ is given by

$$\begin{aligned}
\log q_\mu^*(\mu) &= E_{q_\beta}[\log p(\mu,\beta|D)] \\
&= E_{q_\beta}[\log p(D|\mu,\beta) + \log p(\mu|\beta)] + \text{const} \\
&= E_{q_\beta}\left[-\frac{\beta}{2}\sum_{i=1}^{N}(x_i-\mu)^2 + \frac{\lambda_0\beta}{2}(\mu-\mu_0)^2\right] + \text{const} \qquad (7.223) \\
&= -\frac{E_{q_\beta}[\beta]}{2}\left[\sum_{i=1}^{N}(x_i-\mu)^2 + \lambda_0(\mu-\mu_0)^2\right] + \text{const}
\end{aligned}$$

Completing the square for the μ terms gives

$$\begin{aligned}
&\sum_{i=1}^{N}(x_i-\mu)^2 + \lambda_0(\mu-\mu_0)^2 = N\mu^2 - 2N\mu\bar{x} + \lambda_0\mu^2 - 2\lambda_0\mu_0\mu + \text{const} \\
&= (N+\lambda_0)\mu^2 - 2\mu\left(N\bar{x}+\lambda_0\mu_0\right) = (N+\lambda_0)\left(\mu^2 - \frac{2\mu\left(N\bar{x}+\lambda_0\mu_0\right)}{N+\lambda_0}\right) \qquad (7.224) \\
&= (N+\lambda_0)\left(\mu - \frac{\left(N\bar{x}+\lambda_0\mu_0\right)}{N+\lambda_0}\right)^2 + \text{const}
\end{aligned}$$

Therefore, we have

$$\log q_\mu^*(\mu) = \log N\left(\frac{N\bar{x}+\lambda_0\mu_0}{N+\lambda_0},\ E_{q_\beta}[\beta](N+\lambda_0)\right) \qquad (7.225)$$

Similarly, the posterior over β is given by

$$\begin{aligned}
\log q_\beta^*(\beta) &= E_{q_\mu}[\log p(\mu,\beta|D)] \\
&= E_{q_\mu}[\log p(D|\mu,\beta) + \log p(\mu|\beta) + \log p(\beta)] + \text{const} \\
&= E_{q_\mu}\left[\frac{N}{2}\log(\beta) - \frac{\beta}{2}\sum_{i=1}^{N}(x_i-\mu)^2 - \frac{\lambda_0\beta}{2}(\mu-\mu_0)^2 + (a_0-1)\log(\beta) - b_0\beta\right] \\
&\quad + \text{const}
\end{aligned} \qquad (7.226)$$

Bring terms without μ outside of the integral gives

$$
\begin{aligned}
\log q_\beta^*(\beta) &= E_{q_\mu}[\log p(\mu,\beta|D)] \\
&= E_{q_\mu}[\log p(D|\mu,\beta)+\log p(\mu|\beta)+\log p(\beta)]+\text{const} \\
&= E_{q_\mu}\left[\frac{N}{2}\log(\beta)-\frac{\beta}{2}\sum_{i=1}^{N}(x_i-\mu)^2-\frac{\lambda_0\beta}{2}(\mu-\mu_0)^2+(a_0-1)\log(\beta)-b_0\beta\right]+\text{const} \\
&= E_{q_\mu}\left[\left(\frac{N}{2}+a_0-1\right)\log(\beta)-\beta\left[\frac{1}{2}\sum_{i=1}^{N}(x_i-\mu)^2+\frac{\lambda_0}{2}(\mu-\mu_0)^2+b_0\right]\right]+\text{const}
\end{aligned}
\tag{7.227}
$$

We can rewrite the term b_*, i.e.,

$$
\begin{aligned}
b_* &= b_0+\frac{1}{2}E_{q_\mu}\left[\sum_{i=1}^{N}(x_i-\mu)^2+\lambda_0(\mu-\mu_0)^2\right] \\
&= b_0+\frac{1}{2}E_{q_\mu}\left[-2\mu N\bar{x}+N\mu^2+\lambda_0\mu^2-2\lambda_0\mu_0\mu\right] \\
&+\sum_{i=1}^{N}x_i^2+\lambda_0\mu_0^2 = b_0+\frac{1}{2}\left[(N+\lambda_0)E_{q_\mu}\left[\mu^2\right]-2\left(N\bar{x}+\lambda_0\mu_0\right)E_{q_\mu}[\mu]+\sum_{i=1}^{N}x_i^2+\lambda_0\mu_0^2\right]
\end{aligned}
\tag{7.228}
$$

Finally, we can compute $E_{q_\mu}[\mu]$ and $E_{q_\mu}[\mu^2]$ since $q_\mu(\mu)$ is known from previously.

7.5.4 *Variational Relevance Vector Machine Based on Automatic Relevance Determination Kernel Functions*

In this section, we will introduce an automatic relevance determination kernel (ARDK) function-based variational relevance vector machine (RVM) model (VRVM-ARDK) for embedded feature selection [55]. In this model, an ARDK is used, where each input feature corresponds to an independent scaling kernel parameter. The posterior distributions over these scaling parameters are optimized approximately in the variational inference process. First, we will introduce the preliminaries on variational RVM.

Preliminaries on Variational RVM (VRVM)

Suppose that there is a mapping from an input column vector $\mathbf{x}_i$ to its corresponding output $y_i \in R$, i.e.,

$$y(\mathbf{x}_i|\mathbf{w}) = \sum_{j=1}^{N} w_j K(\mathbf{x}_i, \mathbf{x}_j) + w_0 \mid y(\mathbf{x}_i|\mathbf{w}) = \boldsymbol{\phi}_i^{\mathrm{T}} \mathbf{w} \tag{7.229}$$

where $\boldsymbol{\phi}_i = [1, K(\mathbf{x}_i, \mathbf{x}_1), K(\mathbf{x}_i, \mathbf{x}_2), \ldots, K(\mathbf{x}_i, \mathbf{x}_N)]^{\mathrm{T}}$ and $\mathbf{w} = [w_0, w_1, w_2, \ldots, w_N]^{\mathrm{T}}$ denote the weight vector, $K(\mathbf{x}_i, \mathbf{x}_j)$ is the kernel function, $\mathbf{x}_i = [x_{i1}, x_{i2}, \ldots, x_{ik}, \ldots, x_{ip}]^{\mathrm{T}}$ is the ith input vector with its input dimensionality p, $i = 1,2,\ldots,N$, and N is the number of samples. The observed targets are $\mathbf{t} = [t_1, t_2, \ldots, t_i, \ldots, t_N]^{\mathrm{T}}$, where its ith element $t_i = y(\mathbf{x}_i|\mathbf{w}) + \varepsilon_i$ and ε_i denotes a Gaussian noise with mean zero and variance σ^2. As such, the likelihood function for $\mathbf{t}$ is represented by

$$\begin{aligned} p(\mathbf{t}|\mathbf{w}, \beta) &= N\left(\mathbf{t}|\Phi\mathbf{w}, \beta^{-1}\mathbf{I}\right) \\ &= \frac{(N\beta)^{\frac{1}{2}}}{(2\pi)^{N/2}} \exp\left\{-\frac{\beta}{2}[\mathbf{t} - \Phi\mathbf{w}]^{\mathrm{T}}[\mathbf{t} - \Phi\mathbf{w}]\right\} \end{aligned} \tag{7.230}$$

where $\beta \equiv \sigma^{-2}$ and $\boldsymbol{\Phi} = [\boldsymbol{\phi}_1, \boldsymbol{\phi}_2, \ldots, \boldsymbol{\phi}_i, \ldots, \boldsymbol{\phi}_N]^{\mathrm{T}}$. To infer the weights $\mathbf{w}$, a Gaussian prior distribution with mean zero and variance α_i^{-1} is considered over each weight in $\mathbf{w}$ in the RVM [56], i.e.,

$$p(\mathbf{w}|\boldsymbol{\alpha}) = \prod_{i=0}^{N} N\left(w_i|0, \alpha_i^{-1}\right) \tag{7.231}$$

where $\boldsymbol{\alpha} = [\alpha_0, \alpha_1, \ldots, \alpha_N]^{\mathrm{T}}$ and $\mathbf{A}_{\boldsymbol{\alpha}} = \mathrm{diag}\,(\boldsymbol{\alpha})$. Treating the parameters $\boldsymbol{\alpha}$ and β as random variables, these parameters are associated with the Gamma prior distributions.

$$p(\boldsymbol{\alpha}) = \prod_{i=0}^{N} \Gamma(\alpha_i|a_i, b_i), \quad p(\beta) = \Gamma(\beta|c, d) \tag{7.232}$$

where a_i, b_i, c, and d are the parameters of the Gamma distribution.

According to the variational mean field theory, an arbitrary distribution, $Q(\mathbf{w}, \boldsymbol{\alpha}, \beta) = Q_{\mathbf{w}}(\mathbf{w})Q_{\boldsymbol{\alpha}}(\boldsymbol{\alpha})Q_{\beta}(\beta)$, was used to approximate the true posterior distribution $p(\mathbf{w}, \boldsymbol{\alpha}, \beta|\,\mathbf{t})$. Then, the optimal variational posterior over these variables can be computed by maximizing the lower bound $L = \int Q(\mathbf{w}, \boldsymbol{\alpha}, \beta)\log\frac{p(\mathbf{t}, \mathbf{w}, \boldsymbol{\alpha}, \beta|\mathbf{X})}{Q(\mathbf{w}, \boldsymbol{\alpha}, \beta)} d\mathbf{w}d\boldsymbol{\alpha}d\beta$. Since $Q_{\mathbf{w}}(\mathbf{w})$, $Q_{\boldsymbol{\alpha}}(\boldsymbol{\alpha})$, and $Q_{\beta}(\beta)$ are coupled with each other, one can compute these variational posteriors in an iterative manner, during which the relevant samples can be selected by identifying the non-zero elements within the mean vector of $Q_{\mathbf{w}}(\mathbf{w})$.

Although the posterior distributions over the variables are inferred in the VRVM, the point estimate of the kernel function parameters rather than a posterior distribution has to be determined by cross-validation.

Model Specification for the VRVM-ARDK

Based on the preliminaries of the variational RVM, we will introduce the description of the VRVM-ARDK. Considering an ARDK function in the RVM modeling and setting the ranges of the values of the kernel parameters on the entire real number axis, we use an ARDK in the following form:

$$K_{\text{ard}}\left(\mathbf{x}_i, \mathbf{x}_j\right) = \exp\left\{-\frac{r}{2}\sum_{k=1}^{p}\left(s_k^2 + \varepsilon\right)\left(x_{ik} - x_{jk}\right)^2\right\} \tag{7.233}$$

where $r > 0$ is a predefined constant, $s_k \in R$ ($k = 1,2,\ldots, p$) is the kernel parameter, and $\mathbf{s} = [s_1, s_2, \ldots, s_k, \ldots, s_p]^{\mathrm{T}}$. ε is a fixed very small positive value that guarantees the inequality $s_k^2 + \varepsilon > 0$. To infer the posterior distribution over $\mathbf{s}$, a zero mean Gaussian prior distribution is assigned, i.e.,

$$p(\mathbf{s}|\boldsymbol{\lambda}) = N\left(\mathbf{s}|\mathbf{0}, \mathbf{A}_{\lambda}^{-1}\right) \tag{7.234}$$

where $\mathbf{A}_{\lambda} = \text{diag}\ (\boldsymbol{\lambda})$ and $\boldsymbol{\lambda} = [\lambda_1, \lambda_2, \ldots, \lambda_p]^{\mathrm{T}}$. To complete the hierarchical setting of $\mathbf{s}$, this study fixes the prior over $\boldsymbol{\lambda}$ on the Gamma distribution.

$$p(\boldsymbol{\lambda}) = \prod_{i=1}^{p}\Gamma(\lambda_i|e_i, f_i) \tag{7.235}$$

The prior distributions of the rest of the parameters $\mathbf{w}$, $\boldsymbol{\alpha}$, and β are defined by (7.231)–(7.232), respectively. Therefore, this model is represented by a directed acyclic graph illustrated in Fig. 7.12, where the white circles denote uncertain variables and the shaded ones denote the observed variables.

The uncertain variables in the model consist of $\mathbf{w}$, $\boldsymbol{\alpha}$, β, $\mathbf{s}$, and $\boldsymbol{\lambda}$. According to Bayesian principle, the joint posterior distribution over these variables can be formulated as

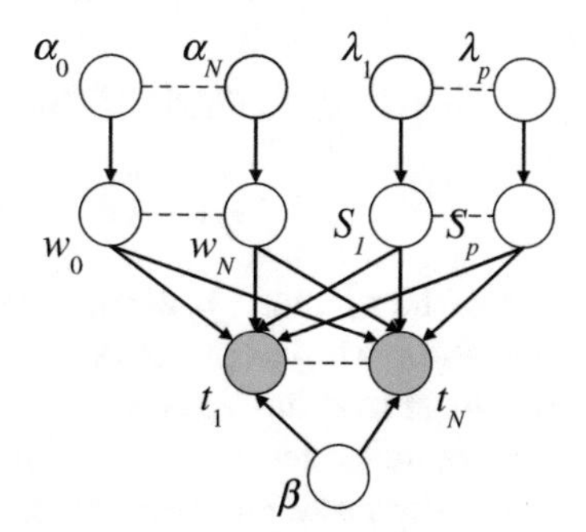

Fig. 7.12 Directed acyclic graphs representing the RVM-ARDK model

$$p(\mathbf{w}, \mathbf{s}, \beta, \boldsymbol{\lambda}, \boldsymbol{\alpha}|\mathbf{t}) = \frac{p(\mathbf{t}|\mathbf{s}, \mathbf{w}, \beta)p(\mathbf{s}|\boldsymbol{\lambda})p(\mathbf{w}|\boldsymbol{\alpha})p(\boldsymbol{\alpha})p(\boldsymbol{\lambda})p(\beta)}{p(\mathbf{t})} \tag{7.236}$$

where $p(\mathbf{t}) = \int p(\mathbf{t}, \mathbf{w}, \mathbf{s}, \beta, \boldsymbol{\lambda}, \boldsymbol{\alpha}) d\mathbf{w} d\mathbf{s} d\beta d\boldsymbol{\lambda} d\boldsymbol{\alpha}$. Since it is hard to compute the integral in the marginal likelihood $p(\mathbf{t})$, the accurate solution of this posterior is not available.

Variational Inference for RVM-ARDK

To compute the joint posterior $p(\mathbf{w}, \mathbf{s}, \beta, \boldsymbol{\lambda}, \boldsymbol{\alpha}|\, \mathbf{t})$, one can combine Bayesian variational inference and Laplace approximation to obtain its approximate solution. According to the mean field variational theory, if an arbitrary decomposable distribution $Q(\mathbf{w}, \boldsymbol{\alpha}, \beta, \mathbf{s}, \boldsymbol{\lambda}) = Q_{\mathbf{w}}(\mathbf{w})Q_{\boldsymbol{\alpha}}(\boldsymbol{\alpha})Q_{\beta}(\beta)Q_{\mathbf{s}}(\mathbf{s})Q_{\boldsymbol{\lambda}}(\boldsymbol{\lambda})$ is employed to approximate the posterior, the explicit form of the posterior of the kernel parameter $\mathbf{s}$ will not be available because it nonlinearly exists in the kernel function.

Here, we carry out a first-order Taylor expansion for local linearization, which produces a quadratic form of the log of the posterior over $\mathbf{s}$ so as to approximate its posterior. As for the other variables, given that their priors are conjugate, the optimal posteriors can be formulated in the same form as their priors. The inference can be addressed as follows:

The lower bound of the logarithm of the joint distribution $\log p(\mathbf{t}, \mathbf{w}, \mathbf{s}, \beta, \boldsymbol{\lambda}, \boldsymbol{\alpha}) = \log p(\mathbf{t}|\, \mathbf{s}, \mathbf{w}, \beta) + \log p(\mathbf{s}|\, \boldsymbol{\lambda}) + \log p(\mathbf{w}|\, \boldsymbol{\alpha}) + \log p(\boldsymbol{\alpha}) + \log p(\boldsymbol{\lambda}) + \log p(\beta)$ is presented by using the Jensen's inequality

$$\begin{aligned} &\log p\,(\mathbf{t}, \mathbf{w}, \mathbf{s}, \beta, \boldsymbol{\lambda}, \boldsymbol{\alpha}) \\ &\geq \mathrm{L}[Q(\mathbf{w}, \boldsymbol{\alpha}, \beta, \mathbf{s}, \boldsymbol{\lambda})] \\ &= \int Q_{\mathbf{w}}(\mathbf{w})\, Q_{\mathbf{s}}(\mathbf{s})\, Q_{\boldsymbol{\alpha}}(\boldsymbol{\alpha})\, Q_{\beta}(\beta)\, Q_{\boldsymbol{\lambda}}(\boldsymbol{\lambda}) \\ &\times \log \frac{p(\mathbf{t}|\mathbf{s}, \mathbf{w}, \beta)p(\mathbf{s}|\boldsymbol{\lambda})p(\mathbf{w}|\boldsymbol{\alpha})p(\boldsymbol{\alpha})p(\boldsymbol{\lambda})p(\beta)}{Q_{\mathbf{w}}(\mathbf{w})\, Q_{\mathbf{s}}(\mathbf{s})\, Q_{\boldsymbol{\alpha}}(\boldsymbol{\alpha})\, Q_{\beta}(\beta)\, Q_{\boldsymbol{\lambda}}(\boldsymbol{\lambda})} \end{aligned} \tag{7.237}$$

1. Posterior over s based on Laplace approximation

   The mean field theory is employed to maximize the lower bound $\mathrm{L}[Q(\mathbf{w}, \boldsymbol{\alpha}, \beta, \mathbf{s}, \boldsymbol{\lambda})]$, finding the log of the variational posterior distribution over $\mathbf{s}$, i.e.,

$$\begin{aligned} \log Q_{\mathbf{s}}(\mathbf{s}) &= \langle \log p(\mathbf{t}, \mathbf{w}, \mathbf{s}, \beta, \boldsymbol{\lambda}, \boldsymbol{\alpha}) \rangle_{\mathbf{w}, \beta, \boldsymbol{\lambda}, \boldsymbol{\alpha}} + \mathrm{const} \\ &= -\frac{1}{2}\langle \beta \rangle_{\beta} \left\langle [\mathbf{t} - \boldsymbol{\Phi}\mathbf{w}]^{\mathrm{T}} [\mathbf{t} - \boldsymbol{\Phi}\mathbf{w}] \right\rangle_{\mathbf{w}} - \frac{1}{2}\mathbf{s}^{\mathrm{T}} \langle \mathbf{A}_{\boldsymbol{\lambda}} \rangle_{\boldsymbol{\lambda}} \mathbf{s} + \mathrm{const} \end{aligned} \tag{7.238}$$

where the symbol $\langle \cdot \rangle_q$ denotes the expectation taken with respect to the variational probability $Q(q)$ over q. As seen in (7.238), $\mathbf{s}$ exists in the matrix $\boldsymbol{\Phi}$. Given a complex nonlinear relationship between $\mathbf{s}$ and $\log Q_{\mathbf{s}}(\mathbf{s})$, it is very difficult to obtain an explicit form of the posterior over $\mathbf{s}$. To avoid finding the optimum of $\mathbf{s}$, which maximizes the log of the posterior distribution, and the computation of the Hessian matrix in the generic Laplace approximation, we carry out a first-order Taylor expansion around $\mathbf{s}_{\mathrm{M}}$ of the term $\mathbf{\Phi w}$ defined by (7.238) to complete local linearization, i.e.,

$$\mathbf{\Phi w} \approx \mathbf{\Phi}_{\mathbf{s}_{\mathrm{M}}}\mathbf{w} + \mathbf{H}_{\mathbf{s}_{\mathrm{M}}}(\mathbf{s} - \mathbf{s}_{\mathrm{M}}) \tag{7.239}$$

where $\mathbf{s}_{\mathrm{M}}$ is the mean vector of $\mathbf{s}$ obtained in the last variational learning update for maximizing the lower bound, $\mathbf{H}_{\mathbf{s}_{\mathrm{M}}} = \left[\frac{\partial \Phi}{\partial s_1}\Big|_{\mathbf{s}=\mathbf{s}_{\mathrm{M}}}\mathbf{w},\ \frac{\partial \Phi}{\partial s_2}\Big|_{\mathbf{s}=\mathbf{s}_{\mathrm{M}}}\mathbf{w},\ \ldots,\ \frac{\partial \Phi}{\partial s_p}\Big|_{\mathbf{s}=\mathbf{s}_{\mathrm{M}}}\mathbf{w}\right]$, $\mathbf{\Phi}_{\mathbf{s}_{\mathrm{M}}} = \mathbf{\Phi}|_{\mathbf{s}=\mathbf{s}_{\mathrm{M}}}$. The dimensionality of the matrix $\mathbf{H}_{\mathbf{s}_{\mathrm{M}}}$ is $N \times p$. Thus, the $\log Q_{\mathbf{s}}(\mathbf{s})$ is approximated as

$$\begin{aligned}
\log Q_{\mathbf{s}}(\mathbf{s}) &\approx -\frac{1}{2}\langle\beta\rangle_\beta\Big\langle\left[\mathbf{t} - \mathbf{\Phi}_{\mathbf{s}_{\mathrm{M}}}\mathbf{w} - \mathbf{H}_{\mathbf{s}_{\mathrm{M}}}(\mathbf{s} - \mathbf{s}_{\mathrm{M}})\right]^{\mathrm{T}} \\
&\quad \left[\mathbf{t} - \mathbf{\Phi}_{\mathbf{s}_{\mathrm{M}}}\mathbf{w} - \mathbf{H}_{\mathbf{s}_{\mathrm{M}}}(\mathbf{s} - \mathbf{s}_{\mathrm{M}})\right]\Big\rangle_{\mathbf{w}} - \frac{1}{2}\mathbf{s}^{\mathrm{T}}\langle\mathbf{A}_\lambda\rangle_\lambda\mathbf{s} + \text{const} \\
&= -\frac{1}{2}\mathbf{s}^{\mathrm{T}}\left(\langle\mathbf{A}_\lambda\rangle_\lambda + \langle\beta\rangle_\beta\left\langle\mathbf{H}_{\mathbf{s}_{\mathrm{M}}}^{\mathrm{T}}\mathbf{H}_{\mathbf{s}_{\mathrm{M}}}\right\rangle_{\mathbf{w}}\right)\mathbf{s} \\
&\quad + \mathbf{s}^T\langle\beta\rangle_\beta\left(\langle\mathbf{H}_s\rangle_{\mathbf{w}}\mathbf{t} - \left\langle\mathbf{H}_{\mathbf{s}_{\mathrm{M}}}^{\mathrm{T}}\mathbf{\Phi}_{\mathbf{s}_{\mathrm{M}}}\mathbf{w}\right\rangle_{\mathbf{w}} + \left\langle\mathbf{H}_{\mathbf{s}_{\mathrm{M}}}^{\mathrm{T}}\mathbf{H}_{\mathbf{s}_{\mathrm{M}}}\right\rangle_{\mathbf{w}}\mathbf{s}_{\mathrm{M}}\right) + \text{const}
\end{aligned} \tag{7.240}$$

One can see that the $\log Q_{\mathbf{s}}(\mathbf{s})$ is a quadratic form of kernel parameter $\mathbf{s}$, representing its posterior by a Gaussian distribution.

$$Q_{\mathbf{s}}(\mathbf{s}) \approx N(\mathbf{s}|\boldsymbol{\mu}_{\mathbf{s}}, \Sigma_{\mathbf{s}}) \tag{7.241}$$

where

$$\begin{aligned}
\boldsymbol{\mu}_{\mathbf{s}} &= \langle\beta\rangle_\beta\Sigma_{\mathbf{s}}\left(\langle\mathbf{H}_s\rangle_{\mathbf{w}}\mathbf{t} - \left\langle\mathbf{H}_{\mathbf{s}_{\mathrm{M}}}^{\mathrm{T}}\mathbf{\Phi}_{\mathbf{s}_{\mathrm{M}}}\mathbf{w}\right\rangle_{\mathbf{w}} + \left\langle\mathbf{H}_{\mathbf{s}_{\mathrm{M}}}^{\mathrm{T}}\mathbf{H}_{\mathbf{s}_{\mathrm{M}}}\right\rangle_{\mathbf{w}}\mathbf{s}_{\mathrm{M}}\right), \quad \Sigma_{\mathbf{s}} \\
&= \left(\langle\mathbf{A}_\lambda\rangle_\lambda + \langle\beta\rangle_\beta\left\langle\mathbf{H}_{\mathbf{s}_{\mathrm{M}}}^{\mathrm{T}}\mathbf{H}_{\mathbf{s}_{\mathrm{M}}}\right\rangle_{\mathbf{w}}\right)^{-1}
\end{aligned} \tag{7.242}$$

The covariance matrix $\Sigma_{\mathbf{s}}$ can easily be proved to be positive definite, which means that (7.241) is a well-defined Gaussian distribution. The explicit computations of the moments $\langle\mathbf{H}_{\mathbf{s}_{\mathrm{M}}}\rangle_{\mathbf{w}}$, $\left\langle\mathbf{H}_{\mathbf{s}_{\mathrm{M}}}^{\mathrm{T}}\mathbf{\Phi}_{\mathbf{s}_{\mathrm{M}}}\mathbf{w}\right\rangle_{\mathbf{w}}$ and $\left\langle\mathbf{H}_{\mathbf{s}_{\mathrm{M}}}^{\mathrm{T}}\mathbf{H}_{\mathbf{s}_{\mathrm{M}}}\right\rangle_{\mathbf{w}}$ are presented as follows:

$$\langle\mathbf{H}_{\mathbf{s}_{\mathrm{M}}}\rangle_{\mathbf{w}} = \left[\frac{\partial \Phi}{\partial s_1}\bigg|_{\mathbf{s}=\mathbf{s}_{\mathrm{M}}}\langle\mathbf{w}\rangle_{\mathbf{w}}, \ldots, \frac{\partial \Phi}{\partial s_u}\bigg|_{\mathbf{s}=\mathbf{s}_{\mathrm{M}}}\langle\mathbf{w}\rangle_{\mathbf{w}}, \ldots, \frac{\partial \Phi}{\partial s_p}\bigg|_{\mathbf{s}=\mathbf{s}_{\mathrm{M}}}\langle\mathbf{w}\rangle_{\mathbf{w}}\right] \tag{7.243}$$

with

$$\frac{\partial \Phi}{\partial s_u}\bigg|_{\mathbf{s}=\mathbf{s}_{\mathrm{M}}} = \left[\frac{\partial \boldsymbol{\phi}_1}{\partial s_u}\bigg|_{\mathbf{s}=\mathbf{s}_{\mathrm{M}}}, \frac{\partial \boldsymbol{\phi}_2}{\partial s_u}\bigg|_{\mathbf{s}=\mathbf{s}_{\mathrm{M}}}, \ldots, \frac{\partial \boldsymbol{\phi}_i}{\partial s_u}\bigg|_{\mathbf{s}=\mathbf{s}_{\mathrm{M}}}, \ldots, \frac{\partial \boldsymbol{\phi}_N}{\partial s_u}\bigg|_{\mathbf{s}=\mathbf{s}_{\mathrm{M}}}\right]^{\mathrm{T}} \tag{7.244}$$

and $\left.\frac{\partial \boldsymbol{\phi}_i}{\partial s_u}\right|_{\mathbf{s}=\mathbf{s}_\mathrm{M}} = \left[0, \left.\frac{\partial K(\mathbf{x}_i, \mathbf{x}_1)}{\partial s_u}\right|_{\mathbf{s}=\mathbf{s}_\mathrm{M}}, \ldots, \left.\frac{\partial K(\mathbf{x}_i, \mathbf{x}_j)}{\partial s_u}\right|_{\mathbf{s}=\mathbf{s}_\mathrm{M}}, \ldots, \left.\frac{\partial K(\mathbf{x}_i, \mathbf{x}_N)}{\partial s_u}\right|_{\mathbf{s}=\mathbf{s}_\mathrm{M}}\right]^\mathrm{T},$

$$\left.\frac{\partial K(\mathbf{x}_i, \mathbf{x}_j)}{\partial s_u}\right|_{\mathbf{s}=\mathbf{s}_\mathrm{M}} = -r \cdot s_u (x_{iu} - x_{ju})^2 K(\mathbf{x}_i, \mathbf{x}_j)\Big|_{\mathbf{s}=\mathbf{s}_\mathrm{M}} \tag{7.245}$$

Besides, $\left\langle \mathbf{H}_{\mathbf{s}_\mathrm{M}}^\mathrm{T} \Phi_{\mathbf{s}_\mathrm{M}} \mathbf{w} \right\rangle_\mathbf{w}$ and $\left\langle \mathbf{H}_{\mathbf{s}_\mathrm{M}}^\mathrm{T} \mathbf{H}_{\mathbf{s}_\mathrm{M}} \right\rangle_\mathbf{w}$ are written in the form

$$\left\langle \mathbf{H}_{\mathbf{s}_\mathrm{M}}^\mathrm{T} \Phi_{\mathbf{s}_\mathrm{M}} \mathbf{w} \right\rangle_\mathbf{w} = \left[\left\langle \mathbf{w}^\mathrm{T} \left.\frac{\partial \Phi}{\partial s_1}\right|_{\mathbf{s}=\mathbf{s}_\mathrm{M}}^\mathrm{T} \Phi_{\mathbf{s}_\mathrm{M}} \mathbf{w} \right\rangle_\mathbf{w}, \ldots, \left\langle \mathbf{w}^\mathrm{T} \left.\frac{\partial \Phi}{\partial s_p}\right|_{\mathbf{s}=\mathbf{s}_\mathrm{M}}^\mathrm{T} \Phi_{\mathbf{s}_\mathrm{M}} \mathbf{w} \right\rangle_\mathbf{w} \right]^\mathrm{T} \tag{7.246}$$

and

$$\left(\left\langle \mathbf{H}_{\mathbf{s}_\mathrm{M}}^\mathrm{T} \mathbf{H}_{\mathbf{s}_\mathrm{M}} \right\rangle_\mathbf{w} \right)_{i,j} = \left\langle \mathbf{w}^\mathrm{T} \left.\frac{\partial \Phi}{\partial s_i}\right|_{\mathbf{s}=\mathbf{s}_\mathrm{M}}^\mathrm{T} \left.\frac{\partial \Phi}{\partial s_j}\right|_{\mathbf{s}=\mathbf{s}_\mathrm{M}} \mathbf{w} \right\rangle_\mathbf{w}, \quad i,j = 1, 2, \ldots, p \tag{7.247}$$

2. Posteriors over $\boldsymbol{\alpha}$, β, $\mathbf{w}$, and $\boldsymbol{\lambda}$

   As for the remaining uncertain variables $\boldsymbol{\alpha}$, β, $\mathbf{w}$, and $\boldsymbol{\lambda}$, one can easily determine the logarithm of their variational posteriors by using the mean field theory. Thus, the posterior of $\mathbf{w}$ can be formulated as a Gaussian distribution form, i.e.,

$$Q_\mathbf{w}(\mathbf{w}) = N(\mathbf{w} | \boldsymbol{\mu}_w, \Sigma_w) \tag{7.248}$$

where

$$\boldsymbol{\mu}_\mathbf{w} = \langle \beta \rangle_\beta \Sigma_w \langle \Phi \rangle_\mathbf{s}^\mathrm{T} \mathbf{t}, \quad \Sigma_\mathbf{w} = \left(\langle \mathbf{A}_\boldsymbol{\alpha} \rangle_\boldsymbol{\alpha} + \langle \beta \rangle_\beta \langle \Phi^\mathrm{T} \Phi \rangle_\mathbf{s} \right)^{-1} \tag{7.249}$$

and the detailed computation of the moments $\langle \Phi \rangle_\mathbf{s}$ and $\langle \Phi^\mathrm{T} \Phi \rangle_\mathbf{s}$ are given as follows. The expectation matrix $\langle \Phi \rangle_\mathbf{s}$ is written by

$$\langle \Phi \rangle_\mathbf{s} = \left[\langle \boldsymbol{\phi}_1 \rangle_\mathbf{s}, \ \langle \boldsymbol{\phi}_2 \rangle_\mathbf{s}, \ldots, \langle \boldsymbol{\phi}_i \rangle_\mathbf{s}, \ldots, \ \langle \boldsymbol{\phi}_N \rangle_\mathbf{s} \right]^\mathrm{T} \tag{7.250}$$

where $\langle \boldsymbol{\phi}_i \rangle_\mathbf{s} = [1, \ \langle K(\mathbf{x}_i, \mathbf{x}_1) \rangle_\mathbf{s}, \ldots, \langle K(\mathbf{x}_i, \mathbf{x}_j) \rangle_\mathbf{s} \ldots, \langle K(\mathbf{x}_i, \mathbf{x}_N) \rangle_\mathbf{s}]^\mathrm{T}$, with its each element represented as

$$\left\langle K(\mathbf{x}_i, \mathbf{x}_j)\right\rangle_{\mathbf{s}} = \int K(\mathbf{x}_i, \mathbf{x}_j)Q(\mathbf{s})d\mathbf{s}$$
$$= \frac{\exp\left\{-\frac{r}{2}\varepsilon\sum_{k=1}^{p}\left(x_{ik}-x_{jk}\right)^2\right\}}{\left|\Sigma_{\mathbf{s}}\Lambda_{1\_ij}+\mathbf{I}\right|^{1/2}} \times \exp\left\{-\frac{1}{2}\boldsymbol{\mu}_{\mathbf{s}}^{T}\left(\Sigma_{\mathbf{s}}^{-1}-\Sigma_{\mathbf{s}}^{-1}\left(\Sigma_{\mathbf{s}}^{-1}+\Lambda_{1\_ij}\right)^{-1}\Sigma_{\mathbf{s}}^{-1}\right)\boldsymbol{\mu}_{\mathbf{s}}\right\} \tag{7.251}$$

with $\Lambda_{1ij} = \mathrm{diag}\left(\left[r\left(x_{i1}-x_{j1}\right)^2, \quad r\left(x_{i2}-x_{j2}\right)^2, \quad \ldots, \quad r\left(x_{ip}-x_{jp}\right)^2\right]\right)$. As for the computation of $\langle\Phi^{\mathrm{T}}\Phi\rangle_{\mathbf{s}}$, its element at location i^*, j^* is written by $\left(\langle\Phi^{\mathrm{T}}\Phi\rangle_{\mathbf{s}}\right)_{i*,j*} = \sum_{m=1}^{N}\left\langle K(\mathbf{x}_{i*-1}, \mathbf{x}_m)K(\mathbf{x}_m, \mathbf{x}_{j*-1})\right\rangle_{\mathbf{s}}$ for i^*, $j^* \neq 1$, and $\left(\langle\Phi^{\mathrm{T}}\Phi\rangle_{\mathbf{s}}\right)_{1,j*} = \left(\langle\Phi^{\mathrm{T}}\Phi\rangle_{\mathbf{s}}\right)_{i*,1} = \sum_{m=1}^{N}\left\langle K(\mathbf{x}_m, \mathbf{x}_{j*})\right\rangle_{\mathbf{s}}$ for $i^* = 1$, or $j^* = 1$, where

$$\left\langle K(\mathbf{x}_i, \mathbf{x}_m)K(\mathbf{x}_m, \mathbf{x}_j)\right\rangle_{\mathbf{s}} = \int K(\mathbf{x}_i, \mathbf{x}_m)K(\mathbf{x}_m, \mathbf{x}_j)Q(\mathbf{s})d\mathbf{s},$$
$$= \frac{\exp\left\{-\frac{r}{2}\varepsilon\sum_{k=1}^{p}\left[\left(x_{ik}-x_{mk}\right)^2+\left(x_{mk}-x_{jk}\right)^2\right]\right\}}{\left|\Sigma_{\mathbf{s}}\Lambda_{2\_imj}+\mathbf{I}\right|^{1/2}} \times \exp\left\{-\frac{1}{2}\boldsymbol{\mu}_{\mathbf{s}}^{\mathrm{T}}\left(\Sigma_{\mathbf{s}}^{-1}-\Sigma_{\mathbf{s}}^{-1}\left(\Sigma_{\mathbf{s}}^{-1}+\Lambda_{2\_imj}\right)^{-1}\Sigma_{\mathbf{s}}^{-1}\right)\boldsymbol{\mu}_{\mathbf{s}}\right\} \tag{7.252}$$

with $\Lambda_{2\_imj} = \mathrm{diag}\left([r(x_{i1}-x_{m1})^2 + r(x_{m1}-x_{j1})^2, \ldots, r(x_{ip}-x_{mp})^2 + r(x_{mp}-x_{jp})^2]\right)$.

Besides, it is noticeable that the posteriors over $\boldsymbol{\alpha}$, β, and $\boldsymbol{\lambda}$ are also Gamma distributions because of their conjugacy characteristics.

$$Q_{\boldsymbol{\alpha}}(\boldsymbol{\alpha}) = \prod_{i=0}^{N}\Gamma\left(\alpha_i|\tilde{a}_i, \tilde{b}_i\right) \tag{7.253}$$

$$Q_{\boldsymbol{\lambda}}(\boldsymbol{\lambda}) = \prod_{i=1}^{p}\Gamma\left(\lambda_i|\tilde{e}_i, \tilde{f}_i\right) \tag{7.254}$$

$$Q_{\beta}(\beta) = \Gamma\left(\beta, |\tilde{c}, |\tilde{d}\right) \tag{7.255}$$

where $\tilde{a}_i = a_i + 1/2$, $\tilde{b}_i = b_i + \left\langle w_i^2\right\rangle_{\mathbf{w}}/2$, $\tilde{e}_i = e_i + 1/2$, $\tilde{f}_i = f_i + \left\langle s_i^2\right\rangle_{s}/2$, $\tilde{c} = c + N/2, \tilde{d}_i = d_i + \frac{1}{2}\sum_{i=1}^{N}t_i^2 - \langle\mathbf{w}\rangle_{\mathbf{w}}^{\mathrm{T}}\langle\Phi\rangle_{\mathbf{s}}^{\mathrm{T}}\mathbf{t} + \frac{1}{2}\left\langle\mathbf{w}^{\mathrm{T}}\langle\Phi^{\mathrm{T}}\Phi\rangle_{\mathbf{s}}\mathbf{w}\right\rangle_{\mathbf{w}}$, and the aforementioned moments are expressed by $\langle\mathbf{w}\rangle = \boldsymbol{\mu}_w$, $\left\langle\mathbf{w}\mathbf{w}^{\mathrm{T}}\right\rangle = \boldsymbol{\mu}_{\mathbf{w}}\boldsymbol{\mu}_{\mathbf{w}}^{\mathrm{T}} + \boldsymbol{\Sigma}_{\mathbf{w}}$, $\left\langle\mathbf{s}\mathbf{s}^{\mathrm{T}}\right\rangle = \boldsymbol{\mu}_{\mathbf{s}}\boldsymbol{\mu}_{\mathbf{s}}^{\mathrm{T}} + \boldsymbol{\Sigma}_{\mathbf{s}}$, $\left\langle\mathbf{w}^{T}\mathbf{R}'\mathbf{w}\right\rangle = \sum_{i=0}^{N}\sum_{j=0}^{N}\langle w_i w_j\rangle\mathbf{R}'_{ij}$, $\langle\mathbf{s}\rangle = \boldsymbol{\mu}_{\mathbf{s}}$, $\langle\alpha_i\rangle = \tilde{a}_i/\tilde{b}_i$, $\langle\log\alpha_i\rangle = \psi\left(\tilde{a}_i\right) - \log\tilde{b}_i$, $\langle\log\lambda_i\rangle = \psi\left(\tilde{e}_i\right) - \log\tilde{f}_i$, $\langle\lambda_i\rangle = \tilde{e}_i/\tilde{f}_i$, $\langle\beta\rangle = \tilde{c}/\tilde{d}$,

$\langle \log\beta \rangle = \psi(\tilde{c}) - \log\tilde{d}$, where $\psi(a) = \frac{d}{da}\log\Gamma(a)$ and $\mathbf{R}'$ is an arbitrary square matrix. Until now, the approximated posterior distributions over the uncertain variables have been completed.

Predictive Distribution and Model Training

In the VRVM-ARDK, the uncertain variables contain $\boldsymbol{\alpha}$, β, $\mathbf{s}$, $\mathbf{w}$, and $\boldsymbol{\lambda}$, of which their posterior distributions will be estimated. Since these posteriors are coupled each other, we resort to an iterative learning algorithm. $L[Q(\mathbf{w}, \boldsymbol{\alpha}, \beta, \mathbf{s}, \boldsymbol{\lambda})]$ is an important term in variational inference. Because of its increase characteristic in the modeling process, it can monitor the iterative learning procedure.

Algorithm 7.12: VRVM-ARDK Algorithm

Initialize: *Since the priors over* $\boldsymbol{\alpha}$, $\boldsymbol{\lambda}$, *and* β *are uninformative, the value of their hyper-parameters including a, b, c, d, e, and f can be here fixed to small values, say,* $a = b = c = d = e = f = 10^{-6}$.

Step 1: *Update* $\boldsymbol{\Sigma}_\mathbf{s}$, $\boldsymbol{\mu}_\mathbf{s}$, $\boldsymbol{\Sigma}_\mathbf{w}$, $\boldsymbol{\mu}_\mathbf{w}$, $\tilde{a}_i$, $\tilde{b}_i$, $\tilde{c}$, $\tilde{d}$, $\tilde{e}_i$, *and* $\tilde{f}_i$ *by using formula (7.241) (7.248), (7.253), (7.254), and (7.255). Defining a threshold* $s_{\min}$*and* $w_{\min}$*(e.g.,* 10^{-6}*), weed the value in* $\boldsymbol{\mu}_\mathbf{s}$*smaller than* $s_{\min}$*, and the value in* $\boldsymbol{\mu}_\mathbf{w}$*smaller than* $w_{\min}$.

Step 2: *Compute the lower bound* $L[Q(\mathbf{w}, \boldsymbol{\alpha}, \beta, \mathbf{s}, \boldsymbol{\lambda})]$ *based on the results produced at Step 1.*

Step 3: *Repeat Step 1–2 until the change of the lower bound value is smaller than the predefined threshold. Finally, the model output contains the parameters defined by (7.241) (7.248), (7.253), (7.254), and (7.255).*

Step 4: *Treat the posterior parameters over* $\boldsymbol{\alpha}$, β, $\mathbf{s}$, $\mathbf{w}$, *and* $\boldsymbol{\lambda}$ *as the model outputs.*

Besides, given a new input $\mathbf{x}_*$, the predictive distribution over its corresponding output t_* is given by using the Jessen's inequality,

$$\begin{aligned} p(t_*|\mathbf{t}) &\geq \int p(t_*|\mathbf{w}, \mathbf{s}, \beta^*) Q_\mathbf{w}(\mathbf{w}) Q_\mathbf{s}(\mathbf{s})\, d\mathbf{w} d\mathbf{s} \\ &= \int p(t_*|\mathbf{s}, \beta^*) Q_\mathbf{s}(\mathbf{s})\, d\mathbf{s} \\ &= \int N\left(t_*|\boldsymbol{\phi}_*^T \boldsymbol{\mu}_\mathbf{w},\ (\beta^*)^{-1} + \boldsymbol{\phi}_*^\mathrm{T} \Sigma_\mathbf{w} \boldsymbol{\phi}_*\right) Q_\mathbf{s}(\mathbf{s}) d\mathbf{s} \end{aligned} \tag{7.256}$$

where $\boldsymbol{\phi}_* = [1, K(\mathbf{x}_*, \mathbf{x}_1),\ K(\mathbf{x}_*, \mathbf{x}_2), \ldots, K(\mathbf{x}_*, \mathbf{x}_N))]^\mathrm{T}$. It is very difficult to obtain an accurate solution of this integral, thus one can resort to the Monte Carlo method for an approximate solution. The approximate predictive distribution is given by

$$p(t_*|\mathbf{t}) \approx \frac{1}{N_s} \sum_{k=1}^{N_s} p(t_*|\mathbf{s}_k, \beta^*) \tag{7.257}$$

where $\{\mathbf{s}_k\}_{k=1}^{N_s}$ are samples of $\mathbf{s}$ drawn from $Q_\mathbf{s}(\mathbf{s})$. Although the accurate predictive density is not Gaussian, its mean and variance can be still calculated.

7.5.5 *Case Study*

To illustrate the performance of the variational inference for parameter estimation, we provide experiments of the VRVM-ARDK method for a practical industrial dataset. Meanwhile, experiments for comparative methods, including the RVM [57], the VRVM [56], the SVM [58], and the GP [59], are conducted.

The size of overflow particle (SOP) of powdered ore in a hydro-cyclone is a significant indicator for judging the production condition in milling grinding process. The workers on site adjust the amount of feed water and feed ore according to the change of the value of SOP so as to balance the entire process. Since the manual measurement mode of SOP is a time-consuming work with high costs, it becomes necessary to predict the corresponding values in real time. Figure 7.13 illustrates a diagrammatic flow chart of a grinding process, and various process variables are listed in Table 7.2. The grinding process aims to obtain powdered ore with its particle size small enough from raw ore. The powdered ore is fed into the hydro-cyclone through the equipment, containing semi-autogenous mill, linear screen, grinding pulp tank, and slurry pump. The particle in hydro-cyclone with bigger size will be returned to ball grinder for secondary grinding, which forms a cycle in the whole grinding process. Therefore, it is a very complex industrial process with the variables coupled each other.

A period of data samples of the aforementioned variables in December 2016 in a practical grinding factory serves as the experimental data in this study. The sampling time is 1 min. Since the time delay between the input variables and the overflow particle is very small, it can be ignored. In this study, 450 samples are randomly selected for training, and 300 samples are used for testing. Note that the input attributes and the output are normalized to the unit interval before training the model.

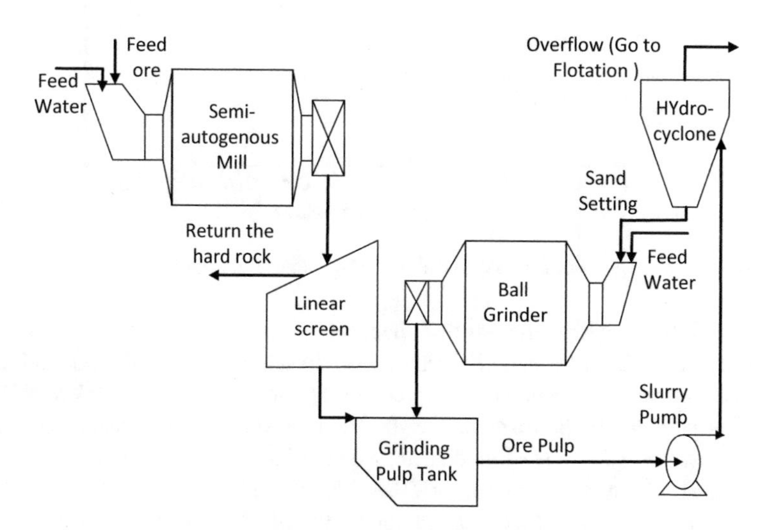

Fig. 7.13 A diagrammatic flow chart of grinding process

Table 7.2 Candidates of input variables and the size of overflow particle

Variables	Meanings of variables	Units
u_1	Feed pressure of cyclone	kPa
u_2	Feed flow of cyclone	t/h
u_3	Feed concentration of cyclone	%
u_4	The sump level	m
u_5	Sand-returning ratio	%
u_6	Amount of the hard stone	t
u_7	Power of ball-milling	kW
u_8	Voice frequency of ball-milling	Hz
u_9	Amount of ore feeding of semi-autogenous mill	t/h
u_{10}	Voice frequency of semi-autogenous mill	Hz
u_{11}	Amount of water feeding of semi-autogenous mill	m^3/h
u_{12}	Power of motor of semi-autogenous mill	kW
y	Size of overflow particle	%

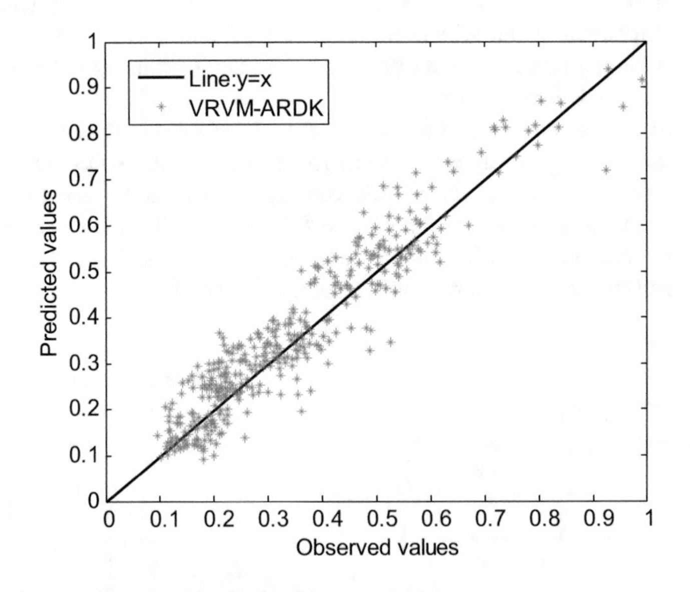

Fig. 7.14 Scatter diagram of the VRVM-ARDK for the industrial dataset

Figure 7.14 provides the scatter diagram of prediction results of the dual estimation-based ESN for such industrial case, in which most of the points locates close to the line $y = x$ showing a good prediction performance of the VRVM-ARDK. To further indicate the performance of the VRVM-ARDK, the statistical results of the prediction error for the testing data are listed in Table 7.3, from which the indices of RMSE, MAPE, and MAE of the VRVM-ARDK are the best than those produced by others. Table 7.4 gives the comparison of the number of relevant samples and

Table 7.3 Statistical results of the methods for the SOP dataset

Methods	RMSE	MAPE	MAE
RVM	0.0850	0.2136	0.0613
VRVM	0.0843	0.2154	0.0611
SVM	0.0888	0.2755	0.0661
GP	0.0775	0.1807	0.0571
VRVM-ARDK	**0.0719**	**0.1791**	**0.0563**

Table 7.4 Comparison of the number of relevant samples and number of relevant features of these methods for the SOP dataset

Methods	Average number of relevant features	Average number of relevant samples
RVM	N/A	14.4
VRVM-ARDK	7.7	10.3

relevant features obtained here. Note that the GP with ARD kernel fails to select relevant features. Besides, the VRVM-ARDK eliminates gradually the irrelevant features during the learning process. The experiments are also repeated 20 times independently, and in these experiments the remaining relevant features with high frequency consists of *the feed pressure of cyclone*, *the feed concentration of cyclone*, *the sump level*, *the amount of the hard stone*, *the voice frequency of semi-autogenous mill*, *the voice frequency of ball-milling*, and *the amount of water feeding of semi-autogenous mill*. These selected features are consistent with those indicated by experts, which demonstrates that the VRVM-ARDK can work well in the industrial setting.

7.6 Parameter Optimization for LS-SVM Based on Noise Estimation

The prediction error of a model can be regarded as the consequence of the samples noise and the inaccuracy of the model. And a prediction model usually has several hyper-parameters which affect the prediction accuracy. The optimization of these hyper-parameters will be interfered by the sample noise. Thereby, the effect of these noises must be considered in the optimization process [60]. Although the framework of the effective noise estimation is introduced in Sect. 5.5.3, we still depict this framework firstly for completeness.

We suppose that var(noise) is the variance of effective noise of the samples. Note that when var(train) > var (noise), it means that the model does not sufficiently learn the nonlinearity of the data, and the model exhibits poor fitting; while, when var (train) < var (noise), the model even learns the dynamic features from the noises, which leads to over-fitting. If var(train) = var (noise), the training error can just reflect the effective noise of the sample data, then the model determines the appropriate hyper-parameters. The variance of training error will be

$$\text{var(train)} = \frac{1}{N}\sum_{i=1}^{N}(e_i - \bar{e}_i)^2 = \frac{1}{N}\sum_{i=1}^{N}e_i^2 \tag{7.258}$$

where e_i is the training error of the ith sample, $\bar{e}_i$ is the mean of e_i of which its value is zero. Besides, var(noise) can be evaluated by the Gamma test technique.

Thereby, we have to minimize the difference between the variance of the training error and that of estimated effective noise. Then, the optimization objective can be formulated as

$$\min d = |\text{var(train)} - \text{var(noise)}| \tag{7.259}$$

We can minimize d with respect to the model parameters to obtain the optimal model parameters. The LSSVM is a promising nonlinear modeling method for regression task, which was introduced in Chap. 3. Using the Gamma test already introduced in Chap. 5, we present an example of parameter optimization for LSSVM based on noise estimation.

7.6.1 Hyper-parameters Optimization Based on the CG Method

In LSSVM regression, the kernel function affects the quality of the model, and the regularization factor establishes a sound trade-off between its complexity and empirical risk. The selection of these hyper-parameters has a direct impact on the prediction accuracy. A CG method based on noise estimation is used to optimize the regularization factor γ and the parameter of Gaussian kernel σ, ($k(\mathbf{x}_i, \mathbf{x}_j) = \exp\left(-\|\mathbf{x}_i - \mathbf{x}_j\|^2/\sigma\right)$) in this section.

As for the LSSVM model, the variance of training error will be written as

$$\text{var(train)} = \frac{1}{N}\sum_{i=1}^{N}e_i^2 = \frac{1}{N}\sum_{i=1}^{N}\frac{\alpha_i^2}{\gamma^2} = \frac{\boldsymbol{\alpha}^{\mathrm{T}}\boldsymbol{\alpha}}{N\gamma^2} \tag{7.260}$$

And we minimize the objective function defined in (7.259). Here, we compute the partial derivatives of objective function, d, with respect to γ and σ to form the directions in the search space of the parameters. First, the partial derivatives of $\mathbf{A}^{-1}$ with respect to γ and σ read as

$$\begin{cases} \dfrac{\partial \mathbf{A}^{-1}}{\partial \sigma} = -\mathbf{A}^{-1}\dfrac{\partial \mathbf{A}}{\partial \sigma}\mathbf{A}^{-1} \\ \quad = -\mathbf{A}^{-1}\left[\exp\left(-\|\mathbf{x}_i - \mathbf{x}_j\|^2/\sigma\right)\|\mathbf{x}_i - \mathbf{x}_j\|^2/\sigma^2\right]_{ij}\mathbf{A}^{-1} \\ \quad = -\mathbf{A}^{-1}\mathbf{B}\mathbf{A}^{-1} \\ \dfrac{\partial \mathbf{A}^{-1}}{\partial \gamma} = -\mathbf{A}^{-1}\dfrac{\partial \mathbf{A}}{\partial \gamma}\mathbf{A}^{-1} = \dfrac{1}{\gamma^2}\mathbf{A}^{-1}\mathbf{A}^{-1} \end{cases} \tag{7.261}$$

Second, the partial derivatives of $b = \dfrac{\vec{\mathbf{1}}^{T}\mathbf{A}^{-1}\mathbf{y}}{\vec{\mathbf{1}}^{T}\mathbf{A}^{-1}\vec{\mathbf{1}}}$ taken with respect to γ and σ is determined as

$$\begin{cases} \dfrac{\partial b}{\partial \sigma} = -\dfrac{\partial\left(\vec{\mathbf{1}}^{\mathrm{T}}\mathbf{A}^{-1}\vec{\mathbf{1}}\right)}{\partial \sigma}\dfrac{\vec{\mathbf{1}}^{\mathrm{T}}\mathbf{A}^{-1}\mathbf{y}}{\left(\vec{\mathbf{1}}^{\mathrm{T}}\mathbf{A}^{-1}\vec{\mathbf{1}}\right)^2} + \dfrac{1}{\vec{\mathbf{1}}^{\mathrm{T}}\mathbf{A}^{-1}\vec{\mathbf{1}}}\dfrac{\partial\left(\vec{\mathbf{1}}^{\mathrm{T}}\mathbf{A}^{-1}\mathbf{y}\right)}{\partial \sigma} \\ \quad = \dfrac{\vec{\mathbf{1}}\dfrac{\partial \mathbf{A}^{-1}}{\partial \sigma}\left(\mathbf{y} - b\,\vec{\mathbf{1}}\right)}{\vec{\mathbf{1}}^{\mathrm{T}}\mathbf{A}^{-1}\vec{\mathbf{1}}} = -\dfrac{\vec{\mathbf{1}}^{\mathrm{T}}\mathbf{A}^{-1}\mathbf{B}\boldsymbol{\alpha}}{\vec{\mathbf{1}}^{\mathrm{T}}\mathbf{A}^{-1}\vec{\mathbf{1}}} \\ \dfrac{\partial b}{\partial \gamma} = --\dfrac{\partial\left(\vec{\mathbf{1}}^{\mathrm{T}}\mathbf{A}^{-1}\vec{\mathbf{1}}\right)}{\partial \gamma}\dfrac{\vec{\mathbf{1}}^{\mathrm{T}}\mathbf{A}^{-1}\mathbf{y}}{\left(\vec{\mathbf{1}}^{\mathrm{T}}\mathbf{A}^{-1}\vec{\mathbf{1}}\right)^2} + \dfrac{1}{\vec{\mathbf{1}}^{\mathrm{T}}\mathbf{A}^{-1}\vec{\mathbf{1}}}\dfrac{\partial\left(\vec{\mathbf{1}}^{\mathrm{T}}\mathbf{A}^{-1}\mathbf{y}\right)}{\partial \gamma} \\ \quad = \dfrac{\vec{\mathbf{1}}\dfrac{\partial \mathbf{A}^{-1}}{\partial \gamma}\left(\mathbf{y} - b\,\vec{\mathbf{1}}\right)}{\vec{\mathbf{1}}^{\mathrm{T}}\mathbf{A}^{-1}\vec{\mathbf{1}}} = \dfrac{\vec{\mathbf{1}}^{\mathrm{T}}\mathbf{A}^{-1}\boldsymbol{\alpha}}{\gamma^2\vec{\mathbf{1}}^{\mathrm{T}}\mathbf{A}^{-1}\vec{\mathbf{1}}} \end{cases} \tag{7.262}$$

Combining (7.261) and (7.262), the partial derivatives of $\boldsymbol{\alpha} = \mathbf{A}^{-1}\left(\mathbf{y} - b\,\vec{\mathbf{1}}\right)$ computed with respect to the hyper-parameters come in the form

$$\begin{cases} \dfrac{\partial \boldsymbol{\alpha}}{\partial \sigma} = \dfrac{\partial \mathbf{A}^{-1}}{\partial \sigma}\left(\mathbf{y} - b\,\vec{\mathbf{1}}\right) + \mathbf{A}^{-1}\dfrac{\partial\left(\mathbf{y} - b\,\vec{\mathbf{1}}\right)}{\partial \sigma} \\ \quad = -\mathbf{A}^{-1}\mathbf{B}\mathbf{A}^{-1}\left(\mathbf{y} - b\,\vec{\mathbf{1}}\right) - \mathbf{A}^{-1}\dfrac{\partial b}{\partial \sigma}\vec{\mathbf{1}} \\ \quad = -\mathbf{A}^{-1}\mathbf{B}\boldsymbol{\alpha} - \mathbf{A}^{-1}\dfrac{\partial b}{\partial \sigma}\vec{\mathbf{1}} \\ \dfrac{\partial \boldsymbol{\alpha}}{\partial \gamma} = \dfrac{\partial \mathbf{A}^{-1}}{\partial \gamma}\left(\mathbf{y} - b\,\vec{\mathbf{1}}\right) + \mathbf{A}^{-1}\dfrac{\partial\left(\mathbf{y} - b\,\vec{\mathbf{1}}\right)}{\partial \gamma} \\ \quad = -\dfrac{1}{\gamma^2}\mathbf{A}^{-1}\mathbf{A}^{-1}\left(\mathbf{y} - b\,\vec{\mathbf{1}}\right) - \mathbf{A}^{-1}\dfrac{\partial b}{\partial \gamma}\vec{\mathbf{1}} \\ \quad = -\dfrac{1}{\gamma^2}\mathbf{A}^{-1}\boldsymbol{\alpha} - \mathbf{A}^{-1}\dfrac{\partial b}{\partial \gamma}\vec{\mathbf{1}} \end{cases} \tag{7.263}$$

Based on the above derivations, the partial derivatives of objective d can be formulated as (7.264).

Thus, the gradient direction for hyper-parameters optimization can be produced. With regard to the above formulas, it is noticeable that all of $\mathbf{A}^{-1}$ in the formulas can be combined in the form of a column vector such as $\vec{\mathbf{1}}$ and $\boldsymbol{\alpha}$, etc., which can transform the task of solving matrix inverse into solving a linear equation $\mathbf{Ax} = \vec{\mathbf{1}}$. In such a way, there is no need for matrix inverse when determining values of the gradients, which largely reduces the computational cost

$$\begin{cases} \dfrac{\partial d}{\partial \gamma} = \dfrac{d}{\mathrm{var(train)} - \mathrm{var(noise)}} \cdot \dfrac{\partial \left(\dfrac{\boldsymbol{\alpha}^{\mathrm{T}}\boldsymbol{\alpha}}{N\gamma^2} \right)}{\partial \gamma} \\ \quad = \dfrac{1}{N\gamma^2 \mathrm{var(train)} - \mathrm{var(noise)}} \left[\left(\left(\dfrac{\partial \boldsymbol{\alpha}}{\partial \gamma} \right)^{\mathrm{T}} \boldsymbol{\alpha} + \boldsymbol{\alpha}^{\mathrm{T}} \dfrac{\partial \boldsymbol{\alpha}}{\partial \gamma} \right) - \dfrac{2}{\gamma} \boldsymbol{\alpha}^{\mathrm{T}}\boldsymbol{\alpha} \right] \\ \dfrac{\partial d}{\partial \sigma} = \dfrac{d}{\mathrm{var(train)} - \mathrm{var(noise)}} \dfrac{\partial \left(\dfrac{\boldsymbol{\alpha}^{\mathrm{T}}\boldsymbol{\alpha}}{N\gamma^2} \right)}{\partial \sigma} \\ \quad = \dfrac{1}{N\gamma^2 \mathrm{var(train)} - \mathrm{var(noise)}} \left(\left(\dfrac{\partial \boldsymbol{\alpha}}{\partial \sigma} \right)^{\mathrm{T}} \boldsymbol{\alpha} + \boldsymbol{\alpha}^{\mathrm{T}} \dfrac{\partial \boldsymbol{\alpha}}{\partial \sigma} \right) \end{cases} \tag{7.264}$$

In this CG algorithm, the line search that satisfies the Wolfe-Powell rule is used for the step length search. Although such type of the algorithm might tend to be trapped in some local minima, it has been experimentally observed that it exhibits sound convergence properties as long as their initial values have been carefully determined.

7.6.2 *Case Study*

The Sinc Function

To quantify the performance of the effective noise estimation-based LSSVM (ENE-LSSVM) model, the Sinc function, which is specified as $\mathrm{sinc}(x) = \sin(x)/x$, is employed. We consider this nonlinear function affected by an additive Gaussian noise

$$y = 5^* \mathrm{sinc}(x) + e \tag{7.265}$$

where e is an additive Gaussian noise with the zero mean and variance equal to 0.2. In this experiment, we first generate 500 inputs in (1.5, 5) and randomly select 300 data to train this model. The corresponding outputs are generated by (7.265). The remaining 200 generated input data are regarded as the testing data x_{test}.

Table 7.5 Average prediction results for the Sinc function with additive noise

Method	Running time(s)	RMSE	NRMSE	MAPE (%)
ENE-LSSVM	0.182	0.0993	0.004537	66.50802
CG-FCV	1.311	0.1006	0.004541	62.86674
BI	11.844	0.1104	0.005135	71.22882
G-LOO	13.485	0.1074	0.004946	65.44225
G-FCV	10.169	0.1037	0.004776	66.70179

The error is computed by comparing the model output with the real value $y = 5^*$ sinc (x_{test}).

A comparative experiment making use of different methods is designed, involving the grid search-fast LOO (G-LOO) [61], the grid search-fast-fold CV (G-FCV) [62], the Bayesian inference (BI) [63], the conjugate gradient-fast-fold CV (CG-FCV) [64], and the ENE-LSSVM method. We randomly chose the initial values of and as 50 and 10 in the ENE-LSSVM, respectively; and set the value of based on experiences of [65] as 10. Here, 20 times independent experiments were repeated for each modeling and predicting, and the quantified results on average are listed as Table 7.5. The calculated noise variance with Gamma test is equal to 0.1971, which is basically consistent with the original variance.

Comparing the results, it is evident that the effective noise estimation-based gradient optimization shows the best performance. Its efficiency is over seven times higher than the one reported for the CG-FCV; furthermore, as 70–90 times as that by grid search-based optimizations or Bayesian inference. Such results show the noticeable advantage of the ENE-LSSVM with regard to the computational cost and usefulness in real-time prediction. As for the prediction accuracy, considering the fact that the function exhibits some additive noise, of which many values are approaching zero, there is no large difference when using these methods to realize prediction.

Industrial Application: Prediction Modeling for Industrial Gas Flow

The gas flow data from a steel plant in China are adopted to establish the effective noise estimation-based LSSVM. The data cover the period from August to October, 2010, and the gas flow types consist of the generation unit and the consumption one. The sampling interval of real-time data from SCADA system is 1 min. The process of modeling and predicting is triggered every 10 points (10 min) to predict the future 60-min gas flow. All in all, the real-time process will be iterated 100 times with time elapsing. Similarly to the experiments with the Sinc function, we repeat the independent experiment 20 times.

First, we perform the analysis of optimal embedded dimensionality (ED). For industrial data, we find that the change of the number of nearest neighbors p in Gamma test significantly impacts the minimal noise variance, while the corresponding minimal ED is basically fixed. Taking the BFG generation flow of

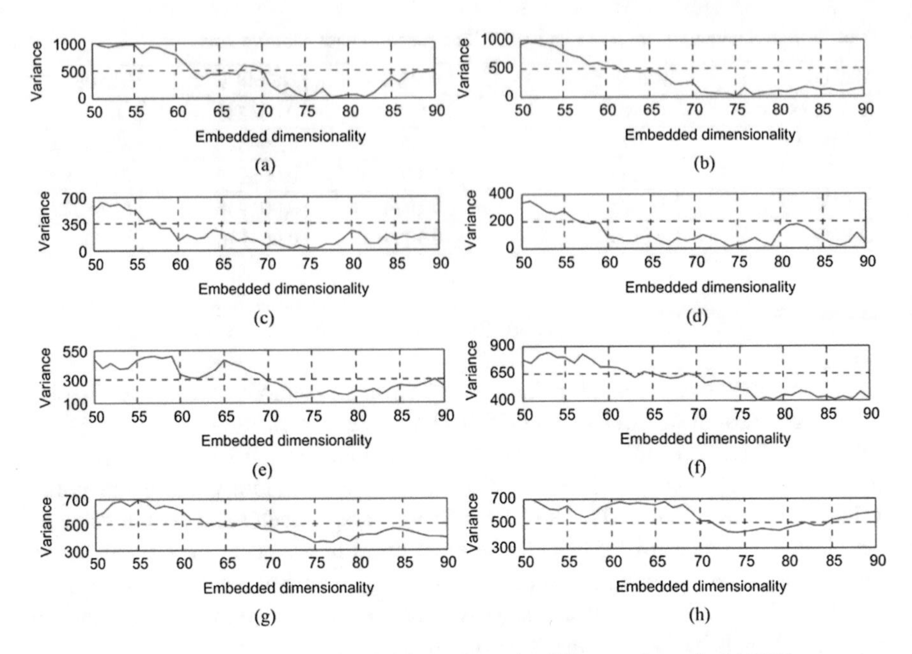

Fig. 7.15 Relationship between ED and noise variance for different values of p for BFG generation flow of #1 blast furnace. (**a**) $p = 10$. (**b**) $p = 15$. (**c**) $p = 20$. (**d**) $p = 25$. (**e**) $p = 30$. (**f**) $p = 35$. (**g**) $p = 40$. (**h**) $p = 45$

Table 7.6 A summary of main results, refer to Fig. 7.15

p	10	15	20	25	30	35	40	45
Optimal ED	75	75	75	74	73	77	75	73
Minimal variance of effective noise	5.91	10.47	15.33	9.74	144.95	398.38	353.61	423.71

#1 blast furnace as an example, as presented in Fig. 3.6, the relationship between the ED and the variance of effective noise with respect to different p are illustrated in Fig. 7.15, refer also to Table 7.6. From Fig. 7.15 and Table 7.6, it can be seen that the optimal ED is relatively insensitive to the change of p; on the contrary, the minimal estimated variance of effective noise has a rather large change that might impact the values of the optimized hyper-parameters. Considering such phenomenon, in this section we can optimize the value of p via off-line cross-validation. We found that once the value of p has been determined with respect to a gas unit, it may be effectively used without change in the online flow prediction. Taking the example of BFG generation by #1 blast furnace, we visualize the relationship between the value of p and the error of cross-validation as shown in Fig. 7.16 when the ED is set to 75. Thus, the optimal value of according to the relationship should be equal to 40 for the generation amount of #1 blast furnace, where the error of cross-validation is minimal. Similarly, the optimal value of p can be experimentally determined off-line.

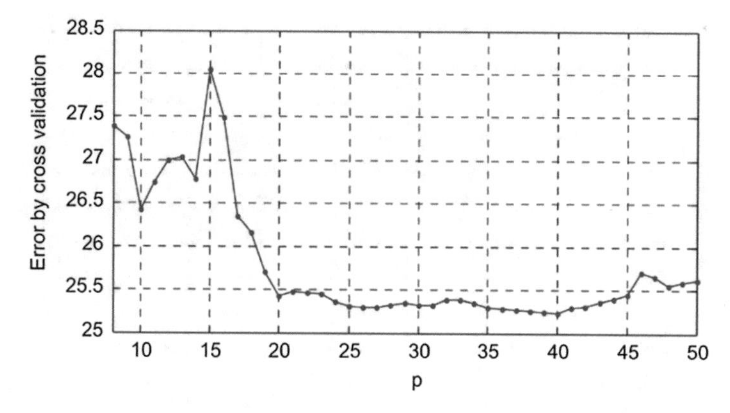

Fig. 7.16 Relationship between and the cross-validation error when ED = 75

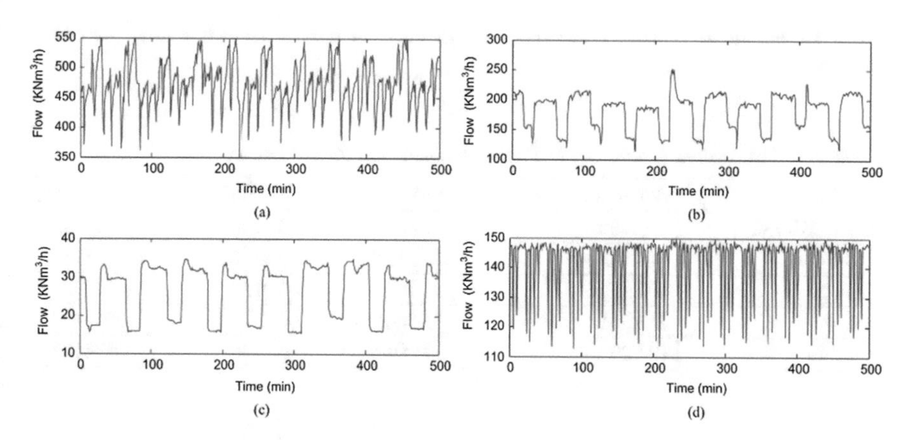

Fig. 7.17 Flow variation of the typical gas generators and users: (**a**) Generation amount of blast furnace; (**b**) BFG consumption of hot blast stove; (**c**) LDG consumption of blast furnace; (**d**) BFG consumption of coke oven

The generation amount of BFG usually exhibits significant fluctuations as presented in Fig. 7.17a. The flow data of #1 blast furnace are considered. To obtain the optimal ED of this time series, we set the initial value of p to be equal to 10. Then, the corresponding ED that takes the minimal variance of noise equal to 75 based on the above calculations, as illustrated as Fig. 7.18. Regarding the width of Gaussian kernel and the regularization factor, we empirically select the initial values of these two parameters set as 100 and 200, respectively. After the grid search and cross-validation, the appropriate value of p corresponding to the minimal training error was found to be equal to 40, see Fig. 7.16. According to the above parameters settings, the γ and α are online optimized by the CG algorithm and the average prediction results using the ENE-LSSVM method and other four methods are listed as Table 7.7, where the 20-time independent experiments are, respectively,

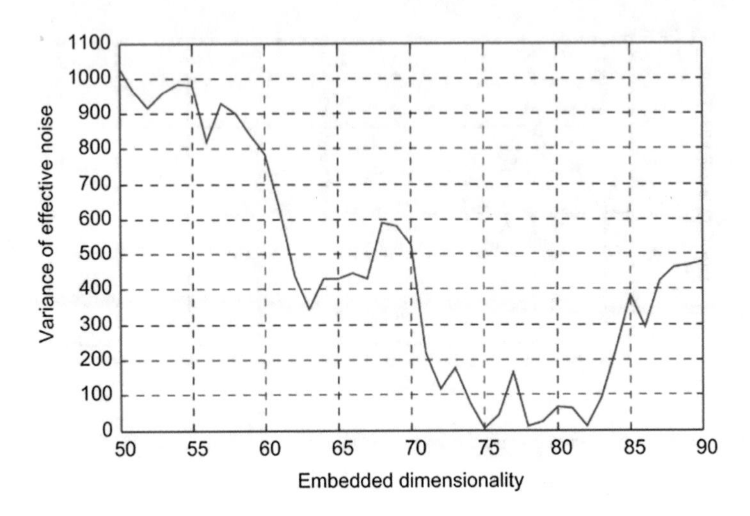

Fig. 7.18 Relationship between ED and noise variance for the gas generation flow of #1 blast furnace

Table 7.7 The prediction accuracy and computational time for BFG generation

Method	Running time(s)	RMSE	NRMSE	MAPE (%)
ENE-LSSVM	4.008	51.8809	0.0125	8.024195
CG-FCV	15.355	52.1851	0.0126	8.073875
G-FCV	102.959	52.1035	0.0125	8.043921
BI	252.861	54.3646	0.0131	8.042392
G-LOO	124.295	57.4911	0.0139	8.834422

performed. For the detailed illustration, the prediction accuracy of each time represented by RMSE and MAPE are shown as Fig. 7.19. It appears that the accuracy of the ENE-LSSVM is the best (on average) compared to the accuracies provided by other methods. In Fig. 7.19, the curve shown red, representing the ENE-LSSVM, shows basically the lowest error indexes. More noticeably, the computational time using the CG algorithm in the ENE-LSSVM is the shortest one. We completed prediction of the 60-min gas flow only within 5 s. This efficiency fully satisfies the real-time requirements for the practical gas flow prediction.

From the experimental results of the above industrial predictions for BFG dataset, the most noticeable issue is the computational efficiency of the ENE-LSSVM method. Based on the results conveyed by Table 7.7, the grid search-based hyper-parameters optimizations, G-FCV and G-LOO, usually expend a lot of computational time to search for the optimal values. This computing overhead does not allow us to meet the industrial real-time demand. Furthermore, although the conjugate gradient-based fast CV (CG-FCV) increases the computational speed over 20–50 times or more compared to the grid search-based methods, there is still a large gap

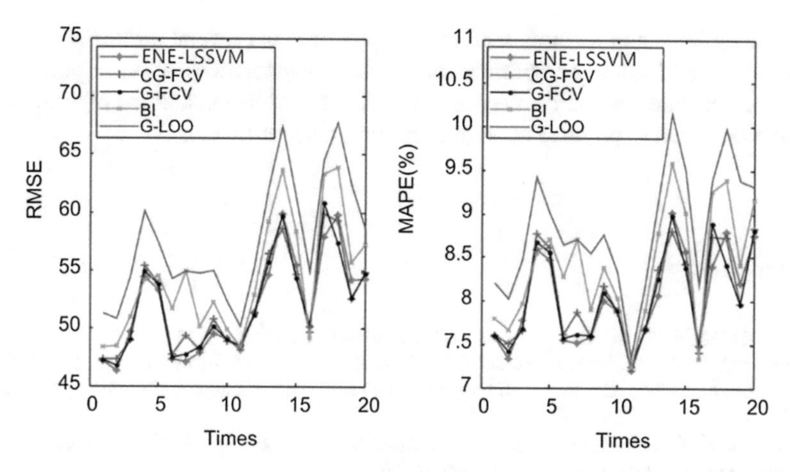

Fig. 7.19 Accuracy indexes of the 20-time prediction for BFG generation flow on #1 blast furnace

between the ENE-LSSVM method based on gradient and variance estimation of noise. Using only several seconds for a 60-min prediction, the ENE-LSSVM method offers a tangible superiority to the other optimization techniques in terms of computing cost, which can fully satisfy the requirements of continuous real-time prediction.

With respect to the prediction accuracy, the performance of the ENE-LSSVM method is also remarkable. Compared the results by G-FCV with that by G-LOO, since the G-FCV leaves many more sample data for validating the fitting error; while, the G-LOO leaves only one of the sample in validation process. As such, although the computational time by G-LOO is somewhat shorter, its prediction performance is worse than that of G-FCV when the sample data is affected by noise, as shown in Table 7.7. Regarding the CV-based methods, the G-FCV and the CG-FCV, the prediction error obtained by the grid search-based G-FCV is usually impacted by the specifications of parameters range and grid density. Thereby, its performance might be better than that obtained by the gradient-based CG-FCV (see Table 7.7).

7.7 Discussion

The parameters contained in a machine learning model can extremely impact the prediction performance of the data-based model. This chapter introduces a series of methods for parameters estimation and optimization, including the gradient-based algorithm, the intelligent optimization algorithm, and the probabilistic methods as well as the nonlinear Kalman-filter estimation for the particular dynamic models. These methods have their own merits and demerits, as described in the corresponding subsections. For different industrial problems, one has to choose a suitable optimization method, then this chapter presents several industrial cases for parameters estimation or optimization. Future development of the optimization methods must depend on the certain industrial characteristic, and the algorithm has to be redesigned to produce good optimization performance.

References

1. Protter, M. H. (2014). Basic elements of real analysis. Springer.
2. Fletcher, R. (2005). On the Barzilai-Borwein method. *Applied Optimization, 96*, 235–256.
3. Bottou, L. (1998). Online algorithms and stochastic approximations. Cambridge University Press.
4. Kiwiel, K. C. (2001). Convergence and efficiency of subgradient methods for quasiconvex minimization. *Mathematical Programming, 90*(1), 1–25.
5. Shanno, D. F. (1970). Conditioning of quasi-Newton methods for function minimization. *Mathematics of Computation, 24*(111), 647–656.
6. Liu, D. C., & Nocedal, J. (1989). On the limited memory BFGS method for large scale optimization. *Springer-Verlag New York Inc*.
7. Dai, Y. H. (2013). A perfect example for the BFGS method. *Mathematical Programming, 138* (1–2), 501–530.
8. Malouf, R. (2002). *A comparison of algorithms for maximum entropy parameter estimation* (pp. 49–55). In Proc. Sixth Conf. on Natural Language Learning (CoNLL).
9. Andrew, G., & Gao, J. (2007). *Scalable training of L-regularized log-linear models*. In Proceedings of the 24th International Conference on Machine Learning.
10. Knyazev, A. V., & Lashuk, I. (2008). Steepest descent and conjugate gradient methods with variable preconditioning. *SIAM Journal on Matrix Analysis and Applications, 29*(4), 1267.
11. Hestenes, M. R., & Stiefel, E. L. (1952). Methods of conjugate gradients for solving linear systems. *Journal of Research of the National Bureau of Standards, 5*(2), 409–432.
12. Fletcher, R., & Reeves, C. (1964). Function minimization by conjugate gradients. *Computer Journal, 7*(1), 149–154.
13. Polak, B., & Ribiere, G. (1969). Note sur la convergence des methods de directions conjuguees. *Rev Francaise Imformmat Recherche Opertionelle, 16*(1), 35–43.
14. Polyak, B. T. (1969). The conjugate gradient method in extreme problems. *USSR Computational Mathematics and Mathematical Physics, 9*(1), 94–112.
15. Fletcher, R. (1987). *Practical methods of optimization, Vol. 1: Unconstrained optimization* (pp. 10–30). New York: Wiley.
16. Liu, Y., & Storey, C. (1991). Efficient generalized conjugate gradient algorithms, Part 1: Theory. *Journal of Optimization Theory and Applications, 69*(1), 129–137.
17. Dai, H. Y., & Yuan, Y. (2000). A nonlinear conjugate gradient method with a strong global convergence property. *SIAM Journal on Optimization, 10*(1), 177–182.
18. Zhang, X. P., Zhao, J., Wei, W., et al. (2010). COG holder level prediction model based on least square support vector machine and its application. *Control and Decision, 25*(8), 1178–1183.
19. Holland, J. H. (1975). *Adaptation in natural and artificial systems*. Ann Arbor: University of Michigan Press.
20. Srinivas, M., & Patnaik, L. (1994). Adaptive probabilities of crossover and mutation in genetic algorithms. *IEEE Transactions on System, Man and Cybernetics, 4*(4), 656–667.
21. Zhang, J., Chung, H., & Lo, W. L. (2007). Clustering-based adaptive crossover and mutation probabilities for genetic algorithms. *IEEE Transactions on Evolutionary Computation, 11*(3), 326–335.
22. Storn, R. (1995). Constrained optimization. *Dr. Dobb's Journal*, 119–123.

23. Das, S., Abraham, A., & Konar, A. (2009). *Differential evolution algorithm: Foundations and perspectives* (Vol. 178, pp. 63–110).
24. Kennedy, J., & Eberhart, R. (1995). *Particle swarm optimization* (pp. 1942–1948). In International Conference on Neural Networks.
25. Eberhart, R., & Kennedy, J. (1995). *A new optimizer using particle swarm theory* (pp. 39–43). In International Symposium on Micro Machine and Human Science.
26. Eberhart, R. C., Shi, Y., & Kennedy, J. (2001). *Swarm intelligence*. Amsterdam: Elsevier.
27. Shi, Y., & Eberhart, R. C. (1998). *Parameter selection in particle swarm optimization* (pp. 591–600). In Evolutionary Programming VI/: Proc. EP98. New York: Springer.
28. Shi, Y., & Eberhart, R. C. (1998). *A modified particle swarm optimizer* (pp. 69–73). In Proceedings of the IEEE International Conference on Evolutionary Computation. Piscataway, NJ: IEEE Press.
29. Kitayama, S., Arakawa, M., & Yamazaki, K. (2006). Penalty function approach for the mixed discrete nonlinear problems by particle swarm optimization. *Structural and Multidisciplinary Optimization, 32*(3), 191–202.
30. Li, D., Wang, B., Kita-Yama, S., Yamazaki, K., & Arakawa, M. (2005). *Application of particle swarm optimization to the mixed discrete non-linear problems* (pp. 315–324). In Artificial intelligence applications and innovations, USA, Vol. 187.
31. Kitayama, S., & Yasuda, K. (2006). A method for mixed integer programming problems by particle swarm optimization. *Electrical Engineering in Japan, 157*(2), 40–49.
32. Chen, W. N., Zhang, J., Chung, H. S. H., et al. (2010). A novel set-based particle swarm optimization method for discrete optimization problems. *IEEE Transactions on Evolutionary Computation, 14*(2), 278–300.
33. Gong, Y. J., Zhang, J., Liu, O., et al. (2012). Optimizing the vehicle routing problem with time windows: A discrete particle swarm optimization approach. *IEEE Transactions on Systems Man & Cybernetics Part C, 42*(2), 254–267.
34. Robinson, D. G. (2005). *Reliability analysis of bulk power systems using swarm intelligence* (pp. 96–102). In Reliability and maintainability symposium, 2005. Proceedings. IEEE.
35. Pampara, G., Franken, N., & Engelbrecht, A. P. (2005). *Combining particle swarm optimisation with angle modulation to solve binary problems* (pp. 89–96). In The 2005 I.E. Congress on Evolutionary Computation, 2005. IEEE.
36. Wu, W. C., & Tsai, M. S. (2011). Application of enhanced integer coded particle swarm optimization for distribution system feeder reconfiguration. *IEEE Transactions on Power Systems, 26*(3), 1591–1599.
37. Kirkpatrick, S., Gelatt, C. D., et al. (1983). Optimization by simulated annealing. *Science, 220*, 671–680.
38. Cerny, V. (1985). Thermodynamical approach to the travelling salesman problem: An efficient simulation algorithm. *Journal of Optimization Theory and Applications, 45*, 41–51.
39. Fleischer, M. A. (1995). *Simulated annealing: Past, present, and future* (pp. 155–161). In Proceedings of the 1995 Winter Simulation Conference, IEEE Press, Arlington, Virginia.
40. Henderson, D., Jacobson, S. H., & Johnson, A. W. (2003). *Handbook of metaheuristics*. Boston: Kluwer.
41. Kumar, P. (2006). A survey of simulated annealing as a tool for single and multiobjective optimization. *Journal of the Operational Research Society, 57*(10), 1143–1160.
42. Sastry, Y. (1971). Decomposition of the extended Kalman filter. *IEEE Transactions on Automatic Control, 16*(3), 260–261.
43. Einicke, G. A. (2012). *Smoothing, filtering and prediction: Estimating the past, present and future*. Rijeka: Intech.
44. Andreasen, M. M. (2013). Non-linear DSGE Models and the central difference Kalman Filter †. *Journal of Applied Econometrics, 28*(6), 929–955.
45. Wan, E. A., & van der Menve, R. (2000). *The unscented Kalman Filter for nonlinear estimation.* In IEEE Conference on Symposium on Adaptive Systems for Signal Processing, Communications, and Control (AS-SPCC).

46. Arasaratnam, I., & Haykin, S. (2009). Cubature Kalman filters. *IEEE Transactions on Automatic Control, 54*(6), 1254–1269.
47. Sheng, C., Zhao, J., Liu, Y., et al. (2012). Prediction for noisy nonlinear time series by echo state network based on dual estimation. *Neurocomputing, 82*(4), 186–195.
48. Venayagamoorthy, G., & Shishir, B. (2009). Effects of spectral radius and settling time in the performance of echo state networks. *Neural Networks, 22*(7), 861.
49. Bishop, C. M. (2006). *Pattern recognition and machine learning (information science and statistics)*. New York: Springer.
50. Bernardo, J. M., & Smith, A. F. M. (1994). *Bayesian theory*. Chichester: Wiley.
51. Gull, S. F. (1989). Developments in maximum entropy data analysis. In J. Skilling (Ed.), *Maximum entropy and Bayesian methods* (pp. 53–71). Dordrecht: Kluwer.
52. MacKay, D. J. C. (1992). The evidence framework applied to classification networks. *Neural Computation, 4*(5), 720–736.
53. Berger, J. O. (1985). *Statistical decision theory and Bayesian analysis* (2nd ed.). New York: Springer.
54. Parisi, G. (1988). *Statistical field theory*. New York: Addison-Wesley.
55. Zhao, J., Chen, L., Pedrycz, W., & Wang, W. (in press). Variational inference based automatic relevance determination kernel for embedded feature selection of noisy industrial data, *IEEE Transactions on Industrial Electronics*. https://doi.org/10.1109/TIE.2018.2815997.
56. Bishop, C. M., & Tipping, M. E. (2000). *Variational relevance vector machines*. In Conference on uncertainty in artificial intelligence.
57. Tipping, M. E. (2001). Sparse Bayesian learning and the relevance vector machine. *Journal of Machine Learning Research, 1*(3), 211–244.
58. Vapnik, V. N. (1995). *The nature of statistical learning theory*. New York: Springer.
59. Rasmussen, C., & Williams, C. (2006). *Gaussian processes for machine learning*. MIT Press.
60. Zhao, J., Liu, Q., Pedrycz, W., et al. (2012). Effective noise estimation-based online prediction for byproduct gas system in steel industry. *IEEE Transactions on Industrial Informatics, 8*(4), 953–963.
61. Zhao, Y., & Keong, K. C. (2004). *Fast leave-one-out evaluation and improvement on inference for LS-SVM* (pp. 1051–4651). In Proc. IEEE Int. Conf. Pattern Recognit., Cambridge, U.K.
62. An, S., Liu, W., & Venkatesh, S. (2007). Fast cross-validation algorithms for least squares support vector machine and kernel ridge regression. *Pattern Recognition, 40*, 2154–2162.
63. Chi, M. V., Wong, P. K., & Li, Y. P. (2006). Prediction of automotive engine power and torque using least squares support vector machines and Bayesian inference. *Engineering Applications of Artificial Intelligence, 19*(3), 277–287.
64. Rubio, G., Pomares, H., Rojas, I., et al. (2009). *Efficient optimization of the parameters of LS-SVM for regression versus cross-validation error* (pp. 406–415). In International Conference on Artificial Neural Networks. Springer.
65. Jones, A. J. (2004). New tools in non-linear modelling and prediction. *Computational Management Science, 1*(2), 109–149.

Chapter 8
Parallel Computing Considerations

Abstract This chapter discusses the computational cost of machine learning model. To reduce its training time is a requisite of its industrial applications since a production process usually requires real-time responses. The commonly used method to accelerate the training process is to develop a parallel computing framework. In literature, two kinds of popular methods speeding up the training involves the one with a computer equipped with graphics processor unit (GPU) and the one with computer cluster including a number of computers. This chapter firstly introduces the basic ideas of GPU acceleration (e.g., the compute unified device architecture (CUDA) created by NVIDIA™) and the computer cluster framework (e.g., the MapReduce framework), then gives some specified examples of them. When training an EKF-based Elman network, the inversion operation of a Jacobian matrix is the most time-consuming procedure; a parallel computing strategy for such an operation is therefore proposed by using the CUDA-based GPU acceleration. Besides, with regard to the LSSVM modeling, a CUDA-based parallel PSO is then introduced for its hyper-parameters optimization. As for the computer cluster version, we design a parallelized EKF based on ESN by using MapReduce framework for acceleration. At the end, we also present a series of experimental analysis by using the practical energy data in steel industry to validate the performance of the accelerating approaches.

8.1 Introduction

In the previous chapters, we introduced different kinds of prediction models for various industrial demands and situations, including time series prediction, factors-based prediction, prediction intervals, and long-term predictions, etc. Actually, for practical industrial application, the training time of a prediction model should be not very large as the industrial operations must be performed in real time. However, most of the algorithms aforementioned in the previous chapters do not meet this demand since there are many time-consuming computation tasks such as the inverse

J. Zhao et al., *Data-Driven Prediction for Industrial Processes and Their Applications*, Information Fusion and Data Science,
https://doi.org/10.1007/978-3-319-94051-9_8

of a matrix and computing loops, especially with large datasets. Thereby, to speed up the algorithms becomes a request in industrial applications.

In practice, we always find that there are lots of computation tasks which can be performed independently in an algorithm. Thus, we can resort to some parallel computing strategy for these computations rather than serial version so as to speed up the training process. Currently, there are various parallel computing strategies, including two kinds of methods: speeding up with graphics processor unit (GPU) within one computer and computing with computer cluster which contains more than one computer.

The GPU-based parallel computing both uses the CPU and GPU to speed up certain computing task collaboratively. A GPU contains a large amount of computing units which can perform a small sub-computation task in parallel. Particularly, the compute unified device architecture (CUDA), created by the NVIDIA Corporation, is a parallel computing platform and application programming interface (API) model. It allows software developers and software engineers to use a CUDA-enabled GPU for general purpose processing. In recent years, GPU technology is experiencing a rapid development, whose floating point operation and programmable capability provide a better platform for general calculating in contrast to its original image processing functions. Therefore, GPU technology lays the foundation for compute-intensive applications in areas such as molecular dynamics, computational chemistry, medical imaging, and seismic exploration.

One of the commonly used computer cluster-based parallel computing approaches is Apache Hadoop. The Apache Hadoop software library is a framework that allows for the distributed processing of large datasets across clusters of computers using simple programming models. It is designed to scale up from single server to thousands of machines, each offering local computation and storage. Besides, there are also other parallel computing architecture based on computer cluster, including Apache Storm and Spark. Apache Storm is a free and open source distributed real-time computation system, which makes it easy to reliably process unbounded streams of data. In addition, Apache Spark is a fast and general purpose cluster computing system. It provides high level APIs in Java, Scala, Python, and R and an optimized engine that supports general execution graphs. It also supports a rich set of higher level tools including Spark SQL for SQL and structured data processing, MLlib for machine learning, GraphX for graph processing, and Spark streaming.

In this chapter, we will introduce the CUDA-based parallel computing and the computer cluster-based Hadoop in details, where their respective computing architectures are also described. Next we shall introduce two specific parallel computing algorithms which are accelerated by CUDA and Hadoop, respectively; moreover, the results of case studies show that these parallel computing algorithms can significantly decrease the training time of prediction models, which can be applied to the industrial situations for online learning.

8.2 CUDA-Based Parallel Acceleration

Modern GPUs are suitable for many general purpose tasks and have emerged as inexpensive high performance co-processors due to their tremendous computing power, which allows us to develop new models of parallel programming [1]. Typically, a GPU device consists of multiple multiprocessors. A multiprocessor manages the creation, handling, and execution of the current threads in the hardware, supporting hundreds of threads under the single instruction multiple threads approach. In a computer, the GPU and its local memory are referred to be the device, and the CPU and its system memory are as the host. To maximize the program efficiency, we have the logical transaction and the serial processing performed on CPU and have the parallelized calculation on GPU for their reasonable cooperation. As described in the introduction, CUDA is a programmable model taking GPU as data parallel computing device, which will cooperate with CPU to perform a specific computing task. In this subsection, we will introduce the CUDA-based computing framework.

8.2.1 What's CUDA?

CUDA developed by the NVIDIA Corporation is a parallel computing platform and API model based on single instruction multithreading (SIMT), which allows the developer to use a CUDA-enabled GPU for general purpose processing. At present, the equipment supporting CUDA is only the product of the NVIDIA manufacturer, which can be divided into two categories, the video card and the acceleration card. Compared with the former, the latter mainly provides data processing rather than the graphical display. The series of video card supporting CUDA, including GeForce 8, GeForce 9, GeForce 200, GeForce FX and GeForce Quadro, etc., and the acceleration card of CUDA is mainly the Telsa series, which is generally used to establish a dedicated computing server with multiple GPUs.

According to the structure of composition, CUDA can be divided into three parts, CUDA driver, CUDA runtime, and CUDA library [2]. CUDA driver provides an access interface to the GPU device at the level of abstraction, which makes it possible to run correctly on different hardware. The salient property of CUDA driver greatly enhances the applicability and vitality of CUDA. CUDA runtime includes the definition of basic data types, all kinds of computation functions, memory management, device access, and execution scheduling, which provides the development interfaces and runtime components of CUDA. CUDA library can be viewed as a tool to assist the developer to save the development time of program. Equipped with these three components, CUDA enables the developer to write programs that can run on a GPU.

8.2.2 Computing Architecture of CUDA

CUDA programs typically consist of a component that runs on the CPU, and a smaller but computationally intensive component called the kernel that runs in parallel on the GPU. A kernel generally refers to a function that is invoked by a CPU host and is executed on the device of GPU. All memories used by the kernel must be pre-allocated. The CPU's main memory is not directly allowed to be accessed by the kernel. The input data for a kernel must be copied to the GPU's onboard memory. The corresponding output data must firstly be written to the GPU's memory, and then they are copied to the CPU's memory. In addition, the kernel cannot use recursion, but loops are allowed.

The computing tasks executed on the GPU are organized in parallel by multiple threads. The kernel function runs with a thread. Several threads can constitute a block, and a grid is made up of a set of blocks. Figure 8.1 gives a detailed description of this organization in the structure of CUDA, showing a combination of the CPU and GPU to perform a specific parallel computing task. When the code of the kernel function is encountered, the main thread of the program in the CPU will allocate the required resources for GPU to execute the parallel computing. The same code for the parallel task can be divided into multiple blocks, each of which will be executed by

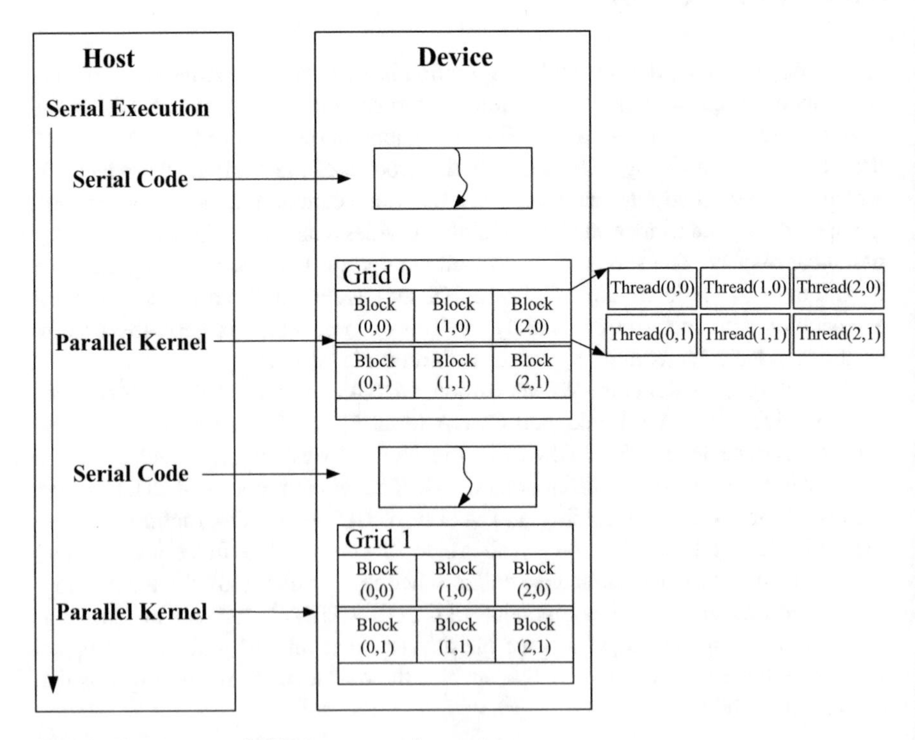

Fig 8.1 The structure of CUDA programming model

its threads in parallel. In addition, each thread consists of a global and local register, and each block has a piece of shared memory.

8.2.3 CUDA Libraries

CUDA libraries created by NVIDIA lay the foundation for compute-intensive applications [2]. Whether you are building a new application or trying to speed up an existing application, the regularly extended and optimized CUDA libraries will provide the easiest way to highly efficient implementations of algorithms. Here, we list a series of commonly used CUDA libraries in Table 8.1.

In the following, we introduce these libraries presented in Table 8.1 in details.

The cuFFT library created by NVIDIA is designed to provide the function of fast Fourier transform (FFT), which is efficient for computing discrete Fourier transforms of complex datasets. It consists of two separate libraries named as cuFFT and cuFFTW, both of which can be widely used for computational physics and general signal processing due to the simple interface for computing FFTs on an NVIDIA GPU.

The cuBLAS library provides a complete implementation of basic linear algebra subprograms (BLAS). This library in CUDA exposes two sets of API, the CUBLASXT API, and the regular cuBLAS API. To use the cuBLAS API, the implementation of accelerated computing is taken place in GPU, and the corresponding results will be transferred from the GPU memory space to the host. However, to use the CUBLASXT API, the library will take care of allocating the required operation to multiple GPUs present in the system.

The cuSPARSE library consists of a set of BLAS used for handling sparse matrices. It can be classified into four categories, operations between a vector in sparse format and a vector in dense format, operations between a matrix in sparse format and a vector in dense format, operations between a matrix in sparse format and a set of vectors in dense format, operations that allow conversion between different matrix formats, and compression of CSR matrices. Although it allows developers to access the computational resources of GPU, it exhibits a poor performance in auto-parallelize across multiple GPUs.

Table 8.1 A collection of supported CUDA library domains

Library name	Domain
cuFFT	Fast Fourier Transforms
cuBLAS	Linear Algebra (BLAS Library)
cuSPARSE	Sparse Linear Algebra
cuSOLVER	Sparse Linear Algebra and Graph Computations
cuRAND	Random Number Generation
NPP	Image and Signal Processing
nvGRAPH	Graph Analytics and Big Data Analytics
Thrust	Parallel Algorithms and Data Structures

The cuRAND library is designed for the simple and efficient generation of high-quality pseudorandom and quasi-random numbers, which has the most of the statistical properties of a random sequence. It contains a library on the host (CPU) side and a device (GPU) header file. The host-side library is similar to that of other CPU library. However, the device (GPU) header file is used for setting up random number generator states and generating sequences of random numbers.

The NPP library is designed for performing CUDA accelerated processing, which is widely applicable for imaging and video processing. Its evolution provides an efficient way to address the task with heavy computational burden.

The nvGRAPH harnesses the sparse linear algebra to deal with the largest graph analytics and big data analytics problems. Its core functionality is the sparse matrix vector product, which uses a semi-ring model with automatic load balancing for any sparsity pattern. It also provides graph construction and manipulation primitives, and a series of useful graph algorithms optimized for GPU.

The Thrust library is based on the standard template library, which allows one to implement high performance parallel applications with minimal programming effort due to a high level interface. It can be composed together to implement complex algorithms with the help of a rich collection of data parallel primitives, such as scan, sort, and reduce.

8.3 Hadoop-Based Distributed Computation

8.3.1 What's Hadoop?

Hadoop, developed by the Apache software foundation, is an open source distributed computing platform. With the Hadoop distributed file system (HDFS) and the MapReduce which is the core of the Hadoop, the distributed infra-structure system with transparent low level details is provided to the users [3]. The advantages of HDFS, such as high fault tolerance and high scalability, allow the users to deploy the Hadoop on cheap hardware, forming a distributed system, and with the MapReduce, the users can develop the parallel applications when being not aware of the underlying distributed system. Therefore, the users can easily organize the computer resources by using the Hadoop, so as to build their own distributed computing platform for processing huge amounts of data.

Why do we need Hadoop? As we all know, with the amount of information of modern society growing, a large amount of data is accumulated, including personal data and industrial data. We need to process and analyze these data in order to get more valuable information. The Hadoop can be used to efficiently store and manage these data, which can improve the speed of reading and writing files by adopting the way of distributed storage. The MapReduce is employed to integrate the distributed file system, guaranteeing the efficiency of processing data. Meanwhile, the Hadoop also ensures the security of data by adopting the way of storing redundancy data.

Hadoop provides a reliable, efficient, and scalable way for processing data. It has the advantages of the following aspects:

High reliability: Since Hadoop assumes that the computation and storage will fail to work, it maintains multiple copies of data, making sure that the failed node can be redistributed again.
Efficiency: As Hadoop works in a parallel way, it can speed up the processing procedure. Hadoop is capable of handling PB bytes of data and can move data between nodes dynamically, ensuring the dynamic balance of each node with very fast processing speed.
High extensibility: Hadoop can allocate the computing tasks using the available computer cluster which can be easily extended to thousands of nodes.
High fault tolerance: Hadoop can automatically save multiple copies of data and can automatically redistribute the failed tasks.
Low cost: Cheap servers can be used to perform Hadoop tasks.
Supporting for multiple programming languages: The Hadoop applications can also be written by other languages, such as C++.

8.3.2 Computing Architecture of Hadoop

Figure 8.2 depicts the system graph of the Hadoop. From this figure, we can see that the Hadoop contains many elements including the Avro, HDFS, HBase, MapReduce, ZooKeeper, Hive, Pig, and Sqoop. The bottom is the HDFS, which stores the Hadoop files on all storage nodes in the computer cluster. The upper layer of HDFS is the engine of MapReduce which is composed of JobTrackers and TaskTrackers.

Avro is the data serialization system. It provides rich data structure types, fast compressible persistent data stored in a binary data format, and set of files. And it can also support the remote procedure call (RPC) protocol. The code generator does not need to read and write files data, and also does not need to use or implement RPC protocol, which is just an optional implementation of the realization of the statically

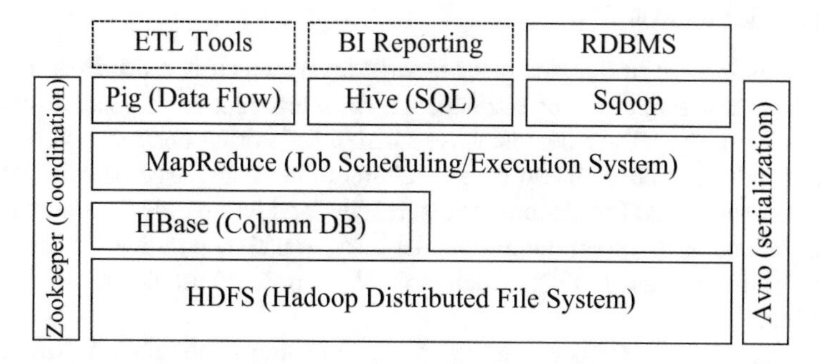

Fig. 8.2 The system graph of the Hadoop

typed languages. Avro system relies on the schema, with the data reading and writing completed under the schema. Thus, one can reduce the time of writing data, so as to improve the speed of serialization. Meanwhile, under the Avro the dynamic scripting languages can be easy to use since the data together with its schema is self-described. In RPC, the client and server of the Avro system exchange the schema through the handshake protocol. Thereby, when the client and server have obtained all their mutual schemas, the problems, (such as the same named fields in different schemas, loss of field, and the consistency of the information,) can be well addressed.

HDFS is a kind of distributed file system, running on a large commercial machine cluster, which provides support of the underlying storage with high reliability for HBase. Since HDFS has the characteristics of high fault tolerance (fault-tolerant), one can design to deploy it on cheap (low-cost) hardware. It can access the data at a high throughput, which is suitable for those applications with very large dataset. HDFS eases the requirements of portable operating system interface (POSIX), thus it can access data in a file system in the form of a flow.

The following is the design goal of HDFS.

1. Realize hardware failure detection and quick recovery. The HDFS system consists of thousands of servers storing the data file, which means high failure rate; therefore, the failure detection and automatic rapid recovery is a key goal of the HDFS.
2. Implement data access with flow form. HDFS makes the applications to have access to their datasets in a streaming form. It is designed to be suitable for batch processing, rather than interactive processing. Therefore, it stresses the throughput of data, rather than the speed of accessing data.
3. Simplify the consistency model. Performing files by most of the HDFS programs need a write operation, and many read operations. Once a file is created, written, and closed, it doesn't need to be modified, so as to avoid the data consistency and high throughput problems.
4. Support communication protocol. All communication protocols are based on the TCP/IP protocol. After a client and an explicitly configured port directory node called the NameNode are connected, their protocol is the client protocol. Communication between the DataNode and the NameNode is realized by the DataNode protocol.

HBase is located in the structured storage layer, which is a distributed storage database. It is a subproject of Hadoop, which is different from general relational database. First, the HBase is a database suitable for storing unstructured data, and second, it has the schema based on the column rather than based on row. Besides, HBase is mainly used for random access, reading and writing big data in real time.

MapReduce is a programming model for parallel computing of large-scale datasets (greater than 1 TB), which will be introduced in details in the next subsection.

ZooKeeper is a centralized service for maintaining configuration information, naming, providing distributed synchronization, and providing group services. All these kinds of services are used in some form or another by distributed applications. Once they are implemented, there is a lot of work that goes into fixing the bugs and race conditions that are inevitable.

Hive was originally designed by Facebook, which is a data warehouse based on Hadoop. It provides some tools for sorting, special query, and analysis of data stored in the Hadoop file set.

Pig is a kind of data flow language and operating environment, whose goal is to retrieve very large datasets, greatly simplifying the common tasks of Hadoop. The Pig can load data, express conversion data and store the final results. The built-in operations in Pig make scene of semi-structured data, e.g., log files. Pig and Hive provide both high level language support for HBase, making the data processing on the HBase very simple.

Sqoop is a tool designed for efficiently transferring bulk data between Apache Hadoop and structured data stores such as relational databases.

8.3.3 MapReduce

MapReduce is the core computing model of Hadoop, which abstracts the complex, parallel computing processes that run on large clusters to two functions: map and reduce [4]. The dataset (or task) which can be suitably handled by the MapReduce needs to meet the basic requirements, that is, the pending datasets can be decomposed into many small datasets and each small dataset can be processed in parallel. The concept of map and reduce, borrowed from functional programming language, contains the characteristics of the vector programming language. The MapReduce brings much convenience for programmers who cannot design distributed parallel programming to run their program on a distributed system.

A MapReduce job usually divides the input dataset into a number of separate data blocks, then the map functions process them in a completely parallel manner. The outputs of the maps will be sorted firstly, and the results are sent to the reduce functions. Usually, the input and output of the job are stored in the file system. The entire framework is responsible for scheduling and monitoring tasks, as well as re-executing tasks that have failed. The MapReduce framework and the distributed file system are run on the same set of nodes, that is, computing and storage nodes are allocated together. This configuration allows the framework to efficiently schedule tasks on nodes that already have data, and the network bandwidth of the entire cluster is utilized very efficiently.

The MapReduce consists of a single master named JobTracker and a slave named TaskTracker for each cluster node. This master is responsible for scheduling all tasks that make up an assignment, which is distributed across different slaves, and it monitors their execution and re-executes the failed task. Besides, the slave only performs tasks assigned by the master.

Figure 8.3 illustrates the process of using MapReduce to deal with large datasets. In a nutshell, the calculation process of the MapReduce divides large

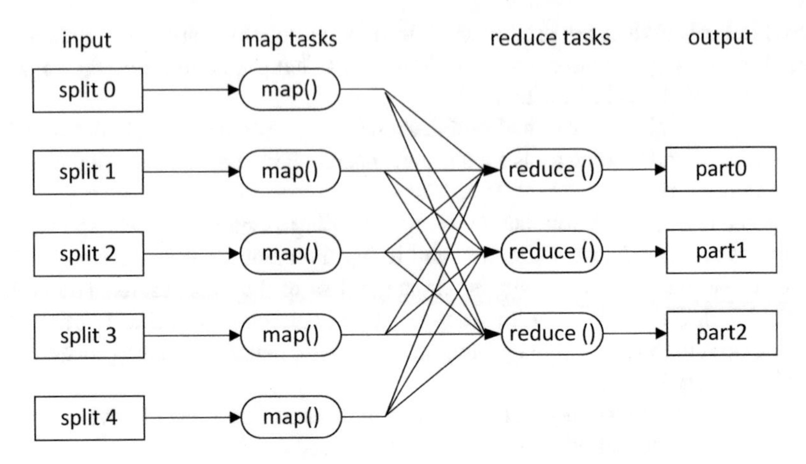

Fig. 8.3 Process of using MapReduce to deal with large datasets

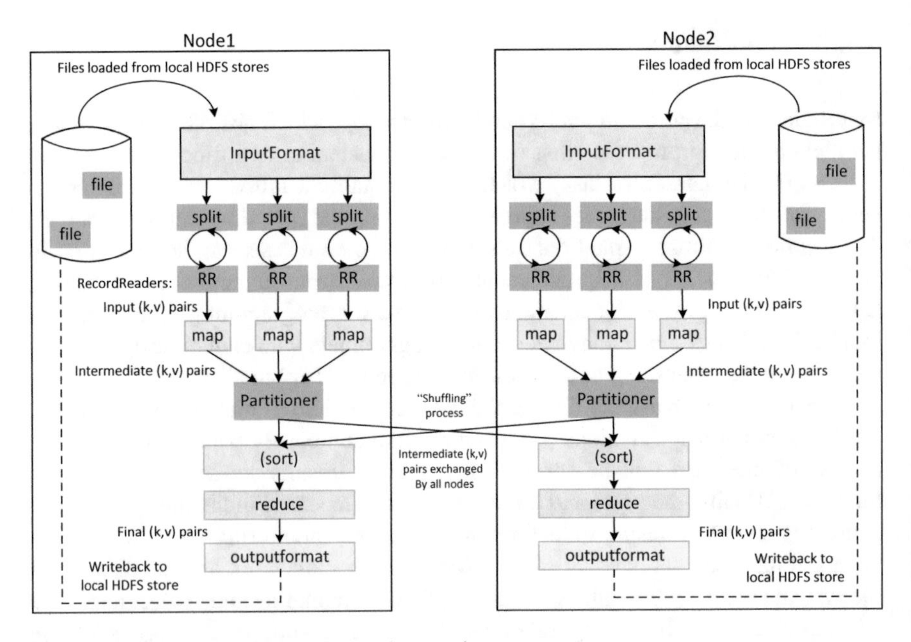

Fig. 8.4 A flow chart of MapReduce task execution process

dataset into thousands of small datasets, with each dataset (or several) being processed by a node (generally a typical computer) in cluster, generating intermediate results. Then, these intermediate results are merged by a large number of nodes, producing the final result.

After introducing the process of dealing with large datasets by MapReduce, we give an overall flow chart of MapReduce task execution process of Hadoop (as shown in Fig. 8.4).

Step 1: Create tasks in client and submit them in a distributed environment by JobTracker.

Step 2: Perform the InputFormat module. It is responsible for preprocessing before the map, which mainly includes the following tasks: 1) verify that whether the input format conforms to the definition of the input of JobConfig. 2) divide the input file into logical inputs (InputSplit). 3) use the RecordReader to process a set of records which are obtained by dividing the input file, and then send the outputs to the map.

Step 3: Consider the results of the RecordReader as the input of the map, then let the map execute the predefined map logic. Store the processed <key, value> in the temporary intermediate file.

Step 4: Execute the "shuffle" and "partitioner." In order to process the results of the map by the reduce in parallel, the outputs of the map should be sorted and divided before sending them to the reduce. The process of organizing and handing outputs of the map to the reduce is called shuffle. The purpose of the partitioner is for selection. Its main function is to specify which reduce is used to progress the result of the Map in the case of multiple reduces. Each reduce will have a separate output file.

Step 5: The reduce performs specific logic task and delivers the produced results to the OutputFormat.

Step 6: Use the OutputFormat to verify whether the output directory already exists and whether the type of output conforms to the configuration type, and if both are true, the result after performing the reduce is produced.

8.4 GPU Acceleration for Training EKF-Based Elman Networks

In this section, we will introduce a GPU acceleration case of the EKF-based Elman networks (EKF-ENs) for time series prediction [5]. The basics of the EKF-ENs have already been introduced in Chap. 7.

Although the EKF-ENs modeling improves the convergence rate, its computational complexity is still rather high at the cost of $O(G^2)$; furthermore, the dynamic feature of the real-world industrial data is usually complicated, where a large-scale network and a large number of sample data might be required. In this case, it is very necessary for practice to accelerate the modeling process. Considering the GPU structure and that a large number of operations in Elman network training based on EKF have high fine grain parallelism, a series of parallelized strategies based on GPU are presented in this section to enhance the efficiency. First, different from the studies that only paid attention to the parallelization related to general matrix computation [6]; this section puts the direct solving of the Jacobian matrix on GPU, which can decrease the time cost of data transfer between CPU and GPU. Second, a series of parallelized kernel functions are developed, such as the parallelized versions of calculating the hyperbolic tangent function and the inverse matrix, in which the shared memory is used to accelerate the data access, and the thread synchronization function is further adopted to coordinate the multiple threads allocated on GPU.

Through a quick visual inspection on (3.72)–(3.76) in Sect. 3.5.1, one can notice that a great deal of matrix operations are involved, and the matrix multiplication and the matrix inversion are the most time-consuming tasks, especially when the network scale increases. Taking the solution to matrix inversion as an example, its calculation usually consumes much more time compared to the matrix multiplication. In this section, a Gaussian elimination method is adopted, which can be expressed as $\mathbf{B} = (\mathbf{A}|\mathbf{I}) \sim (\mathbf{I}|\mathbf{A}^{-1})$. The elimination process is illustrated in Fig. 8.5, taking an example $\mathbf{A}$ with 512×512. The elimination mainly consists of two steps, transforming the original matrix $\mathbf{A}$ into a diagonal one and transforming the calculated diagonal matrix into the identity one. Here, two parallelized kernel functions are developed considering the advantages of shared memory in accessing speed compared to the global memory. In the parallelization, the shared memory serves to access the data of main row (see Fig. 8.5) in the Gaussian elimination, each block is responsible for computing one row of the matrix $\mathbf{B}$, and each thread in a block is in charge of computing one element of the row. Each of these elimination operators is executed by broadcasting the data of main row from the global memory to the shared one, and then one element of the elimination operator is computed by each thread. After each parallelized operations, a synchronization function is

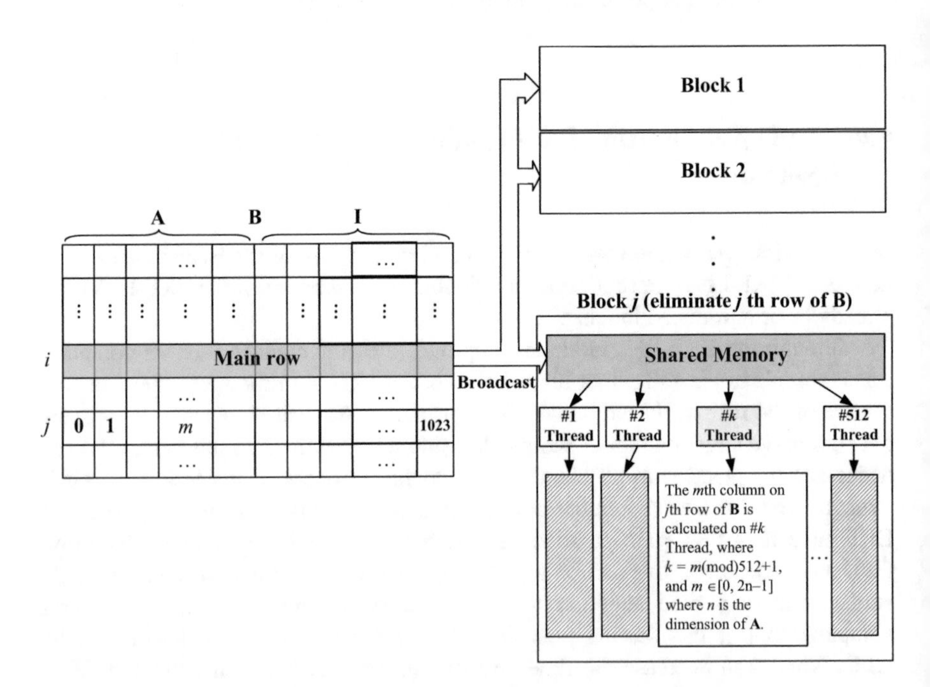

Fig. 8.5 Solving the inverse matrix by parallel acceleration on GPU

adopted to coordinate the multiple threads. To further improve the computational efficiency, some optimizations for CUDA program are applied. For example, the texture memory, viewed as an advantageous alternative to read the device memories, is used to store the values of **Q** and **R** considering they are invariable in iteration. And, the memory alignment is employed to avoid the bank conflict in GPU. As such, one can outline the parallelization of Elman network training based on EKF as follows:

Algorithm 8.1: CUDA-Based Parallel EKF-ENs Modeling

Step 1: Allocate the Host memory for **P**, **Q**, **R**, *and* **w** *and initialize them.*
Step 2: Allocate the GPU local memory for **P**, **Q**, **R**, *and* **w**, *and transfer the data from CPU to GPU.*
Step 3: Build the training sample set $(\mathbf{u}^T(k), \mathbf{d}^T(k))$, $(k = 1, 2, 3, \ldots, M)$ *on GPU and transfer the data from CPU to GPU.*
Step 4: Execute the time update equations of EKF by (3.72)–(3.73) *on GPU using cuBLAS functions.*
Step 5: Calculate the Jacobian matrix **H** *on GPU and the value of* $\mathbf{y}(k) - \widehat{\mathbf{y}}(k)$ *on CPU and transport the result from CPU memory to GPU one.*
Step 6: Compute (3.74)–(3.76) *on GPU by using cuBLAS library functions and the developed kernel functions with CUDA.*
Step 7: Repeat Step 3–6 until the optimal **w** *is obtained.*
Step 8: Transfer the obtained **w** *from GPU memory to CPU memory, then calculate* $\mathbf{W}^{\text{in}}$, **W** *and* $\mathbf{W}^{\text{out}}$ *on CPU to perform the prediction by using* (3.71).

8.4.1 Case Study

To indicate the effectiveness of the CUDA-based parallel computing method, we apply it to an energy system prediction of steel industry. In this section, the data of the real-world gas flows coming from a steel plant of China, ranged between April 1st and April 8th, 2012, are employed. Without loss of generality, we randomly choose the BFG generation flow, the BFG consumption flow on coke oven, and the COG consumption flow on blast furnace as the studied units. A large number of comparative experiments are conducted to clarify the approach effectiveness on the accuracy and the computing cost.

BFG System Prediction

The BFG generation flow of #1 blast furnace and the BFG consumption flow of #1–2 coke oven are randomly selected to conduct the experiments. Based on the experimental approach presented previously, the initial parameters are set as $\mathbf{P}(0|0) = 1000\mathbf{I}$, $\mathbf{Q}(0|0) = 0.0001\mathbf{I}$, and $\mathbf{R}(0|0) = 100\mathbf{I}$. For a fair competition, the optimal parameters in each modeling are also determined by the experimental method. We select 1000 continuous data points from the online database as the training samples, and the next 60 data points as the testing data. The statistic prediction results are reported in Table 8.2, and one of the random selected performances including the predicted values and the errors are illustrated as Figs. 8.6 and 8.7.

From Figs. 8.6 and 8.7, it can be clearly seen that the proposed EKF-ENs model presents the highest prediction accuracy compared to the others. On the statistics perspective, Table 8.2 lists the multiple quantitative evaluation indices for the prediction accuracy, which can comprehensively indicate the advantages of the

Table 8.2 Comparisons of prediction error and training time for the data of the BFG units

Objects	Method	MAPE	RMSE	NRMSE	Computational time (s)
BFG generation flow of #1 blast furnace	BP	7.729	50.484	0.012	53.566
	ESN	8.203	52.860	0.013	0.566
	Kernel learning-based	9.428	63.148	0.016	3.810
	EKF with BPTT	6.504	41.856	0.010	90.140
	EKF-ENs	5.797	39.344	0.010	70.813 (Without GPU) 27.063 (With GPU)
BFG consumption flow of #1–2 coke oven	BP	1.1307	2.0251	0.0018	14.774
	ESN	0.6730	1.2420	0.0011	0.612
	Kernel learning-based	0.6003	1.0482	0.0009	2.016
	EKF with BPTT	0.6262	1.1701	0.0010	42.970
	EKF-ENs	0.5373	0.9821	0.0008	28.312 (Without GPU) 12.201 (With GPU)

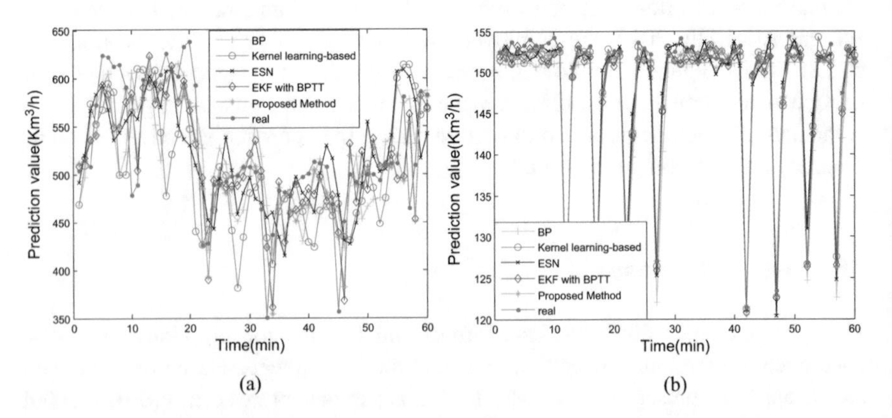

Fig. 8.6 Comparisons of the prediction results produced by the five methods. (**a**) BFG generation flow of # 1 blast furnace. (**b**) BFG consumption flow of # 1–2 coke oven

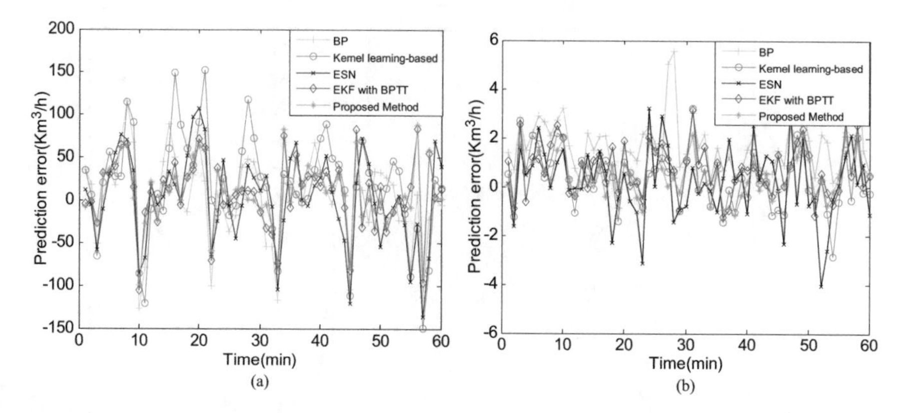

Fig. 8.7 Comparisons of absolute error predicted by the five methods. (**a**) BFG generation flow of # 1 blast furnace. (**b**) BFG consumption flow of # 1–2 coke oven

EKF-ENs. For the industrial data-based experiments, the accuracies of the EKF-ENs are obviously higher than that by the EKF with BPTT. In addition, in respect of the computational time, the parallelized Elman network modeling (using GPU) gives a great improvement on computational efficiency. The real-time demand of online application can be completely satisfied because the required prediction interval is usually set as 1–3 min in practice. By analyzing Table 8.2, although the computing efficiency by the EKF-ENs is lower than those by the ESN and the kernel learning-based approach, its accuracies present great improvements, which will be an important concern for the practical energy scheduling.

COG System Prediction

For a further validation, the COG consumption on #1 blast furnace is selected as another instance. The initial parameters are set as $\mathbf{P}(0|0) = 1000\mathbf{I}$, $\mathbf{Q}(0|0) = 0.0001\mathbf{I}$, and $\mathbf{R}(0|0) = 100\mathbf{I}$. Similarly, 50 times independent experiments are also conducted for comparison. The comparative results that include a segment of the gas flow are presented in Figs. 8.8 and 8.9, and the average statistic results are reported in Table 8.3.

It is apparently from Figs. 8.8 and 8.9 that the performance of the EKF-ENs is also the best, and the statistic results of the multiple experiments also draw the same conclusion, see Table 8.3. From this table, the errors of the EKF-ENs are definitely lower than those by the others as for the overall evaluation indices. Given that the practical sampling interval of the gas flow is 1 min, the parallelization in the Elman networks training can be completed within 10 s, which fully meets the real-time requirement of the industrial application. In addition, the BP and the ESN might generate the prediction failure more or less in the experiments; while, the EKF-ENs can reach the stable prediction results. With long-term application in steel plant, the EKF-ENs model definitely exhibits a good stability, which guarantees the feasibility

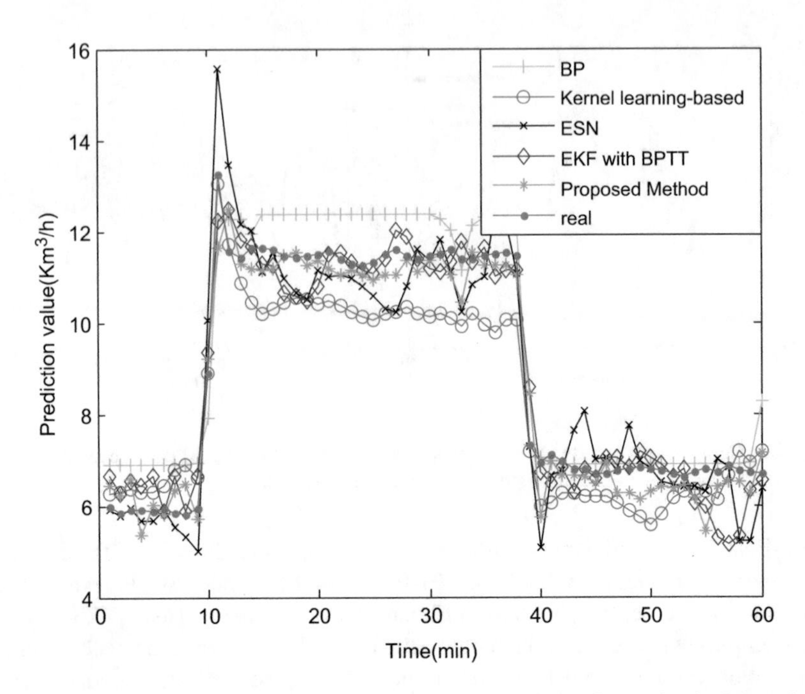

Fig. 8.8 Comparisons of prediction for COG consumption on #1 blast furnace by the five methods

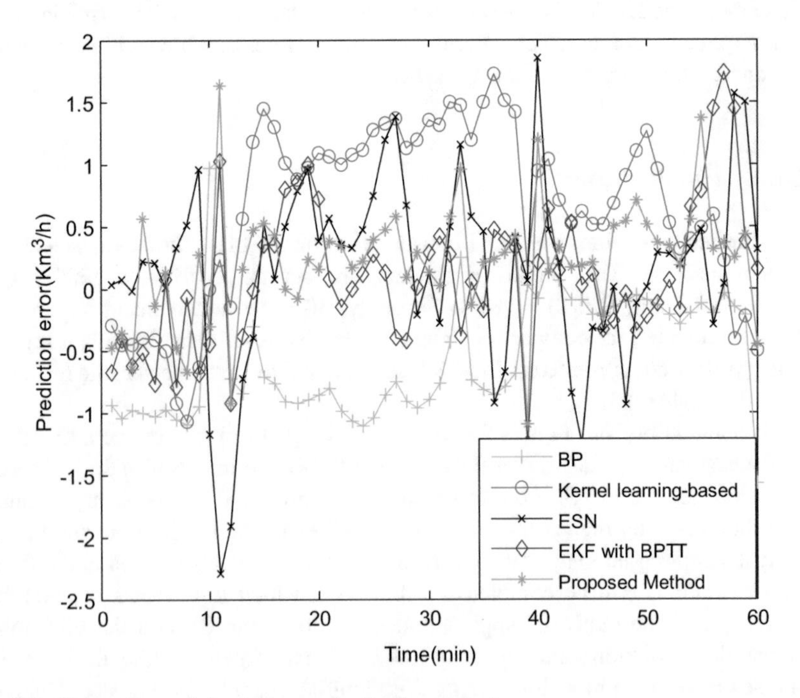

Fig. 8.9 Comparisons of absolute errors by the five methods

Table 8.3 Comparisons of prediction error and computing time for COG consumption on #1 blast furnace

Methods	MAPE	RMSE	NRMSE	Computational time (s)
BP	6.6715	0.6707	0.0094	8.827
ESN	5.9436	0.5778	0.0081	0.580
Kernel learning-based	7.4398	0.9511	0.0133	1.967
EKF with BPTT	5.3572	0.5638	0.0078	28.641
EKF-ENs	5.1094	0.5188	0.0072	23.797 (Without GPU)
				9.962 (With GPU)

of the prediction for the industrial application. All in all, compared to the existing methods, the EKF-ENs approach exhibits the highest prediction accuracies and the applicable stability, which greatly results in the scientific energy scheduling guidance for steel enterprise. By using GPU parallelization, the prediction process is largely accelerated, which enables the EKF-ENs to be applied to the industrial energy prediction.

8.5 Online Parameter Optimization-Based Prediction by GPU Acceleration

In the LSSVM regression model described in Sect. 3.6.3, the kernel function relative to σ determines the expression of the model, and the penalty factor γ is used to establish a sound trade-off between its complexity and empirical risk. If γ and σ are fixed in advance, the phenomenon of over-fitting or poor fitting might occur when the dynamics of sample data changes. Because of this, the model cannot adaptively capture the tendency of the sample data, which ultimately leads to the decrease of the resulting prediction accuracy. To solve this problem, we present an online modeling method to adaptively establish the LSSVM by parallel strategies for prediction.

In this section, we show how to accelerate the modeling process by parallel strategies when dealing with the two aspects of modeling, namely the parameters selection procedure and the validation procedure. Here, a parallel PSO version based on multiple sub-swarms is present to substitute the traditional grid search, and a parallel computing for matrix inversion is used to further speed up the validation phase [7].

The parameter optimization based on cross-validation (CV) has been widely applied to modeling of SVM [8], where the grid search traverses the designated parameter set to reach the optimum. The optimal parameters exhibit a close relationship with the grid partition. If the partition granularity is rather coarse, the optimum might be skipped over. When the partition granularity is too fine, the computational cost of traversing the grid will be excessively high, or even unavailable in an allowable time period. Also, the "traditional" CV needs to solve the inverse matrix associated with the sample data again and again, which consumes a large amount of time. Although [9] and [10] proposed the validation techniques of fast leave-one-out (LOO) and fast CV, a low computational efficiency is still encountered, especially when the dimensionality of the training sample or the number of grid points largely increase. As a consequence, in case of the online prediction for certain industrial case, it can hardly meet the practical requirements.

PSO introduced in Chap. 7 is a swarm intelligence search technique that originates from the ideas of simulating a colony society. When solving an optimization problem, a number of particles are positioned in the search space. Each particle, characterized by its velocity and position, completes search by establishing some collaborative linkages. The pertinent computations are described as follows:

$$v_k^{t+1} = wv_k^t + c_1 r_1 \left(P_{kb} - P_k^t\right) + c_2 r_2 \left(G_b - P_k^t\right) \tag{8.1}$$

$$P_k^{t+1} = P_k^t + v_k^{t+1} \tag{8.2}$$

In (8.1), w is the inertia weight factor; c_1 and c_2 are learning rates on the optima, which are nonnegative constants; r_1 and r_2, are the two random numbers drawn from a uniform distribution in [0,1]. They are combined with c_1 and c_2 to reflect the stochastic influence of local and global optima on the updated velocity of the particle. From this formulation, the new position of kth particle in time $t + 1$, P_k^{t+1}, is determined by its position in t and the new velocity in time $t + 1$, i.e., it is determined by the best-so-far of kth particle, P_{kb}, the best-so-far of the swarm G_b and the velocity of kth particle on time t. In such a way, the new position of the particle is guided by the two components, i.e., its own experience P_{kb} and the swarm cognition level G_b.

In the parallelized PSO version used in this section, to realize the cognition mode of general swarm society, we propose the concept of sub-swarm that is viewed as a part of the entire swarm to describe a neighbor colony, by which the search trace of each particle can be further affected through the local cognition level.

8.5.1 Initialization and Sub-swarm Separation

The parallel PSO generates N particles in the search space at random. For the diversity of individuals in the colony, the following method helps to separate the N particles into M sub-swarms. First, we calculate the value of the objective function for all of the particle positions and arrange the results into a sequence starting from higher fitness to moving to the lower one. Then, the qth particle is placed into the sub-swarm numbered as $q(mod)M + 1$, where $q \in [1, N]$ (mod stands for the modulo operator).

Update of Velocity and Position

Regarding the swarm society, the cognition level of individual is usually affected not only by the individual and the swarm experience, but the learning resulting from the neighbor experiences is also considered. Enlightened by such thought, the local collaboration in a sub-swarm is further considered here when separating the swarm into the sub-swarms. Here, we modify the velocity update rule in the parallel PSO as follows:

$$v_k^t = wv_k^t + c_1r_1\left(P_{kb} - P_k^t\right) + c_2r_2\left(P_{jb} - P_k^t\right) + c_3r_3\left(G_b - P_k^t\right) \tag{8.3}$$

where P_{jb} is the best-so-far of jth sub-swarm that kth particle belongs to. Different from (8.1), the local collaboration, $c_2r_2\left(P_{jb} - P_k^t\right)$, is further added into the process of particle learning in order to affect the their cognition. Thus, the velocity of each particle is updated based on three positions, i.e., particle's own best-so-far, global best-so-far, and the local best-so-far in sub-swarm.

Parameter Adaptation and Process Communication

The inertia coefficient w reflects the impact of the previous velocity on the current one. The larger its value is, the more easily the search process can jump out of local optimum. Low values of the coefficient are beneficial to the convergence of the algorithm. To achieve a sound trade-off in the search process, we proceed with a dynamic selection of w, that is

$$w = w_{\max} - \frac{w_{\max} - w_{\min}}{S_{\max}} \times s \tag{8.4}$$

where $S_{\max}$ denotes the total number of iteration for each particle, s denotes the current iteration; and $w_{\max}$ and $w_{\min}$ are the maximum and minimum of the parameter.

In this parallel PSO, the local best-so-far of sub-swarm and the global best-so-far need updating. We set two zones in the GPU structure for each particle to store the local and the global best-so-far, respectively, in which the matrix inversion are performed based on the Gaussian elimination by CUDA presented in Fig. 8.5. As for the global one, it needs to be used by all of the particles; while as for the neighbor one, it is used by the inner particles in sub-swarm.

The underlying algorithm can be described as a sequence of steps.

Algorithm 8.2: CUDA-Based Parallel LS-SVM

Step 1: *Initiate the velocities and the positions of N particles, the number of sub-swarms M, the iteration number $S_{\max}$, and set the values of the coefficients.*

Step 2: *Calculate the fitness of the N positions, partition sub-swarm by the mentioned method, and record the neighbor and the global best-so-far.*

Step 3: *Asynchronously update velocities and positions of particles in each sub-swarm, and memorize the two best-so-far values into the storage zones.*

Step 4: *Check whether all the sub-swarms complete the iteration. If so, the algorithm finishes, and the optimum is obtained from the storage zone; otherwise, go back to Step 3.*

8.5.2 Case Study

For verifying the feasibility of this parallel strategies for the online parameter optimization of LS-SVM, we predict the variation of the gasholder level in energy system of a steel plant using the available experimental data. Moreover, we also model the LDG consumption flow of blast furnace based on time series for further testing the generalization of the online prediction. All the simulations presented in this section are run on the platform of CUDA. The configuration of hardware involves CPU of Intel Core Q8200 whose main clock frequency is 2.33 GHz, and GPU is GeForce GTX260 by NVIDIA.

Parallelization for Parameter Validation

For achieving the online prediction efficiency as high as possible, we parallelize three validation methods based on the GPU acceleration technique including the generic CV [8], Fast CV [10], and Fast LOO [9], where ten-fold CV is used in the generic CV and the Fast CV. The comparative results are listed in Table 8.4, where n is the number of samples and Acc denotes the acceleration ratio of the parallel version. We develop the prediction models for gasholder level and the LDG consumption flow of blast furnace; the latter employs a time series-based model. In practice, the scheduling workers usually estimate the flow variation of LDG consumption considering the historical data recorded within 1–2 h. Since the sample interval of the time series-based model is 1 min, to achieve a sound compromise between the computational complexity and the prediction accuracy, we set the embedded dimension of the model equal to 90, which amounts to the previous 1.5-h flow data.

It is seen in Table 8.4 that these parallel versions run on GPU are faster than the serial ones, and the acceleration ratios become larger along with the growth of the number of sample data. As for the same number of sample data, the parallelization using Fast LOO leads to the shortest computational time. The validation process took 0.4 s even if the training samples size is 1000. The generic CV produced the worst results. From the acceleration perspective, the acceleration ratio using the Fast CV rapidly increases to the highest in the three validation methods when the amount of sample data grows. However, its serial version spends a rather long time compared to that of the Fast LOO, so the results by Fast CV are somewhat worse. Anyway, using the Fast LOO for parameter validation can completely satisfy the

Table 8.4 Comparison of computing times using three validation methods

Prediction objective	n	Generic CV (s)			Fast CV (s)			Fast LOO (s)		
		Serial	Parallel	Acc	Serial	Parallel	Acc	Serial	Parallel	Acc
Gasholder level	100	0.29	0.045	6.4	0.079	0.013	6.1	0.042	0.011	3.8
	200	2.45	0.103	23.8	0.542	0.022	24.6	0.311	0.017	18.3
	450	24.7	0.545	45.4	6.603	0.070	94.3	2.600	0.053	49.1
	800	119.7	1.911	62.6	35.6	0.261	136.4	17.40	0.22	79.3
	1000	263.6	3.733	70.6	79.6	0.452	176.2	38.30	0.39	99.6
LDG flow on blast furnace	180	2.052	0.214	9.6	0.451	0.02	22.5	0.199	0.015	13.3
	320	10.88	0.812	13.4	2.555	0.047	54.4	1.033	0.032	32.3
	450	26.15	1.29	20.3	7.308	0.080	91.4	2.609	0.057	45.8

Table 8.5 Comparison of prediction accuracy of Fast CV and Fast LOO using MAPE

Prediction objective	Fast CV (%)	Fast LOO (%)
#1 Gasholder level	5.30	4.59
#2 Gasholder level	5.25	4.62
LDG flow on #2 blast furnace	4.12	3.50
LDG flow on #4 blast furnace	6.57	5.90

demand of online prediction for the gasholder level and the LDG users on the perspective of computational efficiency. With respect to the prediction accuracy, we give a further comparison between the Fast CV and the Fast LOO also using the practical energy data from the practice. The comparative experiments were carried out independently 50 times for different data. The prediction horizon was 60 min while the number of points in the grid search was set to 50.

Table 8.5 lists the average prediction results by the Fast CV and the Fast LOO using the MAPE defined by (2.67). From Table 8.5, the prediction accuracies on average by the Fast LOO are somewhat better than that by the Fast CV. It is possible that the prediction precision would be enhanced if the number of girds increases, but the validation time cost will be largely increased as well.

Parallelization for Parameter Selection

In this section, the parallel PSO based on multiple sub-swarms is employed to speed up the process of parameter selection, where each sub-swarm is run at one block of GPU structure, and each particle is run at one thread. The local best-so-far in iteration is stored in shared memory of GPU, and the global best-so-far is stored in global memory. Fifty groups of practical data in different period from a steel plant in China are used to carry out the simulation experiments, and the traditional grid search (GS) is also parallelized for comparison. Considering that the training data has been translated and rescaled component-wise to achieve zero mean and standard deviation before proceeding with learning, we set the ranges of the parameters as $r \in [1, 1000]$, $2\sigma^2 \in [1, 100]$, and the number of grids is also set as 50, i.e., there are 2500 combinations of $r$and σ to be searched in GS. In the parallel PSO, the colony size is 20, and the iteration number is 50. For presenting the effect of the PSO, we give a comparison including the traditional GS, regular PSO, and the parallel PSO. The embedded dimension when predicting the LDG flow on blast furnace based on time series is also set as 90 as the process of parameter validation. The Fast LOO is used in the parameter optimization phase, and the MAPE is also to evaluate the prediction quality. In addition, we report the MSE and the computational time (CT) listed in Table 8.6. As seen from this table, when optimizing the parameters, the prediction accuracy using the two PSO is better than that obtained by GS; furthermore, the computational time by the parallel PSO versions needs only about one third of GS. Although the speed of regular PSO is somewhat higher than of the parallel one, such difference can be completely accepted considering the

Table 8.6 Prediction comparison of parameter optimization by the GS and the PSO

Prediction objective	Method	Number of sample	Results (on average)		
			MSE	MAPE (%)	CT (s)
#1 LDG gasholder level	GS	450	27.6039	4.40	123
	Regular PSO	450	26.9136	4.31	34
	Parallel PSO	450	26.6737	4.26	36
#2 LDG gasholder level	GS	450	27.9649	4.98	123
	Regular PSO	450	27.4891	4.84	34
	Parallel PSO	450	26.9106	4.78	36
LDG flow on #2 blast furnace	GS	320	1.3017	3.56	66
	Regular PSO	320	1.2803	3.53	20
	Parallel PSO	320	1.2451	3.49	21
LDG flow on #4 blast furnace	GS	180	8.5832	5.82	22
	Regular PSO	180	8.5727	5.76	6
	Parallel PSO	180	8.5486	5.69	6

requirements coming from the practice of the problems in the gas system. All in all, the solving efficiency by the parallel PSO is quite sound for online prediction of LDG system.

Online Prediction for LDG System

As for the practical online prediction problem, we first show a comparative experiment to analyze the efficiency of model learning between the serial LS-SVM and the parallel one, where the optimized parameters σ and γ have been determined in advance. We randomly select the various objectives in LDG system with different number of sample data and input dimension to complete the experiments whose results are illustrated in Table 8.7. From this table, the acceleration ratio gradually increases along with the increase of the number of sample data toward a certain of input dimension. We also report the model training time cost realized by the two methods as shown in Fig. 8.10. Apparently, the computing time growth of the parallel learning process is substantially slower than that of the serial one. To complete the gasholder level prediction, we generally select the number of sample data to be equal to 450. Under such condition, the acceleration ratio exceeds a factor of 40 (see Table 8.7).

Table 8.7 Comparison of the training time using different number of samples and input dimension

Number of sample data	Input dimension	Computational time for training (s)		Acc
		Serial LSSVM	Parallel LSSVM	
100	2	0.035	0.004	8.9
200	2	0.204	0.01	20.4
450	2	2.25	0.051	44.0
800	2	12.48	0.205	60.9
1000	2	24.61	0.374	65.8
180	60	0.249	0.043	5.8
320	60	1.067	0.093	11.5
450	60	2.67	0.135	19.8
1000	60	25.98	0.555	46.8
180	80	0.283	0.054	5.2
320	80	1.15	0.107	10.7
450	80	2.85	0.161	17.7
1000	80	26.67	0.639	41.7
320	160	1.51	0.192	7.8
450	160	3.46	0.287	12.0
1000	160	29.61	0.902	32.8

To further demonstrate an overall effectiveness of the parallel strategies for parameters optimization of the LSSVM, we consider 50 groups of energy data to verify the prediction model. We predict a 60-min variation of the objectives and compare the obtained results when using the LSSVM with the fixed parameters, back propagation neural network (BP-NN), and SVM [11]. We take the NRMSE and MAPE to quantify the quality of prediction.

As seen from Table 8.8 that the BP-NN based on the principle of experience risk minimum gives the worst results, and its computational time is also the longest because the network weights are determined by GD algorithm that tends to slow down the learning process which might also be trapped in possible local optimum. Although the prediction precision is evidently enhanced, the generic SVM solving a convex quadratic programming has to require a quantity of computational time period, which is hardly to meet the requirement of real-time prediction for the LDG system. The prediction based on LSSVM solves a series of linear equations to train the regression model. Although the version with fixed parameters uses the least modeling time owing to getting rid of the validation process, it cannot adaptively adjust the kernel-based model according to the dynamics of gas flow. Therefore, its prediction error is obviously larger than that of the online optimized LSSVM. As for the computing time, in general practice, it is acceptable to predict the LDG flow online within one minute, so we propose the real-time optimized LSSVM for the industry application. In application, the modeling process is carried

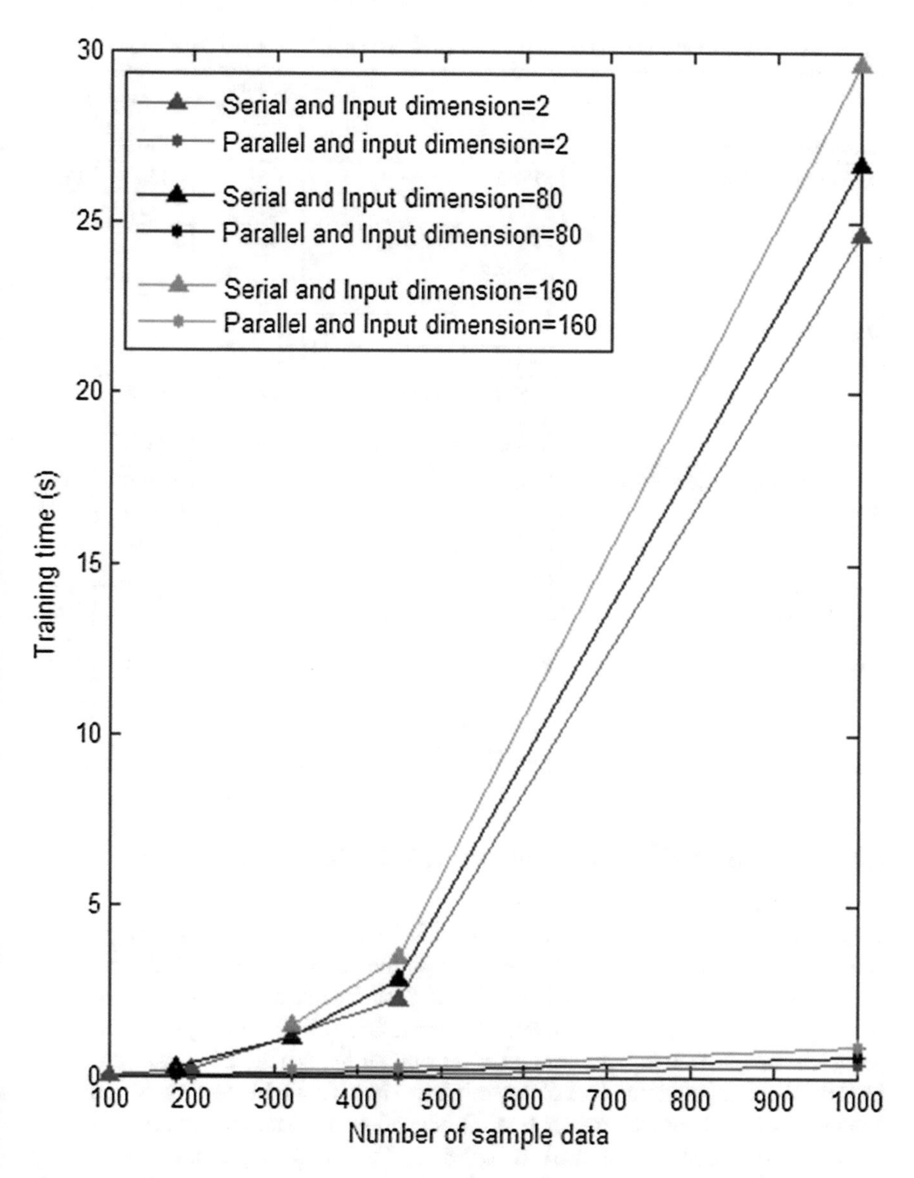

Fig. 8.10 Training time by the serial and the parallel LS-SVM

out at designated time interval in order to meet the demand of online prediction. To clearly visualize the prediction effect, we present the results in Fig. 8.11. The online LSSVM presented in this section can well track the real energy variation for not only the gasholder level but also the LDG consumption amount of gas users.

Table 8.8 Results of comparison using the parallel online optimization and other three methods

Prediction objective	Method	Results		
		MAPE (%)	NRMSE	CT (s)
#1 LDG gasholder level	BP-NN	11.61	0.1352	156.1
	SVM	9.21	0.1128	363.8
	LSSVM with fixed parameters	8.98	0.0929	2.2
	Parallel LSSVM	4.36	0.0588	38.0
#2 LDG gasholder level	BP-NN	13.33	0.1604	120.7
	SVM	11.86	0.1425	363.4
	LSSVM with fixed parameters	10.12	0.1235	2.2
	Parallel LSSVM	4.68	0.0540	38.1
LDG consumption of #2 Blast Furnace	BP-NN	10.59	0.1164	150.5
	SVM	7.35	0.0989	132.8
	LSSVM with fixed parameters	7.26	0.0686	1.1
	Parallel LSSVM	3.49	0.0400	22.2
LDG consumption of #4 Blast Furnace	BP-NN	13.43	0.1674	46.4
	SVM	11.85	0.1524	24.4
	LSSVM with fixed parameters	11.84	0.1174	0.3
	Parallel LSSVM	5.79	0.0898	6.4

8.6 Parallelized EKF Based on MapReduce Framework

In this section, we develop a parallelized EKF based on ESN by using MapReduce framework for acceleration [12]. According to the description of the EKF based on ESN introduced in Sect. 3.5.1, the use of EKF is to achieve a series of inner and outer loops, where the outer one completes the operations of all training samples. For a sufficient sample training, the amount of the outer loops is generally large, which has to make the computational cost of the training very high. In this section, MapReduce framework-based parallelized EKF is introduced to estimate the output weights of the ESN, see the structural chart in Fig. 8.12. Two MapReduce-based models are adopted, each of which is composed of a set of mapper and reducer functions, in which the mapper receives a training sample and generates the updates of the internal states or the output weights, while the reducer merges all updates associated with the same key to produce an average value.

The following is the basic step of the parallel computing method:

Algorithm 8.3: MapReduce-Based Parallel EKF Modeling

Step 1: Receive training samples from the training set.
Step 2: Randomly generate the initial internal states and the output weights.
Step 3: Update the internal states. Execute the mapper tasks, in which each mapper receives one training sample and then computes all updated values for the internal states. And, execute N (the amount of neurons of the reservoir) reducer

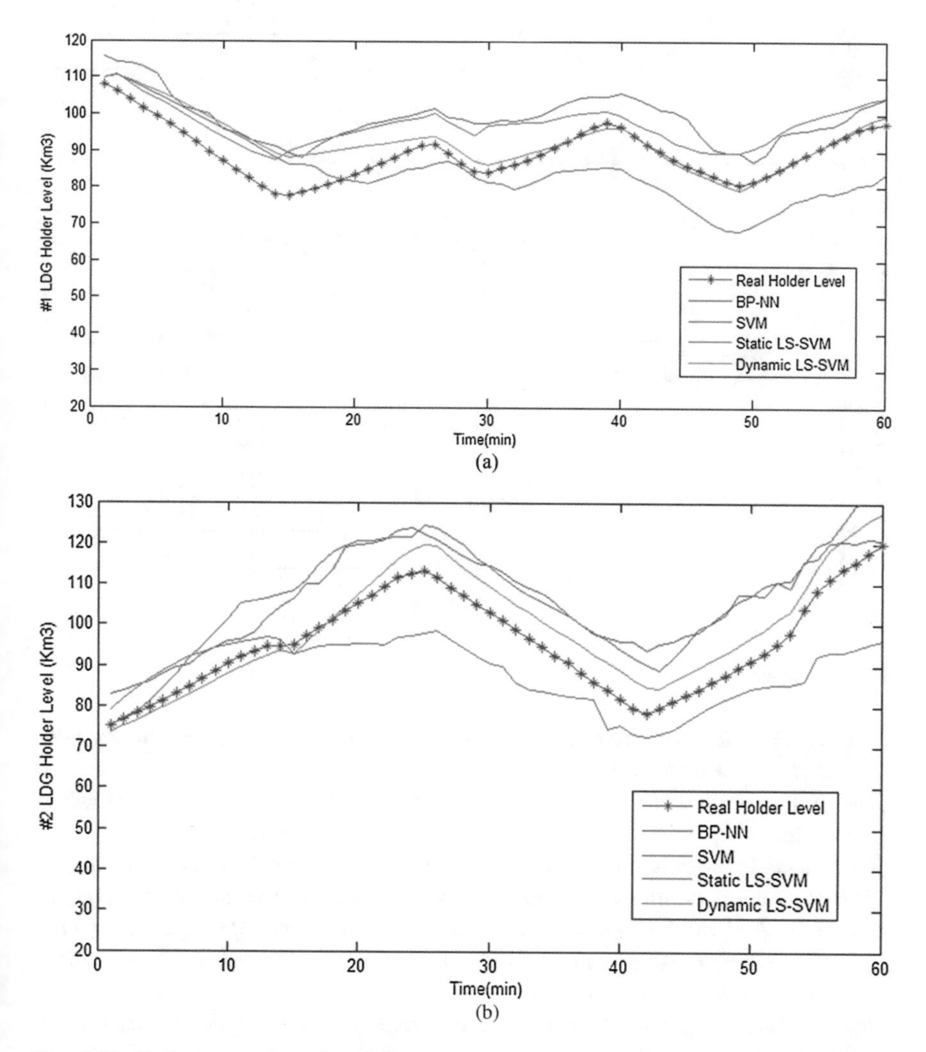

Fig. 8.11 Prediction results using different machine learning-based methods—a comparative analysis. (**a**) #1 gasholder level. (**b**) #2 gasholder level. (**c**) LDG consumption flow of #2 Blast Furnace. (**d**) LDG consumption flow of #4 Blast Furnace

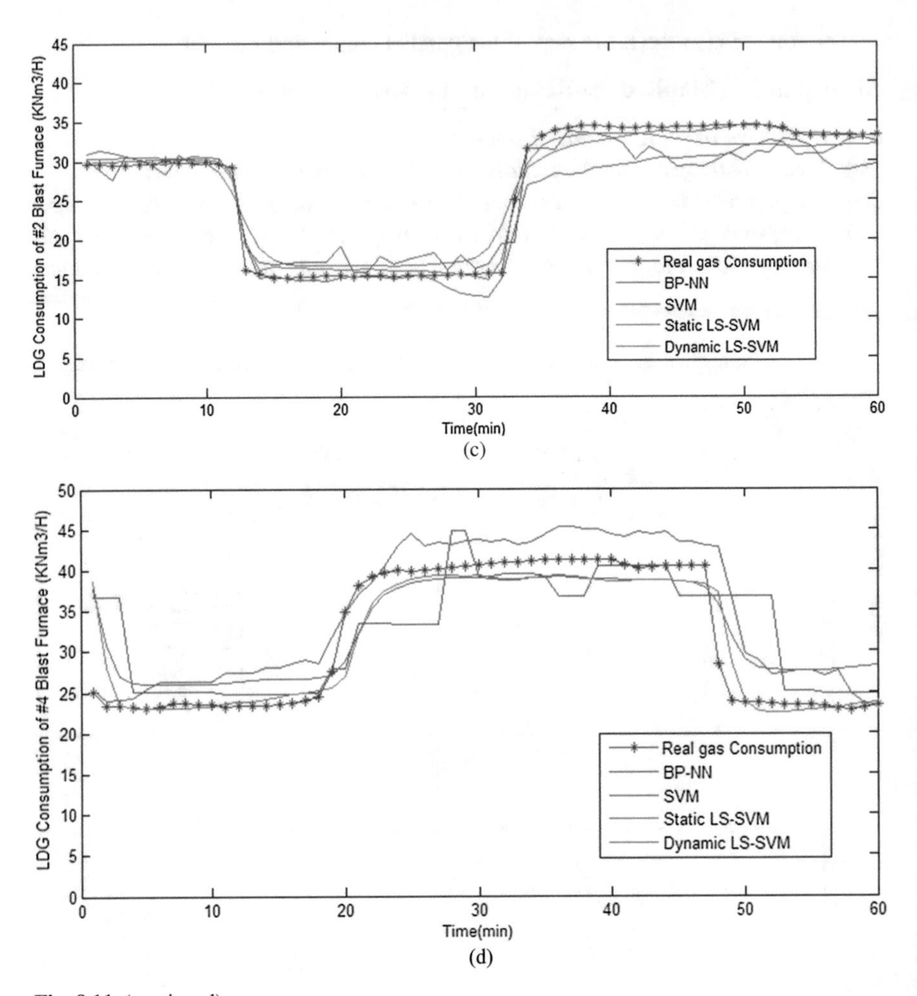

Fig. 8.11 (continued)

tasks, in which each reducer gathers the updated values and calculates their average value as the internal states.

Step 4: Calculating the posterior of the internal states and the corresponding covariance.

Step 5: Update the output weights. Execute the mapper tasks, in which each mapper receives one training sample and then computes all updated values for the output weights. And, execute $N + m$ reducer tasks (m is the number of exogenous inputs), in which each reducer gathers the updated values and calculates their average value as the output weight.

Step 6: Calculating the posterior of the output weights and the corresponding covariance.

Step 7: Return to Step 3, 4, 5, and 6 until the expected precision is achieved.

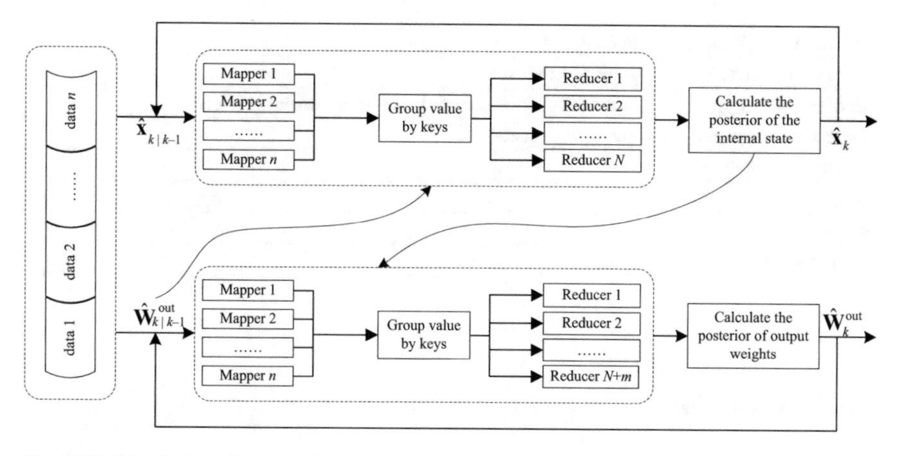

Fig. 8.12 MapReduce framework-based EKF

The details of the mapper and reducer updating the internal states and the output weights can be listed in Procedure 1 and Procedure 2, respectively.

Procedure 1

Mapper for updating the internal states
Input key/value pair <key = the offset in bytes, value=one train item>
Calculating the prior value $\hat{\mathbf{x}}_{k\|k-1}$ of the internal state and the prior covariance $\mathbf{P}_{\mathbf{x}_{k\|k-1}}$;
Calculating the Kalman gain $\mathcal{K}_k$;
Using the priori of the internal states and the output weights, calculating the expected output $\hat{y}_{k\|k-1} = \hat{\mathbf{W}}^{out}_{k\|k-1} \cdot [\mathbf{u}_k, \hat{\mathbf{x}}_{k\|k-1}]$;
For every internal state x, get its update Δx;
Emit intermediate key/value <key=x, value=Δx>
Reducer for merging the updates associated with the same internal state
Input key/value pair <key=x, value=Δx>
count=0,sum=0;
sum = sum + value;
count=count+1;
Output key/value pair <key, value=sum/count>

Procedure 2

Mapper for updating the output weights
Input key/value pair <key=the offset in bytes, value=one train item>
Calculating the prior value of the output weights $\hat{\mathbf{W}}^{out}_{k\|k-1}$ and its prior covariance $\mathbf{P}_{\mathbf{W}^{out}_{k\|k-1}}$;
Calculating the Kalman gain K_k based on the prior covariance;
Using the priori of the output weights and the posteriori of the internal states, calculating the expected output $\hat{y}_{k\|k-1} = \hat{\mathbf{W}}^{out}_{k\|k-1} \cdot (\mathbf{u}_k, \hat{\mathbf{x}}_k)$;

(continued)

For each weight w, get its update Δw;
Output key/value pair <key=w, value=Δw>
Reducer for merging the updates associated with the same output weight
Input key/value pair <key=w, value=Δw>
count=0,sum=0;
sum = sum + value;
count=count+1;
Output key/value pair <key, value=sum/count>

8.6.1 Case Study

To verify the performance of the MapReduce-based EKF modeling, 20 ordinary computers are employed to conduct the experiments of this section, whose configurations are: Inter E7500, 2.94 GHz, 2G RAM. One can take one of the computers as NameNode, JobTracker, DataNode, and TaskTracker, and the others as DataNode and TaskTracker. The version of Hadoop we adopt is 1.0.1.

Here, we employ the real-world gas flow data coming from a steel plant in China and carry out the industrial time series data prediction by using this parallelization method. We randomly select one of the blast furnace gas amounts, the consumption flow of a coke oven as the instance, and empirically set the dimensionality of the reservoir is 60, and the number of the training samples is 1000. Then, the comparative experimental results are illustrated in Fig. 8.13 by using the different methods. From Fig. 8.13a, it is clear that this parallelization method exhibits the best prediction performance for the gas flow data; while, the absolute error comparison presented in Fig. 8.13b also shows that this parallelization method obtains the higher prediction accuracy compared to the other methods.

To provide a further analysis on the performance of the proposed MapReduce-based EKF method, the statistic experimental results by repeating 100 times are listed in Table 8.9, where the RMSE and the MAPE are viewed as the error evaluation criterion. From Table 8.9, it is apparent that the accuracy by using the parallelization method is definitely higher than those of the others. With respect to the computational cost, although the linear regression and the least squares estimation present less computing time than the others, the corresponding accuracies of both of them are relatively low. Besides, the fatal flaw of them is that the ill-condition fails to be avoided in the training process. Among the 100 times experiments, the ill-conditioned phenomenon generate 8 and 6 times, respectively, for the linear regression and the least squares.

To exhibit the advantage of the MapReduce-based EKF, we increase the amount of the training samples and the dimensionality of the reservoir to be 2000 and 100, respectively. From Table 8.9, it can be seen that the prediction accuracy is greatly improved along with the increase of the reservoir dimensionality and the amount of the training samples. As for the computational cost, the linear regression

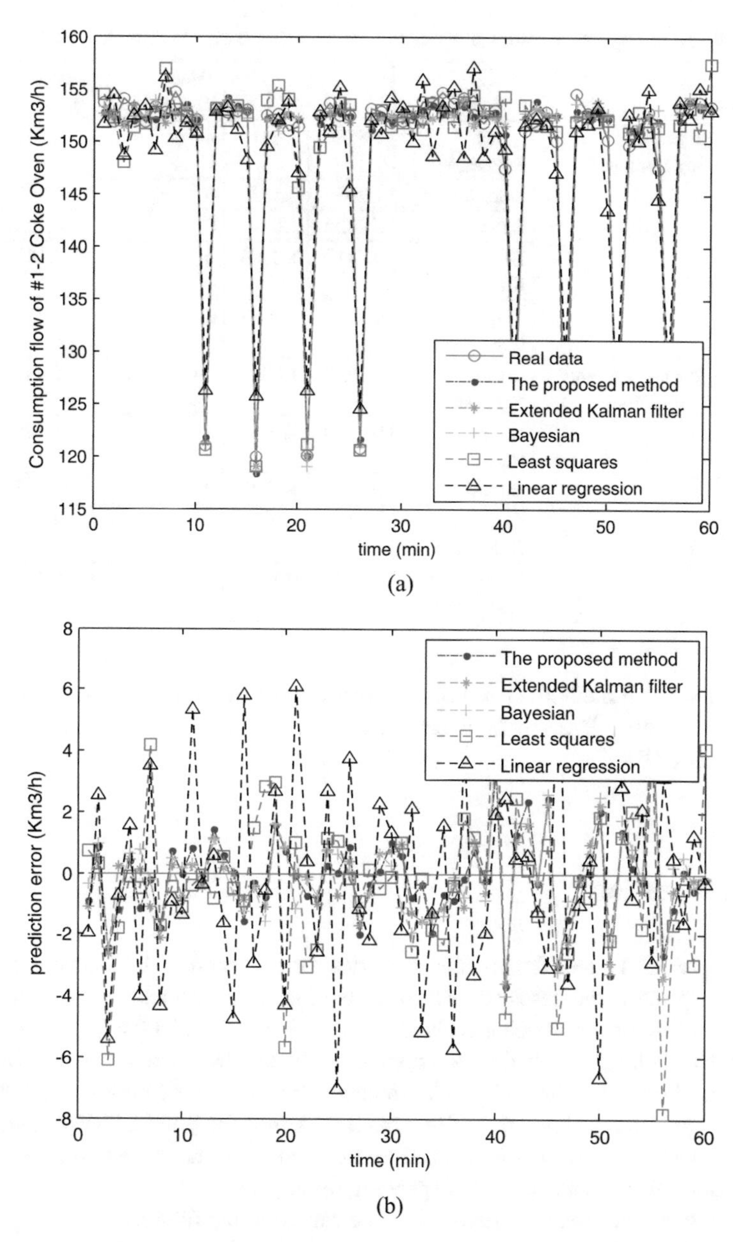

Fig. 8.13 (**a**) The comparative prediction results for gas consumption flow of Coke Oven. (**b**) The comparative errors by using the different methods

and the least squares estimation still present low computational cost than the others. But the times of the ill-condition occurred in the process of prediction also increase. The MapReduce-based EKF costs a much lower computation than the generic EKF

Table 8.9 The comparative results of prediction performance based on different methods

Parameters	Methods	MAPE	RMSE	Times of ill-condition	Computational cost (s)
60 reservoir neurons and 1000 training samples	Linear regression	1.8586	3.2511	8	0.917
	Least squares	1.265	2.563	6	0.9574
	Bayesian	0.8422	1.679	0	134.153
	Extended Kalman-filter	0.7816	1.5283	0	117.169
	MapReduce-based EKF	0.7514	1.4937	0	8.371
100 reservoir neurons and 2000 training samples	Linear regression	0.9985	1.8908	10	1.68
	Lease squares	0.6160	1.195	8	1.696
	Bayesian	0.5755	1.1282	0	362.015
	Extended Kalman-filter	0.5183	0.942	0	569.179
	MapReduce-based EKF	0.4950	0.88	0	35.41

and the Bayesian method. Based on an overall consideration of various factors, the MapReduce-based EKF method performs the comprehensively best to learn the parameters of ESN.

8.7 Discussion

This chapter introduces the parallel strategies for accelerating the algorithms applicable to the industrial cases so as to meet the industrial demand on computational speed. The parallel approaches include the GPU-based parallel computing, e.g., the CUDA by NVIDIA, as well as the computer cluster-based ones, e.g., the Hadoop platform. These two kinds of parallel methods have their own merits and demerits. We also apply the CUDA to the Elman network modeling and LSSVM modeling, and the MapReduce technique is used to accelerate the EKF based on ESN. These applications all have advantage of speeding up the algorithms.

While there are some improvements one can do in the future.

1. For the EKF-ENs modeling, first, in order to reflect more realistic dynamic behaviors of the networks and improve the prediction accuracy, the condition that missing measurements should be considered for the practical application. As such, the methods to reconstruct the networks with missing data, spares connectivity, and time delays could be combined with the EKF-ENs method so as to

extend the its application quality. Second, the use of GPU acceleration mode and its cooperation with serial code may require more attention and further improvements.

2. For the MapReduce-based EKF, it will be extended to other more robust filters such as unscented Kalman-filter (UKF) and particle filters (PF). The MapReduce nonlinear filters-based training will then be generalized as a big data machine learning technique for different neural networks.

References

1. Locans, U., Adelmann, A., Suter, A., et al. (2017). Real-time computation of parameter fitting and image reconstruction using graphical processing units. *Computer Physics Communications, 215*, 71–80.
2. CUDA toolkit, develop, optimize and deploy GPU-accelerated apps. Retrieved from https://developer.nvidia.com/cuda-toolkit
3. Apache Hadoop 3.0.0. Retrieved from http://hadoop.apache.org/docs/current/
4. Ramírez-Gallego, S., Fernández, A., García, S., et al. (2018). Big data: Tutorial and guidelines on information and process fusion for analytics algorithms with MapReduce. *Information Fusion, 42*, 51–61.
5. Zhao, J., Zhu, X., Wang, W., et al. (2013). Extended Kalman filter-based Elman networks for industrial time series prediction with GPU acceleration. *Neurocomputing, 118*(6), 215–224.
6. Heeswijk, M. V., Miche, Y., Oja, E., & Lendasse, A. (2011). GPU-accelerated and parallelized ELM ensembles for large-scale regression. *Neurocomputing, 74*, 2430–2437.
7. Zhao, J., Wang, W., Pedrycz, W., et al. (2012). Online parameter optimization-based prediction for converter gas system by parallel strategies. *IEEE Transactions on Control Systems Technology, 20*(3), 835–845.
8. Chapelle, O., & Vapnik, V. (2000). Model selection for support vector machines. In *Advances in neural information processing systems*. Cambridge, MA: MIT Press.
9. Van, G. T., Suykens, J., Baesens, B., et al. (2004). Benchmarking least squares support vector machine classifiers. *Machine Learning, 54*(1), 5–32.
10. An, S., Liu, W., & Venkatesh, S. (2007). Fast cross-validation algorithms for least squares support vector machine and kernel ridge regression. *Pattern Recognition, 40*, 2154–2162.
11. Scholkopf, B., & Smola, A. J. (2002). *Learning with kernels: Support vector machines, regularization, optimization, and beyond*. Cambridge, MA: MIT Press.
12. Sheng, C., Zhao, J., Leung, H, et al. (2013). Extended Kalman filter based echo state network for time series prediction using MapReduce framework. In *IEEE Ninth International Conference on Mobile Ad-Hoc and Sensor Networks* (pp. 175–180). IEEE.

Chapter 9
Data-Based Prediction for Energy Scheduling of Steel Industry

Abstract Based on the results of a number of different forecasting modes introduced in the previous chapters, this chapter provides a practical case study related to the optimal scheduling for energy system in steel industry based on the prediction outcomes. As for the by-product gas scheduling problem, a two-stage scheduling method is introduced here. On the prediction stage, the states of the optimized objectives, the consumption of the outsourcing natural gas and oil, the power generation, and the gas holder levels are forecasted by using the previous data-driven learning methods. On the optimal scheduling stage, a rolling optimization procedure is performed by employing the predicted results. More typically, with respect to the scheduling task for the oxygen/nitrogen system in steel industry, a similar two-stage method is also developed, in which a granular-computing (GrC)-based prediction model is firstly established on the stage of a long-term prediction, and the scheduling solution is also optimized later. Furthermore, the results of the scheduling system applications also indicate the effectiveness of the real-time prediction and scheduling optimization.

9.1 Introduction

Iron and steel industry is usually associated with high energy consumption and gas emissions. Due to the shortage of coal, oil and other primary sources of energy nowadays, it is very necessary to utilize efficiently the secondary energy sources generated during iron and steel-making procedure in order to reduce the manufacturing cost as well as the emission of the environmental pollution [1]. For example, as for by-product gases, to improve their utilization efficiency, the scheduling workers need to monitor the fluctuation of the gas flow of generation and consumption, as well as the amount of storage in real time, then change the consumption amounts of the adjustment users so as to balance the amount of gas generation and consumption. Besides, the consumption of by-product gas is also associated with other energy sources (oil, steam, etc.), which are coupled each other, making the energy scheduling of the by-product gas rather challenging.

J. Zhao et al., *Data-Driven Prediction for Industrial Processes and Their Applications*, Information Fusion and Data Science,
https://doi.org/10.1007/978-3-319-94051-9_9

- Currently, there were already a number of energy optimization technologies almost modeled by using the physical mechanism-based approaches. A concept of the energy opportunity window was utilized in [2] to allow the machine to be turned off at set periods of time without any throughput loss. Also in [3], an energy consumption reduction in production systems was proposed by scheduling machine startup and shutdown. Besides, a serial production line with finite buffers was considered. However, there are also a large number of applications failed to be modeled by using mechanism-based approaches. Here, we also take the by-product gas systems as an example, due to their complexity, their physical mechanism-based models are very hard to be constructed, and on the contrary, a large amount of real-time energy data has been accumulated by the well-known SCADA system. Thereby, more and more data-driven prediction and optimization methods, applicable to the by-product gas system, are developed. For example, a two-phase data-based forecasting and adjusting method was reported in [4] to complete the operation optimization of coke oven gas. Besides, a data-driven dynamic scheduling strategy was reported in [5] for the BFG system, in which a probability relationship described by a Bayesian network was modeled to determine the adjustable gas users and their scheduling amounts. Overall, data-driven approaches are becoming more and more popular in the field of scheduling of the energy system.

In the previous chapters, we have introduced many data-based prediction methods for different industrial demands. Based on the predictive results of these prediction methods, we can build the scheduling methods for real-time adjustment of energy in steel industry. In this chapter, we will introduce some industrial cases of data-based prediction for energy scheduling of steel industry, including the by-product gas scheduling and the oxygen/nitrogen scheduling.

9.2 A Prediction and Adjustment Method for By-product Gas Scheduling

Within the whole procedure of steel production, from preparation of raw materials and ore, iron-making, steel-making, rolling, heat treatment, coating to product shipment, more than 20 types of energies are involved. At the same time, some useful secondary energy is recycled. Since by-product gas has the advantages of easy combustion, high caloric, and so on, they are mostly reused as the crucial secondary energy. Their utilization can reduce the purchase of oil and natural gas, and increase inner electric power supply. Therefore, to maintain the balance of by-product gasholder level is an important task in optimal scheduling of by-product energy in steel industry. However, this is often influenced by many factors and difficult to obtain a precise mechanism model for analysis.

Here, let us first introduce the by-product gas systems in steel industry in details. Although we have introduced the flow chart of the by-product gas of steel industry

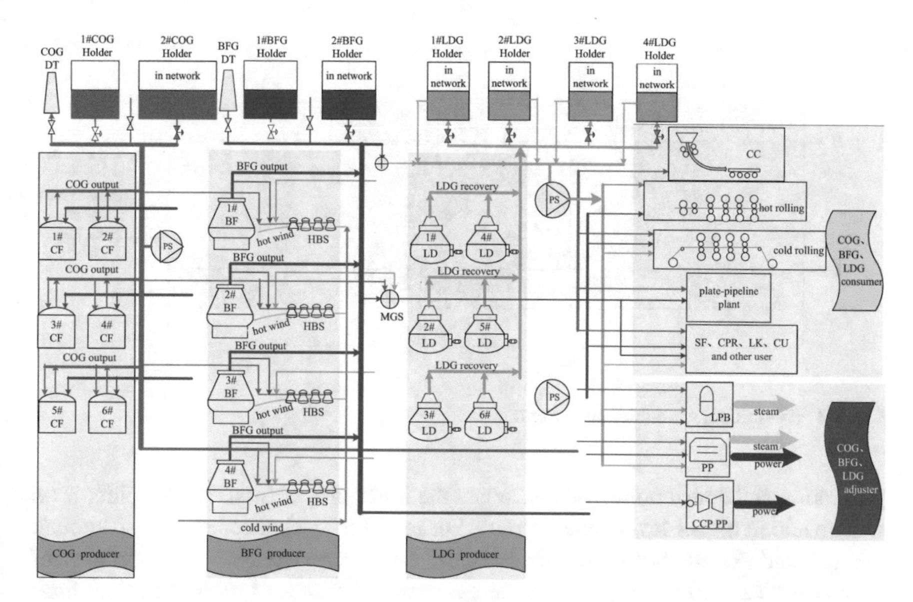

Fig. 9.1 Network structure of by-product gas system in steel industry

including the BFG, COG, and LDG in the previous chapters, for completeness, we still recall these gas systems in brief and point out the main concerns of the by-product gas scheduling.

We show the network of by-product gas production and consumption in a typical steel enterprise in Fig. 9.1. From this figure, we can see that there are mainly three kinds of by-product gases in steel industry: COG (marked in green), BFG (marked in blue), and LDG (marked in orange), of which their corresponding products contain coke, pig iron, and crude steel, respectively. These by-product gases are consumed by many consumption units including coke furnace (CF) and hot blast stove (HBS) of blast furnace (BF). Moreover, after pressured by pressure station (PS), they are indirectly supplied to other consumers containing limekiln (LK), sinter plant (SF), continuous casting (CC), hot rolling, cold rolling, plate-pipe-line plant, chemical products recovery (CPR), and civil user (CU). Besides, there are the adjustable consumers such as power plant (PP) and low pressure boiler (LPB) which can consume the excess gases. The remainders are stored in the corresponding gas-holders connected to the pipe network (marked by in network). While, when there is too much of gas in the pipe net, the diffusing towers (DT) will be opened up extemporaneously to diffuse redundant gases in few cases. Due to the irregular production rhythm of steel manufacturing, the imbalance between the production and consumption of by-product gases often occurs, which is reflected by the gasholder level.

Figure 9.2 shows the trend of gasholder level (marked by solid line) in a certain period t. In general, the gasholder can lie in three different level zones. Safe level operation zone between H and L means that there is a balance between the

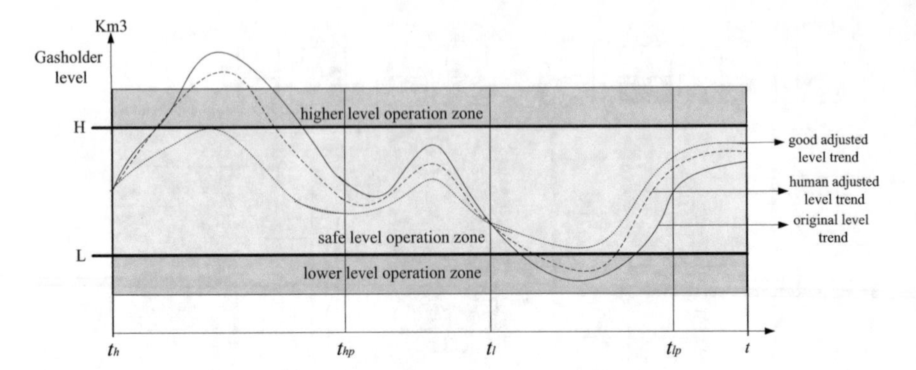

Fig. 9.2 By-product gasholder level trend

generation and consumption of gas. When the imbalance occurs, the gasholder level will be laid in higher level zone beyond H or in lower level zone below L. Although the gasholders can buffer this imbalance, temporary excesses or shortages of by-product gases frequently attack the gas system because of the mechanical limitation of the gasholder. To ensure gasholder safety, the scheduling workers need to continuously monitor the gasholder level and adjust the gas consumption of adjustable users. The accurate estimation and optimal adjustment at time t_h and t_l can obtain a good adjusted level trend within the safe level zone. But manual prediction and adjustment is usually ineffective and inaccurate, thus it cannot always make the adjusted level lie in the safe level. Under this condition, the workers have to diffuse the excessive gases or introduce more oil and natural gas from outside, which will lead to the degradation of environment and an increase of the energy cost.

9.2.1 Holder Level Prediction Models and Adjustment Method

In this section, we apply an optimal method for prediction and adjustment on by-product gasholder [1]. Both single and multiple gasholders level prediction models are established by machine learning methodology considering the different operation styles of gasholders. And, we also develop a hybrid parameter optimization algorithm to optimize the model for pursuing high prediction accuracy. The whole solution of the gasholder prediction and adjustment can be obtained within two stages. The system structure diagram is presented in Fig. 9.3. First, the holder level trend in the near future is predicted by the LSSVM models which are established by the machine learning methodology and optimized by a new hybrid parameter optimization algorithm. Second, the optimal adjustment quantity for each holder is then given based on the predicted level using a novel reasoning calculation.

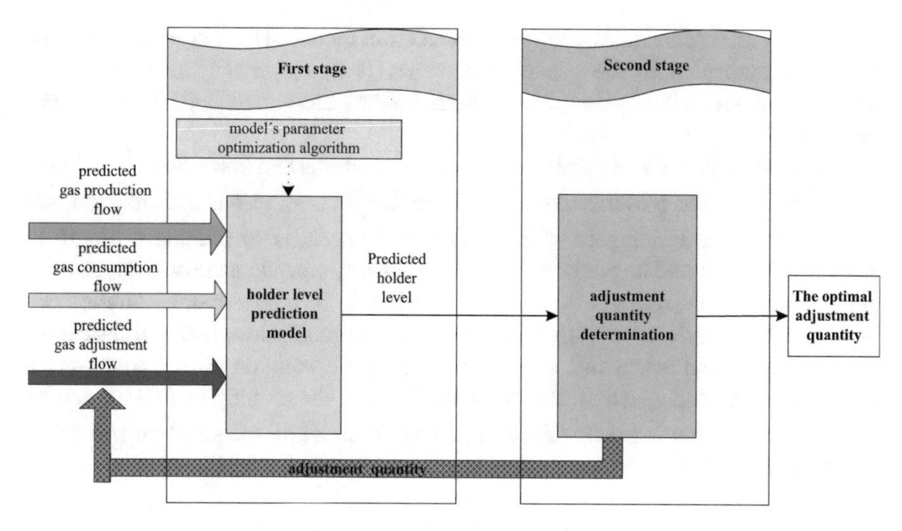

Fig. 9.3 The structure diagram for the prediction and adjustment of gasholder level

Holder Level Prediction Model

In practice, the holder level of each gas system is influenced by all of the producers and consumers through pipeline networks. There are mechanism relationships between them, which are characterized by complex nonlinearity (see Han et al. [6]). Unfortunately, the relationship between them is unknown. Since the machine learning method has been successful in learning from data, we choose to use the LSSVM model to learn the relationship. Here, we assume the amount of production and consumption of gas suppliers and users can be obtained in advance. Taking the practical gasholder operation styles into account, a single gasholder prediction and a multiple gasholder prediction model are established, respectively.

Given the training sample set $S^G(t)$ at time t: $S^G(t) = \{(X_i^G(t - d_{i'}), \mathrm{HL}_i^G(t)), G \in \{\mathrm{COG, BFG, LDG}\}, i = 1, 2, \ldots, n^G, i'=1, 2, \ldots, m^G\}$. where $X_i^G(t - d_{i'}) \in R^{m^G}$, $\mathrm{HL}_i^G(t) \in R^{h^G}$ is m^G inputs and h^G gasholder level outputs of the model; $d_{i'}$ is time delay reflecting the influence of m^G inputs on holder level. And m^G inputs include four kinds of inputs in pipe network, in which m_p^G inputs are gas production flows; m_c^G inputs are gas consumption flows; m_a^G inputs are gas adjustment flows and m_h^G inputs are gasholder levels of previous time. The regression function formulation between the h^G holder level outputs and m^G inputs of the holder level prediction model are as follows:

$$\mathrm{SHL}^G\left(X^G, w^G, b^G\right) = w^G \varphi\left(X^G\right)^{\mathrm{T}} + b^G \tag{9.1}$$

$$\mathrm{MHL}_j^G\left(X^G, w_j^G, b_j^G\right) = w_j^G \varphi_j^G\left(X^G\right)^{\mathrm{T}} + b_j^G, j = 1, \ldots, h^G \tag{9.2}$$

where (9.1) represents the single-output prediction model SHL^G, in which only one holder is connected to the gas pipe network. (9.2) represents the multi-output prediction model $\mathrm{MHL}^G{}_j$, in which several holders are simultaneously in the gas network.

Since MHL_j^G with generalization is more complicated than SHL^G, we take MHL_j^G to make the presentation in details in this section. In our optimization, not only every single training error but also the comprehensive training error of all gasholder level should be punished, which sufficiently considers the property of each holder and the interactions between them. As a result, the whole optimal solution can be guaranteed. And, we employ a hyper-spherical equality loss function instead of the hyper-spherical insensitive one [7] to simplify the solution of the model into a linear equations and speed up the calculation. Thus, the training of MHL_j^G can be viewed as the following constrained optimization based on the structure risk minimization principle.

$$\begin{aligned}
\min\ J_{\mathrm{MHL}^G}(w,E,e) \quad &= \sum_{j=1}^{h^G}\left\|w_j^G\right\|^2 + \sum_{i=1}^{n^G}(C_i^0)^G E_i^G + \sum_{j=1}^{h^G} C_j^G \sum_{i=1}^{n^G}\left(\left(e_i^G\right)_j\right)^2 \\
\text{s.t.} \quad &\left\|\mathrm{HL}_i^G - w^G\Phi^G(X_i^G)^{\mathrm{T}} - b^G\right\|^2 = E_i^G \\
&\left(\mathrm{HL}_i^G\right)_j = w_j^G\varphi_j^G(X_i^G)^{\mathrm{T}} + b_j^G + \left(e_i^G\right)_j
\end{aligned} \tag{9.3}$$

In (9.3), $w = [w_1, w_2, \ldots, w_h]^{\mathrm{T}}$, $b = [b_1, b_2, \ldots, b_h]^{\mathrm{T}}$; $w_j^G \in R^m$ and $b_j^G \in R$ are the weight and bias terms of the model MHL_j^G; $\left(e_i^G\right)_j$ and E_i^G are the single and comprehensive training errors in the ith training sample; $C_j^G > 0$ and $\left(C_i^0\right)^G > 0$ are their corresponding single and comprehensive training regularization parameters; $\Phi^G(X_i) = \left[\varphi_1^G(X_i), \varphi_2^G(X_i), \ldots, \varphi_h^G(X_i)\right]^{\mathrm{T}}$, $\varphi_j^G(X_i) \in R^m$ is the mapping function of input. Then, the solving process of constrained optimization (9.3) is presented as follows:

First, the Lagrange multipliers α^G and $(\beta^G)_j$ $\left(\alpha^G = \left[\alpha_1^G, \alpha_2^G, \ldots, \alpha_{n^G}^G\right]^{\mathrm{T}}, (\beta^G)_j = [\right.$ $\left.\beta_1^G, \beta_2^G, \ldots, \beta_{n^G}^G]^{\mathrm{T}}\right)$ are added in (9.3), which converts the problem (9.3) into an unconstrained optimization problem as (9.4):

$$\begin{aligned}
L_{\mathrm{MHL}^G}\left(w_j^G, b_j^G, (e^G)_j, E_i^G, \alpha^G, (\beta^G)_j\right) = &\sum_{j=1}^{h^G}\left\|w_j^G\right\|^2 + \sum_{i=1}^{n^G}(C_i^0)^G E_i^G + \sum_{j=1}^{h^G} C_j^G \sum_{i=1}^{n}\left(\left(e_i^G\right)_j\right)^2 \\
&- \sum_{i=1}^{n^G}\alpha_i^G\left(E_i^G - \left\|\mathrm{HL}_i^G - w^G\Phi^G(X_i^G)^{\mathrm{T}} - b^G\right\|^2\right) \\
&- \sum_{j=1}^{h^G}\sum_{i=1}^{n^G}(\beta_i^G)_j\left[\left(w_j^G\varphi_j^G(X_i^G)^{\mathrm{T}} + b_j^G + \left(e_i^G\right)_j\right) - \left(\mathrm{HL}_i^G\right)_j\right]
\end{aligned} \tag{9.4}$$

To obtain the expressions of w and b, a series of equations are presented based on the Karush-Kuhn-Tucker conditions:

$$\begin{cases} \frac{\partial L}{\partial b_j^G}=0 \Rightarrow 2(\alpha^G)^{\mathrm{T}}\left[(\mathrm{HL}^G)_j - w_j^G\left(\varphi_j^G\right)^{\mathrm{T}} - \vec{1}b_j^G\right] - \vec{1}(\beta^G)_j = 0 \\ \frac{\partial L}{\partial w_j^G}=0 \Rightarrow w_j^G - 2\left(\varphi_j^G\right)^{\mathrm{T}} D_{\alpha^G}\left[(\mathrm{HL}^G)_j - w_j^G\left(\varphi_j^G\right)^{\mathrm{T}} - \vec{1}b_j^G\right] - \left(\varphi_j^G\right)^{\mathrm{T}}(\beta^G)_j = 0 \\ \frac{\partial L}{\partial E_i^G}=0 \Rightarrow (C_i^0)^G - \alpha_i^G = 0 \\ \frac{\partial L}{\partial (e^G)_j}=0 \Rightarrow 2C_j^G\vec{1}(e^G)_j - \vec{1}(\beta^G)_j = 0 \end{cases}$$

$$\begin{cases} \frac{\partial L}{\partial \alpha_i^G}=0 \Rightarrow \left\|\mathrm{HL}_i^G - w^G\Phi^G(X_i^G)^{\mathrm{T}} - b^G\right\|^2 = E_i^G \\ \frac{\partial L}{\partial \beta_j^G}=0 \Rightarrow \vec{1}(\mathrm{HL}^G)_j = w_j^G\left(\varphi_j^G\right)^{\mathrm{T}} + n^G b_j^G + \vec{1}(e^G)_j \end{cases} \tag{9.5}$$

where, $D_\alpha =$ diag $(\alpha_1, \alpha_2, \ldots, \alpha_n)$; $\varphi_j^G = \left[\varphi_j^G(X_1), \varphi_j^G(X_2), \ldots, \varphi_j^G(X_{n^G})\right]^{\mathrm{T}}$; $\vec{1} = [1, 1, \ldots, 1]$; $(\mathrm{HL}^G)_j = [\mathrm{HL}_1, \mathrm{HL}_2, \ldots, \mathrm{HL}_{n^G}]^{\mathrm{T}}$.

Then, based on Representer Theorem [8], we have

$$w_j^G = \sum_{i=1}^{n^G} (r_i^G)_j \varphi_j^G(X_i^G) = \left(\varphi_j^G\right)^{\mathrm{T}}(r^G)_j \tag{9.6}$$

Given the kernel function $k_j^G(X_i^G, X_k^G) = \varphi_j^G(X_i^G)\cdot\varphi_j^G(X_k^G)$, and combining (9.5) and (9.6), a linear equations about $(r^G)_j$, b_j^G can be concisely written in matrix as

$$\begin{bmatrix} K_j^G + \left(D_{\alpha^G} + C_j^G I\right)^{-1} & 1^{\rightharpoonup \mathrm{T}} \\ \left((\alpha^G)^{\mathrm{T}} - C^j 1^{\rightharpoonup}\right)K_j^G & \left((\alpha^G)^{\mathrm{T}} - C^j 1^{\rightharpoonup}\right)1^{\rightharpoonup \mathrm{T}} \end{bmatrix}\begin{bmatrix} (r^G)_j \\ b_j^G \end{bmatrix} = \begin{bmatrix} (\mathrm{HL}^G)_j \\ \left((\alpha^G)^{\mathrm{T}} - C^j 1^{\rightharpoonup}\right)(\mathrm{HL}^G)_j \end{bmatrix} \tag{9.7}$$

where $\alpha_i^G = (C_i^0)^G$, $K_j^G = \left[(k_{ik}^G)_j = k_j^G(X_i^G, X_k^G)\right]_{i,k=1}^{n^G}$, and I is $n^G \times n^G$ identity matrix. The jth holder level prediction model is given by solving the above-mentioned linear equations.

$$\mathrm{MHL}_j^G\left(X^G, w_j^G, b_j^G\right) = \sum_{i=1}^{n^G} \left(r_i^G\right)_j k_j^G\left(X^G, X_i^G\right) + b_j^G \tag{9.8}$$

Remove the comprehensive training error and the first constraint from (9.3), the similar single-output model SHL^G (as (9.1)) can be easily given as (9.10) by solving the linear equations (9.9):

$$\begin{bmatrix} K^G + \dfrac{1}{C^G} I & 1 \\ 1^{\mathrm{T}} & 0 \end{bmatrix} \cdot \begin{bmatrix} \beta^G \\ b^G \end{bmatrix} = \begin{bmatrix} \mathrm{HL}^G \\ 0 \end{bmatrix} \tag{9.9}$$

$$\mathrm{SHL}^G\left(X^G, w^G, b^G\right) = \sum_{i=1}^{n^G} \beta_i^G k^G\left(X^G, X_i^G\right) + b^G \tag{9.10}$$

Optimization of Holder Level Prediction Model

Since the kernel function and its parameter determine the expression of model, and the regularization parameter is the trade-off between the model complexity and the empirical risk, their rational settings have great effect on the model performance. Previously, grid search based on cross-validation [9] was generally adopted to select the parameters, but that method is extraordinarily time-consuming because of a repetitive training and easy to fall into the local grid due to the lack of guiding search ability. Some researchers proposed Bayesian parameter optimization [10], which is also characterized by the complicated computation process. Recently, intelligent evolutionary algorithms [11–13] were developed to optimize the model. Quantum genetic algorithm (QGA) is a probabilistic optimization algorithm [14] based on quantum computation principle. Compared to the conventional evolutionary algorithm, it has small size of population, rapid convergence, and stronger global optimization ability. Combining the directed search ability of gradient descent and the global search ability of QGA, here we propose a new hybrid optimization algorithm to set the parameters of the proposed LSSVM model, where the gradient denoted as the partial derivative of predicted residual error sum of squares (PRSEE) which is also introduced in Sect. 7.2.5, with respect to the parameters is used to guide the QGA; and a self-adaptive update mechanism in QGA is developed to avoid falling into the local optimum. Here, we take the SHL^G model as an example to present this optimization algorithm in details.

Let $M = K^G + \frac{1}{C^G} I, \tilde{K}^G = \begin{bmatrix} K^G & 1 \end{bmatrix}$, at the ith iteration of fast leave-one-out cross-validation (FLOO-CV) [15], the SHL^G model's coefficient $s_i^G\left(s_i^G = \begin{bmatrix} \beta^G \\ b^G \end{bmatrix}\right)$ and PRSEE are denoted as

$$s_i^G = s^G(i^-) - \frac{s_{(i)}^G}{\left(M^G\right)_{(i,i)}^{-1}} \left(M^G\right)^{-1}(i^-, i) \tag{9.11}$$

$$\text{PRSEE}^G = \sum_{i=1}^{n^G} \left(\tilde{K}^G(i, i^-) \cdot s_i^G - \text{HL}_i^G\right)^2 \tag{9.12}$$

where $s^G(i^-)$ is the sub-vector of the original coefficient vector omitting the ith element, $s_{(i)}^G$ is the ith element of it, $(M^G)^{-1}(i^-, i)$ is the sub-vector formed by the ith column of matrix $(M^G)^{-1}$ omitting the ith element, $(M^G)^{-1}{}_{(i,i)}$ is the ith elements of the diagonal elements in $(M^G)^{-1}, \tilde{K}^G(i, i^-)$ is the sub-vector formed by the ith row of matrix $\tilde{K}^G$ omitting the ith element, and HL_i^G is the ith testing holder level value.

Here, we assume the radial base function (RBF) as an instance. The partial derivative of $\tilde{K}^G$ and $k^G\left(X_i^G, X_k^G\right)$ with respect to parameter σ^G is as (9.13); the partial derivative of $(M^G)^{-1}$ with respect to parameter (C^G, σ^G) is as (9.14); and the partial derivative of s_i^G with respect to parameter (C^G, σ^G) is as (9.15):

$$\begin{cases} \dfrac{\partial \tilde{K}^G}{\partial \sigma^G} = \begin{bmatrix} \dfrac{\partial K^G}{\partial \sigma^G} & 0 \end{bmatrix} \\ \dfrac{\partial k^G\left(X_i^G, X_k^G\right)}{\partial \sigma^G} = -\left(\sigma^G\right)^{-3} k^G\left(X_i^G, X_k^G\right) \left\|X_i^G - X_k^G\right\|^2 \end{cases} \tag{9.13}$$

$$\begin{cases} \dfrac{\partial \left(M^G\right)^{-1}}{\partial C^G} = -\left(M^G\right)^{-1} \begin{bmatrix} -\left(C^G\right)^{-2} I & 0 \\ 0^{\mathrm{T}} & 0 \end{bmatrix} \left(M^G\right)^{-1} \\ \dfrac{\partial \left(M^G\right)^{-1}}{\partial \sigma^G} = -\left(M^G\right)^{-1} \begin{bmatrix} \dfrac{\partial K^G}{\partial \sigma^G} & 0 \\ 0^{\mathrm{T}} & 0 \end{bmatrix} \left(M^G\right)^{-1} \end{cases} \tag{9.14}$$

$$\begin{cases} \dfrac{\partial s_i^G}{\partial C^G} = \dfrac{\partial s^G}{\partial C^G}(i^-) - \dfrac{\dfrac{\partial s^G}{\partial C^G}(i) \left(M^G\right)_{(i^-,i)}^{-1} + s_{(i)}^G \dfrac{\partial \left(M^G\right)^{-1}}{\partial C^G}(i^-, i)}{\left(M^G\right)_{(i,i)}^{-1}} + \dfrac{s_{(i)}^G \left(M^G\right)^{-1}(i^-, i)}{\left[\left(M^G\right)_{(i,i)}^{-1}\right]^2} \cdot \dfrac{\partial \left(M^G\right)^{-1}}{\partial C^G}(i, i) \\ \dfrac{\partial s_i^G}{\partial \sigma^G} = \dfrac{\partial s^G}{\partial \sigma^G}(i^-) - \dfrac{\dfrac{\partial s^G}{\partial \sigma^G}(i) \left(M^G\right)_{(i^-,i)}^{-1} + s_{(i)}^G \dfrac{\partial \left(M^G\right)^{-1}}{\partial \sigma^G}(i^-, i)}{\left(M^G\right)_{(i,i)}^{-1}} + \dfrac{s_{(i)}^G \left(M^G\right)^{-1}(i^-, i)}{\left[\left(M^G\right)_{(i,i)}^{-1}\right]^2} \cdot \dfrac{\partial \left(M^G\right)^{-1}}{\partial \sigma^G}(i, i) \end{cases} \tag{9.15}$$

Combining (9.13)–(9.15), we have

$$\begin{cases} \dfrac{\partial \mathrm{PRSEE}}{\partial C^G} = 2\left(\sum_{i=1}^{n^G} (\tilde{K}^G(i,i^-)s_i^G - \mathrm{hl}_i^G\right) \cdot \tilde{K}^G(i,i^-) \cdot \dfrac{\partial s_i^G}{\partial C^G} \\ \dfrac{\partial \mathrm{PRSEE}}{\partial \sigma^G} = 2\left(\sum_{i=1}^{n^G} (\tilde{K}^G(i,i^-)s_i^G - \mathrm{hl}_i^G\right)\left[\dfrac{\partial \tilde{K}^G}{\partial \sigma^G}(i,i^-) \cdot s_i^G + \tilde{K}^G(i,i^-) \cdot \dfrac{\partial s_i^G}{\partial \sigma^G}\right] \end{cases} \tag{9.16}$$

Then, the update of binary solution in QGA by the steepest gradient decent method can be formulated as

$$\begin{aligned} P_{C^G}(\mathrm{new}) &= P_{C^G} - \frac{\partial \mathrm{PRSEE}}{\partial C^G} \\ P_{\sigma^G}(\mathrm{new}) &= P_{\sigma^G} - \frac{\partial \mathrm{PRSEE}}{\partial \sigma^G} \end{aligned} \tag{9.17}$$

where P_C^G and P_σ^G are the binary solution parameter (C^G, σ^G) in QGA, and $P_C^G(g)_{\mathrm{new}}$ and $P_\sigma^G(g)_{\mathrm{new}}$ are the update solution. It can be seen that such update measure is helpful to obtain a good solution steered by the parameter's gradient direction.

At the gth generation of QGA, the self-adaptive quantum gate mutation to update the quantum individual $Q(g)$ of parameter (C^G, σ^G) is given as (9.18)

$$\begin{cases} \gamma_m^g(\mathrm{new}) = \gamma_m^g \cos\left(\theta_m^g(\mathrm{new})\right) - \lambda_m^g \sin\left(\theta_m^g(\mathrm{new})\right) \\ \lambda_m^g(\mathrm{new}) = \gamma_m^g \sin\left(\left(\theta_m^g(\mathrm{new})\right) - \lambda_m^g \cos\left(\theta_m^g(\mathrm{new})\right)\right. \\ \Delta\theta_m^g = 0.05\pi \cdot \dfrac{f_{\mathrm{avg}}^g}{f_{\mathrm{gbest}}} \\ \theta_m^g(\mathrm{new}) = \theta_m^g + \Delta\theta_m^g \cdot d\left(\gamma_m^g, \lambda_m^g\right) \end{cases} \tag{9.18}$$

where $\gamma_m^g(\mathrm{new})$ and $\lambda_m^g(\mathrm{new})$ are the mth quantum bit of the updated $Q(g)$, f_{gbest} is the fitness value of global best solution up to the gth generation, f_{avg}^g is the average fitness value of the current solution, and the ratio $\dfrac{f_{\mathrm{gbest}}}{f_{\mathrm{avg}}^g} < 1$ denotes the solution distribution status. The smaller the ratio is, the more dispersive distribution has. $\Delta\theta_m^g$ and $d\left(\gamma_m^g, \lambda_m^g\right)$ (see [15]) are the change quantity and direction of θ_i. It is obviously that $\theta_m^g(\mathrm{new})$ can be adaptively adjusted by the solution distribution status. If the distribution is dispersed, $\Delta\theta_m^g$ will be amplified to speed up the update of $Q(g)$. Contrarily, a small $\Delta\theta_m^g$ is beneficial to jump out the local search.

Based on the above-mentioned parameter gradient optimization and QGA with adaptive quantum gate update, the solving steps of the new hybrid optimization can be described as follows:

Algorithm 9.1: Hybrid Optimization of the LSSVM for Gasholder Prediction Modeling

Step 1: *Using the value range of parameter (C^G, $_G{}^G$) to decide the bit number of binary solution P(g), then the corresponding bit of number of quantum individual Q(g) can be determined.*

Step 2: *Initialize the quantum individual and iteration generation g = 0. In general, all γ_m^g and λ_m^g are initialized as $1/\sqrt{2}$.*

Step 3: *Observe Q(g) to get the binary solution P(g).*

Step 4: *Calculate the PRSEE gradient of every P(g) using (9.16), select the maximal gradient, and update the P(g) using (9.17).*

Step 5: *Calculate the PRSEE of every P(g) using (9.13) and store the best b(g).*

Step 6: *Judge whether the termination condition is satisfied. If so, the stored best b (g) is the expected binary solution and go to Step 9; otherwise, go to Step 7.*

Step 7: *Update the Q(g) using (9.18).*

Step 8: *g = g + 1, renewedly observe Q(g − 1) to get P(g), and then go to Step 4.*

Step 9: *Convert the best binary solution to the real value, and the best value of regularization parameter C^G and kernel parameter σ^G can be obtained.*

Optimal Adjustment Quantity Determination

Depending on the prediction of gasholder level by the proposed prediction model, the by-product gas balance can be judged. Once the gasholder level breaks through the normal zone (as in Fig. 9.2), the process of adjustment should be triggered. In practice, the scheduler usually approximates a rough quantity according to the difference between the maximal over level and normal level. But, it is difficult to ensure the normal operation of gasholder in the future since the pre-adjustment time t_h or t_l is too late to produce the enough adjustment quantity, and the pre-adjustment quantity Δa_h or Δa_l is not tested whether the global optimized. We present a gas adjustment quantity calculation method based on the gasholder level prediction examination. The calculation is described as follows:

Algorithm 9.2: Gas Adjustment Quantity Calculation Method

Step 1: *Initialize the pre-adjustment time interval t_a, as shown in Fig. 9.2, $t_h = t_{hp} - t_a$, $t_l = t_{lp} - t_a$.*

Step 2: *During $[t_h, t_{hp}]$ and $[t_l, t_{lp}]$, find all the gasholder levels $HL(t_{hi})$ beyond the higher level H and $HL(t_{lj})$ below the lower level L, and record the related time t_{hi} and t_{lj}.*

Step 3: *Calculate the higher gasholder level difference $DHL(t_{hi})$ between $HL(t_{hi})$ and H, the lower gasholder level difference $DHL(t_{lj})$ between $HL(t_{lj})$ and L.*

Step 4: *Based on the $DHL(t_{hi})$ and $DHL(t_{lj})$ at each time t_{hi} and t_{lj}, calculate their corresponding adjustment quantity Δa_{hi} and Δa_{lj} at pre-adjustment time t_h and t_l,* *the calculation formula is denoted by:* $\Delta a_{hi} = \frac{\mathrm{DHL}(t_{hi})}{t_{hi}-t_h-1}^{*} 60, \Delta a_{lj} = \frac{\mathrm{DHL}(t_{lj})}{t_{lj}-t_l-1}^{*} 60.$

***Step** 5: Select the maximal adjustment quantity from all* Δa_{hi} *and* Δa_{lj} *as the rough adjustment quantity.*

***Step** 6: Add the selected adjustment quantity to the adjuster input of prediction model (as in Fig. 9.3).*

***Step** 7: Predict the gasholder level again, and check whether the adjusted level is within the safety zone. If so, the calculation is completed and the optimal adjustment quantity is obtained; otherwise, revise the* $\Delta a_{h\max}$ *and* $\Delta a_{l\max}$ *and return to Step 6.*

9.2.2 Case Study

The prediction and adjustment method for gasholder balance has been successfully verified in the energy center of a steel plant in China. The following are the settings related to parameters of prediction and adjustment method. Table 9.1 lists the input and output dimension, input time delay, training and testing sample number, and the kernel function of each *G* gasholder level prediction model. The sampling frequency is set as 1 min. Since the training samples have a significant influence on prediction accuracy, the sample numbers for COG, BFG, and LDG are set as different from each other. As for COG, who is with stable variation, the number is as 2000; while as for BFG, who is with frequent variation, it is up to 3500 covering various conditions as much as possible. As for LDG, the number of sample set is as 1500. The kernel function of each prediction model is determined by a large amount of experiments. As for COG and LDG, single RBF is adopted; and as for BFG, multiple kernel functions composed of RBF and polynomial are used. The parameters of the hybrid optimization algorithm are as follows: C and σ belong to [1, 1000] and [0.1, 1000], respectively; degree of polynomial kernel function is set as 2; the population size of QGA is 30; and the termination condition is chosen as the MAE less than 1.5 km^3. Table 9.2 shows the gasholder capacity, the related safety operation range, and pre-adjustment interval. The L and H levels are the boundary to ensure the gasholder within the normal zone, as shown in Fig. 9.2.

In order to demonstrate the advantages of the proposed prediction and adjustment method, a comparison result using the practical data is illustrated. First, we give verifications to the model SHL^G for COG and BFG, and MHL^G for LDG, which consist of three classes of tested change tendency of gasholder level, ascending, stationary, and descending, with the prediction time of 30 min.

In the COG and BFG system, only one gasholder is to store the gas in most cases. In this section, four different approaches are used including manual reasoning (MR), BP network [16] with two hidden layer of 20 and 10 neurons using the scaled conjugate gradient back propagation network training function, standard LSSVM with ten-fold cross-validation (10fc-LSSVM) and the proposed method with hybrid

Table 9.1 Modeling parameters for COG, BFG, and LDG gasholder level prediction model

G gas	Input and output dimension	Input time delay (min)	Training sample number	Testing sample number	Kernel function
COG	$m^{\mathrm{COG}} = 39\left(m_p^{\mathrm{COG}} = 3, m_c^{\mathrm{COG}} = 26, m_a^{\mathrm{COG}} = 9, m_h^{\mathrm{COG}} = 1, h^{\mathrm{COG}} = 1\right)$	$\left[d_{i'}^{\mathrm{COG}}\right]_{i'=1}^{m^{\mathrm{COG}}} = 1$	2000	30	RBF
BFG	$m^{\mathrm{BFG}} = 29\left(m_p^{\mathrm{BFG}} = 4, m_c^{\mathrm{BFG}} = 12, m_a^{\mathrm{BFG}} = 12, m_h^{\mathrm{BFG}} = 1, h^{\mathrm{BFG}} = 1\right)$	$\left[d_{i'}^{\mathrm{BFG}}\right]_{i'=1}^{m^{\mathrm{BFG}}} = 1$	3500	30	a * RBF + b * Polynomial ($a = b = 0.5$)
LDG	$m^{\mathrm{LDG}} = 19\left(m_p^{\mathrm{LDG}} = 3, m_c^{\mathrm{LDG}} = 10, m_a^{\mathrm{LDG}} = 6, m_h^{\mathrm{LDG}} = 1, h^{\mathrm{LDG}} = 2\right)$	$\left[d_{i'}^{\mathrm{LDG}}\right]_{i'=1}^{m^{\mathrm{LDG}}} = 1$	1500	30	RBF

Table 9.2 The constraints for COG, BFG, and LDG gasholder

	Gasholder capacity (km^3)		L (km^3)		H (km^3)		
G gas	#1	#2	#1	#2	#1	#2	Pre-adjustment interval (min)
COG	120	300	70	120	100	230	20
BFG	150	150	90	51	120	120	20
LDG	80	80	28	35	65	65	20

optimization. Note that the MR [6] was widely used at this corporation now. The BP network was commonly adopted to solve the practical regression before LSSVM was presented. The 10fc-LSSVM is now often used for modeling many practical problems. The average results based on 50 groups of testing samples randomly selected are listed in Table 9.3, and the graphical illustration comparison of a certain group of sample are shown as Figs. 9.4 and 9.5. Obviously, in Table 9.3 and Figs.9.4 and 9.5, the proposed model predicts the three different tendency of COG and BFG gasholder level with best performance than MR, BP network, and 10fc-LSSVM. Firstly, it is difficult for MR to establish a reasonable model when many practical influence factors are ignored. In BP network, a model effectively reflecting the practical gasholder level with high complexity and nonlinearity cannot be established because the weight coefficient computed by gradient descent usually falls into local optimum. And based on the empirical minimization principle, the BP model's generalization performance is very poor. However, taking into account the influence of gas production and consumption on gasholder level, the latter two methods based on structure risk minimization principle can transform the nonlinearity to linearity in high dimension by a mapping function, and make a compromise between model complexity and generalization in order to avoid the over-fit and under-fit of gasholder level with parameter optimization. Therefore, the latter two models have better prediction performance than the former models. Nevertheless, 10fc-LSSVM usually spends a lot of computation time since the kernel matrix is composed of the different training samples in each validation, and its inverse need to be computed repeatedly. Additionally, such process is hard to provide an almost unbiased estimation to the prediction model, and usually obtains a sub-optimal parameter by the blind search mechanism. By contrast, the proposed hybrid algorithm adopts QGA's diverse individual representation to enrich the parameter space, and uses the FLOO-CV as selection criterion to accelerate the validation process. Moreover, the developed parameter update further guides to find the better parameter quickly through combining the parameter gradient optimization and adaptive quantum gate update. Therefore, our proposed model can give the better results than 10fc-LSSVM in real time.

In view of the performance statistic (see Table 9.3) for COG and BFG prediction, our proposed model has less MAE, uncorrelated coefficient (UC) and computing time (CT) than the 10fc-LSSVM, and such results can be obtained within 1 min. The difference between the predicted and the practical level are all less than 1.5 km^3 (the required MAE). Although 10fc-LSSVM seems to have a similar performance sometimes, it is difficult to meet the requirement of real time and high accuracy.

Table 9.3 Comparison of COG, BFG gasholder level prediction results with five models

Model	Gasholder	Tendency	MAE	UC	CT(s)
Manual reasoning	#1 COG	Ascending	10.4660	2.0541	11.543
		Stationary	10.1919	2.3781	
		Descending	9.8491	2.0932	
	#2 COG	Ascending	9.5285	2.0367	
		Stationary	10.7655	1.6755	
		Descending	13.9610	2.0266	
	#1 BFG	Ascending	10.9928	2.0011	10.7472
		Stationary	9.5447	1.5368	
		Descending	12.6753	3.5775	
	#2 BFG	Ascending	11.2511	1.9031	
		Stationary	8.1723	2.1516	
		Descending	17.0247	1.3235	
BP network	#1 COG	Ascending	7.2332	1.448	127.894
		Stationary	5.8391	1.4079	
		Descending	6.4763	1.1573	
	#2 COG	Ascending	5.5815	1.1074	
		Stationary	6.0979	1.1448	
		Descending	7.5818	1.1038	
	#1 BFG	Ascending	8.6060	1.2142	116.1070
		Stationary	5.0002	1.3540	
		Descending	6.1905	1.2201	
	#2 BFG	Ascending	6.2162	1.1334	
		Stationary	6.3137	1.2134	
		Descending	9.5645	1.1010	
10kc-LSSVM	#1 COG	Ascending	4.1184	1.0109	302.5475
		Stationary	4.0908	1.2269	
		Descending	3.5062	1.0545	
	#2 COG	Ascending	2.8987	1.0139	
		Stationary	3.5740	1.0184	
		Descending	3.4789	1.0126	
	#1 BFG	Ascending	4.5571	1.0142	327.425
		Stationary	3.1226	1.0971	
		Descending	3.7085	1.0434	
	#2 BFG	Ascending	4.3785	1.0318	
		Stationary	2.7310	1.0415	
		Descending	4.1506	1.0160	
The proposed method	#1 COG	Ascending	0.6148	1.0043	64.5640
		Stationary	0.7344	1.0320	
		Descending	0.7883	1.0068	
	#2 COG	Ascending	1.0321	1.0037	
		Stationary	1.3136	1.0074	
		Descending	1.1190	1.0006	
	#1 BFG	Ascending	1.4123	1.0016	60.0219
		Stationary	1.3831	1.0184	
		Descending	0.8423	1.0030	
	#2 BFG	Ascending	0.9518	1.0005	
		Stationary	0.6286	1.0036	
		Descending	1.0353	1.0004	

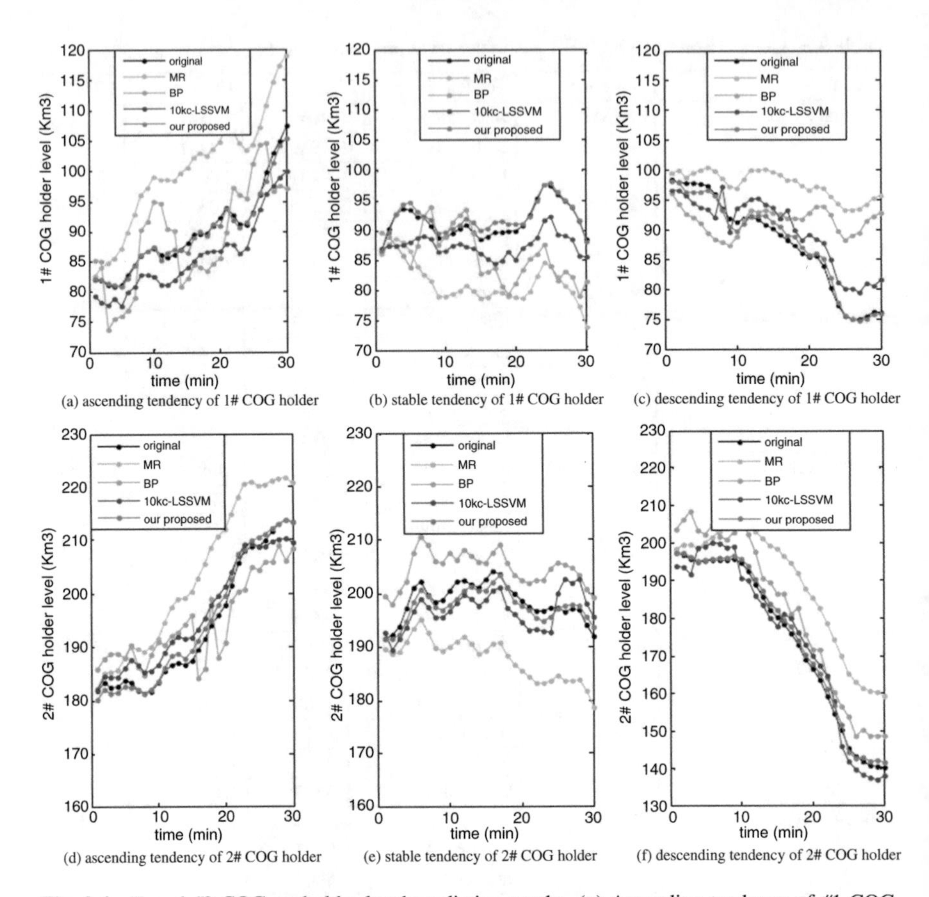

Fig. 9.4 #1 and #2 COG gasholder level prediction results. (**a**) Ascending tendency of #1 COG holder. (**b**) Stable tendency of #1 COG holder. (**c**) Descending tendency of #1 COG holder. (**d**) Ascending tendency of #2 COG holder. (**e**) Stable tendency of #2 COG holder. (**f**) Descending tendency of #2 COG holder

After all, it spends a lot of time to finish the model training (about 5 min). The MAE by BP network is less than manual reasoning, but larger than 10fc-LSSVM and the proposed model. And the fluctuant prediction results with larger UC cannot track the gasholder level's change tendency, especially the ascending and descending tendency, as shown in Figs. 9.4 and 9.5. The spending time is too long, up to 2 min.

In the LDG system, two gasholders are simultaneously connected to the pipe network, and the model MHL^G is adopted. We use three different approaches, MR, MSVMR [7] whose parameter is selected by five-fold cross-validation and our proposed model to compare the prediction performance, in which MSVMR is usually to establish the multi-output model with varying weight iterative training. A set of comparable average results obtained from 50 groups of samples randomly

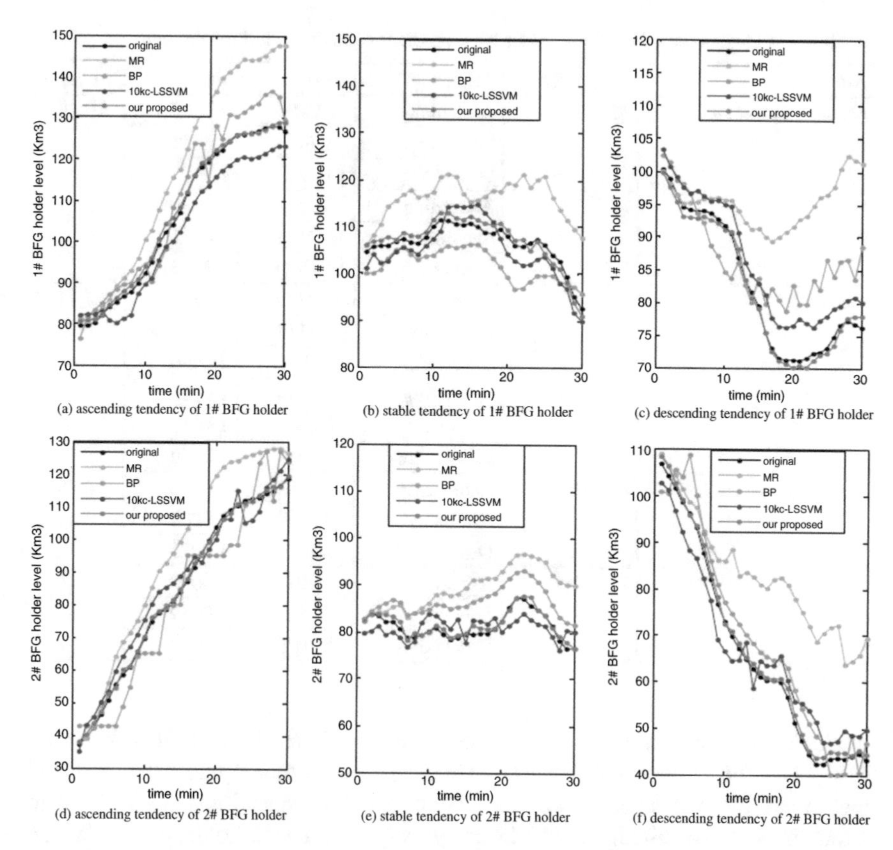

Fig. 9.5 #1 and #2 BFG gasholder level prediction result. (**a**) Ascending tendency of #1 BFG holder. (**b**) Stable tendency of #1 BFG holder. (**c**) Descending tendency of #1 BFG holder. (**d**) Ascending tendency of #2 BFG holder. (**e**) Stable tendency of #2 BFG holder. (**f**) Descending tendency of #2 BFG holder

selected are listed in Table 9.4, and a graphical comparison is in Fig. 9.6. From Table 9.4 and Fig. 9.6, it can be easily seen that the proposed model can successfully predict the LDG holder level with ascending, stationary and descending tendency, which is better than MR and MSVMR. The latter two models consider the practical influence factors and are trained based on the structure risk minimization principle. So their prediction results are better than MR. However, MSVMR's optimization only punishes the whole training error and adopts the same model parameter for different gasholder, which ignores the property of each holder and cannot guarantee the global optimum. And due to the inequality constraint, the optimization becomes a complex QP and is solved by the varying weight iterative training algorithm, which is very time-consuming. As for the performance statistics (see Table 9.4) for LDG

Table 9.4 Comparison of LDG gasholder level prediction results with three models

Model	Gasholder	Tendency	MAE	UC	CT (s)
Manual reasoning	#1 LDG	Ascending	8.8954	1.3402	10.1642
		Stationary	8.7473	2.0288	
		Descending	11.2955	2.5977	
	#2 LDG	Ascending	10.7732	1.1056	
		Stationary	13.2193	2.0081	
		Descending	11.1600	1.2575	
MSVMR	#1 LDG	Ascending	4.9755	1.0275	367.1498
		Stationary	4.5117	1.1737	
		Descending	6.0957	2.1460	
	#2 LDG	Ascending	6.1045	1.0125	
		Stationary	7.0102	1.4615	
		Descending	4.0684	1.0639	
The proposed method	#1 LDG	Ascending	0.7563	1.0046	60.7573
		Stationary	0.8657	1.0002	
		Descending	1.2051	1.0073	
	#2 LDG	Ascending	1.2913	1.0020	
		Stationary	0.8966	1.0049	
		Descending	1.1059	1.0027	

holder level prediction, the proposed model has less MAE, UC, and CT than that by MSVMR. The smaller MAE and UC by our proposed model make the predicted level track the real level without derivation, and time spending is also less than 1 min. In contrast, although MSVMR can grossly track the gasholder level's change tendency compared with MR, as shown in Fig. 9.6, the MAE larger than 4 km^3 and the CT longer than 5 min cannot satisfy the practical requirement.

Based on the accurate predicted level, we give verifications to the adjustment process, which consist of two tested change tendency of ascending, and descending, with the prediction time of 30 min. In this experiment, the manual method is used for comparison. The adjustment quantity for #2 COG, #2 BFG, and #1~#2 LDG gasholder are listed in Table 9.5, and the graphical illustrations are shown in Figs. 9.7, 9.8, 9.9, and 9.10. From the figures, the optimal adjustment quantity given by the proposed method can make the adjusted level within the safety zone; while, the manual adjusting results are always out of the safety zone. For holder's safety, some gas might be diffused or not be effectively supplied to power plant in manual method, which leads to a degradation to the environment and an increase of energy cost. It can be seen from Table 9.5 that our proposed method based on the holder level prediction examination can give more feasible quantity than manual adjusting method.

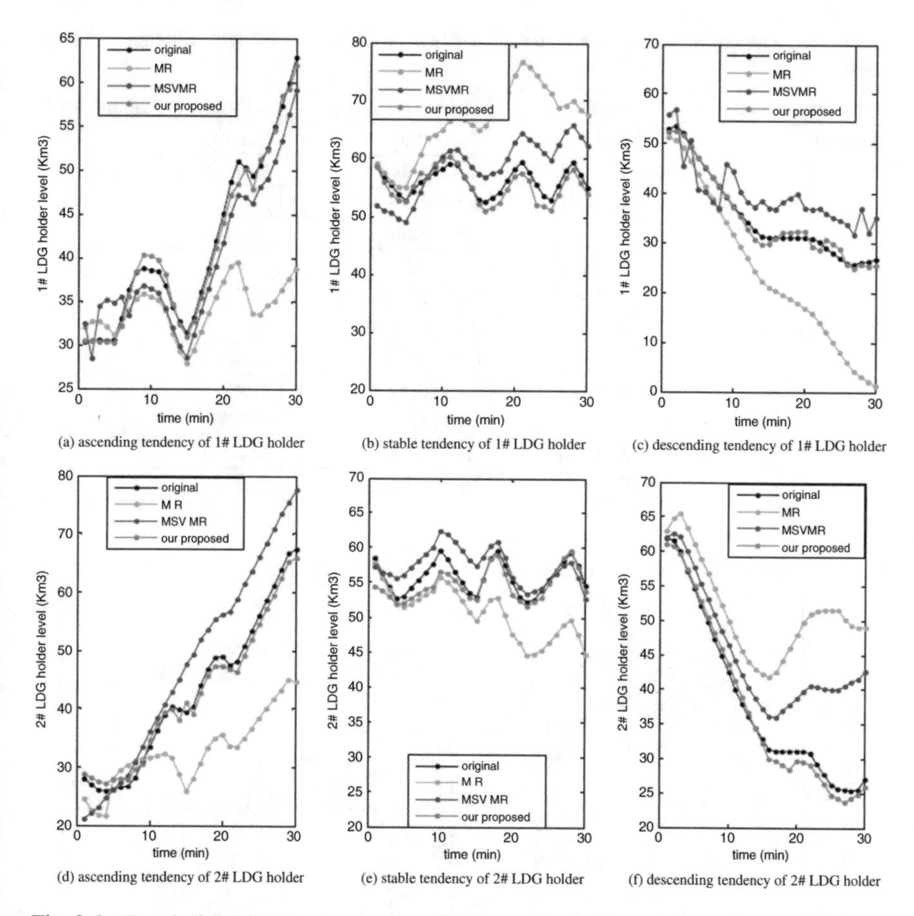

Fig. 9.6 #1 and #2 LDG gasholder level prediction results. (**a**) Ascending tendency of #1 LDG holder. (**b**) Stable tendency of #1 LDG holder. (**c**) Descending tendency of #1 LDG holder. (**d**) Ascending tendency of #2 LDG holder. (**e**) Stable tendency of #2 LDG holder. (**f**) Descending tendency of 2# LDG holder

9.3 Interval Predictive Optimization for By-product Gas System in Steel Industry

In light of significant complexity of the by-product gas system in steel industry (which limits an ability to establish its physics-based model), this section introduces a data-based predictive optimization (DPO) method to carry out real time adjusting for the gas system [17]. Two stages of the method, namely the prediction modeling and real-time optimization, are involved. Different from the point-oriented predicting approach used in Sect. 9.2, here we provide not only the estimated values of the predicted targets but also the indication of their reliability, i.e., PIs.

Table 9.5 The adjustment quantity of COG, BFG, and LDG holder with different adjusting method

Adjusting method	Gasholder	Change tendency	Adjustment quantity (km^3)
Manual	#2 COG	Ascending	25.6809
		Descending	13.8115
	#2 BFG	Ascending	17.1240
		Descending	9.7171
	#1 LDG #2 LDG	Ascending	6.0901
		Descending	5.3216
The proposed method	#2 COG	Ascending	52.1673
		Descending	27.3657
	#2 BFG	Ascending	71.4971
		Descending	64.9709
	#1 LDG #2 LDG	Ascending	22.8379
		Descending	21.6534

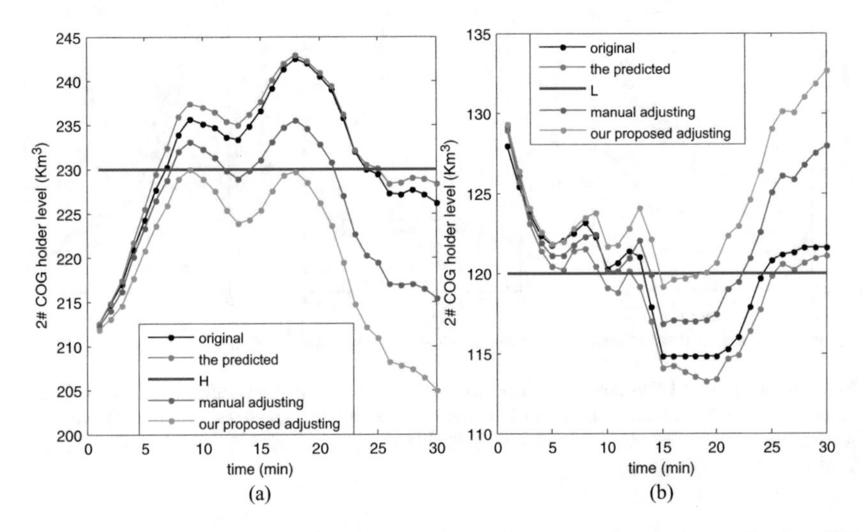

Fig. 9.7 The adjusting results for #2 COG gasholder level. (**a**) Ascending tendency adjusting of #2 COG. (**b**) Descending tendency adjusting of #2 COG

Specifically, at the prediction stage, the states of the optimized objectives, the consumption of the outsourcing natural gas and oil, the power generation, and the holder levels are forecasted based on a proposed mixed Gaussian kernel-based PIs construction model. The Jacobian matrix of this model is represented by a kernel matrix through derivation, which greatly facilitates the subsequent calculation. At the second stage, a rolling optimization based on a mathematical programming technique involving continuous and integer decision-making variables is developed via the PIs.

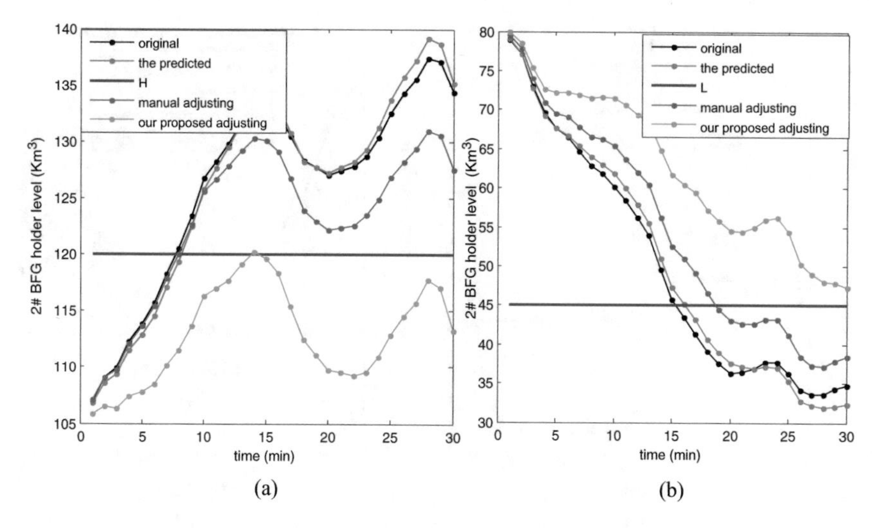

Fig. 9.8 The adjusting results for #2 BFG gasholder level. (**a**) Ascending tendency adjusting of #2 BFG. (**b**) Descending tendency adjusting of #2 BFG

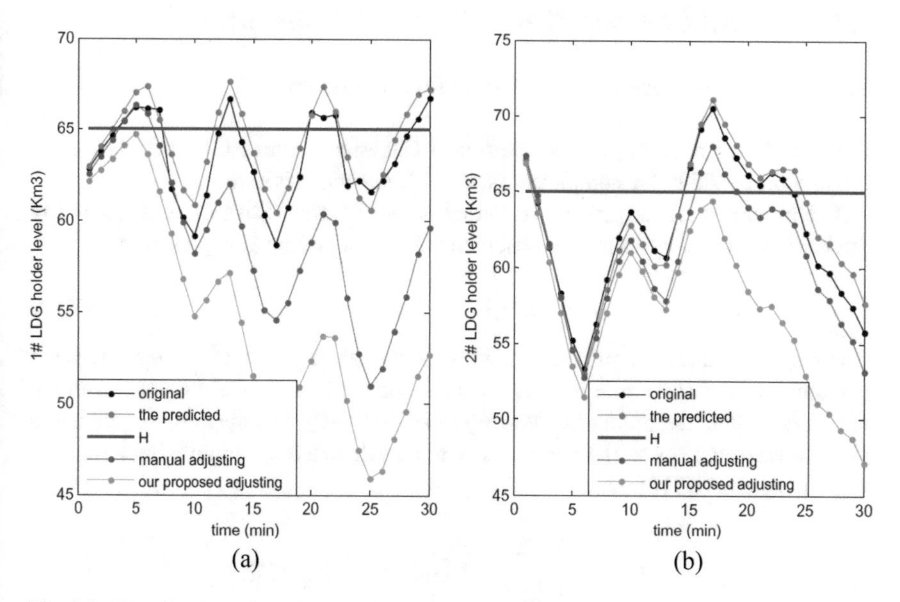

Fig. 9.9 The adjusting results for #1 and #2 LDG gasholder level with ascending tendency

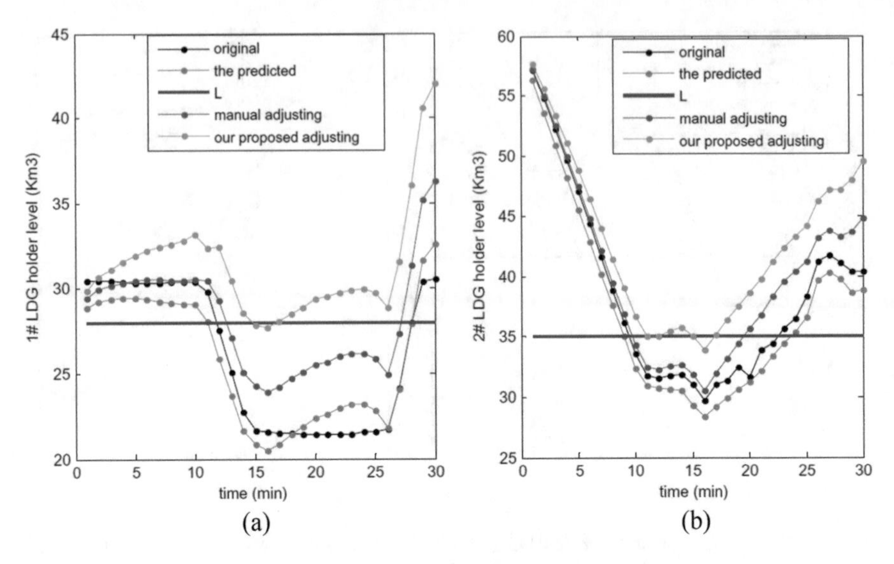

Fig. 9.10 The adjusting results for #1 and #2 LDG gasholder level with descending tendency

9.3.1 Mixed Gaussian Kernel-Based PIs Construction

Mixed Gaussian Kernel-Based Method for Regression

In Chap. 5, we have introduced the mixed Gaussian kernel-based method for PIs construction. While, for completeness, we still introduce it here.

Let us first recall a generalized kernel-based learning. Given the noise widely present in practical systems, the observed value of a system is expressed as

$$t_n = f(\mathbf{x}_n, \mathbf{w}) + \gamma_n \tag{9.19}$$

where t_n is the observed output, $f(\mathbf{x}_n, \mathbf{w})$ is the model output, $\mathbf{x}_n$ is the input, $\mathbf{w}$ is the parameter of the regression model, and γ_n denotes a Gaussian white noise with the zero mean value. Assuming that $f(\mathbf{x}_n, \mathbf{w}) = \mathbf{w}^{\mathrm{T}}\phi(\mathbf{x}_n)$, $\phi(\cdot)$ is a mapping function to a high-dimensional space, the parameters can be calculated by minimizing a regularization expression [18], i.e.,

$$J(\mathbf{w}) = \frac{1}{2}\sum_{n=1}^{N}[\mathbf{w}^{\mathrm{T}}\phi(\mathbf{x}_n) - t_n]^2 + \frac{\lambda}{2}\mathbf{w}^{\mathrm{T}}\mathbf{w} \tag{9.20}$$

where $\lambda \geq 0$ represents a penalty coefficient and N is the number of the sample data. One computes the partial derivative of (9.20) with respect to $\mathbf{w}$ and makes it equal to zero. Then, the dual expression of the original objective function reads as

$$J(\mathbf{a}) = \frac{1}{2}\mathbf{a}^{\mathrm{T}}\mathbf{K}\mathbf{K}\mathbf{a} - \mathbf{a}^{\mathrm{T}}\mathbf{K}\mathbf{t} + \frac{1}{2}\mathbf{t}^{\mathrm{T}}\mathbf{t} + \frac{\lambda}{2}\mathbf{a}^{\mathrm{T}}\mathbf{K}\mathbf{a} \tag{9.21}$$

where $\mathbf{t} = (t_1, t_2, \ldots t_N)^{\mathrm{T}}$, $\mathbf{K}$ is a matrix, its each element reads as $k(\mathbf{x}_n, \mathbf{x}_m) = \phi(\mathbf{x}_n)^{\mathrm{T}}$ $\phi(\mathbf{x}_m)$, and, $\mathbf{a} = (\mathbf{K} + \lambda\mathbf{I}_N)^{-1}\mathbf{t}$. The regression model is formulated as

$$f(\mathbf{x}, \mathbf{w}) = \mathbf{w}^{\mathrm{T}}\phi(\mathbf{x}) = \sum_{n=1}^{N} a_n \phi(\mathbf{x}_n)\phi(\mathbf{x}) = \mathbf{k}(\mathbf{x})^{\mathrm{T}}\mathbf{a} \tag{9.22}$$

where $\mathbf{k}(\mathbf{x})$ is a N-dimensional vector (N kernel functions), each of which can be expressed as $k_n(\mathbf{x}) = k(\mathbf{x}_n, \mathbf{x})$. And, the commonly used kernel function is the Gaussian one in various applications [19]. However, the assumed regression $f(\mathbf{x}_n, \mathbf{w}) = \mathbf{w}^{\mathrm{T}}\phi(\mathbf{x}_n)$ lacks the bias term when compared to the generic SVM [20], i.e., the accuracy could be largely impacted.

We can incorporate two different Gaussian kernels to form a new mixed Gaussian kernel for constructing the regression model, each of which can be regarded as a coordinator of another one. That is,

$$k(\mathbf{x}_i, \mathbf{x}_j) = \beta k_1(\mathbf{x}_i, \mathbf{x}_j) + (1 - \beta)k_2(\mathbf{x}_i, \mathbf{x}_j) \tag{9.23}$$

where $0 < \beta < 1$ is used to exhibit the different kernel widths of the sub-kernels $k_1(\mathbf{x}_i, \mathbf{x}_j)$ and $k_2(\mathbf{x}_i, \mathbf{x}_j)$, i.e.,

$$k_l(\mathbf{x}_i, \mathbf{x}_j) = \exp\left(-\frac{\left\|\mathbf{x}_i - \mathbf{x}_j\right\|^2}{2\sigma_l^2}\right), \quad l = 1, 2. \tag{9.24}$$

Construction of Mixed Gaussian Kernel-Based PIs

Given that the prediction reliability is significantly concerned by the energy schedulers or operators, a PIs construction method is proposed in this section based on the mixed Gaussian kernel model. Here, one can convert (9.22) into a new function $y(\cdot)$ with respect to $\mathbf{a}$. Assuming the estimated parameters are denoted by $\hat{\mathbf{a}}$, the real ones of the system are expressed as $\mathbf{a}^*$, and a Gaussian white noise, ε_n, indicating the difference between the real system output and the model one, then we have a new model of the system as shown below

$$y(\mathbf{x}_n, \mathbf{a}^*) = y(\mathbf{x}_n, \hat{\mathbf{a}}) + \varepsilon_n \tag{9.25}$$

One can still use noise γ_n to specify the difference between the observed value and the real output, see (9.25). Then, the observed system's output reads as follows:

$$t_n = y(\mathbf{x}_n, \widehat{\mathbf{a}}) + \varepsilon_n + \gamma_n \tag{9.26}$$

If $y(\mathbf{x}_0, \widehat{\mathbf{a}})$ (with respect to an input $\mathbf{x}_0$) comes as a result of the first-order Taylor series expansion around $\mathbf{a}^*$, then

$$y(\mathbf{x}_0, \widehat{\mathbf{a}}) \approx y(\mathbf{x}_0, \mathbf{a}^*) + \mathbf{g}_0^{\mathrm{T}}(\widehat{\mathbf{a}} - \mathbf{a}^*) \tag{9.27}$$

where

$$\mathbf{g}_0^{\mathrm{T}} = \left[\frac{\partial y(\mathbf{x}_0, \mathbf{a}^*)}{\partial a_1^*}, \frac{\partial y(\mathbf{x}_0, \mathbf{a}^*)}{\partial a_2^*}, \ldots, \frac{\partial y(\mathbf{x}_0, \mathbf{a}^*)}{\partial a_N^*}\right] \tag{9.28}$$

and $\mathbf{a}^* = (a_1^*, a_2^*, \ldots a_N^*)^{\mathrm{T}}$. It is apparent that if the system is modeled by a kernel-based method, see (9.22), then $\partial y(\mathbf{x}_0, \mathbf{a}^*)/\partial a_1^* = k(\mathbf{x}_0, \mathbf{x}_1)$. Also, assuming $\mathbf{k}(\mathbf{x}_0) = [k(\mathbf{x}_0, \mathbf{x}_1), k(\mathbf{x}_0, \mathbf{x}_2), \ldots k(\mathbf{x}_0, \mathbf{x}_N)]$, one can simplify the expression $\mathbf{g}_0^{\mathrm{T}}$ which reads as $\mathbf{g}_0^{\mathrm{T}} = \mathbf{k}(\mathbf{x}_0)$. As such, integrating the formulas (9.19), (9.26) and (9.27), we obtain

$$t_0 - \widehat{y}_0 \approx \gamma_0 - \mathbf{g}_0^{\mathrm{T}}(\widehat{\mathbf{a}} - \mathbf{a}^*) \tag{9.29}$$

Furthermore, assuming that the value of $\widehat{\mathbf{a}}$ can be considered as the unbiased estimate of $\mathbf{a}^*$, the expectation of the difference comes in the form

$$E(t_0 - \widehat{y}_0) \approx E(\gamma_0) - \mathbf{g}_0^{\mathrm{T}} E[(\widehat{\mathbf{a}} - \mathbf{a}^*)] \approx 0 \tag{9.30}$$

Based on the principle of statistical independence [21], its variance

$$\mathrm{var}[t_0 - \widehat{y}_0] \approx \mathrm{var}[\gamma_0] + \mathrm{var}[\mathbf{g}_0^{\mathrm{T}}(\widehat{\mathbf{a}} - \mathbf{a}^*)] \tag{9.31}$$

Considering that $\gamma_0 \sim N(0, \sigma_\gamma^2)$, the distribution of $\widehat{\mathbf{a}} - \mathbf{a}^*$ is governed by $N\left(0, \sigma_\gamma^2 \left[\mathbf{F}(\widehat{\mathbf{a}})^{\mathrm{T}} \mathbf{F}(\widehat{\mathbf{a}})\right]^{-1}\right)$, where $\mathbf{F}(\widehat{\mathbf{a}})$ is the Jacobian matrix of $y(\mathbf{x}_0, \widehat{\mathbf{a}})$ taken with respect to $\widehat{\mathbf{a}}$, i.e.,

$$\mathbf{F}(\widehat{\mathbf{a}}) = \frac{\partial \mathbf{y}(\mathbf{x}, \widehat{\mathbf{a}})}{\partial \widehat{\mathbf{a}}} = \begin{bmatrix} \left(\frac{\partial y_1(\mathbf{x}_1, \widehat{\mathbf{a}})}{\partial \widehat{a}_1}\right) & \left(\frac{\partial y_1(\mathbf{x}_1, \widehat{\mathbf{a}})}{\partial \widehat{a}_2}\right) & \cdots & \left(\frac{\partial y_1(\mathbf{x}_1, \widehat{\mathbf{a}})}{\partial \widehat{a}_N}\right) \\ \left(\frac{\partial y_2(\mathbf{x}_2, \widehat{\mathbf{a}})}{\partial \widehat{a}_1}\right) & \left(\frac{\partial y_2(\mathbf{x}_2, \widehat{\mathbf{a}})}{\partial \widehat{a}_2}\right) & \cdots & \left(\frac{\partial y_2(\mathbf{x}_2, \widehat{\mathbf{a}})}{\partial \widehat{a}_N}\right) \\ \vdots & \vdots & \vdots & \vdots \\ \left(\frac{\partial y_N(\mathbf{x}_N, \widehat{\mathbf{a}})}{\partial \widehat{a}_1}\right) & \left(\frac{\partial y_N(\mathbf{x}_N, \widehat{\mathbf{a}})}{\partial \widehat{a}_2}\right) & \cdots & \left(\frac{\partial y_N(\mathbf{x}_N, \widehat{\mathbf{a}})}{\partial \widehat{a}_N}\right) \end{bmatrix} \tag{9.32}$$

Besides, given $\partial y(\mathbf{x}_i, \mathbf{a}^*)/\partial a_j^* = k(\mathbf{x}_i, \mathbf{x}_j)$, (9.32) can be formulated as a kernel matrix, i.e., $\mathbf{F}(\widehat{\mathbf{a}}) = \mathbf{K}$. Thus, the advantage of the proposed kernel-based method becomes apparent, which greatly reduces the computational complexity and satisfies the real-time demand of applications. The variance of (9.31) can be further reformulated as

$$\mathrm{var}\left[t_0 - \widehat{y}_0\right] \approx \sigma_\gamma^2 + \sigma_\gamma^2 \mathbf{g}_0^T \left(\mathbf{K}^T\mathbf{K}\right)^{-1} \mathbf{g}_0 \tag{9.33}$$

where the first term comes from the inherent noise of the sample data and the second one denotes the variance of model estimation. Thus, one can form the PI with confidence degree $(1 - \alpha)\%$ of the predicted value $\widehat{y}_0$ as

$$\widehat{y}_0 \pm t_{n-p}^{\alpha/2}\left(\sigma_\gamma^2 + \sigma_\gamma^2 \mathbf{g}_0^{\mathrm{T}} \left(\mathbf{K}^{\mathrm{T}}\mathbf{K}\right)^{-1} \mathbf{g}_0\right)^{1/2} \tag{9.34}$$

where α is the quantile of a t-distribution and $n - p$ is the freedom degree of the t-distribution.

9.3.2 A Novel Predictive Optimization Method

In the predictive optimization, three parameters have to be discussed in advance, see Fig. 9.11, the prediction horizon, the optimization triggering horizon and the optimization duration. The prediction horizon represents the predicted time range of the current time, depending on the predicting accuracy and the application demand; the optimization triggering horizon represents that the adjusting solution will be formed if the system unbalance happens in such a period; and the optimization duration represents the evaluation region of the calculated solution.

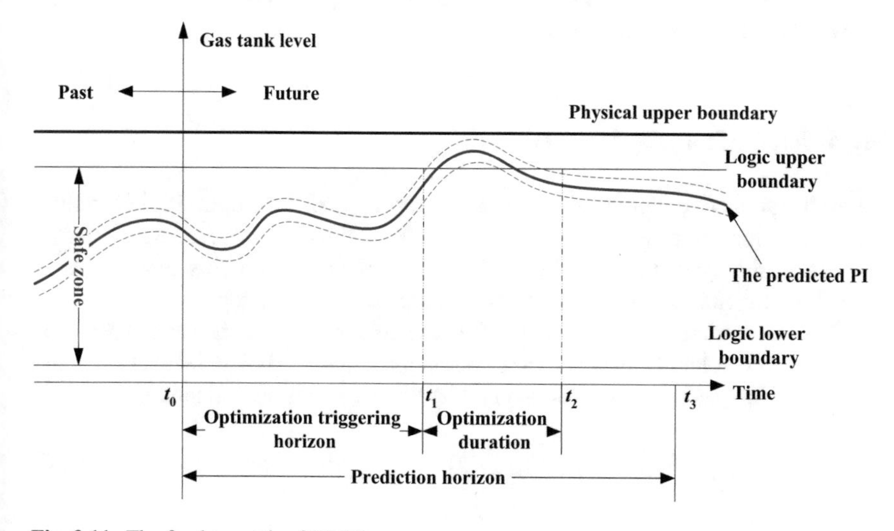

Fig. 9.11 The fundamentals of DPOS

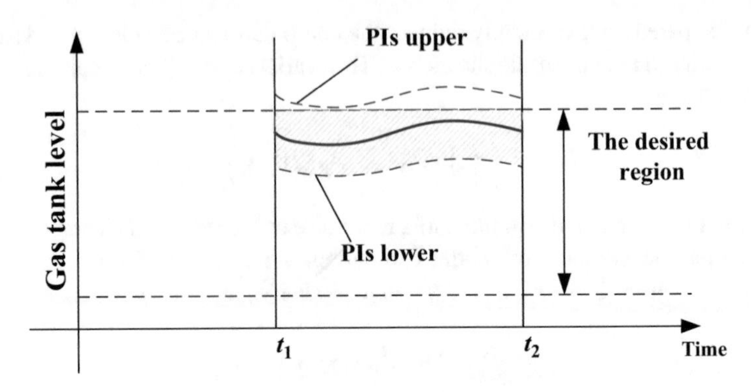

Fig. 9.12 The measure of scheduling solution on the desired region

In this section, a data-driven predictive optimization model is established to optimize the operation of the gas system, of which the first objective is to minimize the purchasing for outsourcing natural gas or oil, the second one ensures the holder levels to run within the desired region and the gas diffusion can be avoided, and the third one is to maximize the benefit from the production of electricity in power stations.

$$\mathrm{Min}\, J = C^{\mathrm{PE}} \sum_{t=t_1}^{t_2} \sum_{i=1}^{\mathrm{NB}} f_{i,t}^{\mathrm{PE}} - C^{\mathrm{gh}} \sum_{G} S^{G} - C^{\mathrm{elec}} \sum_{t=t_1}^{t_2} \left(\mathrm{pw}_{\mathrm{gen},t} - \mathrm{PD}_t \right) \tag{9.35}$$

where $f_{i,t}^{\mathrm{PE}}$ is the flow of the outsourcing energy of i-th boiler at t; NB is the total number of the boilers; S^G denotes the overlap area between the PIs and the desired region, see Fig. 9.12. $\mathrm{pw}_{\mathrm{gen},t}$ is the power generation at t, and PD_t is the internal consumption of the power users. The coefficients C^{gh}, C^{PE}, and C^{elec} serve to adapt the weights of the sub-objectives.

Prediction for Optimized Objectives

As for the predictive optimization, it is necessary to predict the optimized objectives for monitoring their variations. In this section, three objectives are the most concerned, the system safety, the minimization of the outsourcing natural gas or oil and the maximization of the electrical energy of power system.

First, the overlap area between the PIs of the holder level and the desired region is predicted. The holder level at t has a great relationship to the holder level at $t-1$ and the flow of the gas users at t that can be identified by a regression model.

$$h_{M,t}^{G} = \widehat{F}\left[h_{t-1}^{G}, f_{p,t}^{G}(1), f_{p,t}^{G}(2), \ldots, f_{p,t}^{G}(m) \right], \quad \text{for } \forall \text{ G,t} \tag{9.36}$$

where $h^G_{M,t}$ is the predicted holder level of gas G at t. Based on the mixed Gaussian kernel-based model proposed, the PIs for holder level can be constructed, in which the width can be denoted by $2 \times d^G_{p,t}$. Then, the PIs for holder level at t can be

$$h^{G+}_{p,t} = h^G_{M,t} + d^G_{p,t}, h^{G-}_{p,t} = h^G_{M,t} - d^G_{p,t} \tag{9.37}$$

where $h^{G+}_{p,t}$ and $h^{G-}_{p,t}$ is the upper and lower bounds, respectively. And, the overlap area can be computed by using

$$S^G = \begin{cases} \sum_{t=t_1}^{t_2} \left(hh^{G+} - h^{G-}_{p,t}\right)t, h^{G+}_{p,t} \geq hh^{G+} \text{ and } h^{G-}_{p,t} \geq hh^{G-} \\ \sum_{t=t_1}^{t_2} \left(h^{G+}_{p,t} - h^{G-}_{p,t}\right)t, h^{G+}_{p,t} \leq hh^{G+} \text{ and } h^{G-}_{p,t} \geq hh^{G-} \\ \sum_{t=t_1}^{t_2} \left(h^{G+}_{p,t} - hh^{G-}\right)t, h^{G+}_{p,t} \leq hh^{G+} \text{ and } h^{G-}_{p,t} \leq hh^{G-} \end{cases} \tag{9.38}$$

where hh^{G+} and hh^{G-} are the upper and lower bounds of the desired region, respectively.

Besides, the consumption of outsourcing natural gas or oil and the electrical generation of power system still need to be predicted. Considering that the burning mode of boilers remains unchanged without intervention, the consumption of the outsourcing energy can be computed as

$$J^{\mathrm{PE}} = \sum_{t=t_1}^{t_2} \sum_{i=1}^{\mathrm{NB}} f^{\mathrm{PE}}_{i,t} \tag{9.39}$$

where $f^{\mathrm{PE}}_{i,t}$ is the flow of the outsourcing energy of i-th boiler at t that will be a constant value in a period; NB is the total number of the boilers. The electricity generation of the power system can be predicted based on the energy balances for boilers and turbines. The sum of the energy supplied to each boiler multiplied by the efficiencies of each boiler is equal to the total enthalpy change from water to steam in all boilers.

$$\sum_G f^G_{i,t}\mathrm{Cp}^G + f^{\mathrm{PE}}_{i,t}\mathrm{Cp}^{\mathrm{PE}} = \frac{H^{\mathrm{stm}}_{i,t} f^{\mathrm{stm}}_{i,t} - H^{\mathrm{water}}_{i,t} f^{\mathrm{water}}_{i,t}}{\eta^b_i} \tag{9.40}$$

where $f^G_{i,t}$ denotes the gas G flow of i-th boiler at t that will also be a constant value in a period; Cp^G is the caloric value of the gas G; $\mathrm{Cp}^{\mathrm{PE}}$ is the caloric value of outsourcing energy; $f^{\mathrm{stm}}_{i,t}$ is the steam flow of i-th boiler; and, $f^{\mathrm{water}}_{i,t}$ is the water flow. $H^{\mathrm{stm}}_{i,t}$ denotes the enthalpy of steam; $H^{\mathrm{water}}_{i,t}$ is the enthalpy of water; η^b_i is the efficiency of i-th boiler. Based on (9.40), the generation amount of the steam can be computed by

$$f_{i,t}^{\mathrm{stm}} = f_{i,t}^{\mathrm{ps}} + f_{i,t}^{\mathrm{tb}} \tag{9.41}$$

where (9.41) indicates that the generated steam of i-th boiler at t should be the sum of the required steam in the production and the amount transformed into the power. Assuming that the steam for the production process is known, the steam provided for the turbines can also be computed. And then, the power generation can be easily computed as

$$\mathrm{pw}_{\mathrm{gen},i,t} = f_{i,t}^{\mathrm{tb}} H_{i,t}^{\mathrm{stm}} \eta_i^{\mathrm{Tb}} \quad \text{for } \forall\ i,t \tag{9.42}$$

$$J^{\mathrm{pw}} = \sum_{t=t_1}^{t_2} \left(\mathrm{pw}_{\mathrm{gen},t} - \mathrm{PD}_t \right) \tag{9.43}$$

where $\mathrm{pw}_{\mathrm{gen},\,t}$ is the power generation at t, and PD_{t} is the internal consumption of the power users. The three optimized objectives can be predicted before developing a reasonable scheduling strategy.

Rolling Optimization for Adjusting Solution

Sometimes, the burners switching and the boilers adjustment can also lead to some economic losses, so the objective function should also contain the penalties of these two operations besides the objectives defined in (9.35). Therefore, the objective function for optimization reads as

$$\mathrm{Min}\ J^* = J + \sum_{t=t_1}^{t_2} \sum_{i=1}^{\mathrm{NB}_d} \sum_{G} W_{\mathrm{SW}}^{G} \Delta n_{i,t}^{G} + \sum_{t=t_1}^{t_2} \sum_{G} W^{G} \Delta b_t^{G} \tag{9.44}$$

where the second and third terms are the penalties of the burner switching and the boiler adjustment, respectively. $\Delta n_{i,t}^{G}$ is the number of the switched burners of i-th boiler between the times of t to $t+1$, and Δb_t^{G} is the amount of the continuously adjusted boilers at t point.

Apart from constraints on the energy balances and the holder levels stated in (9.40)–(9.42), some other constraints are also considered for the objective of the optimization.

The measures of the energy scheduling are mainly the adjustable gas users, such as the boilers in the power stations and the LPB. The scheduling process is to switch the burners' status or tune the gas input valve of the boilers so as to adjust the gas consumption of the users. Then, the operations should satisfy the following expressions:

$$\Delta n_{i,t}^{G} = n_{i,t}^{G} - n_{i,t-1}^{G} = \mathrm{sw}_{i,t}^{G+} + \mathrm{sw}_{i,t}^{G-} \tag{9.45}$$

$$\mathrm{sw}_{i,t}^{G+}, \mathrm{sw}_{i,t}^{G-} \geq 0 \tag{9.46}$$

where $\mathrm{sw}_{i,t}^{G+}$ denotes the number of the burners that are turned on to use the gas G of ith boiler at t, $\mathrm{sw}_{i,t}^{G-}$ denotes that number of the turned off burners, and $n_{i,t}^{G}$ is the number of the burners that are operated at t. The gas consumption variation of the boilers by the burners switch can be formulated as

$$\Delta f_{i,t}^{G} = U_i^G \Delta n_{i,t}^{G} \Delta t \tag{9.47}$$

where U_i^G denotes the gas flow in an unit period on one burner of ith boiler, and $\Delta f_{i,t}^{G}$ the gas variation between $t-1$ and t points of ith boiler. Then, the gas consumption of ith boiler at t can be represented as

$$f_t^G(i) = f_{t-1}^G(i) + \Delta f_{i,t}^{G} \tag{9.48}$$

Due to the different operational constraints of each boiler or steam turbine, the following formulation must be satisfied:

$$F_{i,t}^{\mathrm{Minstm}} \leq f_{i,t}^{\mathrm{stm}} \leq F_{i,t}^{\mathrm{Maxstm}} \tag{9.49}$$

$$F_{i,t}^{\mathrm{Mintb}} \leq f_{i,t}^{\mathrm{tb}} \leq F_{i,t}^{\mathrm{Maxtb}} \tag{9.50}$$

where (9.48) indicates that the generated steam of ith boiler at t should be the sum of the required steam in the production and the amount transformed into the power. $F_{i,t}^{\mathrm{Minstm}}$ and $F_{i,t}^{\mathrm{Maxstm}}$ are the minimum and the maximum of the generated steam at t; $F_{i,t}^{\mathrm{Mintb}}$ and $F_{i,t}^{\mathrm{Maxtb}}$ are, respectively, those of the steam that enter into the turbines.

The precondition of the energy scheduling should ensure the required energy supplies for production, such as the steam and power, therefore,

$$\sum_{i=1}^{\mathrm{NB}} f_{i,t}^{\mathrm{ps}} \geq F_t^{\mathrm{SD}} \text{ for } \forall\ i,t \tag{9.51}$$

$$\sum_{i=1}^{\mathrm{NB}} \mathrm{pw}_{\mathrm{gen},i,t} \geq \mathrm{PD}_t \text{ for } \forall\ i,t \tag{9.52}$$

where F_t^{SD} denotes the required amount of the steam, and PD_t denotes that of the power.

$$F_i^{G,\min} \leq f_{i,t}^{G} \leq F_i^{G,\max} \text{ for } \forall\ i,t \tag{9.53}$$

where $F_i^{G,\min}$ and $F_i^{G,\max}$ are the minimum and maximum limits for by-product gases inlet.

9.3.3 Case Study

To verify the performance of the proposed DPOS method, all of the experimental data in this section came from the practical energy data center of a steel plant in China. The SCADA system is employed by the plant for the energy data acquisition, and the sampling interval should be less than or equal to 1 min. Furthermore, it is necessary for the sample data to complete the preliminary processing such as data imputation if needed since there are usually a large number of possible missing data points existed in the SCADA system of the production practice. Because such preliminary processing for the sample data belongs to a class of generic methods, this section avoids the redundant technical introduction.

In this section, the by-product gas system of the plant is composed of a BFG system, a COG system and a LDG system (see Fig. 9.1), where the LDG system is split into two subsystems, named #1 and #2 LDG subsystems. In this plant, there are six boilers and five steam turbines. As for each of the gas systems, since there are two holders that would not be meanwhile connected to the pipeline network, this section regards the two holders in each system as one unit.

Figure 9.13 illustrates the PIs of a random selected holder by using the DPOS method, the PIs of BFG holder , in which the PIs can completely cover the real values. Furthermore, a series of statistic comparisons based on the LSSVM [20] and the bootstrap neural network [22] are reported in Table 9.6, where 100 times independent experiments were conducted in order for the fair performance. And, the parameters of these methods are repeatedly tuned by trail-and-error for the best

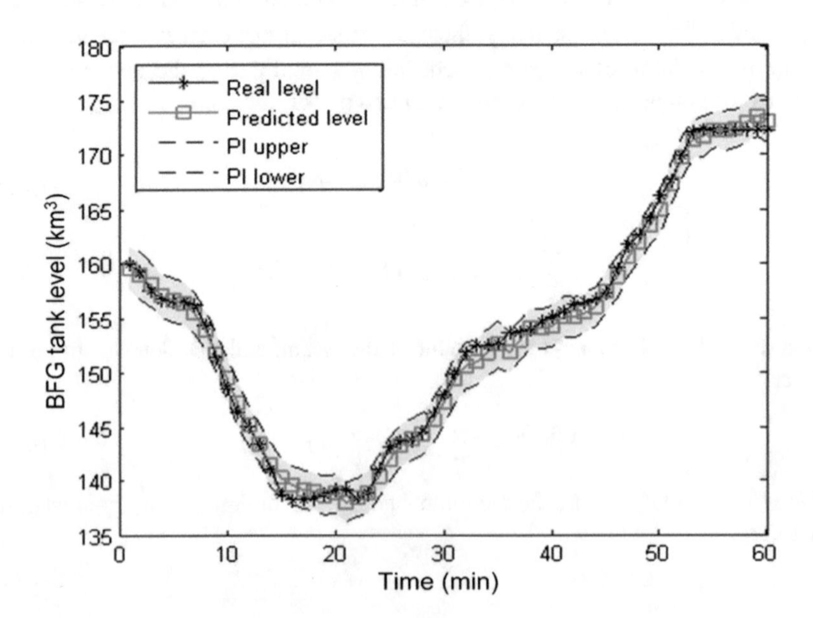

Fig. 9.13 The constructed PIs of the BFG system

Table 9.6 Comparison of the PIs of the by-product gas system constructed by different methods

Gasholders	Methods	RMSE	CWC	CT (s)
BFG holder	LSSVM	0.8757	8.9145	24.093
	Bootstrap NN	0.8104	4.3120	29.122
	DPOS	**0.7646**	**3.9299**	**16.201**
COG holder	LSSVM	0.1215	2.1922	24.047
	Bootstrap NN	0.1144	1.0783	22.230
	DPOS	**0.1058**	**1.0215**	**15.237**
LDG holder of #1 subsystem	LSSVM	0.9752	11.1254	24.063
	Bootstrap NN	0.9073	7.2240	37.492
	DPOS	**0.8953**	**5.9341**	**15.819**
LDG holder of #2 subsystem	LSSVM	0.8065	6.3521	24.072
	Bootstrap NN	0.7939	6.2893	36.643
	DPOS	**0.5778**	**2.8206**	**14.568**

quality of these methods. Here, the RMSE is utilized to quantify the accuracy, and one can refer to the CWC to evaluate the performance of the PIs. As for the computing time, the efficiency improvements of the proposed method are also remarkable, it completely satisfies the practical demand for the online rolling-horizon scheduling encountered in the energy system.

Generally, given that the operational mode of the burners in the boilers is fixed if without manual intervention, the consumption of the outsourcing natural gas and oil, and the generation of the electrical power can be predicted based on the current mode of the burners. Therefore, at the stage of prediction, only the holder levels should be monitored to decide whether a scheduling optimization is required. At the stage of energy scheduling, we offer the scheduling instances when the holder levels exceed the boundary, including two categories of scenarios. The first one deals with a situation when only the BFG holder level exceeds the boundary, and the second one simultaneously involves the BFG and COG holder levels. In this section, the mechanical and logical bounds of the gasholders are listed in Table 9.7 as well as the prediction horizon and frequency. The consumption abilities of different boilers are given in Table 9.8, where five boilers in the power station and one boiler in the lower pressure boiler are considered.

As for the solution of the optimization problem, the IBM software CPLEXTM Optimization Studio is employed for the mathematical programming, which takes only a few seconds to obtain the solution, which can greatly satisfy the real-time demand.

Take the BFG system unbalance as an example, the situation of BFG system that exceeds the boundary is shown, see Fig. 9.14. One can empirically designate the scheduling triggering horizon to be 30 min. The scheduling duration is set as 20 min. Figure 9.14 presents the results, in which the predicted level after scheduling completely run in the desired region and the overlap area between the PIs and the desired region is the largest. And, the system can be continuously traced and

Table 9.7 The mechanical and prediction parameters of gasholders

Gasholders	Mechanical upper bound (km^3)	Logical upper bound (km^3)	Logical lower bound (km^3)	Mechanical lower bound (km^3)	Prediction horizon (min)	Prediction frequency (min)
BFG holder	250	230	120	100	60	5
COG holder	420	358	189	120	60	5
LDG holder of #1 subsystem	130	110	40	25	60	5
LDG holder of #2 subsystem	140	128	48	40	60	5

Table 9.8 The gas consumption measures of the adjustable users

	Power stations					LPB
	Adjustable amount (km^3/h) × The number of burners				Adjustable range (km^3/h)	Adjustable range
Gas	#1	#2	#3	#4	#5	
BFG	20 × 8	20 × 5	15 × 8	600~800	320~380	20~100
COG	4 × 8	6 × 5	N.A.	0~12	N.A.	0.8~1.2
LDG	N.A.	N.A.	N.A.	N.A.	0~6	0~1.6

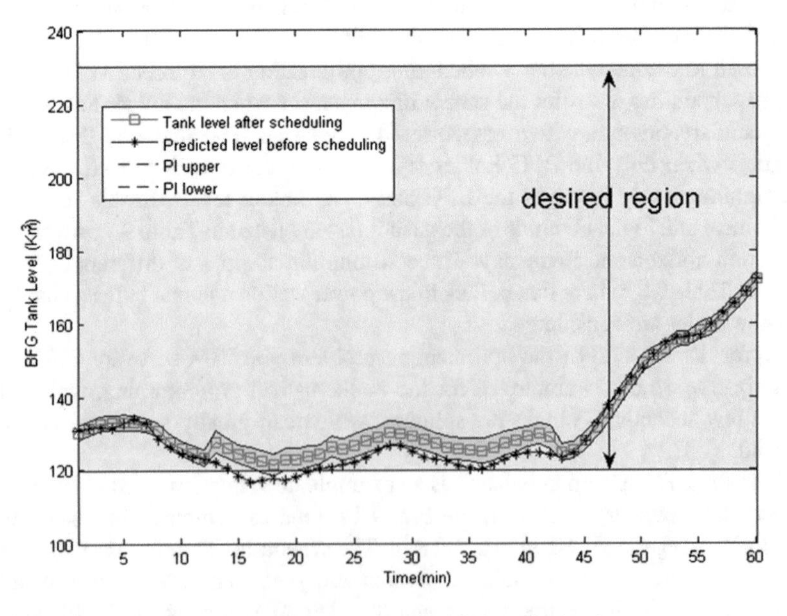

Fig. 9.14 The prediction and scheduling results of the BFG holder level

predicted for the next scheduling, i.e., the rolling-horizon operation. Furthermore, a comparative analysis between the point-oriented prediction and the proposed one is performed, in which the former maintains the holder level at a certain value. Figure 9.15 illustrates the result by the point-oriented prediction, and the setting value of the level is the mean of the desired region. The solutions are listed in Table 9.9, and the detailed performance is in Table 9.10. Although the point-oriented one tends to stablize the level at an exact value, its solution could be inflexibile and extra-rigorous, which leads to a worse performance. Here, the increased energy efficiency indicates the difference of energy transformation ratio compared to that produced by the manual scheduling.

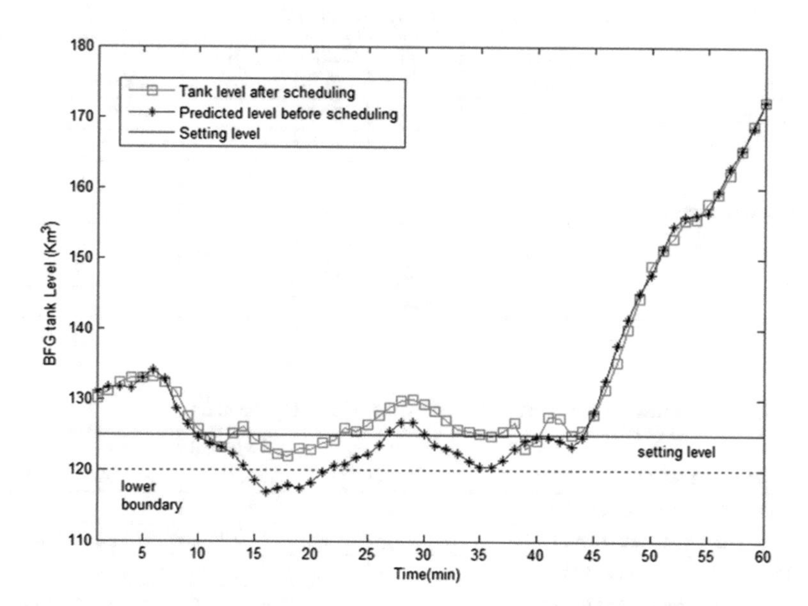

Fig. 9.15 The BFG holder level tendencies by using the point-oriented scheduling method

Table 9.9 The solutions comparison between the proposed DPOS and the point-oriented prediction methods

Scheduling mode	Energy units	DPOS (km^3/h)	Point-oriented prediction (km^3/h)
The burners number for switching	#1 PS	-2×20	-3×20
	#2 PS	0	0
	#3 PS	0	0
Continuous adjustment	#4 PS	0	0
	#5 PS	−29.33	−19.33
	LPB	−13.31	−14.27
Total (km^3/h)		−82.64	−93.8

Table 9.10 The performance using two prediction methods—a comparative analysis

Optimized criterions	DPOS	Point-oriented prediction
The number of burner switching	2	3
Reduced gas consumed by boilers	17.91 km^3	20.32 km^3
Reduced steam generated by boilers	4588.26 kg	5126.53 kg
Reduced power generated by turbines	1455.6 kwh	1818.11 kwh
Increased energy efficiency	0.19	0.11
Total optimization results (saved standard coal equivalent)	**73.06 kg**	56.03 kg

Table 9.11 The scheduling solutions produced by several methods

Scheduling mode	Energy units	DPOS (km^3/h)	Method of [23] (km^3/h)	Manual scheduling (km^3/h)
The burners number for switching	#1 PS	−2×20	−2×20	−3×20
	#2 PS	0	−1×20	0
	#3 PS	0	0	0
Continuous adjustment	#4 PS	0	−19.96	−30
	#5 PS	−29.33	0	0
	LPB	−13.31	−13.75	−15
Total (km^3/h)		**−82.64**	−93.71	−105

Table 9.12 The statistics of results produced by the scheduling methods

Optimized indexes	DPOS	Method of [23]	Manual scheduling
The number of burners switching	2	3	3
Reduced gas consumed by boilers	17.91 km^3	20.30 km^3	22.75 km^3
Reduced steam generated by boilers	4588.26 kg	5047.97 kg	5411.21 kg
Reduced power generated by turbines	1455.6 kwh	1811.93 kwh	2024.57 kwh
Increased energy efficiency	0.19	0.12	0
Total optimization results (saved standard coal equivalent)	**73.06 kg**	57.91 kg	0

In addition, the comparison by using the method in [23] and the widely used manual scheduling were also conducted. The solution and the performance are listed in Tables 9.11 and 9.12, respectively. From Table 9.11, the total scheduling amount by the manual method leads to over-adjustment, which makes the cost of burner switching high. Besides, the over-reduction of the gas resource by the manual method will bring about the decrease of the power generation. From Table 9.12, the results by the DPOS are superior to those by [23] and the manual scheduling. It is obvious that the DPOS can save 73.06 kg standard coal equivalent compared to that when using the manual mode. The reason of such results lies in that the manual scheduling is only carried out based on the experiences of the operator, which is lack of effective comprehensive judgment on the system. And the model built in [23] provides some inapplicable results compared to the DPOS due to the loss of the forecasting function. Thus, the DPOS achieves the best performance of the optimization.

Some theoretical improvements related to the data-based method presented in this section deserve further studies. For instance, when developing a PIs construction model, it could be more effective method that has a higher efficiency under the condition of unbiased estimation for the system. While this is a practically viable option, in the future, it would be beneficial to carry out some theoretical analysis on the PIs-based optimized scheduling. Therefore, the energy optimization technology can be further extended.

9.4 A Two-Stage Method for Predicting and Scheduling Energy in an Oxygen/Nitrogen System of the Steel industry

Oxygen/nitrogen system in steel industry, which contains a series of energy media such as oxygen and nitrogen is one of most fundamental energy resources for a steel enterprise. In daily routine production, the related processes such as steel-making by converters, iron-making by blast furnace, and non-ferrous metal smelting always require continuous supplying of oxygen or nitrogen. Therefore, oxygen and nitrogen are widely utilized and on greatly demand in steel industry, and its scheduling works will be significantly beneficial for the production efficiency, energy saving, and even be indirectly related to economic profits of steel enterprises.

Regarded as a significant energy resource system in steel industry, a typical structure of an oxygen/nitrogen system in a plant can be illustrated as Fig. 9.16, in which eight ASUs, the liquefying plant (LP), the liquid tanks, the consumption units, and the transportation system are involved. Acting as the energy generation units, the ASUs supply oxygen and nitrogen mostly in gaseous state, and then a partition of the generated energy are stored in the tanks with liquid state. These liquid tanks are named by its related devices. Take "LTASU1234(O_2)" as an example, it stores liquidized oxygen coming from #1 to #4 ASU. The compressors and the pipeline networks are the energy transportation system delivering the oxygen/nitrogen in various pressure levels to the consumption units such as blast furnaces, converters, and cold/hot rolling. Besides, there are a number of redundant compressors available if a compressor falls down. Under such mechanism, the production process and the energy utilization will not be interrupted.

Thus, one can summarize the characteristics of the oxygen/nitrogen system as follows: 1) Due to various load capacity of the ASUs, multiple categories of pressure levels of the energy, and long distributed pipeline networks, it is extremely difficult to establish a physics-based model for describing the oxygen/nitrogen system. 2)

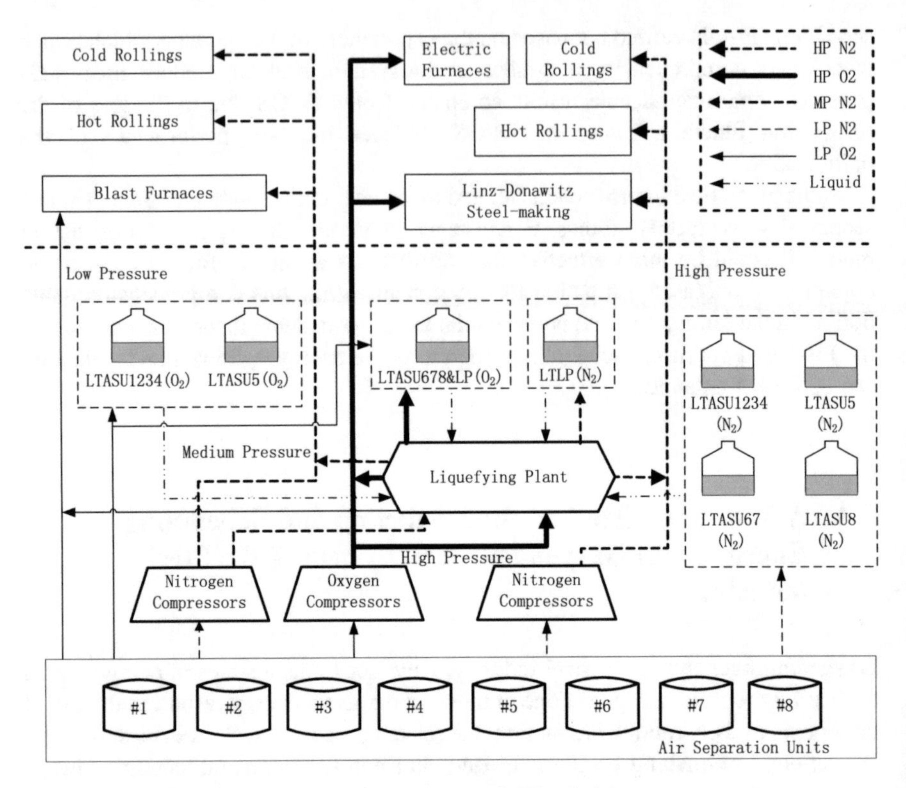

Fig. 9.16 Structural chart of oxygen/nitrogen system in a steel plant

The adjustable capacity (the scheduling measures) of the oxygen/nitrogen system is relatively limitative, i.e., only a few of liquid tanks with small volume can be used for temporary storage, and the load changing of ASU could also be available for scheduling.

In production practice, it is noticeable that the oxygen/nitrogen utilization process always involves some regular operational modes or phases, which can be reflected by a series of similar data segments. It can be obviously depicted from Fig. 9.17 that there are generally three categories of operational modes of an oxygen consumption unit. One can use the feature modes of the following semantically different information granules, see Mode #1–#3, to describe the operational characteristics of the equipment. Although there are some similarities between these modes, the amplitude and the duration of each mode are substantially different. As such, the practical features should be reasonably and generally considered in the model establishment.

Considering a practical oxygen/nitrogen system of a steel plant in China, this section introduces a two-stage predictive scheduling method for the energy scheduling [24]. Given that the scheduling operation (e.g., ASU load change procedure) requires a relative long time period, at the prediction stage a long-term prediction model is established by using a GrC-based method for oxygen/nitrogen consumption demand prediction. Based on industrial semantics directly related to energy devices

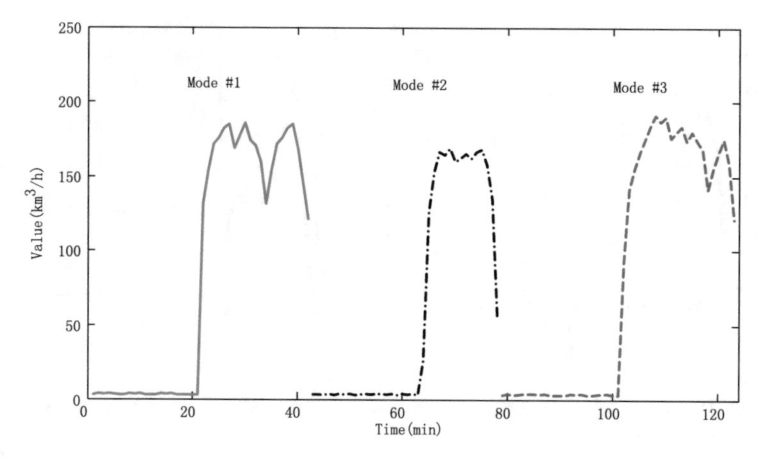

Fig. 9.17 Three operational modes of an oxygen consumption unit in the oxygen/nitrogen system

operations, data granulation is performed so that the predictive horizon can be reasonably extended. Then, an MILP model, considering the adjusting capacity of the energy network and the energy conversion relationship between gaseous and liquid state, is established at the scheduling stage, where the constraints of the scheduled equipment amount is also taken into account.

9.4.1 GrC-Based Long-Term Prediction Model

Since the energy scheduling, which primarily refers to formulation of scheduling scheme and procedures of device operation (e.g., load change of ASUs), usually expends a long period of time, a long-term prediction is on high demand for sufficiently providing helpful guidance on the decision-making of the energy system in steel industry. As described in Chap. 6, GrC becomes a popular paradigm on the subject of data-driven method [25–27] by deeply merging with fuzzy modeling, rough sets, and the related theories [28, 29]. It takes data granule, i.e., data segments with a series of points, as the analysis unit instead of single data point, which enables to extend the prediction horizon (long-term prediction). First, we will introduce the data granulation procedure, and then present the long-term prediction modeling scheme.

Data Granulation

As an essential preparation for GrC modeling, a reasonable data granulation always considers prior knowledge for making the model more human centric. Given that the energy utilization (data flow) of the studied oxygen/nitrogen system usually corresponds to a specific device operation status, e.g., continuous consumption and short

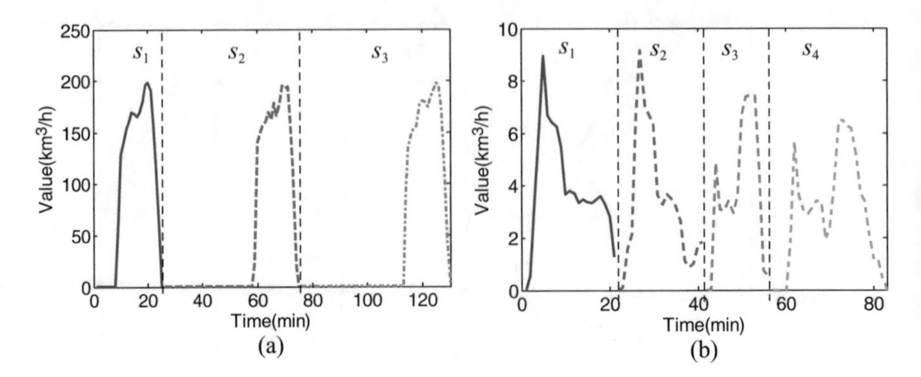

Fig. 9.18 Data granulation results. (**a**) oxygen consumption of #1 converter (**b**) nitrogen consumption of #2 blast furnace

pauses, mentioned as the above illustrative description, it is obviously useful to adopt such practical semantic meaning for data granulation process.

Here, to clarify the granulation process considering the operational feature of industrial data, two examples are illustrated in Fig. 9.18, where Fig. 9.18a is the oxygen consumption flow of #1 converter and Fig. 9.18b is the nitrogen consumption of #2 blast furnace. It is obviously depicted that the oxygen consumption of #1 converter takes a heating round as the basic unit, which involves a short temporary break and variable oxygen consumption. As such, one can segment such a data interval as a data granule. Similarly, the nitrogen consumption flow of #2 blast furnace can also be granulated by its quasi-periodic features. In such a way, the data granules considering practical semantic meaning are finally obtained.

Long-Term Prediction Modeling

In this section, a long-term prediction model is established for forecasting the oxygen/nitrogen requirements of the energy system. One can consider that FCM always deals with the data segment [30, 31], and this unsupervised algorithm is with a low computation cost. Here, a FCM algorithm is used to compute the membership value so as to perform the fuzzy inference. The objective of FCM is denoted as

$$Q = \sum_{i=1}^{c} \sum_{j=1}^{N} u_{ij}^{m} d_{ij}^{2} \tag{9.54}$$

$$\text{s.t.} \quad \sum_{i=1}^{c} u_{ij} = 1 \quad j = 1, 2, \ldots N \tag{9.55}$$

where N refers to the number of data granules, u_{ij} denotes the membership of the jth data with respect to the ith cluster, $1 < c \leq N$ is the number of clusters, $m > 1$ is the fuzzy coefficient, and d_{ij} denotes the distance of the jth granule with respect to the ith clustering center, i.e.,

$$d_{ij} = \left\| \boldsymbol{s}_j - \boldsymbol{v}_i \right\| \quad i = 1, 2, \ldots, c \quad j = 1, 2, \ldots, N \tag{9.56}$$

where $\boldsymbol{v}_i$ is the ith clustering center. With the help of Lagrange multiplier [21], the clustering center can be computed by

$$v_{ij} = \frac{\sum_{k=1}^{N} u_{ik}^{m} S_{kj}}{\sum_{k=1}^{N} u_{ik}^{m}} \quad i = 1, 2, \ldots, c \quad j = 1, 2, \ldots, w \tag{9.57}$$

Based on the above computation, the corresponding fuzzy rules can be determined by fuzzy inference [32, 33]. Here, we utilize a concept of "which cluster the granule to the most extent belongs to" to construct the industrial linguistic variables. If assuming the input amount as 3, the fuzzy rules can be formulated by

$$R: \text{If } s_{k-3} \text{ is } c_3, s_{k-2} \text{ is } c_2, s_{k-1} \text{ is } c_1, \text{ then } s_k \text{ is } c_0 \tag{9.58}$$

where c_3, c_2, c_1, c_0 are the tags of the clusters, e.g., c_3 will be tagged as 2 if the maximal membership of s_{k-3} is toward 2nd cluster. Therefore, a rule base involving $N - 3$ fuzzy rules is then formed. Then, the forecasted membership function can be calculated by

$$\boldsymbol{p} = \sum_{j=1}^{N} h_j u_{rj}^{L} \tag{9.59}$$

where $_r$ denotes the cluster that h_j tags, and u_{rj}^{L} is the maximal value in the jth column of the partition matrix which indicates the most belonging grade toward certain cluster. Then, centroid defuzzification [34, 35] is utilized here, i.e.,

$$\widehat{\boldsymbol{p}} = \boldsymbol{p}^T \boldsymbol{V} \tag{9.60}$$

where $\boldsymbol{V}$ is the prototype matrix, and $\widehat{\boldsymbol{p}}$ is the final prediction result.

For clarifying the modeling process presented in this section, a list of detailed procedures is shown as follows:

Algorithm 9.3: Procedure of the Long-Term Prediction Modeling

Step 1: *Generate the clusters for a given dataset with (9.54), (9.55), (9.56), and (9.57).*

Step 2: *Establish fuzzy rule basement as (9.58) utilizing fuzzy inference tricks.*

Step 3: *Predict the membership grade by using (9.59).*

Step 4: *Obtain the long-term prediction result by the defuzzification of (9.60).*

9.4.2 *MILP-Based Scheduling Model for Oxygen/Nitrogen System*

Industrial scheduling can be deemed as an optimization problem with multiple constraints [36], which were usually formulated by mathematical programming models [37, 38]. Based on the practical concerns, a minimization of the oxygen/nitrogen diffusion can be regarded as the objective function in this section. And, the restrictions on scheduling capacity of the equipment as well as the relationships for unit conversion are practically considered in the constraints. In addition, the energy scheduling aims at minimizing the number of the operated devices as few as possible for the purpose of reducing the correlative impacts on the other energy media or the manufacturing process. Therefore, the number of the scheduled devices is generally limitative in practice.

Objective

As one of the most representative indexes in practical production, the energy diffusion flow plays a crucial rule for measuring the effect of scheduling works in steel industry. Taking oxygen as an example, it is a quotient percentage $O_{2_{dr}}$ which is defined as follows:

$$O_{2_{dr}} = \frac{O_{2_g} - O_{2c}}{O_{2_g}} \times 100\% \tag{9.61}$$

where O_{2_g} refers to the energy generation and O_{2_c} the consumption. The objective of the energy scheduling here can be formulated by

$$\min(O_{2_{dr}} + n_{2_{dr}}) \tag{9.62}$$

where $n_{2_{dr}}$ denotes the diffusion rate of nitrogen.

Constraints

In this section, one can summarize the device-related constraints into three parts, i.e., the ASUs, the liquefying plant and the liquid tanks. Besides, the practical concerned constraints are further established such as restriction on the amount of scheduling devices, etc.

1. ASUs

   Each of ASU has its upper and lower limitations on oxygen/nitrogen generation, which are approximately 80% and 105% of the capacity, respectively. Assuming the variables of the ASU load as $O_{2_{AUSi}}$ and $n_{2_{ASUi}}$, the related constraints can be described as

$$\begin{array}{l} \mathrm{LB}_{\mathrm{ASU}i}^{\mathrm{o}_2} \leq \mathrm{o}_{2_{\mathrm{ASU}i}} \leq \mathrm{UB}_{\mathrm{ASU}i}^{\mathrm{o}_2} \\ \mathrm{LB}_{\mathrm{ASU}i}^{\mathrm{n}_2} \leq \mathrm{n}_{2_{\mathrm{ASU}i}} \leq \mathrm{UB}_{\mathrm{ASU}i}^{\mathrm{n}_2} \end{array} \tag{9.63}$$

where $\mathrm{LB}_{\mathrm{ASU}i}^{\mathrm{o}_2}$, $\mathrm{UB}_{\mathrm{ASU}i}^{\mathrm{o}_2}$ denotes the lower, upper bound of the oxygen generation, $\mathrm{LB}_{\mathrm{ASU}i}^{\mathrm{n}_2}$, $\mathrm{UB}_{\mathrm{ASU}i}^{\mathrm{n}_2}$ for the nitrogen, and $i = 1, 2, \ldots, n_{\mathrm{ASU}}$ refers to the number of the ASU. Besides, the decision process of restarting/shutting down an ASU requires careful planning and complex evaluation work, which means that the status of them cannot be arbitrarily changed. Therefore, owing to the burdensome work and complicated situation, the units that are out of service are not specially addressed in this model.

2. Liquefying Plant

   Similar to the ASUs, the liquefying plant also has a load range which is typically from 60 to 100% of its capacity, see

$$\begin{array}{l} \mathrm{LB}_{\mathrm{LP}}^{\mathrm{o}_2} \leq \mathrm{o}_{2_{\mathrm{LP}}} \leq \mathrm{UB}_{\mathrm{LP}}^{\mathrm{o}_2} \\ \mathrm{LB}_{\mathrm{LP}}^{\mathrm{n}_2} \leq \mathrm{n}_{2_{\mathrm{LP}}} \leq \mathrm{UB}_{\mathrm{LP}}^{\mathrm{n}_2} \end{array} \tag{9.64}$$

where $\mathrm{o}_{2_{\mathrm{LP}}}$ and $\mathrm{n}_{2_{\mathrm{LP}}}$ are the variables for the load of the liquefying plant, $\mathrm{LB}_{\mathrm{LP}}^{\mathrm{o}_2}$ and $\mathrm{UB}_{\mathrm{LP}}^{\mathrm{o}_2}$ refers to lower and upper bound of the oxygen liquefying amount, and $\mathrm{LB}_{\mathrm{LP}}^{\mathrm{n}_2}$, $\mathrm{UB}_{\mathrm{LP}}^{\mathrm{n}_2}$ for the nitrogen.

3. Liquid Tanks

   Considering the equipment safety, the practical storage volumes of the liquid tanks are restricted at 10–95% of theirs rated capacity. Furthermore, two sets of coefficients are required for unit normalization, of which $\mathrm{O}_{\mathrm{t2m}}$ means the unit of this variable is transformed from ton to cubic meter, and $\mathrm{O}_{\mathrm{l2g}}$ for the one from liquid state to gaseous state.

   As such, the liquid tank related constraints are then established. Some examples are given as follows:

$$\begin{array}{l} \mathrm{O}_{\mathrm{t2m}} \times \mathrm{O}_{\mathrm{l2g}} \times \mathrm{LB}_{\mathrm{LTASU1234}}^{\mathrm{o}_2} \leq \mathrm{O}_{2_{\mathrm{LTASU1234}}} \leq \mathrm{O}_{\mathrm{t2m}} \times \mathrm{O}_{\mathrm{l2g}} \times \mathrm{UB}_{\mathrm{LTASU1234}}^{\mathrm{o}_2} \\ \mathrm{O}_{\mathrm{l2g}} \times \mathrm{LB}_{\mathrm{LTASU5}}^{\mathrm{O}_2} \leq \mathrm{o}_{2_{\mathrm{LTASU5}}} \leq \mathrm{O}_{\mathrm{l2g}} \times \mathrm{UB}_{\mathrm{LTASU5}}^{\mathrm{o}_2} \end{array} \tag{9.65}$$

Note that the subscript denotes the devices that are attached to the liquid tanks. For example, LTASU1234 in (9.67) refers to the liquid tank that is connected with the #1–#4 ASUs, LTASU5 refers to the liquid tank with the #5 ASU.

4. Other constraints

   Two sets of constraints are still needed for making the scheduling solution applicable. First, excessive scheduling should be avoided. Taking a situation that oxygen is on shortage as an example, here the energy supply increase is needed, whereas a sufficient solution should not over-schedule the energy as surplus for this could further cause unnecessary scheduling with needless waste of time and manpower. Bearing this in mind, two constraints are established

$$\begin{aligned}&\sum_{i=1}^{n_{\mathrm{ASU}}}\Delta o2_{\mathrm{ASU}i}+\Delta o2_{\mathrm{LTASU1234}}+\Delta o2_{\mathrm{LTASU5}}+\Delta o2_{\mathrm{LTASU678}}-\Delta o2_{\mathrm{LP}}\geq\delta_{\mathrm{o}_2}\\&\sum_{i=1}^{n_{\mathrm{ASU}}}\Delta n2_{\mathrm{ASU}i}+\Delta n2_{\mathrm{LTASU1234t}}+\Delta n2_{\mathrm{LTASU1234m}}+\Delta n2_{\mathrm{LTASU5}}\\&\quad+\Delta n2_{\mathrm{LTASU6}}+\Delta n2_{\mathrm{LTASU78}}-\Delta n2_{\mathrm{LP}}\geq\delta_{\mathrm{n}_2}\end{aligned}\tag{9.66}$$

where δ_{o_2} and δ_{n_2} refer to the amount of surplus/shortage for oxygen and nitrogen. The subscripts denote the liquefying plant, ASUs or the attached liquid tanks. All the variables on the left side refer to delta values, i.e., subtraction of the value before and after scheduling.

Second, a reasonable scheduling solution should not involve too much devices with consideration of the costs on time and manpower. As a result, a set of binary variables are utilized here as the constraints for the number of scheduled units. Then,

$$\begin{aligned}&\sum_{i=1}^{n_{\mathrm{ASU}}}\mathrm{o}_{2\mathrm{ASU}i}^{\mathrm{B}}+\mathrm{o}_{2\mathrm{LTASU1234}}^{\mathrm{B}}+\mathrm{o}_{2\mathrm{LTASU5}}^{\mathrm{B}}+\mathrm{o}_{2\mathrm{LTASU678}}^{\mathrm{B}}-\mathrm{o}_{2\mathrm{LP}}^{\mathrm{B}}\leq N_{\mathrm{o}_2}\\&\sum_{i=1}^{n_{\mathrm{ASU}}}\mathrm{n}_{2\mathrm{ASU}i}^{\mathrm{B}}+\mathrm{n}_{2\mathrm{LTASU1234t}}^{\mathrm{B}}+\mathrm{n}_{2\mathrm{LTASU1234m}}^{\mathrm{B}}+\mathrm{n}_{2\mathrm{LTASU1234}}^{\mathrm{B}}+\mathrm{n}_{2\mathrm{LTASU6}}^{\mathrm{B}}\\&\quad+\mathrm{n}_{2\mathrm{LTASU78}}^{\mathrm{B}}-\mathrm{n}_{2\mathrm{LP}}^{\mathrm{B}}\leq N_{\mathrm{n}_2}\end{aligned}\tag{9.67}$$

where N_{o_2} and N_{n_2} denotes the amount of adjustable devices. And, all of the variables on the left side are binary values. It should be clarified that the liquid tanks for nitrogen which connected with #1, #2, #3, and #4 are actually consisted of two sets, one with the unit of ton and another cubic meter. As a result, we have the related subscripts as LTASU1234t and LTASU1234m, respectively.

Here, one can take #1 ASU as an example to describe how (9.66) and (9.67) are actually applied. Assuming that the current load of #1 ASU is 30000 and take the upper bound 31,500, (see Table 9.13) into account, then the corresponding delta variable in (9.66) can be computed as $\Delta o2_{\mathrm{ASU1}}=31,500-30,000=1500$. At the meantime, the binary variable $\mathrm{o}_{2\mathrm{ASU1}}^{\mathrm{B}}$ in (9.67) is set at 1. The parameters δ_{o_2}, δ_{n_2}, N_{o_2}, and N_{n_2} are determined in view of practical application which will be provided in experimental analysis of next section.

Table 9.13 Capacity of the air separation units and the associated liquid tanks (1 Nm^3 means 1 m^3 at normal state with 101.325 kPa and 20 °C, "t" refers to the unit "ton" for which 1 t = 1000 kg)

ASU	Oxygen (Nm^3/h)	Liquid tank (LO_2)	Nitrogen (Nm^3/h)	Liquid tank (LN_2)
#1	30,000	650 t	45,000	100 t + 500 m^3
#2	30,000		45,000	
#3	27,000		30,000	
#4	30,000		40,000	
#5	50,000	2500 m^3	95,000	2500 m^3
#6	60,000	2000 m^3 (also connect with liquefying plant)	64,000	100 m^3 × 2
#7	60,000		64,000	2000 m^3
#8	60,000		64,000	

9.4.3 Case study

Long-Term Prediction on Oxygen/Nitrogen Requirements

Since the flows of the oxygen consumption of #1 converter and the nitrogen consumption of #2 blast furnace are relatively typical in this energy system, they are selected as practical illustrative examples for long-term prediction in this section. To demonstrate the performance of the proposed approach, one can also designate two classes of data-driven modeling methods as the comparative ones. Here, the ESN and LSSVM are selected due to their pointwise prediction advantage for industrial application [39, 40]. The length of prediction is set at 24 h, i.e., 1440 data points. The MAPE and the RMSE are adopted as the evaluation criterions to quantify the prediction quality. The following results are randomly selected from 6-month online running application.

Figure 9.19 shows the prediction results of the oxygen flow of #1 converter, and the error statistics and the CT can also be listed as Table 9.14. It is clear that the proposed method gives more accurate estimation compared to the other methods. For such a long prediction period, although some deviations between the predicted

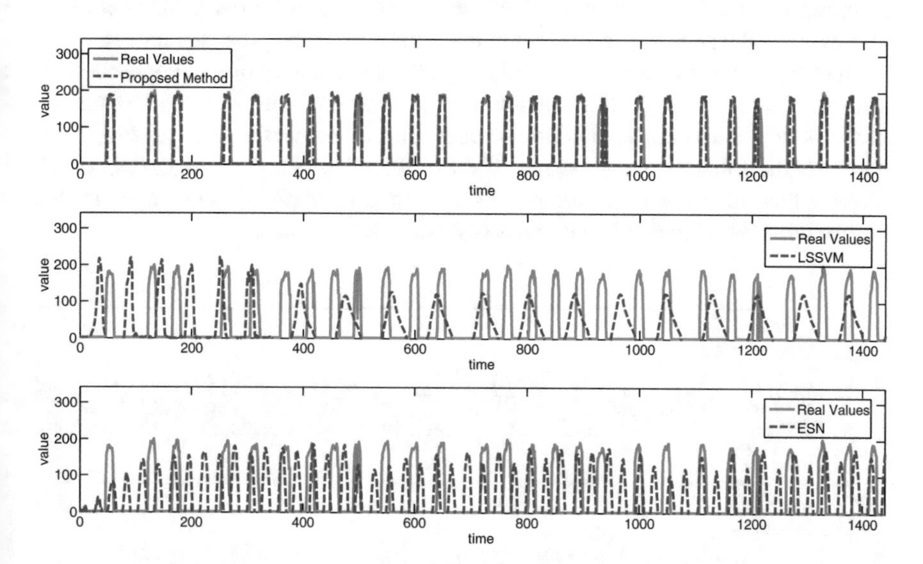

Fig. 9.19 Prediction results of oxygen consumption of #1 converter with different methods

Table 9.14 Error statistics and time costs on the prediction of oxygen consumption of #1 converter with different methods

Methods	MAPE	RMSE	CT (s)
ESN	18.47	64.2827	0.098
LSSVM	15.37	51.1434	7.998
Proposed method	6.14	28.2563	0.005

values and the real ones exist, the predicted tendency of the production status switching is fairly apparent for real application. Based on the results which reflect the future trend of energy requirements, the scheduling scheme can be then scientifically formulated. In contrast, LSSVM fails on forecasting the data on both tendency and exact values. And, the result of ESN is also unsatisfactory which exhibits unstable oscillation. In addition, the numerical results from this table also explicitly indicate that ESN and LSSVM present severe error on MAPE and RMSE, and LSSVM also costs long computing time. Thereby, with regard to the accuracy and the computing cost, the proposed approach behaves outstanding performance.

Compared with the oxygen consumption which exhibits quasi-periodic characteristics, the nitrogen consumption of #2 blast furnace here behaves somewhat irregular oscillation with respect to both width and height of each granule. A set of long-term prediction result with different methods on this object is given in Fig. 9.20. For clarifying the detailed predicting quality of the methods, one can refer to the illustrative results of first 120, see Fig. 9.21. It is apparently that the LSSVM can only provide an acceptable result in such a short period. With the iteration process going on, the accuracy becomes worse. For ESN, the amplitude of the data is wrongly estimated. As a result, the comparative two methods are inappropriate for long-term prediction for the practical industrial data. Table 9.15 also exhibits the superior accuracy and low time cost of the proposed model.

In order to demonstrate the feasibility and the applicability of the proposed long-term prediction method, this section further provides here a 3-month running statistics on the average prediction accuracy after implementing it in practice compared to the other two approaches. Table 9.16 gives the error statistics, which indicate that the accuracy of the proposed method is relatively stable and capable of providing the remarkable outcomes compared to the others.

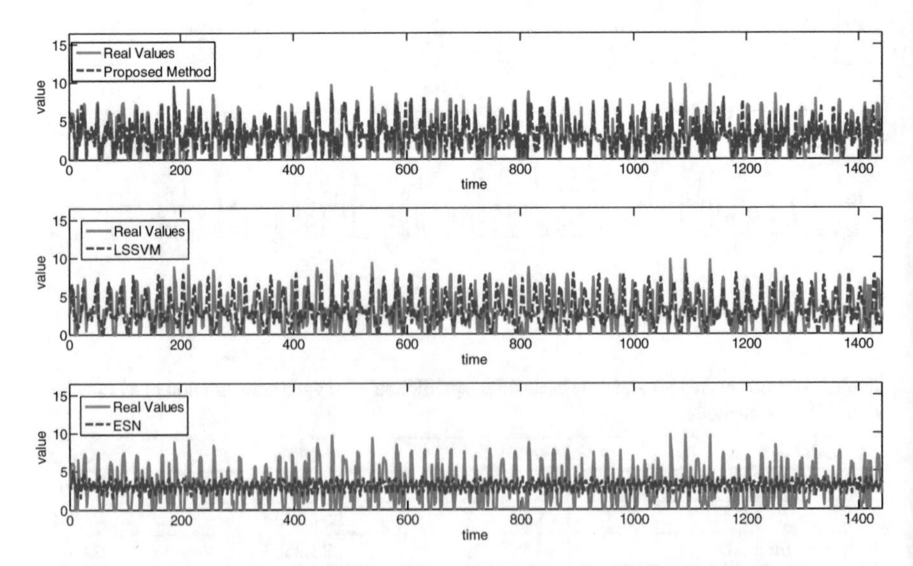

Fig. 9.20 Prediction results of nitrogen consumption of #2 blast furnace with different methods

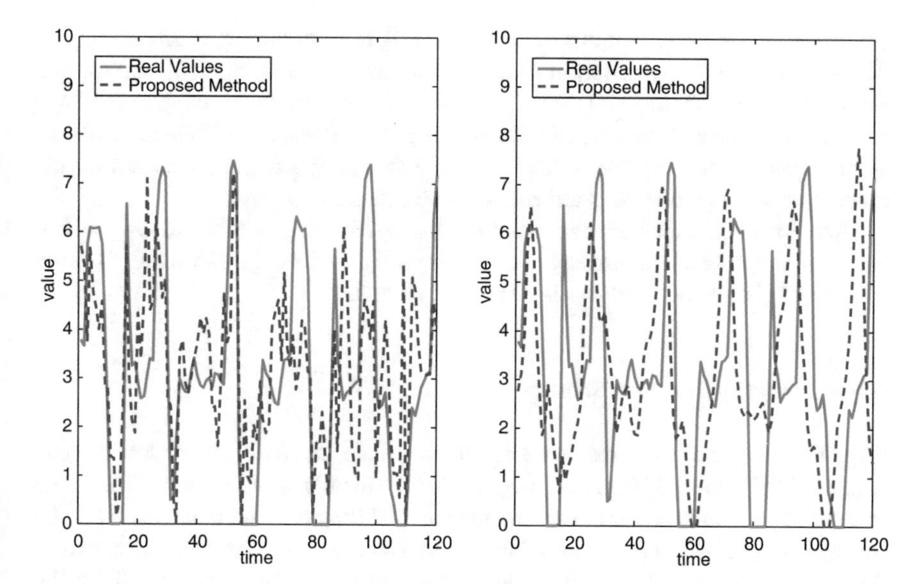

Fig. 9.21 Prediction results of first 120 points on the nitrogen consumption of #2 blast furnace with different methods

Table 9.15 Error statistics and time costs on the prediction of oxygen consumption of #2 blast furnace with different methods

Methods	MAPE	RMSE	CT (s)
ESN	10.98	4.827	1.144
LSSVM	12.85	7.689	8.247
Proposed method	4.56	2.256	0.013

Table 9.16 Error statistics (average of the 3-month results)

Items	Methods	MAPE	RMSE	CT (s)
Oxygen consumption of #1 converter	ESN	20.74	62.1452	1.926
	LSSVM	14.63	55.7945	9.323
	Proposed method	7.01	30.0013	0.644
Nitrogen consumption of #2 blast furnace	ESN	11.63	5.223	1.662
	LSSVM	14.34	7.969	10.015
	Proposed method	5.08	2.531	0.429

Energy Scheduling and Optimization

Based on the predicted requirements of oxygen and nitrogen, the objective of the energy scheduling work is formulated by (9.61). In the perspective of the scheduling practice, four typical categories of circumstances are involved in this section, i.e., oxygen surplus/nitrogen surplus $[O_{dr} > 0, N_{dr} > 0]$, oxygen shortage/nitrogen surplus $[O_{dr} < 0, N_{dr} > 0]$, oxygen surplus/nitrogen shortage $[O_{dr} > 0, N_{dr} < 0]$, and oxygen shortage/nitrogen shortage $[O_{dr} < 0, N_{dr} < 0]$. And two of them, the oxygen shortage/nitrogen shortage and the oxygen surplus/nitrogen shortage, are addressed here to be as the representative instances. The manual scheduling scheme conducted by the scheduling workers on-site is utilized for performance comparison.

Here, the coefficients are computed as $O_{12g} = 700$, $N_{12g} = 800$ and $O_{t2m} = 1/1.14$, $N_{t2m} = 1/0.808$ (at standard atmospheric pressure). δ_{o_2} and δ_{n_2} in (9.66) are all set at $1000\ km^3/h$, while N_{o_2} and N_{n_2} in (9.67) are all at 3.

Oxygen Shortage/Nitrogen Shortage

Assuming a practical circumstance, the shortage of 8000 Nm^3 for oxygen ($o_{2_{dr}}: -7.5\%$) and 10,000 Nm^3 ($n_{2_{dr}}: -6.3\%$) for nitrogen, Table 9.17 lists the scheduling solutions coming from the manual and the proposed methods. And, the corresponding effects on the diffusion rate by using a Gantt chart are illustrated in Fig. 9.22. The real and the predicted energy diffusion rates (computed by the proposed GrC-based long-term prediction model) are shown in the above two sub-plots. Because of the inaccurate estimation on the capacity of the oxygen/nitrogen system, the manual one results in a worse solution in respect of both time cost and the scheduling effects, which leads to an excessive oxygen generation such that a further scheduling has to be performed for avoiding the energy diffusion. Taking the related equipment of the energy network into account, the proposed one presents a practicable solution involving far less units. When the diffusion rates are closing to

Table 9.17 Scheduling solution of different method (oxygen ($-8000\ Nm^3$)/nitrogen ($-10,000\ Nm^3$), 1 Nm^3 means 1 m^3 at normal state with 101.325 kPa and 20 °C)

Method	Device	Energy	Current load (Nm^3)	Scheduling tendency	Scheduling quantity (Nm^3)	CT (approx.)
Manual	#3 ASU	O_2	25,000	+	3000	6.5 h
		N_2	26,000	+	5000	
	#4 ASU	O_2	27,000	+	4000	
		N_2	41,000	+	6000	
	#7 ASU	O_2	60,000	+	2000	
	Liquid tank (in total)	O_2	3800	Liquidization	1500	
		N_2	4600	Liquidization	2000	
	#5 ASU	O_2	51,000	+	500	
		N_2	98,000	+	1000	
Proposed	#6 ASU	O_2	55,000	+	5000	3.5 h
		N_2	60,000	+	6500	
	#7 ASU	O_2	60,000	+	3000	
		N_2	63,000	+	3500	

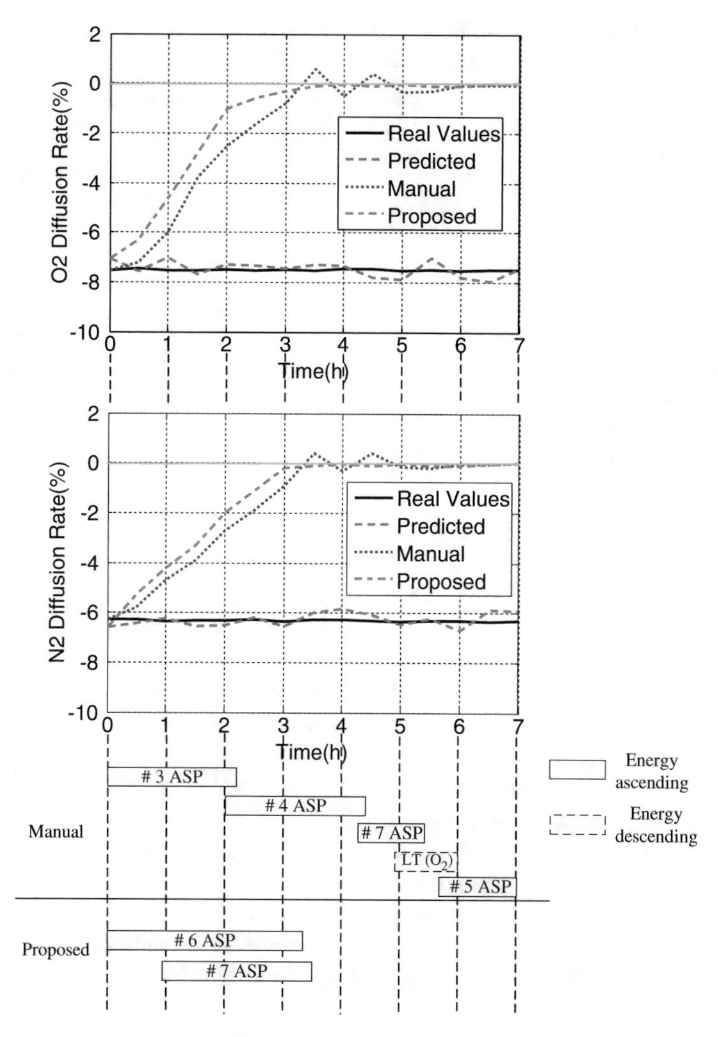

Fig. 9.22 Diffusion rate and Gantt chart when applying the scheduling solutions (oxygen (-8000 Nm^3)/nitrogen ($-10{,}000$ Nm^3))

0%, the scheduling work tends to be more cautious. Therefore, the plots in Fig. 9.22 after 3 h are varying little. And, the Gantt chart manifests that the scheduling operation can be simultaneously carried out. As a result, the diffusion rate is also operated in an acceptable range with the help of the accurate prediction and the scheduling solution computation.

Table 9.18 Scheduling solution of different method (oxygen (+3000 Nm^3)/nitrogen (−4000 Nm^3), 1 Nm^3 means 1 m^3 at normal state with 101.325 kPa and 20 °C)

Method	Device	Energy	Current load (Nm^3)	Scheduling tendency	Scheduling quantity (Nm^3)	CT (approx.)
Manual	#1 ASU	O_2	29,000	−	2000	6 h
	#2 ASU	O_2	29,000	−	2500	
	Liquid tank (in total)	O_2	3600	Liquidization	1500	
	#6 ASU	N_2	63,000	+	3000	
	Liquid tank (in total)	N_2	3350	Evaporation	1000	
Proposed	Liquid tank (in total)	N_2	3350	Evaporation	1500	2.5 h
	#5 ASU	O_2	49,000	+	2500	
	#6 ASU	O_2	55,000	−	6000	

Oxygen Surplus/Nitrogen Shortage

Another typical situation that is with redundant oxygen (3000 Nm^3, $o_{2_{dr}}$: 8.0%) and insufficient nitrogen (4000 Nm^3, $n_{2_{dr}}$: −5.0%) is taken here as an illustrative example.

Although the shortage/surplus has been successfully supplemented, the details of the manual and the proposed scheduling scheme are obviously different (see Table 9.18). Owing to the long scheduling time and the burdensome work of even 3 ASUs and several liquid tanks, the manual solution is not suitable for real application compared to the proposed one. While the proposed scheme operates only 2 ASUs and 1 liquid tank which is more convenient for realization and also consumes far less time. Besides, the following diffusion rate curves and Gantt chart (see Fig. 9.23) also indicate that the proposed approach scheduled the energy with less fluctuations and better accuracy. Note that due to the closeness to 0% of the diffusion rate, the scheduling is going to be temperate which leads to little fluctuation of the plots after 2.5 h in Fig. 9.23.

Similarly, here this section provides a 3-month running statistics on diffusion rate reduction as Table 9.19 for demonstrating the superiority of the proposed scheduling approach.

9.5 Discussion

In this chapter, we developed several applications of data-based prediction for energy scheduling of steel industry. First, single and multiple outputs prediction models based on LSSVM with hybrid optimization are established. Steered by the prediction results of the flow of the by-product gas, the gasholder level is adjusted to be maintained at a safe range. Similarly, we also introduced a predictive optimization

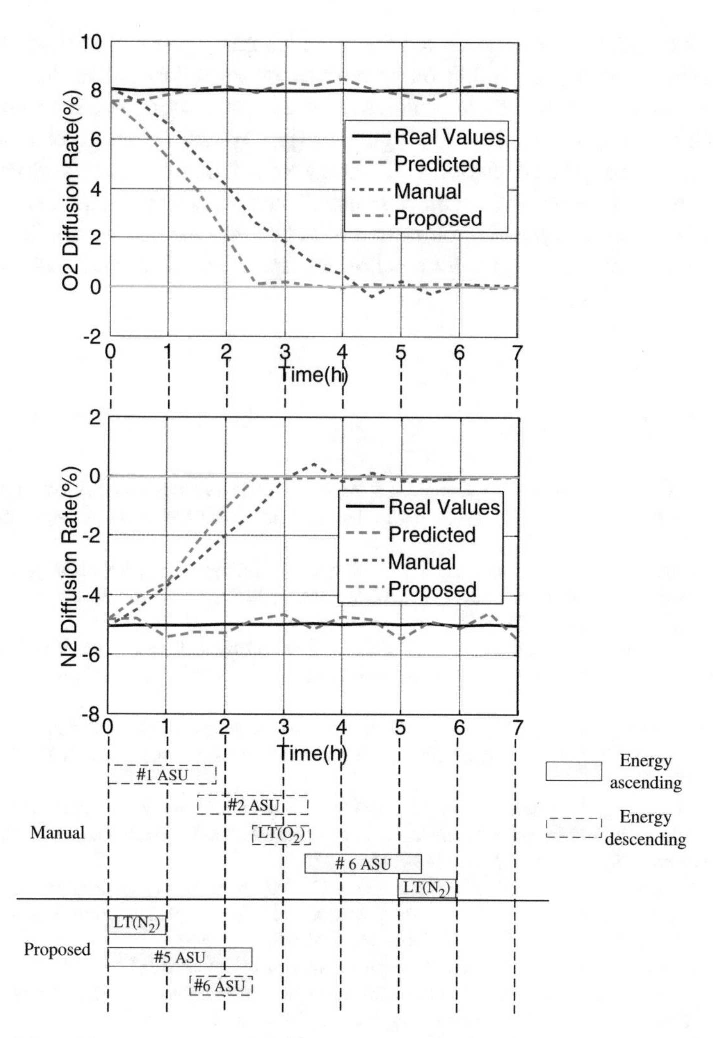

Fig. 9.23 Diffusion rate and Gantt chart when applying the scheduling solutions (oxygen (+3000 Nm^3)/nitrogen (−4000 Nm^3))

Table 9.19 Statistics of diffusion rate reduction with different methods (average of 3-month results)

Method	Diffusion rate reduction (%)
Manual	1.357
Proposed	2.015

method with prediction intervals for the real-time gas adjusting problem, in which the scheduling is a process of rolling-horizon optimization online, and then the tentative solutions can also be validated by the prediction model to ensure its applicability in practice. As for oxygen/nitrogen system of the steel industry, a two-stage method for prediction and energy scheduling is also developed. The experiments and online application demonstrate that accurate prediction is necessary for real-time scheduling and optimization of industrial system. Overall, these aforementioned methods may give illumination for researchers to perform scheduling in other process industries.

References

1. Zhang, X., Zhao, J., Wang, W., Cong, L., & Feng, W. (2011). An optimal method for prediction and adjustment on byproduct gas holder in steel industry. *Expert Systems with Applications, 38* (4), 4588–4599.
2. Brundage, M. P., Chang, Q., Li, Y., et al. (2013). Energy efficiency management of an integrated serial production line and HVAC system. *IEEE Transactions on Automation Science and Engineering, 11*(3), 789–797.
3. Chen, G., Zhang, L., Arinez, J., et al. (2012). Energy-efficient production systems through schedule-based operations. *IEEE Transactions on Automation Science and Engineering, 10*(1), 27–37.
4. Zhao, J., Wang, W., Liu, Y., et al. (2011). A two-stage online prediction method for a blast furnace gas system and its application. *IEEE Transactions on Control Systems Technology, 19* (3), 507–520.
5. Zhao, J., Wang, W., Sun, K., et al. (2014). A Bayesian networks structure learning and reasoning-based byproduct gas scheduling in steel industry. *IEEE Transactions on Automation Science and Engineering, 11*(4), 1149–1154.
6. Han, C., Chu, Y. H., Kim, J. H., et al. (2004). Control of gasholder level by trend prediction based on time-series analysis and process heuristics. In *The Seventh International Symposium on Advanced Control of Chemical Processes, Hong Kong, China.*
7. Sanchez-Fernandez, M., de-Prado-Cumplido, M., Arenas-Garcia, J., et al. (2004). SVM multiregression for nonlinear channel estimation in multiple-input multiple-output systems. *IEEE Transactions on Signal Processing, 52*(8), 2298–2307.
8. Scholkopf, B., & Smola, A. J. (2001). *Learning with kernels*. Cambridge, MA: MIT Press.
9. An, S., Liu, W., & Venkatesh, S. (2007). Fast cross-validation algorithms for least square support vector machine and kernel ridge regression. *Pattern Recognition, 40*(8), 2154–2162.
10. Van Gestel, T., Suykens, J. A. K., et al. (2001). Financial time series prediction using least squares support vector machines within the evidence framework. *IEEE Transactions on Neural Networks, 12*(4), 809–821.
11. Friedrichs, F., & Igel, C. (2005). Evolutionary tuning of multiple SVM parameters. *Neurocomputing, 64*, 107–117.
12. Pai, P. F., & Hong, W. C. (2005). Support vector machines with simulated annealing algorithms in electricity load forecasting. *Energy Conversation Management, 46*, 2669–2688.
13. Guo, X. C., Yang, J. H., Wu, C. G., et al. (2008). A novel LS-SVM's hyper-parameter selection based on particle swarm optimization. *Neurocomputing, 71*(16), 3211–3215.
14. Han, K. H., & Kim, J. H. (2002). Quantum–inspired evolutionary algorithm for a class of combinatorial optimization. *IEEE Transaction Evolution Computation, 6*(6), 580–593.
15. Cawley, G., & Talbot, N. (2004). Fast exact leave–one–out cross–validation of sparse least–squares support vector machines. *Neural Networks, 17*(10), 1467–1475.

16. Haykin, S. (1999). *Neural networks: A comprehensive foundation* (2nd ed.). New Jersey: Prentice-Hall.
17. Zhao, J., Sheng, C., Wang, W., et al. (2017). Data-based predictive optimization for byproduct gas system in steel industry. *IEEE Transactions on Automation Science and Engineering, 14*(4), 1761–1770.
18. Liu, Y., Liu, Q., Wang, W., et al. (2012). Data-driven based model for flow prediction of steam system in steel industry. *Information Sciences, 193*, 104–114.
19. An, S., Liu, W., & Venkatesh, S. (2007). Fast cross-validation algorithms for least squares support vector machine and kernel ridge regression. *Pattern Recognition, 40*, 2154–2162.
20. Brabanter, K. D., Brabanter, J. D., Suykens, J. A. K., et al. (2011). Approximate confidence and prediction intervals for least squares support vector regression. *IEEE Transactions on Neural Networks, 22*(1), 110–120.
21. Bishop, C. (2006). *Pattern recognition and machine learning*. New York: Springer.
22. Sheng, C., Zhao, J., Wang, W., et al. (2013). Prediction intervals for a noisy nonlinear time series based on a bootstrapping reservoir computing network ensemble. *IEEE Transactions on Neural Networks and Learning Systems, 24*(7), 1036–1048.
23. Kim, J. H., Yi, H., & Han, C. (2002). Optimal byproduct gas distribution in the iron and steel making process using mixed integer linear programming. In *Proc. Int. Symp. Adv. Control Ind. Process, Kumamoto, Japan* (pp. 581–586).
24. Han, Z., et al. (2016). A two-stage method for predicting and scheduling energy in an oxygen/ nitrogen system of the steel industry. *Control Engineering Practice, 52*, 35–45.
25. Bargiela, A., & Pedrycz, W. (2003). *Granular computing: An introduction*. Berlin: Springer.
26. Bargiela, A., & Pedrycz, W. (2008). Toward a theory of granular computing for human-centered information processing. *IEEE Transactions on Fuzzy Systems, 16*(2), 320–330.
27. Pedrycz, W., Skowron, A., & Kreinovich, V. (Eds.). (2008). *Handbook of granular computing*. New York: Wiley.
28. Pawlak, Z., & Skowron, A. (2007). Rough sets: Some extensions. *Information Sciences, 177*(1), 28–40.
29. Pedrycz, W. (2009). From fuzzy sets to shadowed sets: Interpretation and computing. *International Journal of Intelligent Systems, 24*(1), 48–61.
30. Bai, C., Dhavale, D., & Sarkis, J. (2014). Integrating Fuzzy C-Means and TOPSIS for performance evaluation: An application and comparative analysis. *Expert Systems with Applications, 41*(9), 4186–4196.
31. Daneshwar, M. A., & Noh, N. M. (2015). Detection of stiction in flow control loops based on fuzzy clustering. *Control Engineering Practice, 39*, 23–34.
32. Boyacioglu, M. A., & Avci, D. (2010). An adaptive network-based fuzzy inference system (ANFIS) for the prediction of stock market return: The case of the Istanbul stock exchange. *Expert Systems with Application, 37*(12), 7908–7912.
33. Chang, F. J., & Chang, Y. T. (2006). Adaptive neuro-fuzzy inference system for prediction of water level in reservoir. *Advances in Water Resources, 29*(1), 1–10.
34. Van Broekhoven, E., & De Baets, B. (2006). Fast and accurate center of gravity defuzzification of fuzzy system outputs defined on trapezoidal fuzzy partitions. *Fuzzy Sets and Systems, 157*(7), 904–918.
35. Wang, Y. M. (2009). Centroid defuzzification and the maximizing set and minimizing set ranking based on alpha level sets. *Computers & Industrial Engineering, 57*(1), 228–236.
36. Price, L., Sinton, J., Worrell, E., et al. (2002). Energy use and carbon dioxide emissions from steel production in China. *Energy, 27*(1), 429–446.
37. Borghetti, A., D'Ambrosio, C., Lodi, A., et al. (2008). An MILP approach for short-term hydro scheduling and unit commitment with head-dependent reservoir. *IEEE Transactions on Power Systems, 23*(3), 1115–1124.
38. Karimi, A., Kunze, M., & Longchamp, R. (2007). Robust controller design by linear programming with application to a double-axis positioning system. *Control Engineering Practice, 15* (2), 197–208.

39. Han, Z., Liu, Y., Zhao, J., et al. (2012). Real time prediction for converter gas tank levels based on multi-output least square support vector regressor. *Control Engineering Practice, 20*(12), 1400–1409.
40. Zhao, J., Liu, Q., Wang, W., et al. (2012). Hybrid neural prediction and optimized adjustment for coke oven gas system in steel industry. *IEEE Transactions on Neural Networks and Learning Systems, 23*(3), 439–450.

Index

J. Zhao et al., *Data-Driven Prediction for Industrial Processes and Their Applications*, Information Fusion and Data Science,
https://doi.org/10.1007/978-3-319-94051-9